T0360426

MATHEMATICAL METHODS FOR THE NATURAL AND ENGINEERING SCIENCES

Second Edition

Series on Advances in Mathematics for Applied Sciences – Vol. 87

MATHEMATICAL METHODS FOR THE NATURAL AND ENGINEERING SCIENCES

Second Edition

Ronald E Mickens

Clark Atlanta University, USA

 World Scientific

NEW JERSEY · LONDON · SINGAPORE · BEIJING · SHANGHAI · HONG KONG · TAIPEI · CHENNAI · TOKYO

Published by

World Scientific Publishing Co. Pte. Ltd.

5 Toh Tuck Link, Singapore 596224

USA office: 27 Warren Street, Suite 401-402, Hackensack, NJ 07601

UK office: 57 Shelton Street, Covent Garden, London WC2H 9HE

Library of Congress Cataloging-in-Publication Data

Names: Mickens, Ronald E., 1943–

Title: Mathematical methods for the natural and engineering sciences /
by Ronald E. Mickens (Clark Atlanta University, USA).

Description: 2nd edition. | New Jersey : World Scientific, 2017. |
Series: Series on advances in mathematics for applied sciences ; vol. 87 |
Includes bibliographical references and index.

Identifiers: LCCN 2016042801 | ISBN 9789813202702 (hardcover : alk. paper)

Subjects: LCSH: Mathematical analysis.

Classification: LCC QA300 .M5195 2017 | DDC 515--dc23

LC record available at https://lccn.loc.gov/2016042801

British Library Cataloguing-in-Publication Data

A catalogue record for this book is available from the British Library.

Printed in Singapore

This book is dedicated to my teachers:

James Williamson, my grandfather

Florence Walker and *LaVerne Wright Goodridge*, my mathematics
    teachers at Peabody High School

Nelson Fuson and *James Lawson*, my physics professors at Fisk
    University

Wendell Holladay, my research advisor at Vanderbilt University

Preface

First Edition

The main purpose of this book is to introduce and explore a variety of mathematical methods useful for analyzing equations arising in the modeling of phenomena from the natural and engineering sciences. The book is intended for students beginning their third year of undergraduate work in one of these sciences. Students will reap maximum benefit from the book if they bring to it knowledge gained from having taken the following courses: one year of general physics, including an introduction to the essential issues of modern physics; a standard three semester course in calculus; an introductory course in differential equations; and a one semester course in linear algebra. This book is based on my experiences related to teaching courses of "mathematical physics" and "quantum mechanics" for over twenty-five years at Fisk University, Atlanta University, and Clark Atlanta University; and from presenting the results of my research activities to both undergraduate and graduate students at numerous workshops and seminars. These experiences led me to the following conclusion: It is generally of greater utility to make science and engineering students aware of the existence of a broad range of mathematical techniques, needed to solve the equations arising in their work, both academically and professionally, than it is to give the intimate details of a few methods within the framework of a rigorous theorem and proof format. Clearly, the huge success of the natural and engineering sciences has been based largely on this philosophy.

Most textbooks on the subject of mathematical methods and their applications tend to be encyclopedic in both size and coverage, and, in general, have much more material than can be presented in one year. The topics of this book can be covered in two semesters, and most chapters can be read independently of each other. However, the order given is usually the one I use in my teaching this course.

While we present many theorems and related results in rigorous mathematical form, in general, few proofs are explicitly given. However, we do provide for each chapter a list of books and related publications which give such proofs and/or alternative ways of stating the theorems.

An important aspect of this book is the inclusion of a rather large number of worked examples. They are worked in sufficient detail such that the reader can fully understand the concept or method under discussion. In general, the reader/student should work all of the problems listed at the end of each chapter. Their solution not only allows the filling in of certain gaps, but often they show how a particular concept or method can be generalized or extended to other situations.

Another feature of the book is that it gives equal emphasis to both qualitative and quantitative methods. In particular, much of the presentation concerns the use of various "asymptotic methods" to determine analytical approximations to the solutions for both differential equations and certain functions defined by integrals. A good starting point for a course, based on this book, is to begin with a general, but, quick review of the material presented in the two appendices.

The writing of this book has been both a learning and "fun" experience for me. It will be of great value to me and persons who may use this book if you, as a reader/user, report to me any "errors" that you find.• In particular, I welcome any comments as to how the presentation on any topic can be improved (within the framework of keeping the book finite). My contact information is

Ronald E. Mickens
Clark Atlanta University
Physics Department-Box 172
Atlanta, Georgia 30314
Telephone: (404) 880-6923
Email: rohrs@math.gatech.edu

I thank my students and colleagues for both their interest in this book and for the many suggestions as to how it could be improved. To Ms. Annette Rohrs, I am truly grateful for all her efforts in seeing that the many handwritten pages got transformed to a final typed manuscript. She has also served as illustrator and preliminary copy editor. I also thank the Department of Energy and the NIH/MBRS-SCORE Program at Clark Atlanta University for support of my research and providing funds for release time to write this book. Finally, my great appreciation and thanks go to the publisher, World Scientific, for accepting this book for publication and giving excellent service in all matters related to its production.

Atlanta, Georgia   Ronald E. Mickens
August 2003

Second Edition

This second edition corrects a number of typos and minor errors which appeared in the original book. Further, it contains a number of additional topics covering mathematical methods of relevance to investigations on the modeling analysis of systems in the natural and engineering sciences. The three new chapters include introductions to and applications on the following subjects:

- generalizations of the standard (trigonometric) periodic functions
- Lambert-W function (the newest addition to the class of "elementary functions")
- dimensional analysis and its use to derive important relations among the variables and parameters for physical systems
- functional equations
- calculation of $\operatorname{Re} F(z)$ given $\operatorname{Im} F(z)$
- exponential functions
- principle of dynamic consistency
- method of dominant balance

This book, as was the first edition, is not based on rigorous mathematical arguments. Its major goals include the introduction of useful mathematical concepts, definitions, and functions, and the list of relevant references to research journals and books, for which, upon study, will lead the user of this volume to a deeper understanding of the particular subject of interest.

Finally, I would like to especially thank three individuals who have played important roles in the conceptualization, production, and writing of this second edition:

- Ms. E. H. Chionh, editor, at World Scientific, for her suggesting the need for a second edition, and for her efficient professional aid in seeing that this volume would come to a speedy conclusion ...
- Ms. Annette Rohrs, for doing the really hard work needed to transfer my hand-written pages to the book you now behold ...

- Maria (my wife, partner, friend, queen and ruler of it all) for providing me with many kindnesses which allowed the orderly writing of this book.

Atlanta, Georgia   Ronald E. Mickens
August 2016

Contents

MATHEMATICAL METHODS FOR THE NATURAL AND ENGINEERING SCIENCES

Second Edition

Chapter 1

Introduction

The need to study various mathematical techniques grows out of the requirement to understand, solve and/or evaluate the different types of equations that arise from the mathematical modeling of phenomena in the natural, social, and engineering sciences. The first equations written down for a particular system involve a number of parameters appearing in the equations relating variable quantities. What is generally required, at this point, is to convert these equations into structurally similar equations, but having the important property that all of the new parameters and variables have no physical dimensions. One of the purposes of this chapter is to provide some indication as to how to do this and illustrate the basic method by means of examples.

The final section briefly introduces the concept of nonlinearity and the important role that it plays in the sciences. The bibliography gives a short listing of books and related materials on the main issues raised in this chapter.

1.1 Mathematical Modeling

Suppose there is a system that you wish to understand. This understanding may include the correlation of its past behavior with the current state of the system, along with the ability to predict its future evolution. An important tool for obtaining such information is to construct a mathematical model for the relevant phenomena related to the system. The particular mathematical formalism used will depend on the nature of the system. It should be pointed out that the same system can have various mathematical representations with the one selected dependent upon what questions need to be answered relative to the system. The actual equations occurring in the construction of a particular mathematical model will depend on what broad principles can be applied to the system to restrict its dynamical behavior. Examples of such requirements include the conservation laws (energy, momentum, charge, mass, etc.). However, in general, we define mathematical modeling and the associated model as follows:

Definition *Mathematical modeling* is the process of constructing appropriate equations to represent the relevant aspects of a system. This activity includes the various methods needed to analyze these equations, the obtaining of exact and approximate solutions, possible numerical evaluations, and simulations.

Definition A *mathematical model* is the set of equations arising from the modeling process, along with any restrictions or auxiliary conditions that must be imposed on the solutions to the equations.

While there cannot be a general prescription which, if followed, leads to a successful mathematical model for a given system, the following is a minimal list of issues and steps that clearly must be included in any such effort:

(i) The underlying science should be as fully understood as possible.

(ii) As many simplifying assumptions, as needed, should be made to reduce the complexities of the system to a form where suitable mathematical equations can be derived, while at the same time assuring that all relevant behavior and properties of the system are captured by the equation.

(iii) The equations of the mathematical model should be checked for both consistency within a given equation and consistency among the various equations.

(iv) The mathematical analysis of the model equations should begin with a detailed study of the qualitative properties of its possible solution behaviors. Possible methods to be used include examination of trajectories in the phase-space of the dynamical variables of the system, and the application of dimensional analysis and scaling to reduce the number of free parameters.

(v) If possible, special, but exact solutions should be found and their properties examined.

(vi) If exact solutions cannot be obtained, then analytical approximations to these solutions should be constructed by use of appropriate mathematical techniques.

(vii) It is often of great value to apply numerical methods to generate numerical solutions. A deep understanding of the system may sometimes be obtained using this technique.

(viii) Finally, all of the above elements should be combined to form a grand synthesis that hopefully will allow a full and deep understanding of those aspects of the system which were of original interest.

Finally, it should be realized that these steps may have to be iterated several times before a suitable mathematical model is reached. Also, it must be understood that difference systems may lead to similar, if not identical mathematical models. Consequently, an understanding of the dynamics of one system may provide additional insight into the mechanisms of other (mathematically) related processes.

The bibliography to the chapter provides a listing of books that treat mathematical modeling in detail, including general questions on philosophical aspects of the general modeling process. Several present a variety of examples as to how this procedure is to be applied to various areas of the sciences.

1.2   Mathematical versus Physical Equations

Mathematical models in the sciences generally are expressed in terms of differential equations [1]. In the standard case, the variables and parameters appearing in these equations have magnitudes determined by the system of "units" used for their expression [2]. However, the system of units used has no absolute significance and other units can be applied to the same problem. Often, the nature of the units used is dependent on what questions the modeler has in mind when considering a given problem. That is, different questions can lead the modeler to use different sets of units to characterize the same problem. For example, consider a mechanical system composed of two subsystems, each of which are electrically charged. It is always possible to study the full system using the basic set of standard units $(M, L, T) \equiv$ (mass, length, time). But, it will prove to be more useful, for this case, if an additional physical unit is added to the standard units, i.e., a unit of electrical charge, Q. Thus, the more appropriate units are now (M, L, T, Q) [2].

Assuming that the original modeling equations are differential equations, then they will take the general form

$$\frac{\partial u}{\partial t} = f(x, t, u, p), \tag{1.2.1}$$

$$\begin{cases} u^T \equiv (u_1, u_2, \ldots, u_N), \\ x^T \equiv (x_1, x_2, \ldots, x_M), \\ p^T \equiv (p_1, p_2, \ldots, p_K), \end{cases} \qquad (1.2.2)$$

and u represents the N-components of the dependent variable; x, the M-components of independent the "space" variable; t is the time variable; and p represents the K-parameters which appear in the modeling equations. In general, Eq. (1.2.1) is a short-hand notation for N-coupled differential equations and they comprise the mathematical model.

Equation (1.2.1) is a *physical equation* in the sense that, in general, all of the variables and parameters, (u, x, t, p), are expressed in terms of a given set of physical units such as mass, length, time, etc. Now what is most desirable is to have a set of equations for which all the new variables and parameters are dimensionless. This new or derived set of equations are the mathematical equations and they possess the important property of not having the values of their variables and parameters depend on any particular system of units [2].

1.3 Dimensionless Variables and Characteristic Scales

Given a set of "physical" differential equations, the primary issue is how does one construct from them the related "mathematical equations"? As we outline below, the method for their determination is relatively direct, but does not lead, in general, to a unique set of mathematical equations. However, this non-uniqueness can often be used by the investigator to study different aspects of the same system.

Consider, for example, a system characterized by k parameters, $p = (p_1, p_2, \ldots, p_k)$. All possible length scales will take the form

$$L_i = \prod_{j=1}^{k} (p_j)^{\ell_j}, \qquad i = 1, 2, \ldots, I; \qquad (1.3.1)$$

where the ℓ_i can be explicitly calculated and there are I scales. Note that it is unlikely that value of I will be known prior to the actual determination of the set of length scales $\{L_i : 1, 2, \ldots, I\}$. Similar forms also hold for the other scales involving mass, time, dependent variables, etc. [2]; for example,

$$M_r = \prod_{j=1}^{k} (p_j)^{v_j}, \qquad r = 1, 2, \ldots, R; \qquad (1.3.2)$$

$$T_s = \prod_{j=1}^{k} (p_j)^{w_j}, \qquad s = 1, 2, \ldots, S; \tag{1.3.3}$$

where the $\{v_j, w_j : j = 1, 2, \ldots, K\}$ are calculable.

To form a dimensionless or mathematical set of equations, we select a member from each different type of scale, i.e.,

$$\begin{cases} L : \{L_1, L_2, \ldots, L_I\}, \\ M : \{M_1, M_2, \ldots, M_R\}, \\ T : \{T_1, T_2, \ldots, T_S\}, \\ u^* : \{u_1^*, u_2^*, \ldots, u_Q^*\}, \\ \qquad \text{etc.,} \end{cases} \tag{1.3.4}$$

where (I, R, T, Q) are positive integers and the scales for the dependent variables are indicated by u^*, and define new dimensionless variables $\bar{u}, \bar{x}, \bar{t}$, etc., as follows,

$$u = u^* \bar{u}, \quad x = L\bar{x}, \quad t = T\bar{t}, \text{ etc.} \tag{1.3.5}$$

Substitution of these quantities into Eq. (1.2.1) and simplifying, gives an equation having the structure

$$\frac{d\bar{u}}{d\bar{t}} = F(\bar{x}, \bar{t}, \bar{u}, \lambda), \tag{1.3.6}$$

where the function F is determined from a knowledge of f, and the new dimensionless set of parameters $\{\lambda_j : 1, 2, \ldots, \bar{K}\}$ is usually fewer in number than the original set of k-parameters, i.e., $\bar{K} \leq K$. Since Eq. (1.3.6) is now expressed only in terms of dimensionless variables and parameters, it is the required mathematical equation. This is the equation that can now be studied using all available tools of applied mathematics.

Finally, an issue indicated above should be reiterated. Suppose we have a problem for which only the scales (L, M, T, u^*) arise. Then it follows, from the results stated in Eq. (1.3.4), that there are $IQRS$ possible combinations of scalings that can be constructed and used to transform the physical equations to mathematical equations. Thus, the selection of a particular set of scales should be based on both an understanding of the dynamics of the system and what questions are being asked. In few cases does this nonuniqueness cause difficulties. A very useful way to proceed is to construct the mathematical equation by beginning with generic, *a priori* unspecified, scales $(T_1, L_1, \text{etc.})$ that are later calculated to make the coefficients of the various terms in this equation reflect their magnitude [2].

In the next section, we illustrate the general method by applying it to a number of mathematical models for a variety of systems.

1.4   Construction of Mathematical Equations

1.4.1   *Decay Equation*

The following first-order, linear differential equation plays a fundamental role in the study of systems that can transform their state by decaying [3],

$$\frac{dc}{dt} = -\lambda c, \qquad c(0) = c_0 > 0. \tag{1.4.1}$$

If we measure c in terms of particle density, then the various quantities have the dimensions,

$$\begin{cases} [c] \equiv \# = \text{particles per unit volume}, \\ [t] \equiv T = \text{time}, \\ [\lambda] \equiv T^{-1}, \end{cases} \tag{1.4.2}$$

where the square bracket $[\cdots]$ around a symbol denotes its units. Now for this system, only two units occur, namely, $\#$ and T. Also, only two parameters are at our disposal: the measure of the decay rate, given by λ and the initial concentration, c_0. Thus, the scales for time and concentration are

$$T_1 = \frac{1}{\lambda}, \qquad c^* = c_0, \tag{1.4.3}$$

and the corresponding dimensionless quantities are

$$c \to \bar{c} = \frac{c}{c_0}, \qquad t \to \bar{t} = \frac{t}{T_1} = \lambda t. \tag{1.4.4}$$

Substitution of these results into Eq. (1.4.1) gives the mathematical equation

$$\frac{d\bar{c}}{d\bar{t}} = -\bar{c}, \qquad \bar{c}(0) = 1. \tag{1.4.5}$$

Observe that the ODE for the dimensionless quantities only involves a single, definite value of the initial condition and no free parameters occur in Eq. (1.4.5).

In summary, the initial physical equation was characterized by one parameter, λ, and one initial value for the concentration, and the initial value, c_0, could take on a range of values, i.e.,

$$c_0 > 0. \tag{1.4.6}$$

However, the mathematical equation had no free parameters and only one initial value needed to be considered.

1.4.2 Logistic Equation

The logistic equation provides a model for population growth of a single species, but with self-interaction that places a limit on the ultimate size of the population [4]. Let $P(t)$ be population number, then the modeling "physical" equation is

$$\frac{dP}{dt} = \lambda_1 P - \lambda_2 P^2, \qquad P(0) = P_0 > 0, \qquad (1.4.7)$$

where λ_1 is a parameter related to the birthrate at low values of P and λ_2 is related to the strength of the self-interaction in the population. The units of the various quantities for this system are

$$[P] = \#, \quad [P_0] = \#, \quad [\lambda] = \frac{1}{T}, \quad [\lambda_2] = \frac{1}{\#T}, \quad [t] = T. \qquad (1.4.8)$$

From these, two characteristic times, T_1 and T_2, can be constructed, as well as one characteristic population value, P^*. They are given by

$$T_1 = \frac{1}{\lambda_1}, \qquad T_2 = \frac{1}{P_0 \lambda_2}, \qquad P^* = \frac{\lambda_1}{\lambda_2}. \qquad (1.4.9)$$

The following dimensionless variables can now be constructed

$$t \to \bar{t} = \frac{t}{T_1} = \lambda_1 t, \qquad P \to \bar{P} = \frac{P}{P^*} = \left(\frac{\lambda_2}{\lambda_1}\right) P, \qquad (1.4.10)$$

and with their substitution into Eq. (1.4.7), the logistic equation becomes

$$\frac{d\bar{P}}{d\bar{t}} = \bar{P}(1 - \bar{P}), \qquad \bar{P}(0) = \frac{P(0)}{P^*} > 0. \qquad (1.4.11)$$

Note that the mathematical equation contains no arbitrary parameters, while the original physical equation had two parameters, λ_1 and λ_2. However, both the physical and mathematical equations have to satisfy a single initial condition.

A more detailed study of the logistic equation [4] allows the following general conclusions to be reached with regard to the qualitative properties of its solutions:

(i) For any positive initial condition, $\bar{P}(0) > 0$, all solutions monotonically approach the value one, i.e.,

$$\lim_{t \to \infty} \bar{P}(t) = 1, \qquad \bar{P}(0) > 0. \qquad (1.4.12)$$

(ii) Starting from an initial value $\bar{P}(0) > 0$, then the time, T_∞, (in dimensionless units) to "almost" reach the population value, $\bar{P}(\infty) = 1$, is approximately given by the expression

$$T_\infty \simeq \frac{\lambda_1}{P_0 \lambda_2}. \qquad (1.4.13)$$

1.4.3 *The Fisher Equation*

Many interesting and important phenomena in the sciences can be modeled by the nonlinear Fisher partial differential equation [5]. This equation takes the form

$$\frac{\partial u}{\partial t} = D \frac{\partial^2 u}{\partial x^2} + \lambda_1 u - \lambda_2 u^2, \qquad (1.4.14)$$

where all three parameters $(D, \lambda_1, \lambda_2)$ are non-negative. If the unit of u is particle density, then

$$[u] = \#, \quad [D] = \frac{L^2}{T}, \quad [\lambda_1] = \frac{1}{T}, \quad [\lambda_2] = \frac{1}{\#T}. \qquad (1.4.15)$$

The corresponding scales are given by the expressions

$$u^* = \frac{\lambda_1}{\lambda_2}, \quad T_1 = \frac{1}{\lambda_1}, \quad L_1 = \left(\frac{D}{\lambda_1}\right)^{1/2}. \qquad (1.4.16)$$

With the dimensionless variables defined by

$$\bar{u} = \frac{u}{u^*} = \left(\frac{\lambda_2}{\lambda_1}\right) u, \quad \bar{t} = \frac{t}{T_1} = \lambda_1 t, \quad \bar{x} = \frac{x}{L_1} = \left(\frac{\lambda_1}{D}\right)^{1/2} x, \qquad (1.4.17)$$

the physical Fisher equation is transformed into the following dimensionless mathematical equation,

$$\frac{\partial \bar{u}}{\partial \bar{t}} = \frac{\partial^2 \bar{u}}{\partial \bar{x}^2} + \bar{u}(1 - \bar{u}). \qquad (1.4.18)$$

Thus, in contrast to the original Fisher equation, where three parameters appeared, our transformed equation contains no free parameters.

Since there exists a characteristic length and a characteristic time, it is now possible to calculate a characteristic velocity; it is given by

$$c^* \equiv \frac{L_1}{T_1} = (\lambda_1 D)^{1/2}. \qquad (1.4.19)$$

This c^* is an estimate of how fast phenomena modeled by the Fisher equation propagate. In fact, it is known that the minimum speed of propagation for the Fisher equation is [5],

$$c = 2(\lambda_1 D)^{1/2}. \qquad (1.4.20)$$

Thus, our estimate, c^*, is in excellent agreement with this value of c.

1.4.4 *Duffings' Equation*

A variety of mechanical oscillating systems [6; 7] can be approximated by the following second-order, nonlinear differential equation,

$$m \frac{d^2 x}{dt^2} + kx + k_1 x^3 = 0, \tag{1.4.21}$$

where the initial conditions are selected to be

$$x(0) = A, \qquad \frac{dx(0)}{dt} = 0. \tag{1.4.22}$$

This equation is called Duffings' equation and relates the oscillatory motions of a mass, m, subject to both linear and nonlinear elastic forces. As presented above the equation depends on three parameters (m, k, k_1) along with one essential initial condition, $x(0) = A$.

Since each term in Eq. (1.4.21) corresponds to a force, they each have units of MLT^{-2}. Thus, the parameters and essential initial condition have the units

$$[m] = M, \quad [k] = MT^{-2}, \quad [k_1] = MT^{-2}L^{-2}, \quad [A] = L, \tag{1.4.23}$$

and from them the following scales can be constructed,

$$T_1 = \left(\frac{m}{k}\right)^{1/2}, \qquad L_1 = \left(\frac{k}{k_1}\right)^{1/2}, \qquad L_2 = A. \tag{1.4.24}$$

The time scale T_1 is related to the period of the free oscillations of the linear system, i.e., when $k_1 = 0$. The two length scales have interpretations first as a measure of the intrinsic size of the system, i.e., L_1, and second as the scale associated with the (external to the system) initial condition. Two sets of time and length scales can be formed; they are (T_1, L_1) and (T_2, L_2), and give, respectively, two difference forms of dimensionless Duffings' equation. (T_1, L_1) :

$$\bar{x} = \frac{x}{L_1}, \qquad \bar{t} = \frac{t}{T_1}, \tag{1.4.25}$$

$$\frac{d^2 \bar{x}}{d\bar{t}^2} + \bar{x} + \bar{x}^3 = 0, \qquad \bar{x}(0) = \frac{A}{L_1}, \qquad \frac{D\bar{x}(0)}{d\bar{t}} = 0; \tag{1.4.26}$$

(T_2, L_2) :

$$\bar{x} = \frac{x}{L_2}, \qquad \bar{t} = \frac{t}{T_1}, \tag{1.4.27}$$

$$\frac{d^2 \bar{x}}{d\bar{t}^2} + \bar{x} + \epsilon \bar{x}^3 = 0, \qquad \bar{x}(0) = 1, \qquad \frac{d\bar{x}(0)}{d\bar{t}} = 0; \tag{1.4.28}$$

$$\epsilon = \left(\frac{L_2}{L_1}\right)^2. \tag{1.4.29}$$

The dimensionless parameter ϵ is the square of the ratio of the initial displacement to the intrinsic size of the system. Thus, the scales (T_1, L_1) corresponds to the preparation of the system in an initial state where the value of $x(0)$ is close to the intrinsic length scale of the system. On the other hand, Eq. (1.4.28) is the equation to use when the displacements in x are small compared to the intrinsic length scale. This is the form to be used when applying a perturbation method to calculate solutions for Duffings equation.

It is true, in general, that the intrinsic scales of a system are determined entirely by the parameters appearing in the physical equation. They can consequently be considered as the quantities relevant for a full understanding of the dynamics of the system. This characterization of the scales is also helpful in determining conditions for small changes in the dynamics of the system. For example, if a system has a time scale associated with it of magnitude $\bar{T}$, then the dynamics will not change much if the interval of time over which the system is observed is small compared to $\bar{T}$. Similarly, if initial conditions take values that are small compared to those of the corresponding intrinsic scales, then the system can be considered as a perturbed state of one for which either the initial conditions are zero or the values of some of the parameters are zero.

1.4.5 *Budworm Population Dynamics*

An interesting problem in mathematical ecology is to model the dynamics of budworm populations [7]. These pesky insects attack the leaves of a certain type of tree and when an outbreak takes place, they defoliate and eventually kill most of the trees that they feed on in several years. The following first-order ordinary differential equation has been used for the initial study of the budworm-tree interaction,

$$\frac{dN}{dt} = RN\left(1 - \frac{N}{K}\right) - \frac{BN^2}{A^2 + N^2}, \tag{1.4.30}$$

where N is the budworm population. Note that this equation only models the budworm population.

The models contain four parameters (A, B, R, K) and our task is to determine a dimensionless mathematical equation equivalent to Eq. (1.4.30). An examination of this equation indicates that the units of the various quantities appearing in it are

$$[N] = \#, \quad [R] = T^{-1}, \quad [K] = \#, \quad [A] = \#, \quad [B] = \#T^{-1}. \quad (1.4.31)$$

From the four parameters, the following population and time scales can be constructed,

$$N_1^* = A, \quad N_2^* = K, \quad N_3^* = \frac{B}{R}, \quad (1.4.32)$$

$$T_1 = \frac{1}{R}, \quad T_2 = \frac{K}{B}, \quad T_3 = \frac{A}{B}, \quad T_4 = \left(\frac{KA}{B^2}\right)^{1/2}. \quad (1.4.33)$$

Out of the possible twelve sets of (N^*, T) scales, we will use the one containing (N_1^*, T_3). Therefore, we have

$$\bar{N} = \frac{N}{A}, \quad \bar{t} = \frac{t}{T_3} = \left(\frac{B}{A}\right) t. \quad (1.4.34)$$

Substitution of these into Eq. (1.4.30) and simplifying the resulting expression gives

$$\frac{d\bar{N}}{d\bar{t}} = q\bar{N}\left(1 - \frac{\bar{N}}{p}\right) - \left(\frac{\bar{N}^2}{1 + \bar{N}^2}\right), \quad (1.4.35)$$

where

$$p = \frac{K}{A}, \quad q = \frac{RA}{B}. \quad (1.4.36)$$

In summary, we have transformed the original physical equation for the budworm problem into the equivalent dimensionless mathematical equation with the number of free parameters reduced from four to two.

1.5 Nonlinearity

Much of science, in particular physics, up to about the mid-twentieth century was centered on the construction and analysis of linear or linearized models for the various phenomena studied. However, the creation of digital computers allowed the beginning of a better understanding of nonlinear systems and the associated mathematical structures. This included processes for which realistic theories already existed, but for which no set of analytical techniques were available to provide solutions and other systems of interest for which no fundamental theories existed, yet knowledge of their dynamic behavior and related properties were needed.

A critical feature of systems modeled by linear differential equations is the property of linear superposition [9]. This means that if two different solutions, $x_1(t)$ and $x_2(t)$, are found, then their sum

$$x(t) = c_1 x_1(t) + c_2 x_2(t), \tag{1.5.1}$$

is a solution for arbitrary constants c_1 and c_2. In many cases, a knowledge of the actual explicit functional form of one solution could then allow the calculation of a second solution. Of interest is the fact that for a linear system, the output or reaction to a given input is directly proportional to the input to the system. Thus, small changes in the initial data generally give rise to some perturbations in the future evolution of the system.

However, such nice features do not hold true for nonlinear systems: small perturbations in the input data (initial conditions, parameters, etc.) can have a major impact on the future states of the systems. Often this aspect of nonlinear systems is characterized by the phrase "sensitive dependence on initial conditions" [10]. Related to this phenomena is the possible sensitive dependence on changes in parameter values; this area falls under the general topic of bifurcation theory [11; 12].

Nonlinear systems have many other properties that do not appear in linear systems. Examples include chaotic behavior, with the associated structure of "strange attractor" [11], and solitary waves which have many of the properties of classical particles [13]. Special nonlinear systems can even possess normal modes and have localized solutions [14]. It is expected that other types of solution behaviors will be found from future research on these systems.

While the central theme of this book is not concerned with the above indicated systems and the mathematical techniques needed to study them, the various topics included here are needed to begin such an effort. In particular, the genesis of most investigations on the nonlinear differential equations modeling nonlinear systems begins with the determination of the equilibrium solutions, i.e., fixed-points, and proceeds then to study the stability properties of the system equations linearized about these constant solutions. Thus, the techniques to be presented in this book form the basis of the broad background of resources needed to understand the dynamics of these complex systems.

Problems

Section 1.3

1.3.1 Explain why the $(\ell_j : j = 1, 2, \ldots, k)$ in Eq. (1.3.1) are unique for a particular L_i. The same reasoning applies to the results of Eqs. (1.3.2) and (1.3.3).

2.3.2 Determine $F(\bar{x}, \bar{t}, \bar{u}, \bar{\lambda})$ in Eq. (1.3.6) explicitly in terms of the $f(x, t, u, p)$ given by Eq. (1.2.1).

Section 1.4

1.4.1 Rescale the budworm population equation using (N_2^*, T_1).

1.4.2 Which scales should be used and why if the following conditions hold for the budworm population problem:

(a) $A \gg K$,

(b) $A \ll K$.

1.4.3 Construct a dimensionless mathematical equation for the system of differential equations:

$$\frac{dx_1}{dt} = rx_1 \left(1 - \frac{x_1}{K}\right) - \frac{\beta x_1 x_2}{\alpha + x_2},$$

$$\frac{dx^2}{dt} = sx_2 \left(1 - \frac{x_2}{\nu x_1}\right),$$

where all the parameters $(r, s, K, \beta, \alpha, \nu)$ are positive. These coupled, nonlinear, first-order differential equations model an animal predator (x_2)-prey (x_1) system [8].

1.4.4 The Rayleigh equation [6]

$$m\frac{d^2x}{dt^2} - \left[\alpha - \left(\frac{\beta}{3}\right)\left(\frac{dx}{dt}\right)^2\right]\frac{dx}{dt} + ky = 0$$

models nonlinear oscillations of a mass m acted upon by a linear elastic restoring force and a nonlinear energy source. Construct the required scales and transform it into a dimensionless equation. Can all of the original parameters (m, α, β, k) be transformed away in the dimensionless equation?

Section 1.5

1.5.1 Do there exist nonlinear differential equations having the property that if $x(t)$ is a solution, then $cx(t)$ is also a solution for any constant c?

1.5.2 Can a linear system of ordinary differential equations have chaotic solutions?

Comments and References

[1] Differential equations are not the only mathematical structures that can be used to represent such models. See the books of Aris, Brauer and Castillo-Chávez, Rosenblatt and Bell, and Shier and Wallenius.

[2] See for example the books of Logan, Mickens and Szirtes.

[3] J. W. Rohlf, *Modern Physics from α to Z^0* (Wiley, New York, 1994).

[4] J. D. Murray, *Mathematical Biology I: An Introduction*, 3rd ed. (Springer-Verlag, New York, 2003).

[5] See, for example,
 * J. D. Murray, *Mathematical Biology* (Springer-Verlag, Berlin, 1989).
 * L. Debnath, *Nonlinear Partial Differential Equations* (Birkhäuser, Boston, 1997).

[6] R. E. Mickens, *Nonlinear Oscillations* (Cambridge University Press, New York, 1981); section 1.3.

[7] D. Ludwig, D. D.Jones, and C. S. Holling, *J. Animal Ecology* **47**, 315 (1978).

[8] R. May, *Model Ecosystems* (Princeton University Press; Princeton, NJ; 1973).

[9] R. E. Williamson, *Introduction to Differential Equations and Dynamical Systems* (McGraw-Hill, New York, 1996).

[10] D. Gulick, *Encounters with Chaos* (McGraw-Hill, New York, 1992).

[11] S. Wiggins, *Global Bifurcations and Chaos: Analytical Methods* (Springer-Verlag, New York, 1988).

[12] G. Iooss and D. Joseph, *Elementary Stability and Bifurcation Theory* (Springer-Verlag, Berlin, 1980).

[13] E. Infeld and G. Rowlands, *Nonlinear Waves, Solitons, and Chaos* (Cambridge University Press, Cambridge, 1990).

[14] A. F. Vakakis, et al., *Normal Models and Localization in Nonlinear Systems* (Wiley-Interscience, New York, 1996).

Bibliography

Dimensionless Variables and Scaling

G. Birkhoff, *Hydrodynamics – A Study in Logic, Fact, and Similitude*, 2nd ed. (Princeton University Press; Princeton, NJ; 1960).

H. E. Huntley, *Dimensional Analysis* (Dover, New York, 1967).

E. de St. Q. Isaacson and M. de St. Q. Isaacson, *Dimensional Methods in Engineering and Physics* (Wiley, New York, 1975).

S. J. Kline, *Similitude and Approximation Theory* (McGraw-Hill, New York, 1965).

J. D. Logan, *Applied Mathematics* (Wiley, New York, 1977).

R. E. Mickens, *Oscillations in Planar Dynamic Systems* (World Scientific, Singapore, 1996).

L. A. Segel, *Society for Industrial and Applied Mathematics* **14**, 547 (1992).

T. Szirtes, *Applied Dimensional Analysis and Modeling* (McGraw-Hill, New York, 1998).

Mathematical Modeling

F. R. Adler, *Modeling the Dynamics of Life* (Brooks/Cole, New York, 1998).

R. Aris, *Mathematical Modeling Techniques* (Dover, New York, 1994).

D. Basmadjian, *Mathematical Modeling of Physical Systems* (Oxford University Press, New York, 2003).

E. Beltrami, *Mathematics for Dynamic Modeling* (Academic Press, New York, 1987).

F. Brauer and C. Castillo-Chávez, *Mathematical Models in Population Biology and Epidemiology* (Springer-Verlag, New York, 2001).

M. Gitterman and V. Halpern, *Qualitative Analysis of Physical Problems* (Academic Press, New York, 1981).

G. Godfrey, *Compartmental Models and Their Applications* (Academic Press, New York, 1983).

R. Haberman, *Mathematical Models* (Prentice-Hall; Englewood Cliffs, NJ; 1977).

J. W. Haefner, *Modeling Biological Systems* (Chapman and Hall, New York, 1996).

C. C. Lin and L. A. Segel, *Mathematics Applied to Deterministic Problems in the Natural Sciences* (New York, Macmillan, 1974).

D. G. Luenberger, *Introduction to Dynamic Systems: Theory Models and Applications* (Wiley, New York, 1979).

D. Mooney and R. Swift, *A Course in Mathematical Modeling* (The Mathematical Association of America; Washington, DC; 1999).

J. Rosenblatt and S. Bell, *Mathematical Analysis for Modeling* (CRC Press, New York, 1999).

L. A. Segel, *Modeling Dynamic Phenomena in Molecular and Cellular Biology* (Cambridge University Press, New York, 1984).

D. R. Shier and K. T. Wallenius, *Applied Mathematical Modeling: A Multidisciplinary Approach* (Chapman and Hall/CRC, New York, 2000).

J. A. Spriet and G. C. Vansteenkiste, *Computer-aided Modeling and Simulation* (Academic Press, London, 1982).

Nonlinearity

K. Alligood, T. Sauer, and J. A. Yorke, *Chaos: An Introduction to Dynamical Systems* (Springer-Verlag, New York, 1997).

R. Bellman, *Stability Theory of Differential Equations* (McGraw-Hill, New York, 1954).

A. Blaquiere, *Nonlinear System Analysis* (Academic Press, New York, 1966).

T. V. Davies and E. M. Jones, *Nonlinear Differential Equations* (Addison-Wesley; Reading, MA; 1966).

R. C. Hilborn, *Chaos and Nonlinear Dynamics: An Introduction for Scientists and Engineers* (Oxford University Press, New York, 1994).

V. Nemytskii and V. Stepanov, *Qualitative Theory of Differential Equations* (Princeton University Press; Princeton, NJ; 1959).

G. Nicolis, *Introduction to Nonlinear Science* (Cambridge University Press, Cambridge, 1995).

A. Perlmutter and L. F. Scott, editors, *The Significance of Nonlinear in the Natural Sciences* (Plenum, New York, 1977).

R. H. Rand and D. Armbruster, *Perturbation Methods Bifurcation Theory and Computer Algebra* (Springer-Verlag, New York, 1987).

T. L. Saaty and J. Bram, *Nonlinear Mathematics* (McGraw-Hill, New York, 1964).

R. Z. Sagdeev, D. A. Usikov, and G. M. Zaslavsky, *Nonlinear Physics* (Harwood, New York, 1988).

D. R. Shier and K. T. Wallenius, *Applied Mathematical Modeling: A Multidisciplinary Approach* (Chapman and Hall, New York, 2000).

D. Siljak, *Nonlinear System: The Parameter Analysis and Design* (Wiley, New York, 1969).

S. H. Strogatz, *Nonlinear Dynamics and Chaos – with Applications to Physics, Biology, Chemistry and Engineering* (Addison-Wesley; Reading, MA; 1994).

P. N. V. Tu, *Dynamical Systems with Applications in Economics and Biology* (Springer-Verlag, Berlin, 1994).

B. J. West, *An Essay on the Importance of Being Nonlinear* (Springer-Verlag, Berlin, 1985).

S. Wiggins, *Introduction to Applied Nonlinear Dynamical Systems and Chaos* (Springer-Verlag, New York, 1990).

Chapter 2

Trigonometric Relations and Fourier Analysis

2.1 Introduction

Many systems in the natural and engineering sciences, if their motions are bounded, oscillate. Often the oscillations are periodic or eventually become periodic. Thus there is a need to understand how oscillatory functions can be represented. The general area of mathematics concerned with this and related issues is Fourier analysis. The main purpose of this chapter is to introduce the elements of Fourier series and also include some interesting results on the various functional relations from trigonometry, and the Fourier and Laplace transforms. We generally only state the relevant theorems and provide no proofs of them. Such details are provided by the books listed in the bibliography given at the end of the chapter.

An illustration as to how the techniques of this chapter can be applied is given by the example of the following differential equation which models a very nonlinear oscillator,

$$\frac{d^2x}{dt^2} + x^3 = 0. \tag{2.1.1}$$

Defining, $y = dx/dt$, and using

$$\frac{d^2x}{dt^2} = \frac{d}{dt}\left(\frac{dx}{dt}\right) = \frac{dy}{dt} = \frac{dy}{dx}\frac{dx}{dt} = y\frac{dy}{dx}, \tag{2.1.2}$$

we obtain

$$y\frac{dy}{dx} + x^3 = 0. \tag{2.1.3}$$

Integrating once gives,

$$\frac{y^2}{2} + \frac{x^4}{4} = E \geq 0. \tag{2.1.4}$$

In the (x, y) phase-plane, $y(x)$ is a closed curve and, as will be seen in later chapters of this book, such a closed curve corresponds to $x(t)$ and $y(t)$ being periodic functions of t. Several questions and issues naturally arise:

(i) Can an explicit solution to Eq. (2.1.1) be found?

(ii) If no simple explicit solution exists, can a suitable analytical approximation be constructed?

(iii) Related to (ii) is the question: How, in fact, can such an analytical approximation be actually constructed? If there are several such methods, what mathematical procedures are required for their application?

(iv) Can the period be calculated without the need to have an actual explicit solution?

The topics in this chapter provide some of the background mathematics needed later in the book to answer these questions.

2.2   Euler's Formula and DeMoivre's Theorem

Define the function $E(\theta)$ as

$$E(\theta) \equiv \cos\theta + i\sin\theta, \qquad i = \sqrt{-1}. \tag{2.2.1}$$

Using the fact that

$$\frac{d\cos\theta}{d\theta} = -\sin\theta, \qquad \frac{d\sin\theta}{d\theta} = \cos\theta, \tag{2.2.2}$$

we have

$$\frac{dE(\theta)}{d\theta} = -\sin\theta + i\cos\theta$$
$$= i^2\sin\theta + i\cos\theta$$
$$= i(\cos\theta + i\sin\theta)$$
$$\frac{dE(\theta)}{d\theta} = iE(\theta). \tag{2.2.3}$$

This last expression is a separable, first-order differential equation,

$$\frac{dE}{E} = id\theta, \tag{2.2.4}$$

which can be solved to give

$$E(\theta) = ce^{i\theta}, \tag{2.2.5}$$

where c is the integration constant. Since, from the definition of $E(\theta)$, it follows that $E(0) = 1$, we have $c = 1$ and

$$E(\theta) = e^{i\theta} = \cos\theta + i\sin\theta. \tag{2.2.6}$$

Taking the complex conjugate of Eq. (2.2.6) gives

$$e^{-i\theta} = \cos\theta - i\sin\theta. \tag{2.2.7}$$

Now Eqs. (2.2.6) and (2.2.7) constitute two equations linear in $\cos\theta$ and $\sin\theta$. Solving them gives

$$\cos\theta = \frac{e^{i\theta} + e^{-i\theta}}{2}, \qquad \sin\theta = \frac{e^{i\theta} - e^{-i\theta}}{2i}. \tag{2.2.8}$$

The relations given in Eqs. (2.2.6) and (2.2.8) are called the *Euler's formulas*.

Note that if θ is set equal to π in Eq. (2.2.6), then

$$e^{i\pi} + 1 = 0. \tag{2.2.9}$$

This remarkable expression contains seven of the most important symbols in mathematics: $1, e, \pi, i, 0, +, =$.

Another significant relation can be constructed by taking the mth power of each side of the expression in Eq. (2.2.6); doing this gives,

$$(e^{i\theta})^m = (\cos\theta + i\sin\theta)^m. \tag{2.2.10}$$

Now

$$(e^{i\theta})^m = e^{im\theta} = \cos(m\theta) + i\sin(m\theta), \tag{2.2.11}$$

and we conclude that

$$(\cos\theta + i\sin\theta)^m = \cos(m\theta) + i\sin(m\theta). \tag{2.2.12}$$

This last result is known as *DeMoivre's theorem*.

2.3 Derivation of Trigonometric Relations

We now show that the use of Euler's formulas will allow the direct determination of the various trigonometric relations derived in elementary courses.

To begin, consider the product of two cosine functions and replace them by their Euler formulas, i.e.,

$$
\begin{aligned}
\cos\theta_1\cos\theta_2 &= \left(\frac{e^{i\theta_1} + e^{-i\theta_1}}{2}\right) \cdot \left(\frac{e^{i\theta_2} + e^{-i\theta_2}}{2}\right) \\
&= \left(\frac{1}{4}\right)\left[e^{i(\theta_1+\theta_2)} + e^{-i(\theta_1+\theta_2)} + e^{i(\theta_1-\theta_2)} + e^{-i(\theta_1-\theta_2)}\right] \\
&= \left(\frac{1}{2}\right)\left[\frac{e^{i(\theta_1+\theta_2)} + e^{-i(\theta_1+\theta_2)}}{2}\right] + \left(\frac{1}{2}\right)\left[\frac{e^{i(\theta_1-\theta_2)} + e^{-i(\theta_1-\theta_2)}}{2}\right]
\end{aligned}
$$

$$= \left(\frac{1}{2}\right) [\cos(\theta_1 + \theta_2) + \cos(\theta_1 - \theta_2)]. \qquad (2.3.1)$$

For $\theta_1 = \theta_2 = \theta$, we obtain

$$(\cos\theta)^2 = \left(\frac{1}{2}\right) [1 + \cos(2\theta)]. \qquad (2.3.2)$$

Using, $(\sin\theta)^2 + (\cos\theta)^2 = 1$, gives

$$(\sin\theta)^2 = \left(\frac{1}{2}\right) [1 - \cos(2\theta)]. \qquad (2.3.3)$$

Higher powers of $\cos\theta$ can be easily calculated. For example,

$$(\cos\theta)^3 = (\cos\theta)(\cos\theta)^2 = (\cos\theta) \left(\frac{1}{2}\right) [1 + \cos(2\theta)]$$

$$= \left(\frac{1}{2}\right) \cos\theta + \left(\frac{1}{2}\right) \cos\theta \cos 2\theta$$

$$= \left(\frac{1}{2}\right) \cos\theta + \left(\frac{1}{2}\right)\left(\frac{1}{2}\right) [\cos(3\theta) + \cos\theta]$$

$$= \left(\frac{3}{4}\right) \cos\theta + \left(\frac{1}{4}\right) \cos(3\theta), \qquad (2.3.4)$$

and

$$(\cos\theta)^4 = (\cos\theta)^2 (\cos\theta)^2 = \left(\frac{1}{4}\right) [1 + \cos(2\theta)]^2$$

$$= \left(\frac{1}{4}\right) [1 + 2\cos(2\theta) + \cos^2(2\theta)]$$

$$= \left(\frac{1}{4}\right) [1 + 2\cos(2\theta) + \frac{1}{2} + \left(\frac{1}{2}\right) [\cos(4\theta)]$$

$$= \frac{3}{8} + \left(\frac{1}{2}\right) \cos(2\theta) + \left(\frac{1}{8}\right) \cos(4\theta). \qquad (2.3.5)$$

Applying the same technique to the product of two sines gives,

$$\sin\theta_1 \sin\theta_2 = \left(\frac{e^{i\theta_1} - e^{-i\theta_2}}{2i}\right) \left[\frac{e^{i\theta_2} - e^{-i\theta_2}}{2i}\right]$$

$$= \left(\frac{-1}{4}\right) \left[e^{i(\theta_1+\theta_2)} - e^{i(\theta_1-\theta_2)} - e^{-i(\theta_1-\theta_2)} + e^{-i(\theta_1+\theta_2)}\right]$$

$$= \left(\frac{1}{2}\right) \cos(\theta_1 - \theta_2) - \left(\frac{1}{2}\right) \cos(\theta_1 + \theta_2), \qquad (2.3.6)$$

and for $\theta_1 = \theta_2 = \theta$, the formula of Eq. (2.3.3) is obtained. Likewise,

$$(\sin\theta)^3 = (\sin\theta)(\sin\theta)^2 = (\sin\theta) \left(\frac{1}{2}\right) [1 - \cos(2\theta)]$$

$$= \left(\frac{1}{2}\right)\sin\theta - \left(\frac{1}{2}\right)\sin\theta\cos(2\theta), \qquad (2.3.7)$$

and we must stop because we do not yet know how to calculate $\sin\theta\cos(2\theta)$. However, we can determine $(\sin\theta)^4$; it is,

$$(\sin\theta)^4 = (\sin\theta)^2(\sin\theta)^2 = \left(\frac{1}{4}\right)[1 - \cos(2\theta)]^2$$

$$= \left(\frac{1}{4}\right)[1 - 2\cos(2\theta) + \cos^2(2\theta)]$$

$$= \left(\frac{1}{4}\right)\left[1 - 2\cos(2\theta) + \frac{1}{2} + \left(\frac{1}{2}\right)\cos(4\theta)\right]$$

$$= \frac{3}{8} - \left(\frac{1}{2}\right)\cos(2\theta) + \left(\frac{1}{8}\right)\cos(4\theta). \qquad (2.3.8)$$

Let us now calculate the product of sine and cosine functions, i.e.,

$$\sin\theta_1\cos\theta_2 = \left[\frac{e^{i\theta_1} - e^{-i\theta_1}}{2i}\right]\left[\frac{e^{i\theta_2} + e^{-i\theta_2}}{2}\right]$$

$$= \left(\frac{1}{4i}\right)\left[e^{i(\theta_1+\theta_2)} - e^{-i(\theta_1+\theta_2)} + e^{i(\theta_1-\theta_2)} - e^{-i(\theta_1-\theta_2)}\right]$$

$$= \left(\frac{1}{2}\right)[\sin(\theta_1 + \theta_2) + \sin(\theta_1 - \theta_2)]. \qquad (2.3.9)$$

Using this result in the second line of Eq. (2.3.7) gives

$$(\sin\theta)^3 = \left(\frac{1}{2}\right)\sin\theta - \left(\frac{1}{2}\right)\left(\frac{1}{2}\right)[\sin(3\theta) - \sin\theta]$$

$$= \left(\frac{3}{4}\right)\sin\theta - \left(\frac{1}{4}\right)\sin(3\theta). \qquad (2.3.10)$$

It is also easy to determine that

$$\sin\theta\cos\theta = \left(\frac{1}{2}\right)\sin(2\theta). \qquad (2.3.11)$$

From Eqs. (2.3.1), (2.3.6), and (2.3.9), we can determine the sine and cosine of the sum and difference of two angles. This calculation gives

$$\sin(\theta_1 \pm \theta_2) = \sin\theta_1\cos\theta_2 \pm \cos\theta_1\sin\theta_2, \qquad (2.3.12)$$

$$\cos(\theta_1 \pm \theta_2) = \cos\theta_1\cos\theta_2 \mp \sin\theta_1\sin\theta_2. \qquad (2.3.13)$$

At this point, we are able to directly calculate all the other standard trigonometric relations. As a (last) example, let us derive a formula for the tangent of the sum of two angles, i.e.,

$$\tan(\theta_1 + \theta_2) = \frac{\sin(\theta_1 + \theta_2)}{\cos(\theta_1 + \theta_2)} = \frac{\sin\theta_1\cos\theta_2 + \cos\theta_1\sin\theta_2}{\cos\theta_1\cos\theta_2 - \sin\theta_1\sin\theta_2}. \qquad (2.3.14)$$

Now divide both numerator and denominator by $\cos\theta_1\cos\theta_2$; this gives

$$\tan(\theta_1 + \theta_2) = \frac{\tan\theta_1 + \tan\theta_2}{1 - \tan\theta_1\tan\theta_2}. \qquad (2.3.15)$$

2.4   Periodic, Even and Odd Functions

2.4.1   *Periodic Functions*

Assume that a function $f(x)$ is defined on the interval, $-\infty < x < +\infty$. Further assume that there exists a fixed positive T such that

$$f(x + T) = f(x). \tag{2.4.1}$$

Then T is called a period of the function $f(x)$. It follows from this definition that if n is any integer, then

$$f(x + nT) = f(x). \tag{2.4.2}$$

In general, the smallest value T for which Eq. (2.4.1) holds is called the fundamental period for $f(x)$. Note that many periods exist for $f(x)$; in fact, any integer multiple of T is a period. Also, it is easy to verify that the following properties hold:

(i) If $f(x)$ is a periodic function, with fundamental period T, then $cf(x)$, where c is an arbitrary constant, is also a periodic function with fundamental period T.

(ii) Let $f(x)$ and $g(x)$ be periodic functions of period T, then $c_1 f(x) + c_2 g(x)$ is a periodic function, where c_1 and c_2 are arbitrary constants.

(iii) Let $f(x)$ be an integrable function with period T. Then, for any real constant c,

$$\int_c^{c+T} f(x)dx = \int_0^T f(x)dx. \tag{2.4.3}$$

2.4.2   *Even and Odd Functions*

The mathematical modeling of systems generally give rise to differential equations for which initial and/or boundary values must be specified. For example, Eq. (2.1.1), rewritten here again,

$$\frac{d^2x}{dt^2} + x^3 = 0 \tag{2.4.4}$$

has initial conditions $x(0)$ and $dx(0)/dt$. By proper selection of these values, the solution $x(t)$ can be made to be either an even or an odd function of t. In particular,

$$\begin{cases} x(0) \neq 0, & dx(0)/dt = 0 \Rightarrow x(t) \text{ even,} \\ x(0) = 0, & dx(0)/dt \neq 0 \Rightarrow x(t) \text{ odd.} \end{cases} \tag{2.4.5}$$

Thus, we see that there is a fundamental need to understand the properties of even and odd functions. This section summarizes the basic features of these functions.

Our interest will be on functions defined on the symmetric interval $(-a, a)$, $a > 0$, where a may be unbounded.

Definition The function $f(x)$ is an even function on the interval $(-a, a)$ if and only if

$$f(-x) = f(x). \tag{2.4.6}$$

Definition The function $f(x)$ is an odd function on the interval $(-a, a)$ if and only if

$$f(-x) = -f(x). \tag{2.4.7}$$

Let $g(x)$ be defined on the interval $(-a, a)$, then $g(x)$ can be written

$$g(x) = g_{(+)}(x) + g_{(-)}(x), \tag{2.4.8}$$

where

$$g_{(+)}(x) \equiv \frac{g(x) + g(-x)}{2}, \tag{2.4.9}$$

$$g_{(-)}(x) \equiv \frac{g(x) - g(-x)}{2}, \tag{2.4.10}$$

and

$$g_{(+)}(-x) = g_{(+)}(x), \qquad g_{(-)}(-x) = -g_{(-)}(x). \tag{2.4.11}$$

The functions, $g_{(+)}(x)$ and $g_{(-)}(x)$, are called, respectively, the even and odd parts of $g(x)$. This construction shows that the breakup into even and odd parts is unique.

It follows from the definitions of even and odd functions that the following statements are true:

(i) The product of two even functions is an even function, i.e.,

$$f(-x) = f(x), \quad g(-x) = g(x) \Rightarrow \begin{cases} h(-x) = +h(x), \\ \text{where } h(x) = f(x)g(x). \end{cases} \tag{2.4.12}$$

(ii) The product of two odd functions is an even function, i.e.,

$$f(-x) = -f(x), \quad g(-x) = -g(x) \Rightarrow \begin{cases} h(-x) = +h(x), \\ \text{where } h(x) = f(x)g(x). \end{cases}$$
$$\tag{2.4.13}$$

(iii) The product of an even and an odd function is an odd function, i.e.,

$$f(-x) = f(x), \quad g(-x) = -g(x) \Rightarrow \begin{cases} h(-x) = -h(x), \\ \text{where } h(x) = f(x)g(x). \end{cases} \quad (2.4.14)$$

(iv) If $f(x)$ is even and also an integrable function on the interval $(-a, a)$, then

$$\int_{-a}^{a} f(x)dx = 2 \int_{0}^{a} f(x)dx. \quad (2.4.15)$$

This result can be proved by first defining I_- and I_+ as follows,

$$I_+ \equiv \int_{0}^{a} f(x)dx, \qquad I_- \equiv \int_{-a}^{0} f(x)dx, \quad (2.4.16)$$

and writing the left-side integral in Eq. (2.4.15) as

$$\int_{-a}^{a} f(x)dx = I_- + I_+. \quad (2.4.17)$$

Then, using the substitution, $x = -y$, we obtain

$$I_- = \int_{-a}^{0} f(x)dx = -\int_{a}^{0} f(-y)dy = \int_{0}^{a} f(y)dy. \quad (2.4.18)$$

Therefore, $I_- = I_+$ and the result given in Eq. (2.4.15) immediately follows.

(v) If $f(x)$ is an odd function and integrable over the interval $(-a, a)$, then

$$\int_{-a}^{a} f(x)dx = 0. \quad (2.4.19)$$

This result is a consequence of the fact that for this case $I_- = -I_+$.

(vi) If $f(x)$ is an even (odd) function over the interval $(-a, a)$, then its derivative is an odd (even) function on the same interval, i.e.,

$$\begin{cases} f(x) \text{ even} \quad \Rightarrow \quad \dfrac{df(x)}{dx} \text{ odd}, \\ f(x) \text{ odd} \quad \Rightarrow \quad \dfrac{df(x)}{dx} \text{ even}. \end{cases} \quad (2.4.20)$$

We show the correctness of these statements for the case where $f(x)$ is even. Define

$$y(x) \equiv \frac{df(x)}{dx}. \quad (2.4.21)$$

Then,

$$y(-x) = \frac{df(-x)}{d(-x)} = -\frac{df(x)}{dx}, \quad (2.4.22)$$

or

$$y(-x) = -y(x). \quad (2.4.23)$$

Thus, if $f(x)$ is even, its derivative is odd. Similar reasoning can be applied to the second part of the result given in Eq. (2.4.20).

2.5   Fourier Series

A number of important physical systems are modeled by differential equations for which the relevant solutions are only needed over a finite interval, say, $0 \leq x \leq L$. Particular examples include the vibrating mechanical string and the quantum mechanical particle-in-a-box. In both cases, the solutions correspond to linear combinations of sine and cosine functions. More generally, we may have to study systems over the full x-axis where the associated solutions oscillate. In either case, the method of Fourier series solutions can play an important role in representing the periodic behavior of the dynamics of these systems.

This section introduces the elements of Fourier series. Critical concepts, definitions, and theorems are stated without proof. The bibliography lists several books that provide both extensions of our results, proofs of theorems, and applications. Two very nice and concise general, but elementary treatments of Fourier series are:

(i) S. L. Ross, *Differential Equations* (Blaisdell, Waltham, MA; 1964); see section 12.4.

(ii) J. D. Logan, *Applied Mathematics, A Contemporary Approach* (Wiley-Interscience, New York, 1987); see pps. 178–182.

Our treatment of Fourier series is closely based on their presentations.

For future reference, we make note of the following integral properties of the sine and cosine functions:

$$\int_{-L}^{L} \cos\left(\frac{m\pi x}{L}\right) \cos\left(\frac{n\pi x}{L}\right) dx = 0, \qquad m \neq n, \tag{2.5.1}$$

$$\int_{-L}^{L} \sin\left(\frac{m\pi x}{L}\right) \sin\left(\frac{n\pi x}{L}\right) dx = 0, \qquad m \neq n, \tag{2.5.2}$$

$$\int_{-L}^{L} \cos\left(\frac{m\pi x}{L}\right) \sin\left(\frac{n\pi x}{L}\right) dx = 0, \qquad m \neq n, \tag{2.5.3}$$

$$\int_{-L}^{L} \cos^2\left(\frac{n\pi x}{L}\right) dx = \int_{-L}^{L} \sin^2\left(\frac{n\pi x}{L}\right) dx = L, \tag{2.5.4}$$

where m and n are specified by

$$\begin{cases} m = 0, 1, 2, 3 \ldots, \\ n = 1, 2, 3, \ldots. \end{cases} \tag{2.5.5}$$

Note that $n = 0$ is not required, since for this case the associated sine function is zero.

Definition Let $f(x)$ be defined on the interval, $-L < x < L$, and is such that the following two integrals exist,

$$\int_{-L}^{L} f(x) \cos\left(\frac{n\pi x}{L}\right) dx, \qquad \int_{-L}^{L} f(x) \sin\left(\frac{n\pi x}{L}\right) dx, \qquad (2.5.6)$$

for $n = 0, 1, 2, \ldots$. Let (a_n, b_n) be defined as

$$\begin{cases} a_n \equiv \left(\dfrac{1}{L}\right) \displaystyle\int_{-L}^{L} f(x) \cos\left(\dfrac{n\pi x}{L}\right) dx, & (n = 0, 1, 2, \ldots), \\ b_n \equiv \left(\dfrac{1}{L}\right) \displaystyle\int_{-L}^{L} f(x) \sin\left(\dfrac{n\pi x}{L}\right) dx, & (n = 1, 2, \ldots). \end{cases} \qquad (2.5.7)$$

Then the (formal) *Fourier series* of $f(x)$ on, $-L < x < L$, is given by the expression,

$$f(x) \sim \frac{a_0}{2} + \sum_{n=1}^{\infty} \left[a_n \cos\left(\frac{n\pi x}{L}\right) + b_n \sin\left(\frac{n\pi x}{L}\right) \right]. \qquad (2.5.8)$$

Definition Let $f(x)$ be defined on the interval, $0 \le x \le L$, and is such that the following integrals exist,

$$\int_{-L}^{L} f(x) \cos\left(\frac{n\pi x}{L}\right) dx, \qquad (n = 0, 1, 2, \ldots). \qquad (2.5.9)$$

Let a_n be defined as

$$a_n \equiv \left(\frac{2}{L}\right) \int_{0}^{L} f(x) \cos\left(\frac{n\pi x}{L}\right) dx, \qquad (n = 0, 1, 2 \ldots). \qquad (2.5.10)$$

Then the (formal) *Fourier cosine series* of $f(x)$ on, $0 \le x \le L$, is given by the expression,

$$f(x) \sim \frac{a_0}{2} + \sum_{n=1}^{\infty} a_n \cos\left(\frac{n\pi x}{L}\right). \qquad (2.5.11)$$

Definition Let $f(x)$ be defined on the interval, $0 \le x \le L$, and is such that the following integrals exist,

$$\int_{0}^{L} f(x) \sin\left(\frac{n\pi x}{L}\right) dx, \qquad (n = 1, 2, 3, \ldots). \qquad (2.5.12)$$

Let b_n be defined as

$$b_n \equiv \left(\frac{2}{L}\right) \int_{0}^{L} f(x) \sin\left(\frac{n\pi x}{L}\right) dx, \qquad (n = 1, 2, 3 \ldots). \qquad (2.5.13)$$

Then the (formal) *Fourier sine series* of $f(x)$ on, $0 \le x \le L$, is given by the expression,

$$f(x) \sim \sum_{n=0}^{\infty} b_n \sin\left(\frac{n\pi x}{L}\right). \qquad (2.5.14)$$

Note that if $f(x)$ is defined on the interval, $0 \le x \le L$, then it can be continued on to the interval, $-L < x < 0$, in, at least, two important ways. One is to extend it evenly to the symmetric interval, $-L < x < L$; the other is to extend it in such a way as to make $f(x)$ odd over the interval, $-L < x < L$. The following example illustrates how this can be done.

$$\begin{cases} f(x) = x, & 0 \le x \le L; \\ \text{Even extension: } f(x) = |x|, & -L \le x \le L; \\ \text{Odd extension: } f(x) = x, & -L \le x \le L. \end{cases} \qquad (2.5.15)$$

Definition A function $f(x)$ is said to be *piecewise continuous* on a finite interval, $a \le x \le b$, if this interval can be partitioned into a finite number of subintervals such that $f(x)$ is continuous in the interior of each of the subintervals and $f(x)$ has finite limits as x approaches either end point of each subinterval from its interior.

Thus, if $x = x_0$ is a point of discontinuity for $f(x)$ on $a \le x \le b$, then the one-sided limits, denoted by

$$f(x_0^+) = \lim_{x \to x_0^+} f(x), \qquad f(x_0^-) = \lim_{x \to x_0^-} f(x), \qquad (2.5.16)$$

both exist.

Definition A function $f(x)$ is said to be *piecewise smooth* on a finite interval, $a \le x \le b$, if both $f(x)$ and $f'(x)$ are piecewise continuous on $a \le x \le b$.

Theorem 2.5.1 *If $f(x)$ is piecewise smooth on the interval, $-L < x < L$, then its Fourier series, given by*

$$\frac{a_0}{2} + \sum_{n=1}^{\infty} \left[a_n \cos\left(\frac{n\pi x}{L}\right) + b_n \sin\left(\frac{n\pi x}{L}\right) \right], \qquad (2.5.17)$$

where

$$a_n = \left(\frac{1}{L}\right) \int_{-L}^{L} f(x) \cos\left(\frac{n\pi x}{L}\right) dx, \qquad b_n = \left(\frac{1}{L}\right) \int_{-L}^{L} f(x) \sin\left(\frac{n\pi x}{L}\right) dx,$$
$$(2.5.18)$$

converge at every point x to the value

$$\frac{f(x^+) + f(x^-)}{2}, \qquad (2.5.19)$$

where $f(x^+)$ is the right-hand limit of the function f at x and $f(x^-)$ is the left-hand limit of the function f at x.

Note that if f is continuous at x, then the theorem implies that the Fourier series of f at x converges to $f(x)$.

Another Fourier type theorem can be derived for the class of functions that are square integrable on the interval, $-L < x < L$, i.e.,

$$\int_{-L}^{L} [f(x)]^2 dx < \infty. \tag{2.5.20}$$

For this case, the theorem reads:

Theorem 2.5.2 *If $f(x)$ is defined on the interval, $-L \le x \le L$, and is square integrable, then the Fourier series associated with $f(x)$ converges in the mean to $f(x)$, i.e.,*

$$\lim_{N \to \infty} \int_{-L}^{L} [f(x) - f_N(x)]^2 \, dx = 0, \tag{2.5.21}$$

where

$$f_N(x) \equiv \frac{a_0}{2} + \sum_{m=1}^{N} \left[a_m \cos\left(\frac{m\pi x}{L}\right) + b_m \sin\left(\frac{m\pi x}{L}\right) \right]. \tag{2.5.22}$$

Fourier series may be integrated, usually without causing difficulties. The following theorem is useful for many situations.

Theorem 2.5.3 *The Fourier series for $f(x)$ can be integrated term by term from a to x, and the resulting series will converge uniformly to*

$$\int_{a}^{x} f(u) du \tag{2.5.23}$$

provided that $f(x)$ is piecewise continuous on the interval, $[-L, L]$ and both a and x are in this interval, i.e.,

$$-L \le a < x \le L. \tag{2.5.24}$$

However, differentiation of Fourier series may give rise to invalid expressions. The following theorem [1] can often be applied and provides the needed conditions for knowing when this operation can be performed on many Fourier series.

Theorem 2.5.4 *Let $f(x)$ be continuous on $(-\infty, \infty)$, be periodic of period $2L$, and have a derivative $f'(x)$ that is continuous except at a finite number of points on $[-L, L]$. Let either (i) $|f'(x)|$ be bounded, or (ii) $f'(x)$ be integrable. Then both $f(x)$ and $f'(x)$ will have Fourier coefficients and the Fourier series of $f'(x)$ will be obtained from that of $f(x)$ by term by term differentiation. Under the conditions required by this theorem, if $f(x)$ and $f'(x)$ have, respective, Fourier coefficients $\{a_n, b_n : n = 0, 1, 2, \ldots\}$ and $\{\bar{a}_n, \bar{b}_n : n = 1, 2, \ldots\}$, then they are related as follows,*

$$\bar{a}_n = \left(\frac{n\pi}{L}\right) b_n, \qquad \bar{b}_n = -\left(\frac{n\pi}{L}\right) a_n. \tag{2.5.25}$$

2.6 Worked Examples for Fourier Series

2.6.1 *Parseval's Identity*

This relation is given by the expression,

$$\left(\frac{1}{L}\right)\int_{-L}^{L}[f(x)]^2 dx = \frac{a_0^2}{2} + \sum_{n=1}^{\infty}(a_n^2 + b_n^2). \tag{2.6.1}$$

We now derive it starting with the Fourier series for $f(x)$, i.e.,

$$f(x) = \frac{a_0}{2} + \sum_{n=1}^{\infty}\left[a_n\cos\left(\frac{n\pi x}{L}\right) + b_n\sin\left(\frac{n\pi x}{L}\right)\right]. \tag{2.6.2}$$

Therefore,

$$\int_{-L}^{L}[f(x)]^2 dx = \int_{-L}^{L}\left\{\frac{a_0}{2} + \sum_{n=1}^{\infty}\left[a_n\cos\left(\frac{n\pi x}{L}\right) + b_n\sin\left(\frac{n\pi x}{L}\right)\right]\right\}$$

$$\cdot\left\{\frac{a_0}{2} + \sum_{m=1}^{\infty}\left[a_m\cos\left(\frac{m\pi x}{L}\right) + b_m\sin\left(\frac{m\pi x}{L}\right)\right]\right\}dx$$

$$= \left(\frac{a_0^2}{4}\right)\int_{-L}^{L}dx$$

$$+ 2\left(\frac{a_0}{2}\right)\int_{-L}^{L}\sum_{n=1}^{\infty}\left[a_n\cos\left(\frac{n\pi x}{L}\right) + b_n\sin\left(\frac{n\pi x}{L}\right)\right]dx$$

$$+ \int_{-L}^{L}\sum_{n=1}^{\infty}\sum_{m=1}^{\infty}\left[a_n\cos\left(\frac{n\pi x}{L}\right) + b_n\sin\left(\frac{n\pi x}{L}\right)\right]$$

$$\cdot\left[a_m\cos\left(\frac{m\pi x}{L}\right) + b_m\sin\left(\frac{m\pi x}{L}\right)\right]dx$$

$$= \left(\frac{a_0^2}{4}\right)(2L) + 2\left(\frac{a_0}{2}\right)(0)$$

$$+ \sum_{n=1}^{\infty}\sum_{m=1}^{\infty}\left[a_n a_m L\delta_{nm} + b_n b_m L\delta_{nm} + a_n b_m(0) + a_m b_n(0)\right]$$

$$= L\left[\frac{a_0^2}{2} + \sum_{n=1}^{\infty}(a_n^2 + b_n^2)\right], \tag{2.6.3}$$

where δ_{nm} is Kronecker delta function, i.e.,

$$\delta_{nm} = \begin{cases} 1, & \text{if } m = n, \\ 0, & \text{if } m \neq n, \end{cases} \tag{2.6.4}$$

and we have used the integral relations given in Eqs. (2.5.1) to (2.5.4). Dividing by L gives the result of Eq. (2.6.1).

One of the implications of Parseval's identity is that

$$\lim_{n \to \infty} |a_n| = 0, \qquad \lim_{n \to \infty} |b_n| = 0. \tag{2.6.5}$$

2.6.2 $f(x) = |x|, \ -\pi \leq x \leq \pi$

This function is even and the coefficients for the sine terms will be zero, i.e., $b_n = 0$ for $n = 1, 2, 3, \ldots$. Therefore, $L = \pi$ and a_n is to be calculated from the expression,

$$a_n = \frac{1}{\pi} \int_{-\pi}^{\pi} |x| \cos(nx) dx = \left(\frac{2}{\pi} \right) \int_{0}^{\pi} x \cos(nx) dx$$

$$= \left(\frac{2}{\pi} \right) \left[\left(\frac{1}{n^2} \right) \cos(nx) + \left(\frac{1}{n} \right) x \sin(nx) \right]_{0}^{\pi}$$

$$= \left(\frac{2}{\pi n^2} \right) [(-1)^n - 1]. \tag{2.6.6}$$

Therefore,

$$a_n = \begin{cases} 0, & \text{for } n = 2, 4, 6, \ldots, \\ -\left(\dfrac{4}{\pi n^2} \right), & \text{for } n = 1, 3, 5, \ldots, \end{cases} \tag{2.6.7}$$

and

$$a_0 = \left(\frac{2}{\pi} \right) \int_{0}^{\pi} x \, dx = \pi. \tag{2.6.8}$$

Thus, the Fourier series is

$$f(x) = |x| = \frac{\pi}{2} - \left(\frac{4}{\pi} \right) \left[\cos x + \frac{\cos(3x)}{3^2} + \frac{\cos(5x)}{5^2} + \cdots \right]$$

$$= \frac{\pi}{2} - \left(\frac{4}{\pi} \right) \sum_{n=1}^{\infty} \frac{\cos(2n-1)x}{(2n-1)^2}, \qquad -\pi \leq x \leq \pi. \tag{2.6.9}$$

2.6.3 $f(x) = x, \ -\pi < x < \pi$

This function is odd on the interval, $(-\pi, \pi)$. Therefore, all the cosine coefficients will be zero. For the b_n, we have,

$$b_n = \left(\frac{2}{\pi} \right) \int_{0}^{\pi} x \sin(nx) dx$$

$$= \left(\frac{2}{\pi} \right) \left[\left(\frac{1}{n^2} \right) \sin(nx) - \left(\frac{1}{n} \right) x \cos(nx) \right]_{0}^{\pi}$$

$$= \left(\frac{2}{\pi}\right)\left(\frac{-1}{n}\right)\pi\cos(n\pi) = (-1)^{n+1}\left(\frac{2}{n}\right), \tag{2.6.10}$$

and

$$f(x) = x = 2\sum_{n=1}^{\infty}(-1)^{n+1}\left[\frac{\sin(nx)}{n}\right], \qquad -\pi < x < \pi. \tag{2.6.11}$$

2.6.4 The Square Wave Function

The square wave function can be defined as

$$g(x) = \begin{cases} 1, & \text{for } 0 < x < \pi, \\ -1, & \text{for } -\pi < x < 0. \end{cases} \tag{2.6.12}$$

With respect to the interval, $(-\pi, \pi)$, this function is odd and all the cosine coefficients, a_n, are zero. Therefore, we need to calculate the b_n; they are given by

$$b_n = \left(\frac{2}{\pi}\right)\int_0^{\pi}\sin(nx)dx = \left(\frac{2}{\pi}\right)\left[\frac{-\cos(nx)}{n}\right]_0^{\pi}$$

$$= \left(\frac{2}{\pi n}\right)\left[(-1)^{n+1} + 1\right] \tag{2.6.13}$$

and, finally,

$$b_n = \begin{cases} 0, & \text{for } n = 2, 4, 6, \ldots, \\ \frac{4}{\pi n}, & \text{for } n = 1, 3, 5, \ldots. \end{cases} \tag{2.6.14}$$

Therefore, the Fourier series for $g(x)$ is

$$g(x) = \left(\frac{4}{\pi}\right)\left[\sin x + \frac{\sin(3x)}{3} + \frac{\sin(5x)}{5} + \cdots\right]$$

$$= \left(\frac{4}{\pi}\right)\sum_{n=1}^{\infty}\frac{\sin(2n-1)x}{(2n-1)}, \qquad -\pi < x < \pi. \tag{2.6.15}$$

2.6.5 Comparison of Problems 2.6.2 and 2.6.4

In problems 2.6.2 and 2.6.4, we computed the Fourier series, respectively, for the two functions

$$f(x) = |x|, \qquad -\pi < x < \pi, \tag{2.6.16}$$

$$g(x) = \begin{cases} 1, & \text{for } 0 < x < \pi, \\ -1, & \text{for } -\pi < x < 0. \end{cases} \tag{2.6.17}$$

It should be clear that

$$g(x) = \frac{df(x)}{dx}, \qquad -\pi < x < \pi. \tag{2.6.18}$$

Comparing the corresponding Fourier series, it is evident that the Fourier series for $g(x)$ is the derivative of the Fourier series for $f(x)$. These results are consistent with Theorem 2.5.4.

To further illustrate when (or not) the derivative of a Fourier series represents the derivative of the original function, consider the function in problem 2.6.3, i.e.,

$$f(x) = x, \qquad -\pi < x < \pi. \tag{2.6.19}$$

The associated sine Fourier series is, from Eq. (2.6.11),

$$f(x) = \left(\frac{2}{\pi}\right) \sum_{n=1}^{\infty} (-1)^{n+1} \left[\frac{\sin(n\pi x)}{n}\right], \qquad -\pi < x < \pi. \tag{2.6.20}$$

The term by term differentiation of this series gives

$$2 \sum_{n=1}^{\infty} (-1)^{n+1} \cos(n\pi x) = 2[\cos(\pi x) - \cos(2\pi x) + \cdots], \tag{2.6.21}$$

and this corresponds to

$$a_n = (-1)^{n+1}, \qquad b_n = 0. \tag{2.6.22}$$

But,

$$\left(\frac{1}{\pi}\right) \int_{-\pi}^{\pi} [f(x)]^2 dx = \left(\frac{1}{\pi}\right) \int_{-\pi}^{\pi} x^2 dx = \left(\frac{2}{3}\right) \pi^2, \tag{2.6.23}$$

and

$$\sum_{n=1}^{\infty} (a_n^2 + b_n^2) = \sum_{n=1}^{\infty} (1 + 0) = \infty. \tag{2.6.24}$$

Therefore, we conclude that the Parseval identity is not satisfied and the expression given by Eq. (2.6.21) is not a Fourier series. Note that this result is a consequence of the series not converging. From the perspective of Theorem 2.5.4, this result follows from this particular $f(x)$ not being continuous over the interval $(-\infty, \infty)$.

2.6.6 $f(x) = \sin x,\ 0 < x < \pi$

This function has the following even and odd extensions, on the interval, $-\pi < x < \pi$,

$$\begin{cases} \text{Even extension: } f_e(x) = |\sin x|, \\ \text{Odd extension: } f_o(x) = \sin x. \end{cases} \qquad (2.6.25)$$

Note the Fourier series for the odd extension is

$$f_o(x) = \sin x, \qquad -\pi < x < \pi, \qquad (2.6.26)$$

i.e., it's just the function itself. Since $f_o(x)$ is continuous on the interval, $(-\infty, \infty)$, its derivative can be taken to give the Fourier series for $f'_o(x)$,

$$\frac{df_o(x)}{dx} = \cos x, \qquad -\pi < x < \pi. \qquad (2.6.27)$$

The even extension can be calculated by determining the cosine Fourier series, for

$$f(x) = \sin x, \qquad 0 < x < \pi, \qquad (2.6.28)$$

and using the series for $f(x)$ on the interval $(-\pi, \pi)$. Therefore, we have $b_n = 0$ and

$$\begin{aligned} a_n &= \left(\frac{2}{\pi}\right) \int_0^\pi \sin x \cos(nx)dx \\ &= \left(\frac{1}{\pi}\right) \int_0^\pi [\sin(n+1)x - \sin(n-1)x]dx \\ &= \left(\frac{1}{\pi}\right) \left[-\frac{\cos(n+1)x}{n+1} + \frac{\cos(n-1)x}{n-1} \right]_0^\pi \\ &= -\left(\frac{2}{\pi}\right) \left[\frac{1 + \cos(n\pi)}{n^2 - 1} \right], \qquad n \neq 1, \\ &= -\left(\frac{2}{\pi}\right) \left[\frac{1 + (-1)^n}{n^2 - 1} \right], \qquad n \neq 1. \qquad (2.6.29) \end{aligned}$$

The two cases, $n = 0$ and 1, have to be now calculated; they are

$$a_0 = \left(\frac{2}{\pi}\right) \int_0^\pi \sin x\, dx = \left(\frac{2}{\pi}\right) (-\cos x)\Big|_0^\pi = \frac{4}{\pi}, \qquad (2.6.30)$$

$$a_1 = \left(\frac{2}{\pi}\right) \int_0^\pi \sin x \cos x\, dx = \left(\frac{1}{\pi}\right) \sin^2 x\Big|_0^\pi = 0. \qquad (2.6.31)$$

Therefore, the Fourier series for

$$f(x) = \begin{cases} \sin x, & 0 < x < \pi, \\ \text{or} \\ |\sin x|, & -\pi < x < \pi, \end{cases} \qquad (2.6.32)$$

is given by

$$
\begin{aligned}
f(x) &= \left(\frac{2}{\pi}\right) - \left(\frac{2}{\pi}\right) \sum_{n=2}^{\infty} \left[\frac{1 + (-1)^n}{n^2 - 1}\right] \cos(nx) \\
&= \left(\frac{2}{\pi}\right) - \left(\frac{4}{\pi}\right) \left[\frac{\cos(2n)}{2^2 - 1} + \frac{\cos(4x)}{4^2 - 1} + \cdots\right] \\
&= \left(\frac{2}{\pi}\right) - \left(\frac{4}{\pi}\right) \sum_{k=1}^{\infty} \frac{\cos(2kx)}{(2k)^2 - 1}. \qquad (2.6.33)
\end{aligned}
$$

Examination of $f_e(x) = |\sin x|$ shows that it is continuous on the interval, $(-\infty, \infty)$. Applying Theorem 2.5.4 allows us to conclude that the Fourier series of $f'(x)$ can be determined by taking the term by term derivative of Eq. (2.6.33), i.e.,

$$
\begin{aligned}
f'(x) &= \begin{cases} \cos x, & 0 < x < \pi, \\ -\cos x, & -\pi < x < 0, \end{cases} \\
&= \left(\frac{8}{\pi}\right) \sum_{k=1}^{\infty} \frac{k \sin(2kx)}{2k^2 - 1}. \qquad (2.6.34)
\end{aligned}
$$

2.6.7 *Fourier Series of x^2, $-\pi < x < \pi$ by Integration*

From problem 2.6.3, we have

$$x = 2 \sum_{n=1}^{\infty} (-1)^{n+1} \left[\frac{\sin(nx)}{n}\right], \qquad -\pi < x < \pi. \qquad (2.6.35)$$

Integrating both sides gives, on multiplying by two,

$$x^2 = c + 4 \sum_{n=1}^{\infty} (-1)^n \left[\frac{\cos(nx)}{n^2}\right], \qquad (2.6.36)$$

where the integration constant c is given by

$$c = 4 \sum_{n=1}^{\infty} \frac{(-1)^{n+1}}{n^2}. \qquad (2.6.37)$$

(Note that this value of c is gotten by setting $x = 0$, evaluating both sides of Eq. (2.6.36), and then solving for c.) This constant can also be determined by use of the fact that Eq. (2.6.36) is a Fourier cosine series and

$$c = \frac{a_0}{2} = \left(\frac{1}{\pi}\right) \int_0^\pi x^2 dx = \left(\frac{1}{\pi}\right)\left(\frac{\pi^3}{3}\right) = \frac{\pi^2}{3}. \qquad (2.6.38)$$

Therefore, comparing Eqs. (2.6.37) and (2.6.38), we obtain

$$\sum_{n=1}^\infty \frac{(-1)^{n+1}}{n^2} = \frac{\pi^2}{12}. \qquad (2.6.39)$$

Finally the Fourier series for

$$f(x) = x^2, \qquad -\pi < x < \pi, \qquad (2.6.40)$$

is

$$x^2 = \frac{\pi^2}{3} + 4\sum_{n=1}^\infty (-1)^n \left[\frac{\cos(nx)}{n^2}\right]. \qquad (2.6.41)$$

2.6.8 *Riemann-Lebesgue Theorem*

This result states that if $f(x)$ is absolutely integrable over the interval $(-L, L)$, then

$$\lim_{n\to\infty} \int_{-\pi}^\pi f(x)\sin(nx)dx = 0, \qquad \lim_{n\to\infty} \int_{-\pi}^\pi f(x)\cos(nx)dx = 0. \quad (2.6.42)$$

This result follows from the definition of the Fourier coefficients,

$$a_n = \left(\frac{1}{\pi}\right) \int_{-\pi}^\pi f(x)\cos(nx)dx, \quad b_n = \left(\frac{1}{\pi}\right) \int_{-\pi}^\pi f(x)\sin(nx)dx, \quad (2.6.43)$$

and the Parseval identity

$$\left(\frac{1}{\pi}\right) \int_{-\pi}^\pi [f(x)]^2 dx = \frac{a_0^2}{2} + \sum_{n=1}^\infty (a_n^2 + b_n^2). \qquad (2.6.44)$$

If the series of Eq. (2.6.44) is to converge, then

$$\lim_{n\to\infty} a_n = 0, \qquad \lim_{n\to\infty} b_n = 0. \qquad (2.6.45)$$

2.7   Fourier Series for $(\cos\theta)^\alpha$ and $(\sin\theta)^\alpha$

The following nonlinear, second-order differential equation

$$\frac{d^2x}{dt^2} + x^{1/3} = 0, \tag{2.7.1}$$

can be shown to have all of its solutions periodic. Using

$$y = \frac{dx}{dt} \Rightarrow \frac{d^2x}{dt^2} = y\frac{dy}{dx}, \tag{2.7.2}$$

then Eq. (2.7.1) becomes

$$y\frac{dy}{dx} + x^{1/3} = 0. \tag{2.7.3}$$

Integrating once gives

$$\frac{y^2}{2} + \left(\frac{3}{4}\right)x^{4/3} = E \ge 0. \tag{2.7.4}$$

In the (x, y) phase-plane, $y(x)$ is a closed curve for any constant $E > 0$.
Hence, all solutions are periodic. While Eq. (2.7.1) does not have a general
solution that can be expressed in terms of a finite number of the elementary
functions, methods do exist for constructing analytic approximations to
these solutions. For such techniques, a knowledge of the Fourier expansions
of $(\cos\theta)^\alpha$ and $(\sin\theta)^\alpha$ are needed, where α is a fraction satisfying the
constraint [2],

$$\alpha = \frac{2n+1}{2m+1}, \qquad (m, n : 0, 1, 2 \ldots). \tag{2.7.5}$$

Likewise, other related nonlinear oscillatory systems can require a knowledge of the Fourier series of $(\cos\theta)^\beta$ and $(\sin\theta)^\beta$ where [3],

$$\beta = \frac{2n}{2m+1}, \qquad (m, n : 1, 2, 3 \ldots). \tag{2.7.6}$$

While general expressions can be obtained for all of the Fourier coeffi-
cients, we will present explicit results for the cases where $\alpha = \frac{1}{3}$, and $\beta = \frac{2}{3}$
[4]. One place to start is to use the known formula [5],

$$\int_0^{\pi/2} (\cos\theta)^{\nu-1}\cos(a\theta)d\theta = \frac{\pi}{(2^\nu)\nu B\left(\frac{\nu+a+1}{2}, \frac{\nu-a+1}{2}\right)}, \tag{2.7.7}$$

where $B(p, q)$ is the beta function,

$$B(p, q) \equiv \frac{\Gamma(p)\Gamma(q)}{\Gamma(p+q)}, \tag{2.7.8}$$

which is defined in terms of the so-called gamma function. Since $(\cos\theta)^{1/3}$ is an even function of θ, the sine terms in the Fourier series will not occur because their coefficients are zero, i.e.,

$$b_n = 0, \qquad (n = 1, 2, 3, \dots). \tag{2.7.9}$$

The coefficients for the cosine terms are given by the expression

$$
\begin{aligned}
a_n &= \left(\frac{1}{\pi}\right) \int_{-\pi}^{\pi} (\cos\theta)^{1/3} \cos(n\theta)d\theta \\
&= \left(\frac{2}{\pi}\right) \int_{0}^{\pi} (\cos\theta)^{1/3} \cos(n\theta)d\theta \\
&= \left(\frac{2}{\pi}\right) \int_{0}^{\pi/2} (\cos\theta)^{1/3} \cos(n\theta)d\theta \\
&\quad + \left(\frac{2}{\pi}\right) \int_{\pi/2}^{\pi} (\cos\theta)^{1/3} \cos(n\theta)d\theta.
\end{aligned}
\tag{2.7.10}
$$

If we replace θ by $(\pi - \theta)$ in the second integral, we find that

$$a_n = \left(\frac{2}{\pi}\right) [1 + (-1)^{n+1}] \int_{0}^{\pi/2} (\cos\theta)^{1/3} \cos(n\theta)d\theta. \tag{2.7.11}$$

Note that if n is even, then $a_n = 0$. Therefore only odd n coefficients will contribute; we now write them as

$$a_{2p+1} = \left(\frac{4}{\pi}\right) \int_{0}^{\pi/2} (\cos\theta)^{1/3} \cos(2p+1)\theta\, d\theta, \quad (p = 0, 1, 2, \dots). \tag{2.7.12}$$

Using the formula of Eq. (2.7.7), it follows that

$$a_{2p+1} = \frac{3\Gamma\left(\frac{7}{3}\right)}{2^{4/3}\Gamma\left(p + \frac{5}{3}\right)\Gamma\left(\frac{2}{3} - p\right)}, \tag{2.7.13}$$

and, in particular [6],

$$a_1 = \frac{\Gamma\left(\frac{7}{3}\right)}{2^{1/3}\left[\Gamma\left(\frac{5}{3}\right)\right]^2} = \frac{\Gamma\left(\frac{1}{3}\right)}{2^{1/3}\left[\Gamma\left(\frac{2}{3}\right)\right]^2}. \tag{2.7.14}$$

Numerically, a_1 has the value,

$$a_1 = 1.159595266963929. \tag{2.7.15}$$

It follows from Eq. (2.7.13) that

$$\frac{a_{2p+1}}{a_{2p-1}} = -\left(\frac{3p-2}{3p+2}\right). \tag{2.7.16}$$

Two conclusions can be reached from this result:

(i) For $p \geq 1$, the coefficients a_{2p+1}, alternate in sign.

(ii) All of the coefficients can be expressed as the product of a_1 and a known numerical factor that is a rational number.

Using Eq. (2.7.16) repeatedly, we obtain

$$(\cos\theta)^{1/3} = a_1 \left[\cos\theta - \frac{\cos(3\theta)}{5} + \frac{\cos(5\theta)}{10} - \frac{7\cos(7\theta)}{110} \right.$$

$$+ \frac{\cos(9\theta)}{22} - \frac{13\cos(11\theta)}{374}$$

$$\left. + \frac{26\cos(13\theta)}{935} - \frac{494\cos(15\theta)}{21505} + \cdots \right]. \qquad (2.7.17)$$

If we replace θ by $\left(\frac{\pi}{2} - \theta\right)$ in Eq. (2.7.12), we find that

$$\begin{cases} a_{2p+1} = (-1)^p \left(\frac{4}{\pi}\right) \int_0^{\pi/2} (\sin\theta)^{1/3} \sin(2p+1)\theta \, d\theta \\ p = 0,1,2,\ldots, \end{cases} \qquad (2.7.18)$$

and conclude that the Fourier series for $(\sin\theta)^{1/3}$ is,

$$(\sin\theta)^{1/3} = \sum_{p=0}^{\infty} (-1)^p a_{2p+1} \sin(2p+1)\theta. \qquad (2.7.19)$$

In a similar manner, the following expression are obtained for $(\cos\theta)^{2/3}$ and $(\sin\theta)^{2/3}$,

$$(\cos\theta)^{2/3} = a_0 \left[\frac{1}{2} + \frac{\cos(2\theta)}{4} - \frac{\cos(4\theta)}{14} + \frac{\cos(6\theta)}{28} - \frac{2\cos(8\theta)}{91} + \frac{11\cos(10\theta)}{728} \right.$$

$$\left. - \frac{11\cos(12\theta)}{988} + \frac{17\cos(14\theta)}{1976} - \frac{17\cos(16\theta)}{2470} + \cdots \right], \qquad (2.7.20)$$

$$(\sin\theta)^{2/3} = a_0 \left[\frac{1}{2} + \frac{\cos(2\theta)}{4} + \frac{\cos(4\theta)}{14} + \frac{\cos(6\theta)}{28} + \frac{2\cos(8\theta)}{91} + \frac{11\cos(10\theta)}{728} \right.$$

$$\left. + \frac{11\cos(12\theta)}{988} + \frac{17\cos(14\theta)}{1976} + \frac{17\cos(16\theta)}{2470} + \cdots \right], \qquad (2.7.21)$$

where

$$a_0 = \frac{3 \cdot 2^{4/3} \Gamma\left(\frac{2}{3}\right)}{\left[\Gamma\left(\frac{1}{3}\right)\right]^2} = 1.426348256. \qquad (2.7.22)$$

2.8 Fourier Transforms

Thus far, this chapter has only considered periodic functions on an interval $(-L, L)$. These functions can be extended to the whole x-axis by use of the periodicity relation,

$$\begin{cases} f[x \pm n(2L)] = f(x), \\ n = 1, 2, 3, \ldots. \end{cases} \tag{2.8.1}$$

An issue can be raised as to how functions "periodic" over the infinite interval, $-\infty < x < \infty$, can be represented? Problems of this sort occur in systems describing scattering of waves. In particular, the fields of quantum mechanics, optics, and underwater acoustics are important areas where the propagation of waves are studied over an effective unbounded space range.

This chapter briefly introduces an integral transform method called Fourier transforms. After defining the conditions under which the transforms exist, we present a summary of their more significant properties. In the next section, we provide several examples which illustrate the application of the Fourier transforms to several problems of importance in the sciences.

2.8.1 *Definition of Fourier Transforms*

Assume that a function $f(x)$ satisfy the following conditions:
(i) For every finite interval (a, b), $f(x)$ is piecewise smooth.
(ii) $f(x)$ is an absolutely integrable function in $(-\infty, \infty)$, i.e.,

$$\int_{-\infty}^{\infty} |f(x)| dx < \infty. \tag{2.8.2}$$

Under these conditions the Fourier transform of $f(x)$ is defined by the following integral,

$$F(k) = \frac{1}{\sqrt{2\pi}} \int_{-\infty}^{\infty} f(x) e^{ikx} dx. \tag{2.8.3}$$

The major usefulness of the Fourier transform is that $F(x)$ can be written as

$$f(x) = \frac{1}{\sqrt{2\pi}} \int_{-\infty}^{\infty} F(k) e^{-ikx} dk. \tag{2.8.4}$$

Note that the exponential factors in the integrands have opposite signs. Also, observe that

$$e^{\pm ikx} = \cos(kx) \pm i \sin(kx), \tag{2.8.5}$$

and, further, since $-\infty < k < \infty$, the integrals may be interpreted as "sums" over every possible wavelength or period. It is this feature that indicates the nature of the generalization from Fourier series having a given period, to Fourier transforms that involve all periods.

Functions $f(x)$ that are either even or odd can be rewritten to forms reflecting these symmetries:

(i) Let $f(x)$ be an even function, i.e.,

$$f(-x) = f(x); \qquad (2.8.6)$$

then the Fourier transforms are,

$$F_c(k) = \sqrt{\frac{2}{\pi}} \int_0^\infty f(x)\cos(kx)dx, \qquad (2.8.7)$$

$$f(k) = \sqrt{\frac{2}{\pi}} \int_0^\infty F_c(x)\cos(kx)dk. \qquad (2.8.8)$$

These functions are called the Fourier cosine transforms of each other.

(ii) Let $f(x)$ be an odd function, i.e.,

$$f(-x) = -f(x); \qquad (2.8.9)$$

then the Fourier sine transforms are,

$$F_s(k) = \sqrt{\frac{2}{\pi}} \int_0^\infty f(x)\sin(kx)dx, \qquad (2.8.10)$$

$$f(k) = \sqrt{\frac{2}{\pi}} \int_0^\infty F_s(x)\sin(kx)dk. \qquad (2.8.11)$$

2.8.2 *Basic Properties of Fourier Transforms*

The following is a listing of the fundamental properties of the Fourier transforms. In most cases, the proofs are rather direct to obtain. However, the books listed in reference [7] provide the details of the required proofs. We use the following notation to denote, respectively, the operation of obtaining the Fourier transform,

$$F(k) = \mathcal{F}[f(x); k]. \qquad (2.8.12)$$

1. Parseval's relation:

$$\int_{-\infty}^\infty |F(k)|^2 dk = \int_{-\infty}^\infty |f(x)|^2 dx. \qquad (2.8.13)$$

2. Let $f(x)$ and $g(x)$ have Fourier transforms $F(k)$ and $G(k)$, and let (c_1, c_2) be arbitrary constants; then

$$\mathcal{F}[c_1 f(x) + c_2 g(x)] = c_1 F(k) + c_2 G(k). \tag{2.8.14}$$

3. Let $a \neq 0$, be an arbitrary constant, then

$$\mathcal{F}[f(ax); k] = \left(\frac{1}{|a|}\right) F\left(\frac{k}{a}\right). \tag{2.8.15}$$

4. Let a be an arbitrary real constant, then

$$\mathcal{F}[f(x - a); k] = e^{ika} F(k). \tag{2.8.16}$$

5. Let a be an arbitrary real constant, then

$$\mathcal{F}[e^{iax} f(x); k] = F(k + a). \tag{2.8.17}$$

6. Let $n > 0$ be an integer. Let $f(x)$ have an nth derivative that is piecewise continuous and differentiable, and let each of the derivatives

$$f^{(m)}(x), \qquad (m : 0, 1, \ldots, n) \tag{2.8.18}$$

be absolutely integrable for the interval, $-\infty < x < \infty$, then,

$$\mathcal{F}[f^{(n)}(x); k] = (-ik)^n F(k). \tag{2.8.19}$$

7. The Fourier convolution of $f(x)$ and $g(x)$ is defined to be,

$$f * g \equiv \frac{1}{\sqrt{2\pi}} \int_{-\infty}^{\infty} f(x - t) g(t) dt. \tag{2.8.20}$$

The Fourier convolution has the following properties:

$$f * g = g * f; \tag{2.8.21}$$

$$f * (cg) = cf * g, \qquad c = \text{constant}; \tag{2.8.22}$$

$$\begin{cases} f * (c_1 g + c_2 h) = c_1 f * g + c_2 f * h, \\ \quad c_1 \text{ and } c_2 \text{ constants}; \end{cases} \tag{2.8.23}$$

$$\mathcal{F}[f * g; k] = F(k) G(k). \tag{2.8.24}$$

This last expression states that the product of the Fourier transforms of two functions $f(x)$ and $g(x)$ corresponds to the Fourier transform of their convolution; therefore,

$$\mathcal{F}[fg; k] \neq F(k) G(k). \tag{2.8.25}$$

Tables of Fourier transform pairs are compiled in many handbooks. See reference [8] for a listing.

2.9 Application of Fourier Transforms

2.9.1 *Fourier Transform of the Square Pulse*

The square pulse function is

$$f(x) = \begin{cases} 1, & \text{for } |x| < a, \\ 0, & \text{for } |x| > a. \end{cases} \tag{2.9.1}$$

The Fourier transform is calculated by evaluating the expression

$$F(k) = \left(\frac{1}{\sqrt{2\pi}}\right) \int_{-\infty}^{\infty} f(x) e^{ikx} = \left(\frac{1}{\sqrt{2\pi}}\right) \int_{-a}^{a} e^{ikx} dx$$

$$= \left(\frac{1}{\sqrt{2\pi}}\right) \left[\frac{e^{ikx}}{ik}\right]_{-a}^{a} = \sqrt{\frac{2}{\pi}} \frac{\sin(ka)}{k}, \tag{2.9.2}$$

where

$$F(0) = \lim_{k \to 0} F(k) = \sqrt{\frac{2}{\pi}} \cdot a. \tag{2.9.3}$$

2.9.2 *Fourier Transform of the Gaussian Function*

The Gaussian function is

$$f(x) = e^{-x^2}. \tag{2.9.4}$$

The corresponding Fourier transform is

$$F(k) = \left(\frac{1}{\sqrt{2}}\right) \int_{-\infty}^{\infty} e^{-x^2} e^{ikx} dx. \tag{2.9.5}$$

Now, the quantity in the exponent, $x^2 - ikx$, can be rewritten as follows,

$$x^2 - ikx = x^2 - ikx - \frac{k^2}{4} + \frac{k^2}{4} = \left(x - \frac{ik}{2}\right)^2 + \frac{k^2}{4}. \tag{2.9.6}$$

If we now define y as

$$y = x + \frac{ik}{2}, \tag{2.9.7}$$

then the integral of Eq. (2.9.6) becomes

$$F(k) = \left(\frac{1}{\sqrt{2\pi}}\right) e^{-k^2/4} \int_{-\infty}^{\infty} e^{-y^2} dy$$

$$= \left(\frac{1}{\sqrt{2\pi}}\right) \sqrt{\pi} e^{-k^2/4} = \left(\frac{1}{\sqrt{2}}\right) e^{-k^2/4}. \tag{2.9.8}$$

We therefore reach the interesting conclusion that the Fourier transform of a Gaussian function, in x, is a Gaussian function in k.

2.9.3 The Convolution Theorem

Let $f(x)$ and $g(x)$ have, respective, Fourier transforms $F(k)$ and $G(k)$. The convolution of $f(x)$ and $g(x)$ is defined to be the function

$$f * g = c(x) \equiv \left(\frac{1}{\sqrt{2\pi}}\right) \int_{-\infty}^{\infty} f(x-y)g(y)dy. \qquad (2.9.9)$$

Now,

$$f(x-y) = \left(\frac{1}{\sqrt{2\pi}}\right) \int_{-\infty}^{\infty} F(k)e^{-ik(x-y)}dk, \qquad (2.9.10)$$

and substituting this into the integral of Eq. (2.9.9) gives,

$$\begin{aligned}
c(x) &= \left(\frac{1}{2\pi}\right) \int_{-\infty}^{\infty} g(y) \left[\int_{-\infty}^{\infty} F(k)e^{-ik(x-y)}dk\right] dy \\
&= \left(\frac{1}{2\pi}\right) \int_{-\infty}^{\infty} F(k) \left[\int_{-\infty}^{\infty} g(y)e^{iky}dy\right] e^{-ikx}dk \\
&= \left(\frac{1}{\sqrt{2\pi}}\right) \int_{-\infty}^{\infty} F(k)G(k)e^{-ikx}dk. \qquad (2.9.11)
\end{aligned}$$

Comparing the last result of Eq. (2.9.11) with that given in the definition provided by Eq. (2.9.9), we conclude that

$$\mathcal{F}[f * g; k] = F(k)G(k). \qquad (2.9.12)$$

This is just the result stated in Eq. (2.8.24).

2.9.4 The Diffusion Equation

Diffusion of either some material or temperature is governed by the following partial differential equation if the system is linear,

$$\frac{\partial u}{\partial t} = D\frac{\partial^2 u}{\partial x^2}, \qquad (2.9.13)$$

where $D > 0$, is the constant diffusion coefficient, and $u = u(x,t)$ is a function of the space variable x and the time, t. We consider the initial value problem,

$$u(x,0) = f(x), \qquad -\infty < x < \infty, \qquad (2.9.14)$$

where the $f(x)$ is specified. Note that for most problems in the natural and engineering sciences, the following requirements hold,

$$f(x) \geq 0 \quad \text{and} \quad \int_{-\infty}^{\infty} f(x)dx < \infty. \qquad (2.9.15)$$

In more detail, we assume that (u, u_t, u_x, u_{xx}) are continuously differentiable and absolutely integrable on the interval, $-\infty < x < \infty$.

To proceed, we apply the Fourier transform procedure to the x variable and treat t as a parameter. Denoting $U(k, t)$ as the Fourier transform of $u(x, t)$, we obtain

$$U_t(k, t) = \left(\frac{1}{\sqrt{2\pi}}\right) \int_{-\infty}^{\infty} u_t(x, t) e^{ikx} dx, \tag{2.9.16}$$

$$-k^2 U(k, t) = \left(\frac{1}{\sqrt{2\pi}}\right) \int_{-\infty}^{\infty} u_{xx}(x, t) e^{ikx} dx, \tag{2.9.17}$$

where the second expression is a consequence of the result given by Eq. (2.8.18). Therefore, taking the Fourier transform of Eq. (2.9.13) gives

$$U_t(k, t) + Dk^2 U(k, t) = 0, \tag{2.9.18}$$

which is a first-order, linear differential equation in the variable t. Its solution is

$$U(k, t) = C(k) e^{-Dk^2 t}, \tag{2.9.19}$$

where $C(k)$ is a currently unknown function of k. It can be determined by taking the Fourier transform of the initial condition presented by Eq. (2.9.14), i.e.,

$$U(k, 0) = \left(\frac{1}{\sqrt{2\pi}}\right) \int_{-\infty}^{\infty} u(x, 0) e^{ikx} dx$$
$$= \left(\frac{1}{\sqrt{2\pi}}\right) \int_{-\infty}^{\infty} f(x) e^{ikx} dx = F(k). \tag{2.9.20}$$

Since $f(x)$ is specified, then from Eq. (2.9.20), we know $F(k)$. Setting $t = 0$, in Eq. (2.9.19), gives

$$C(k) = F(k), \tag{2.9.21}$$

and

$$U(k, t) = F(k) e^{-Dk^2 t}. \tag{2.9.22}$$

Now $u(x, t)$ is given by the inverse Fourier transform relation,

$$u(x, t) = \left(\frac{1}{\sqrt{2\pi}}\right) \int_{-\infty}^{\infty} F(k) e^{-Dk^2 t} e^{-ikx} dk$$
$$= \left(\frac{1}{2\pi}\right) \int_{-\infty}^{\infty} \left[\int_{-\infty}^{\infty} f(y) e^{iky} dy\right] e^{-Dk^2 t} e^{-ikx} dk$$

$$= \left(\frac{1}{2\pi}\right) \int_{-\infty}^{\infty} f(y) \left[\int_{-\infty}^{\infty} e^{-ik(x-y)-Dk^2t} dk\right] dy. \qquad (2.9.23)$$

The integral in the square bracket can be rewritten to the following form

$$\int_{-\infty}^{\infty} e^{-ik(x-y)-Dk^2t} dk = 2 \int_{0}^{\infty} e^{-Dk^2t} \cos[k(x-y)] dk$$

$$= \left(\sqrt{\frac{\pi}{Dt}}\right) \exp\left[-\frac{(x-y)^2}{4Dt}\right], \qquad (2.9.24)$$

where we used

$$e^{-ik(x-y)} = \cos[k(x-y)] - i\sin[k(x-y)], \qquad (2.9.25)$$

and the fact that the integral with the sine term is zero. Note that the last integral in Eq. (2.9.25) is easily evaluated or looked-up in a standard table of integrals. Therefore, the solution to the linear diffusion equation with the initial condition, given by Eq. (2.9.14) is

$$u(x,t) = \left(\frac{1}{\sqrt{4\pi Dt}}\right) \int_{-\infty}^{\infty} f(y) \exp\left[-\frac{(x-y)^2}{4Dt}\right] dy. \qquad (2.9.26)$$

2.9.5 The Wave Equation

One of the fundamental equations modeling linear waves is the one-dimensional wave equation given by

$$u_{tt} = \left(\frac{1}{c^2}\right) u_{xx}. \qquad (2.9.27)$$

We consider the case for which the following initial conditions are specified,

$$u(x,0) = f(x), \qquad u_t(x,0) = 0. \qquad (2.9.28)$$

For many applications, the requirement

$$\int_{-\infty}^{\infty} |f(x)|^2 dx < \infty, \qquad (2.9.29)$$

is also made. However, in the analysis to come, no use will be made of this restriction.

As in the previous problem, we will take Fourier transforms with respect to x and consider t a parameter. Thus, if we write,

$$u(x,t) = \left(\frac{1}{\sqrt{2\pi}}\right) \int_{-\infty}^{\infty} U(k,t) e^{-ikx} dk, \qquad (2.9.30)$$

and substitute it into Eq. (2.9.27), we obtain

$$\left(\frac{1}{\sqrt{2\pi} \cdot c^2}\right) \int_{-\infty}^{\infty} U_{tt} e^{-ikx} dx = \left(\frac{1}{\sqrt{2\pi}}\right) (-k^2) \int_{-\infty}^{\infty} U e^{ikx} dx, \qquad (2.9.31)$$

and conclude that

$$U_{tt}(k,t) = -c^2 k^2 U(k,t). \tag{2.9.32}$$

The general solution to this ordinary, linear, second-order differential equation in t is,

$$U(k,t) = A(k)e^{ickt} + B(k)e^{-ikct}, \tag{2.9.33}$$

where $A(k)$ and $B(k)$ are two, currently unknown, functions. Now, we have

$$f(x) = u(x,0) = \left(\frac{1}{\sqrt{2\pi}}\right) \int_{-\infty}^{\infty} U(k,0)e^{-ikx}dk, \tag{2.9.34}$$

and therefore,

$$U(k,0) = \left(\frac{1}{\sqrt{2\pi}}\right) \int_{-\infty}^{\infty} f(x)e^{ikx}dx. \tag{2.9.35}$$

It follows from

$$\frac{\partial u(x,0)}{\partial t} = \left(\frac{1}{\sqrt{2\pi}}\right) \int_{-\infty}^{\infty} \frac{\partial U(k,0)}{\partial t} e^{-ikx}dk = 0, \tag{2.9.36}$$

that

$$\frac{\partial U(k,0)}{\partial t} = 0. \tag{2.9.37}$$

Therefore, from Eqs. (2.9.33), (2.9.35), and (2.9.37), we find

$$A(k) = B(k) = \left(\frac{1}{2}\right) U(k,0). \tag{2.9.38}$$

If we let $C(k) = U(k,0)/2$, then $C(k)$ is a known function of k, given by Eq. (2.9.35), and we obtain,

$$U(k,t) = C(k)[e^{ickt} + e^{-ickt}] \tag{2.9.39}$$

and

$$u(x,t) = \left(\frac{1}{\sqrt{2\pi}}\right) \int_{-\infty}^{\infty} C(k)[e^{ik(x+ct)} + e^{ik(x-ct)}]dk. \tag{2.9.40}$$

This latter expression can be written,

$$u(x,t) = \left(\frac{1}{2}\right) [f(x+ct) + f(x-ct)], \tag{2.9.41}$$

and the solution is completely determined from just a knowledge of $u(x,0) = f(x)$.

2.10   The Laplace Transform

Another linear transform is that of taking a given function $f(x)$ and from it construct a function $\mathcal{L}\{f(t); s\}$ defined by the expression,

$$\mathcal{L}\{f(t); s\} = F(s) = \int_0^\infty f(t)e^{-st}dt. \tag{2.10.1}$$

Thus, the function $F(s)$ exists depending on whether the integral exists. For example, if

$$f(t) = e^{t^2}, \tag{2.10.2}$$

then

$$\int_0^\infty e^{t^2 - st}dt = \infty, \tag{2.10.3}$$

and there is no Laplace transform corresponding to this function. To deal with the existence or not of a Laplace transform for a particular function, $f(t)$, we first introduce the concept of exponential order.

Definition   Let $f(t)$ be defined for, $t \geq 0$. Let real constants (T, M, α) exist

$$T > 0, \qquad M > 0, \tag{2.10.4}$$

such that

$$|f(t)| \leq Me^{\alpha t}, \qquad t > T. \tag{2.10.5}$$

Then $f(t)$ is said to be of exponential order.

The application of the following theorem will allow us to determine when a given $f(x)$ has a Laplace transform.

Theorem   *If $f(t)$ is a piecewise continuous function in every finite interval, $0 \leq t \leq T$, and is of exponential order for $t > T$, then $\mathcal{L}\{f(t); s\}$ exists for $s > \alpha$.*

Note that in the general case where s can be complex, a Laplace transform will exist under the condition $\mathrm{Re}(s) > \alpha$.

If $F(s)$ is the Laplace transform of $f(t)$, then $f(t)$ is the inverse Laplace transform of $F(s)$ and we denote this by the symbolic representation,

$$f(t) = \mathcal{L}^{-1}\{F(s); t\}. \tag{2.10.6}$$

Extensive tables of both Laplace and inverse Laplace transforms exist. We refer the reader to the books and handbooks listed in reference [9] for such tables as well as applications of these transforms.

The remainder of this section consists of a summary of the major properties of the Laplace transform. Proofs for most of these results are given in the following book: M. R. Spiegel, *Advanced Mathematics for Engineers and Scientists* (McGraw-Hill, New York, 1999); Chapter 4.

However, before proceeding, let us introduce the *theta function* which also goes under the names, *unit step function* and *Heaviside's unit step function*. This function will be denoted by $\theta(t - a)$ and is defined to be

$$\theta(t - a) = \begin{cases} 0, & t < a, \\ 1, & t > a. \end{cases} \qquad (2.10.7)$$

Note that this function is discontinuous at $t = a$. Its Laplace transform is[1]

$$\mathcal{L}\{\theta(t - a)\} = \int_0^\infty \theta(t - a)e^{-st}dt = \int_a^\infty e^{-st}dt = \frac{e^{-as}}{s}, \qquad s > 0. \qquad (2.10.8)$$

Linearity Condition

Let c_1 and c_2 be arbitrary constants, then

$$\mathcal{L}[c_1 f(x) + c_2 g(x)] = c_1 \mathcal{L}[f(x)] + c_2 \mathcal{L}[g(x)] = c_1 F(s) + c_2 G(x). \qquad (2.10.9)$$

Shift Theorems

$$\mathcal{L}\{e^{at}f(t)\} = F(s - a), \qquad (2.10.10)$$

$$\mathcal{L}\{\theta(t - a)f(t - a)\} = e^{-as}F(s). \qquad (2.10.11)$$

Scaling Condition

$$\mathcal{L}\{f(at)\} = \left(\frac{1}{a}\right) F\left(\frac{s}{a}\right). \qquad (2.10.12)$$

---

[1] For the remainder of our discussion, we drop the indication of the "s" variable in the notation for the Laplace transform, i.e.,

$$\mathcal{L}\{f(t), s\} \equiv \mathcal{L}\{f(t)\}.$$

Differentiation of a Transform

$$\mathcal{L}\{t^n f(t)\} = (-1)^n \frac{d^n F(s)}{ds^n}, \qquad (n = 1, 2, 3 \ldots). \qquad (2.10.13)$$

Transform of an Integral

If

$$\underset{t \to \infty}{\text{Lim}} \left[e^{-st} \int_0^t f(y) dy \right] = 0, \qquad (2.10.14)$$

then

$$\mathcal{L}\left\{ \int_0^t f(y) dy \right\} = \frac{F(s)}{s}. \qquad (2.10.15)$$

Integration of the Transform

If

$$\underset{t \to 0}{\text{Lim}} \left[\frac{f(t)}{t} \right] < \infty, \qquad (2.10.16)$$

then

$$\mathcal{L}\left\{ \frac{f(t)}{t} \right\} = \int_s^\infty F(y) dy. \qquad (2.10.17)$$

Transform of Derivatives

Let $f(t)$ be such that the first $(n-1)$ derivatives are continuous and $f^{(n)}(t)$ is piecewise continuous in every finite interval, $0 \le t \le T$. Also, let $f^{(m)}(t)$, $(m : 0, 1, 2, \ldots, n-1)$, be of exponential order for $t > T$, then

$$\mathcal{L}\{f^{(n)}(t)\} = s^n \mathcal{L}\{f(t)\} - s^{n-1} f(0) - s^{n-2} f'(0) - \cdots - f^{(n-1)}(0). \qquad (2.10.18)$$

Convolution Theorem

Define the Laplace convolution $f * g$ of two functions as follows,

$$f * g \equiv \int_0^t f(x - y) g(y) dy. \qquad (2.10.19)$$

A consequence of this definition is the result,

$$f * g = g * f, \qquad (2.10.20)$$

i.e., the convolution is commutative. It is also true that

$$\mathcal{L}\{f * g\} = F(s) G(s). \qquad (2.10.21)$$

Transform of a Periodic Function

Let $f(t)$ be a piecewise continuous function for $x \geq 0$ and assume that it is periodic with period P, then

$$\mathcal{L}\{f(t)\} = \left[\frac{1}{1 - e^{-sP}}\right] \int_0^P f(y)e^{-sy}dy. \qquad (2.10.22)$$

A comparison of the above theorems for the Laplace transforms with those for the Fourier transforms indicates that there is a close connection between the two transforms.

2.11 Worked Problems Using the Laplace Transform

2.11.1 *Laplace Transform of* $t^{-1/3}$

Consider the function

$$f(t) = \frac{1}{t^{1/3}}. \qquad (2.11.1)$$

Note that it is not continuous in the interval, $0 \leq t \leq T$. However,

$$\mathcal{L}\{t^{-1/3}\} = \int_0^\infty t^{-1/3}e^{-st}dt = \frac{\Gamma\left(\frac{2}{3}\right)}{s^{2/3}}, \qquad (2.11.2)$$

where $\Gamma(x)$ is the gamma function (to be defined in the next chapter). Since $\Gamma\left(\frac{2}{3}\right)$ has a definite numerical value, it follows that while the above $f(t)$ violates a stated condition for the existence of the Laplace transform, nevertheless, the transform exists. Our conclusion is that these conditions are only sufficient.

2.11.2 *The Square Wave Function*

For $t \geq 0$, define the square wave function as

$$f(t) = \begin{cases} 1, & \text{for } 2k \leq t \leq 2k + 1, \\ -1, & \text{for } 2k + 1 \leq t < 2k + 2, \\ k = 0, 1, 2, \ldots. \end{cases} \qquad (2.11.3)$$

This function has period $P = 2$. Therefore, from Eq. (2.11.3) we obtain

$$\begin{aligned}
\mathcal{L}\{f(t)\} &= \left[\frac{1}{1 - e^{-2s}}\right] \left[\int_0^1 e^{-sy}dy - \int_1^2 e^{-sy}dy\right] \\
&= \left[\frac{1}{1 - e^{-2s}}\right] \left\{\left[\frac{1 - e^{-s}}{s}\right] + \left[\frac{e^{-2s} - e^{-s}}{s}\right]\right\} \\
&= \frac{(1 - e^{-s})^2}{s(1 - e^{-2s})}. \qquad (2.11.4)
\end{aligned}$$

2.11.3   The Dirac Delta Function

Assume there exists a "function" $\delta(t)$, such that for any continuous function $f(t)$, the following integral relationship holds

$$\int_0^\infty f(t)\delta(t - t_0)dt = f(t_0), \qquad t_0 \geq 0. \tag{2.11.5}$$

Then,

$$\mathcal{L}\{\delta(t)\} = \int_0^\infty \delta(t)e^{-st}dt = 1, \tag{2.11.6}$$

and

$$\mathcal{L}\{\delta(t - t_0)\} = \int_0^\infty \delta(t - t_0)e^{-st}dt = e^{-st_0}, \qquad t_0 \geq 0. \tag{2.11.7}$$

Let us now assume that it makes "sense" to differentiate both sides of Eq. (2.11.7). If we do this, then the following result is obtained,

$$\frac{d}{dt_0}\left[\int_0^\infty \delta(t - t_0)e^{-st}dt\right] = -se^{-st_0}, \tag{2.11.8}$$

and it follows that

$$\int_0^\infty \delta'(t - t_0)e^{-st}dt = -se^{-st_0}, \tag{2.11.9}$$

where $\delta'(t - t_0) = d\delta(t - t_0)/dt_0$. For the case where $t_0 = 0$, we obtain,

$$\mathcal{L}\{\delta'(t)\} = s. \tag{2.11.10}$$

The $\delta(x)$ is called the Dirac delta function and we will study its properties in more detail in the next chapter.

Problems

Section 2.2

2.2.1 Let $y(\theta) = e^{i\theta}$ be represented as

$$y(\theta) = c(\theta) + is(\theta)$$

where $c(\theta)$ and $s(\theta)$ are assumed to be not known. Show that $c(\theta)$ and $s(\theta)$ have the following properties:

(i) $c^2(\theta) + s^2(\theta) = 1$,

(ii) $\dfrac{dc}{d\theta} = -s, \quad \dfrac{ds}{d\theta} = c$,

(iii) $\dfrac{d^2c}{d\theta^2} + c = 0, \quad \dfrac{d^2s}{d\theta^2} + s = 0$,

(iv) $c(\theta) = \dfrac{e^{i\theta} + e^{-i\theta}}{2}$,

$s(\theta) = \dfrac{e^{i\theta} - e^{-i\theta}}{2i}$.

2.2.1 Does i^i have a definite value? Explain your answer.

Section 2.3

2.3.1 Is $|\cos\theta|^2$ the same as $(\cos\theta)^2$?

2.3.2 Show that the following relations are correct:

   (i) $\tan\theta_1 \pm \tan\theta_2 = \dfrac{\sin(\theta_1 \pm \theta_2)}{\cos\theta_1 \cos\theta_2}$,

   (ii) $\tan\left(\dfrac{\theta}{2}\right) = \dfrac{\sin\theta}{1 + \cos\theta} = \dfrac{1 - \cos\theta}{\sin\theta}$.

Section 2.4

2.4.1 Prove the result stated by Eq. (2.4.2).

2.4.2 Show that the constant function, $f(x) = c$, where c is an arbitrary real constant, is a periodic function, but having no fundamental period.

2.4.3 Prove the result stated in Eq. (2.4.3).

2.4.4 Show that the statements presented in Eq. (2.4.5) are correct. Hint: Construct a Taylor series representation for the required solutions.

Section 2.5

2.5.1 Construct and plot both even and odd extensions of the following functions defined over the indicated intervals:

   (i) $f(x) = \sin\pi x$, $0 \le x \le 1$;

   (ii) $f(x) = 1$, $0 \le x \le \pi$;

   (iii) $f(x) = \frac{1}{x}$, $0 < x < \infty$;

   (iv) $f(x) = \begin{cases} 1, & 0 < x < 1, \\ 0, & 1 < x < 2, \\ 3x - 6, & 2 < x < 3. \end{cases}$

2.5.2 Determine the even and odd parts of the functions for the interval, $(-\pi, \pi)$:

   (i) $f(x) = e^x$,

   (ii) $f(x) = e^{ix}$,

   (iii) $f(x) = \dfrac{x - \pi}{1 + x^2}$,

   (iv) $f(x) = 1 - e^{-2x}$.

2.5.3 Which of the functions in problems 2.5.1 and 2.5.2 have Fourier series representations?

Section 2.6

2.6.1 Show for $0 \leq x \leq \pi$, that $x(\pi - x)$ has the following sine and cosine Fourier series:

(i) $x(\pi - x) = \dfrac{\pi^2}{6} - \displaystyle\sum_{n=1}^{\infty} \dfrac{\cos(2nx)}{n^2}$

(ii) $x(\pi - x) = \left(\dfrac{8}{\pi}\right) \displaystyle\sum_{n=1}^{\infty} \dfrac{\sin(2n-1)x}{(2n-1)^3}$.

2.6.2 Using the results from the previous problem, show that

(i) $\displaystyle\sum_{n=1}^{\infty} \dfrac{1}{n^2} = 2 \sum_{n=1}^{\infty} \dfrac{(-1)^{n+1}}{n^2} = \dfrac{\pi^2}{6}$,

(ii) $\displaystyle\sum_{n=1}^{\infty} \dfrac{(-1)^{n+1}}{(2n-1)^3} = \dfrac{\pi^3}{32}$.

2.6.3 From problem 2.6.1 and the application of the Parseval identity show that the following sums are correct:

(i) $\dfrac{\pi^4}{90} = \displaystyle\sum_{n=1}^{\infty} \dfrac{1}{n^4}$,   (ii) $\dfrac{\pi^6}{945} = \displaystyle\sum_{n=1}^{\infty} \dfrac{1}{n^6}$.

2.6.4 Construct the Fourier series for

$$f(x) = x^2, \qquad 0 \leq x \leq 2\pi,$$

and show that it is

$$\dfrac{4\pi^2}{3} + 4 \sum_{n=1}^{\infty} \left[\dfrac{\cos(nx)}{n^2} - \dfrac{\pi \sin(nx)}{n} \right].$$

Use the Parseval identity and the result from part (i) of problem 2.6.3 to obtain

$$\dfrac{\pi^2}{6} = \sum_{n=1}^{\infty} \dfrac{1}{n^2}.$$

Section 2.7

2.7.1 Obtain the result given by Eq. (2.7.11).

2.7.2 Derive the result given by Eq. (2.7.18) by replacing θ in Eq. (2.7.12) by $\left(\frac{\pi}{2} - \theta\right)$.

2.7.3 What is the Fourier series expansion for $(\cos\theta)^{1/5}$?

Section 2.8

2.8.1 Prove the properties for the Fourier transforms as given in the following equations:

(i) Eqs. (2.8.6), (2.8.7), (2.8.8);
(ii) Eqs. (2.8.9), (2.8.10), and (2.8.11);
(iii) the linearity condition of Eq. (2.8.14);
(iv) the scaling condition of Eq. (2.8.15);
(v) the shift condition of Eq. (2.8.16);
(vi) the k-shift condition of Eq. (2.8.17);
(vii) the differentiation condition of Eq. (2.8.19).

2.8.2 Are there functions $f(x)$ and $g(x)$ for which

$$\mathcal{F}[fg; k] = F(k)G(k)?$$

Section 2.9

2.9.1 Plot for the pulse function, defined by Eq. (2.9.1), both $f(x)$ vs x and $F(k)$ vs k. Where do the zeros of $F(k)$ occur? Where do the minima of $F(k)$ occur?

2.9.2 The limits of the integral in Eq. (2.9.8), given the transformation $y = x + \frac{ik}{2}$, should be $-\infty + \frac{ik}{2}$ to $+\infty + \frac{ik}{2}$. What arguments allow us to justify the use of $-\infty$ to $+\infty$?

2.9.3 Derive the final result shown in Eq. (2.9.24).

2.9.4 Show that for $x_0 = $ constant,

$$u(x, t, x_0) = \left(\frac{1}{\sqrt{4\pi Dt}} \right) \exp\left[-\frac{(x - x_0)^2}{4Dt} \right]$$

is a solution to the diffusion equation

$$u_t = Du_{xx}.$$

Plot $u(x, t, x_0)$ for t small and t large. Further, show that

(i) $\lim_{t \to 0^+} u(x, t, x_0) = 0$, $x \neq x_0$;
(ii) $\lim_{t \to \infty} u(x, t, x_0) = 0$, $-\infty < x < \infty$;
(iii) $\lim_{t \to 0^+} u(x, t, x_0) = +\infty$, $x = x_0$.

Calculate the following integral and give it a suitable interpretation,

$$\int_{-\infty}^{\infty} u(x, t, x_0)dx, \qquad t > 0.$$

Section 2.10

2.10.1 Prove the shift theorems given by Eqs. (2.10.10) and (2.10.11).

2.10.2 Derive the scaling condition of Eq. (2.10.12).

2.10.3 Derive the expression for the differentiation of a transform by directly taking the derivatives of Eq. (2.10.1).

2.10.4 Show that the expressions given by Eqs. (2.10.15) and (2.10.17) correctly give the results, respectively, for the transform of an integral and the integration of the transform.

2.10.5 Derive the convolution theorem given by Eq. (2.10.21).

2.10.6 Show that the transform of a periodic function only requires a knowledge of the function for one period. See Eq. (2.10.22).

Section 2.11

2.11.1 Let n be a non-negative integer. Prove that

$$\mathcal{L}\{t^n\} = \frac{n!}{s^{n+1}}.$$

2.11.2 Does $\mathcal{L}\{t^{-1}\}$ exist? Explain your answer.

2.11.3 The rectified sine wave is the periodic function

$$f(t) = \begin{cases} \sin t, & 0 \le t < \pi, \\ 0, & \pi \le t < 2\pi. \end{cases}$$

Determine its Laplace transform.

2.11.4 Calculate the derivative of $f(t)$ as defined in the previous problem. Determine its Laplace transform.

2.11.5 Does the derivative of the theta function,

$$\theta(t - a) = \begin{cases} 0, & \text{for } t < a, \\ 1, & \text{for } t > a, \end{cases}$$

have a Laplace transform? That is, what is

$$\mathcal{L}\{\theta'(t - a)\}?$$

2.11.6 Use the Laplace transform to solve the boundary value problem [10]

$$u_t = D u_{xx}, \qquad t > 0, \quad x > 0;$$

$$u(x, 0) = 0, \, x > 0 \text{ and } u(0, t) = 1, \, t > 0.$$

Comments and References

[1] D. C. Champeney, *A Handbook of Fourier Theorems* (Cambridge University Press, New York, 1987); see section 15.6.

[2] R. E. Mickens, *J. Sound and Vibration* **259**, 457 (2003).

[3] R. E. Mickens, *J. Sound and Vibration* **261**, 567 (2003).

[4] The results to be given below were derived jointly by both R. E. Mickens and J. Ernst Wilkins, Jr., during April–July 2002. Professor Wilkins has studied the general case and proved a number of very interesting theorems on the properties of the Fourier coefficients, including the relationship of the coefficients for $(\cos\theta)^\alpha$ and $(\sin\theta)^\alpha$, and their asymptotic properties in the labelling index.

[5] See, Formulas 3.631 in: I. S. Gradshteyn and I. M. Ryzhik, *Tables of Integrals, Series and Products* (Academic Press, New York, 1965).

[6] See chapter 3 for a discussion of the properties of the gamma function. For this calculation, we use $\Gamma(x+1) = x\Gamma(x)$.

[7] Proofs of the various properties of Fourier transforms can be found in the books:
  * G. B. Arfken and Hans J. Weber, *Mathematical Methods for Physicists* (Academic Press, New York, 1995).
  * D. C. Champeney, *Fourier Transforms and Their Physical Applications* (Academic Press, New York, 1973).
  * A. Papoulis, *The Fourier Integral and Its Applications* (McGraw-Hill, New York, 1962).
  * I. N. Sneddon, *Fourier Transforms* (McGraw-Hill, New York, 1951).
  * K. B. Wolf, *Integral Transforms in Science and Engineering* (Plenum Press, New York, 1979).

[8] Tables of Fourier transforms appear in the following books:
  * A. Erdélyi et al., *Tables of Integral Transforms, Vols. I and II* (McGraw-Hill, New York, 1954).
  * A. Jeffrey, *Handbook of Mathematical Formulas and Integrals* (Academic Press, New York, 1995).
  * O. I. Marichev, *Handbook of Integral Transforms of Higher Transcendental Functions, Theory and Algorithmic Tables* (Ellis Horwood, Chichester,

1982).

[9] See the listing in [8] above as well as the following publications:
 * G. Doetsch, *Handbuch der Laplace-Transformation, Vols. I–IV* (Birkhäuser, Basel, 1950–1956).
 * G. Doetsch, *Theory and Application of the Laplace Transform* (Chelsea, New York, 1965).
 * F. Oberhettinger and L. Badii, *Tables of Laplace Transforms* (Springer-Verlag, Berlin, 1973).
 * D. V. Widder, *The Laplace Transforms* (Princeton University Press; Princeton, NJ; 1941).

[10] See J. D. Logan, *Applied Mathematics, A Contemporary Approach* (Wiley-Interscience, New York, 1987); see pps. 191–192.

Bibliography

H. S. Carslaw, *Introduction to the Theory of Fourier Series and Integrals*, 3rd ed. (Dover, New York, 1952).

D. C. Champeney, *Fourier Transforms and Their Physical Applications* (Academic Press, New York, 1973).

R. V. Churchill, *Fourier Series and Boundary Value Problems* (McGraw-Hill, New York, 1941).

L. Debnath, *Integral Transforms and Their Applications* (CRC Press, New York, 1995).

J. R. Hanna, *Fourier Series and Integrals of Boundary Value Problems* (Wiley; Somerset, NJ; 1982).

K. B. Howell, *Principles of Fourier Analysis* (Chapman and Hall/CRC; Boca Raton, FL; 2001).

T. W. Körner, *Fourier Analysis* (Cambridge University Press, Cambridge, 1988).

A. Kufner and J. Kadlec, *Fourier Series* (Iliffe, London, 1971).

W. R. LePlage, *Complex Variables and the Laplace Transform for Engineers* (McGraw-Hill, New York, 1961).

P. A. McCollum and B. F. Brown, *Laplace Transform Tables and Theorems* (Holt, Reinhart and Winston, New York, 1965).

J. W. Miles, *Integral Transforms in Applied Mathematics* (Cambridge University Press, Cambridge, 1971).

F. Oberhettinger, *Fourier Expansions, A Collection of Formulas* (Academic Press, New York, 1973).

G. E. Roberts and H. Kaufman, *Table of Laplace Transforms* (Saunders, Philadelphia, 1966).

W. Rogosinski, *Fourier Series* (Chelsea Publishing, New York, 1950).

E. C. Titchmarsh, *Eigenfunction Expansions* (Oxford University Press, Oxford, 1946).

D. V. Widder, *The Laplace Transform* (Princeton University Press; Princeton, NJ; 1946).

A. Zygmund, *Trigonometric Series* (Cambridge University Press, Cambridge, 1977).

Chapter 3

Gamma, Beta, Zeta, and Other Named Functions

3.1 Scope of Chapter

A number of functions, somewhat more complex in nature than the standard elementary functions, arise from the mathematical analysis of many systems in the sciences. These include the gamma, beta, and zeta functions; Dirichlet integrals; the Dirac delta function; and other "named functions" represented by definite integrals such as the exponential, sine, cosine, Fresnel Sine and Cosine integrals. We define these functions, derive several of their important properties, and then show how they can be applied.

We also include a brief section on elliptic integrals and functions. The elliptic functions are a generalization of the trigonometric functions and satisfy many similar relations. The basic properties are derived for these functions along with the nonlinear, second-order differential equations to which they are solution. We also give their Fourier series representations.

The chapter ends with a topic that will be looked at somewhat differently in Chapter 9. We introduce integrals depending on parameters and show that many can be explicitly evaluated using various techniques involving differentiation with respect to these parameters. Application of the method to several examples demonstrates that it is a powerful method, when applicable, for evaluating integrals.

3.2 Gamma Function

Define $f(a)$ to be

$$f(a) \equiv \int_0^\infty e^{-at} dt, \qquad \text{Re}(a) > 0, \qquad (3.2.1)$$

where it is indicated that the integral exists for complex a, provided $\text{Re}(a)$ is positive. In fact, $f(a)$ can be easily computed and is found to be

$$f(x) = \frac{1}{a}.$$ (3.2.2)

If the following expression

$$\frac{1}{a} = \int_0^\infty e^{-at} dt$$ (3.2.3)

is differentiated n-times with respect to a, then we obtain

$$\frac{(-1)^n n!}{a^{n+1}} = (-1)^n \int_0^\infty t^n e^{-at} dt,$$ (3.2.4)

and for $a = 1$,

$$n! = \int_0^\infty t^n e^{-t} dt; \qquad n = 0, 1, 2, \dots .$$ (3.2.5)

Thus, we see that there is a connection between the factorial function, $n!$, and the integral on the right-side of Eq. (3.2.5). The following question immediately comes to mind: How can we generalize this result? In particular, suppose the non-negative integer, n, is replaced by a complex number. Does the integral have a well defined meaning for this case? The answer to this question is Yes! This generalization is the gamma function, defined as

$$\Gamma(z) \equiv \int_0^\infty t^{z-1} e^{-t} dt.$$ (3.2.6)

The integral converges for $\text{Re}(z) > 0$. Note that if $z = n$, where n is a positive integer, then

$$\Gamma(n) = (n-1)!$$ (3.2.7)

It follows from this equation that

$$\Gamma(n+1) = n! = n(n-1)!$$ (3.2.8)

and

$$\Gamma(n+1) = n\Gamma(n).$$ (3.2.9)

This is a simple recurrence relation expressing one gamma function in terms of its immediate successor. We now show that such a relation also holds for $\Gamma(z)$. This follows from integrating by parts $\Gamma(z+1)$, i.e.,

$$\Gamma(z+1) = \int_0^\infty t^z e^{-t} dt$$

$$= \left[-e^{-t}t^z\right]\Big|_{t=0}^{t=\infty} + z\int_0^\infty t^{z-1}e^{-t}dt \qquad (3.2.10)$$

and, finally,

$$\Gamma(z+1) = z\Gamma(z). \qquad (3.2.11)$$

From Eq. (3.2.11), it follows that for n a positive integer

$$\Gamma(z+n) = (z+n-1)(z+n-2)\cdots(z+1)z\Gamma(z), \qquad (3.2.12)$$

and

$$\Gamma(z) = \frac{\Gamma(z+n)}{z(z+1)\cdots(z+n-1)}. \qquad (3.2.13)$$

Note that this expression allows us to extend the definition of $\Gamma(z)$ to $\mathrm{Re}(z) > (-n)$, where n is an arbitrary positive integer. This can be seen by rewriting Eq. (3.2.13) to the form

$$\Gamma(z) = \left[\frac{1}{z(z+1)\cdots(z+n-1)}\right]\int_0^\infty t^{z+n-1}e^{-t}dt. \qquad (3.2.14)$$

It is also clear that for $z = n \geq 1$, then

$$\Gamma(n) > 0. \qquad (3.2.15)$$

Furthermore, $\Gamma(z)$ has simple zeros at the negative integers.

The gamma function has two interesting infinite product representations, namely, the Euler expression,

$$\Gamma(z) = \left(\frac{1}{z}\right)\prod_{k=1}^\infty \left[\left(1+\frac{z}{k}\right)^{-1}\left(1+\frac{1}{k}\right)^z\right], \qquad (3.2.16)$$

and the Weierstrass formula

$$[\Gamma(z)]^{-1} = ze^{\gamma z}\prod_{k=1}^\infty \left[\left(1+\frac{z}{k}\right)e^{-z/k}\right], \qquad (3.2.17)$$

where γ is the s-called Euler constant,

$$\gamma = 0.5772156649\ldots \qquad (3.2.18)$$

The following product formula holds for the gamma function:

$$\Gamma(z)\Gamma(-z) = -\frac{\pi}{z\sin(\pi z)} \qquad (3.2.19)$$

This can be derived by using Eq. (3.2.17) to evaluate the product $\Gamma(z)\Gamma(-z)$ and then using the fact that the sine function has the representation

$$\sin(z) = x\prod_{k=1}^\infty \left(1-\frac{z^2}{\pi^2 k^2}\right). \qquad (3.2.20)$$

If, for z real, the values of the gamma function are known for any interval of length one in $z = x$, then the relationship of Eq. (3.2.11) can be used to calculate the gamma function for any other unit interval. Tables of such values are usually given for $1 \leq x \leq 2$. A selected set of values for $\Gamma(z)$ is presented in Table 3.2.1.

Table 3.2.1 Selected Values for Gamma Function:

$\Gamma(x) = \int_0^\infty t^{x-1}e^{-t}dt$	
x	$\Gamma(x)$
1.00	1.00000
1.01	0.99433
1.05	0.97350
1.10	0.95135
1.15	0.93304
1.20	0.91817
1.30	0.89747
1.40	0.88726
1.50	0.88623
1.60	0.89352
1.70	0.90864
1.80	0.93138
1.90	0.96177
2.00	1.00000

The following asymptotic expansion, i.e., $z \to \infty$, is known as Stirling's formula

$$\Gamma(z+1) = (2\pi z)^{1/2} z^z e^{-z} \left\{ 1 + \frac{1}{12z} + \frac{1}{288z^2} - \frac{139}{51840z^3} + O\left(\frac{1}{z^4}\right) \right\}. \quad (3.2.21)$$

For integer z, this gives very accurate values of the gamma functions for values as small as $z = n = 10$.

3.3   The Beta Function

The beta function, $B(p, q)$, is defined by the integral

$$B(p, q) \equiv \int_0^1 t^{p-1}(1 - t)^{q-1}dt, \quad (3.3.1)$$

where to have the integral exist, we must require

$$\text{Re}(p) > 0, \qquad \text{Re}(q) > 0. \quad (3.3.2)$$

Let $t = 1 - x$, then $B(p, q)$ can be written as

$$B(p, q) = \int_1^0 (1 - x)^{p-1} x^{q-1} (-dx)$$

$$= \int_0^1 x^{q-1} (1 - x)^{p-1} dx = B(q, p). \tag{3.3.3}$$

Thus, $B(p, q)$ is symmetric in (p, q).

We show that $B(q, p)$ can be represented in terms of the gamma function. Using the integral definition of $\Gamma(p)$ and $\Gamma(q)$, it follows that

$$\Gamma(p)\Gamma(q) = \left(\int_0^\infty u^{p-1} e^{-u} du \right) \left(\int_0^\infty v^{q-1} e^{-v} dv \right). \tag{3.3.4}$$

Now replace u and v by

$$u = x^2, \qquad v = y^2, \tag{3.3.5}$$

in Eq. (3.3.4) and obtain

$$\Gamma(p)\Gamma(q) = \left(2 \int_0^\infty x^{2p-1} e^{-x^2} dx \right) \left(2 \int_0^\infty y^{2q-1} e^{-y^2} dy \right)$$

$$= 4 \int_0^\infty \int_0^\infty x^{2p-1} y^{2q-1} e^{-(x^2+y^2)} dx\, dy. \tag{3.3.6}$$

If we convert to plane polar variables, i.e.,

$$x = r \cos\theta, \qquad y = r \sin\theta, \tag{3.3.7}$$

then

$$\Gamma(p)\Gamma(q) = 4 \left(\int_0^\infty r^{2(p+q)-1} e^{-r^2} dr \right)$$

$$\cdot \left(\int_0^{\pi/2} (\cos\theta)^{2p-1} (\sin\theta)^{2q-1} d\theta \right). \tag{3.3.8}$$

Using the change of variables, $t = r^2$, in the first integral gives

$$\int_0^\infty r^{2(p+q)-1} e^{-r^2} dr = \left(\frac{1}{2} \right) \int_0^\infty t^{p-q-1} e^{-t} dt = \left(\frac{1}{2} \right) \Gamma(p + q). \tag{3.3.9}$$

For the second integral, let $x = \cos\theta$ and obtain

$$\int_0^{\pi/2} (\cos\theta)^{2p-1} (\sin\theta)^{2q-1} d\theta = \left(\frac{1}{2} \right) \int_0^1 x^{p-1} (1 - x)^{q-1} dx$$

$$= \left(\frac{1}{2} \right) B(p, q). \tag{3.3.10}$$

Substituting the last two results into the right-side of Eq. (3.3.8) gives

$$\Gamma(p)\Gamma(q) = (4)\left[\left(\frac{1}{2}\right)\Gamma(p+q)\right]\left[\left(\frac{1}{2}\right)B(p,q)\right]$$
$$= \Gamma(p+q)B(p,q) \qquad (3.3.11)$$

or

$$B(p,q) = \frac{\Gamma(p)\Gamma(q)}{\Gamma(p+q)}. \qquad (3.3.12)$$

Note that the original restrictions on p and q, given by Eq. (3.3.2), can now be dropped. The beta function can now be defined for any values of (p,q) for which the right-side of Eq. (3.3.12) exists.

3.4   The Riemann Zeta Function

The Riemann Zeta function is defined by the following infinite series

$$\zeta(z) \equiv \sum_{n=1}^{\infty} \frac{1}{n^z} = 1 + \frac{1}{2^z} + \frac{1}{3^z} + \cdots. \qquad (3.4.1)$$

This series converges provided $\text{Re}(z) > 1$. This function has the following properties:

(i) For z real, i.e., $z = x$, $\zeta(x)$ is positive, i.e.,

$$\zeta(x) > 0, \qquad x > 1; \qquad (3.4.2)$$

(ii)

$$\zeta(x+1) < \zeta(x) < \zeta(x-1), \qquad x > 2; \qquad (3.4.3)$$

(iii)

$$1 < \zeta(x) < 2, \qquad x > 2; \qquad (3.4.4)$$

(iv) $\zeta(x)$ rapidly approaches 1 as x increases, i.e.,

$$\zeta(x) = 1 + O\left(\frac{1}{2^x}\right); \qquad (3.4.5)$$

(v) $\zeta(x)$ is a monotonic decreasing function for $x > 1$;
(vi) the following limit holds

$$\lim_{x \to 1^+} \zeta(x) = \infty. \qquad (3.4.6)$$

(vii) For even integer values of x, the zeta function can be explicitly evaluated to give

$$\zeta(2n) = c_n \pi^{2n}, \qquad (3.4.7)$$

where

$$c_1 = \frac{1}{6}, \quad c_2 = \frac{1}{90}, \quad c_3 = \frac{1}{945}. \tag{3.4.8}$$

Consider the function $\zeta(z; -)$ defined by alternating the signs of terms in $\zeta(z)$,

$$\zeta(z; -) \equiv \sum_{n=1}^{\infty} \frac{(-1)^{n+1}}{n^z} = 1 - \frac{1}{2^z} + \frac{1}{3^z} - \frac{1}{4^z} + \cdots. \tag{3.4.9}$$

This series can be rewritten to the form

$$\begin{aligned}
\zeta(z; -) &= \sum_{n=1}^{\infty} \frac{1}{(2n-1)^z} - \sum_{n=1}^{\infty} \frac{1}{(2n)^z} \\
&= \sum_{n=1}^{\infty} \frac{1}{(2n-1)^z} + \sum_{n=1}^{\infty} \frac{1}{(2n)^z} - \sum_{n=1}^{\infty} \frac{1}{(2n)^n} - \sum_{n=1}^{\infty} \frac{1}{(2n)^z} \\
&= \sum_{n=1}^{\infty} \frac{1}{n^z} - 2 \sum_{n=1}^{\infty} \frac{1}{(2n)^z} \\
&= \sum_{n=1}^{\infty} \frac{1}{n^z} - \left(\frac{1}{2^{z-1}} \right) \sum_{n=1}^{\infty} \frac{1}{n^z}, \tag{3.4.10}
\end{aligned}$$

and, finally,

$$\zeta(z; -) \equiv \sum_{n=1}^{\infty} \frac{(-1)^{n+1}}{n^z} = \zeta(z) \left[1 - \frac{1}{2^{z-1}} \right]. \tag{3.4.11}$$

Another related function is

$$\zeta(z; \text{odd}) \equiv \sum_{n=1}^{\infty} \frac{1}{(2n-1)^z}. \tag{3.4.12}$$

Writing it in the form

$$\zeta(z; \text{odd}) = \sum_{n=1}^{\infty} \left[\frac{1}{n^z} - \frac{1}{(2n)^z} \right], \tag{3.4.13}$$

we can easily obtain the result,

$$\zeta(z; \text{odd}) = \zeta(z) \left(1 - \frac{1}{2^z} \right). \tag{3.4.14}$$

The generalized Riemann zeta function is defined to be

$$\zeta(z; a) \equiv \sum_{n=0}^{\infty} \frac{1}{(n+a)^z}. \tag{3.4.15}$$

For our purposes, we take a to be a real number satisfying

$$0 < a \leq 1, \tag{3.4.16}$$

and, of course, for convergence of the series, we need $\mathrm{Re}(z) > 1$. Comparison of Eqs. (3.4.1) and (3.4.15) gives

$$\zeta(z) = \zeta(z; 1). \tag{3.4.17}$$

From the definition of the gamma function

$$\Gamma(z) = \int_0^\infty t^{z-1} e^{-t} dt, \tag{3.4.18}$$

it follows that

$$\int_0^\infty t^{z-1} e^{-(n+a)t} dt = \frac{\Gamma(z)}{(n+a)^z}, \tag{3.4.19}$$

and

$$\sum_{n=0}^\infty \int_0^\infty t^{z-1} e^{-(n+a)t} dt = \Gamma(z) \sum_{n=0}^\infty \frac{1}{(n+a)^z} = \Gamma(z)\zeta(z; a) \tag{3.4.20}$$

Interchanging the integration and summation (which is legitimate for this problem) gives

$$\sum_{n=0}^\infty \int_0^\infty t^{z-1} e^{-(n+a)t} = \int_0^\infty t^{z-1} e^{-at} \left[\sum_{n=0}^\infty e^{-nt} \right] dt$$

$$= \int_0^\infty \frac{t^{z-1} e^{-at}}{1 - e^{-t}} dt, \tag{3.4.21}$$

and we obtain the result

$$\Gamma(z)\zeta(z; a) = \int_0^\infty \frac{t^{z-1} e^{-at}}{1 - e^{-t}} dt. \tag{3.4.22}$$

The relation can be used to evaluate integrals having the structure on the right-side of this expression.

Finally, we note that the Riemann zeta function has a product representation in terms of prime numbers. (A prime number is a positive integer, divisible only by itself and one.) A heuristic argument to give this expression begins with noting that

$$\zeta(z) \left[1 - \frac{1}{2^z} \right] = 1 + \frac{1}{3^z} + \frac{1}{5^z} + \frac{1}{7^z} + \cdots, \tag{3.4.23}$$

contains no terms where N is a multiple of 2. Likewise

$$\zeta(z) \left[1 - \frac{1}{2^z} \right] \left[1 - \frac{1}{3^z} \right] = 1 + \frac{1}{5^z} + \frac{1}{7^z} + \frac{1}{11^z} + \cdots, \tag{3.4.24}$$

has no terms for which n is a multiple of 3. Continuing this procedure, it is clear that

$$\zeta(z) \left[1 - \frac{1}{2^z} \right] \left[1 - \frac{1}{3^z} \right] \left[1 - \frac{1}{5^z} \right] \cdots \left[1 - \frac{1}{P_m^z} \right],$$

where P_m is the mth prime, contains no terms where n is a multiple of P_m. In the limit as $m \to \infty$, we have

$$\zeta(z) \prod_{P(\text{prime})=2}^{\infty} \left(1 - \frac{1}{P^z} \right) = 1, \tag{3.4.25}$$

and thus the following infinite product representation for the Riemann zeta function,

$$\zeta(z) = \left[\prod_{P(\text{prime})=2}^{\infty} \left(1 - \frac{1}{P^z} \right) \right]^{-1}. \tag{3.4.26}$$

3.5 The Dirac Delta Function

The so-called Dirac delta function was invented by Paul A. M. Dirac [1] within the context of constructing a formal theory of quantum mechanics. This "function" came to play a significant role in many areas of the natural and engineering sciences. While the Dirac delta function is not strictly speaking a function, it can be used and treated in applications as if it were a function. In fact the Dirac delta function was among the first of such objects that are now called generalized functions [2].

Our presentation focuses on issues needed for the use of Dirac delta function in applications. Rigorous proofs will not be given, in general; the details of the mathematical correct procedures are provided in the books by Friedman and Lighthill; also, see the book of Lea [3].

Any function, $\delta(x)$, is said to be a delta function if it satisfies three conditions:

(i)

$$\int_{-\infty}^{\infty} \delta(x)dx = 1, \tag{3.5.1}$$

(ii)

$$\delta(x) = 0, \qquad \text{for } x \neq 0, \tag{3.5.2}$$

(iii)

$$\int_{-\infty}^{\infty} \delta(x)f(x)dx = f(0). \tag{3.5.3}$$

One way to approach the concept of a delta function is to consider it to be the limit of a sequence of functions $\{\delta_n(x) : n = 1, 2, 3, \dots\}$ such that

$$\int_{-\infty}^{\infty} \delta_n(x)dx = 1, \qquad n = 1, 2, 3, \dots, \qquad (3.5.4)$$

$$\int_{-\infty}^{\infty} \delta(x)f(x)dx = \lim_{n \to \infty} \int_{-\infty}^{\infty} \delta_n(x)f(x) = f(0). \qquad (3.5.5)$$

Thus, the delta function is the limit function of this sequence, i.e.,

$$\delta(x) = \lim_{n \to \infty} \delta_n(x). \qquad (3.5.6)$$

Particular sequences that have these properties are:

$$\delta_n(x) = \begin{cases} \sqrt{\dfrac{n}{\pi}}\, e^{-nx^2}, \\[2mm] \left(\dfrac{n}{\pi}\right) \dfrac{1}{1 + n^2x^2}, \\[2mm] \begin{cases} \dfrac{n}{2}, & \text{for } |x| < \frac{1}{n}, \\[2mm] 0, & \text{for } |x| \geq \frac{1}{n}. \end{cases} \end{cases} \qquad (3.5.7)$$

As a check and also interesting exercise, consider the third example of $\delta_n(x)$ in Eq. (3.5.7). Clearly, its integral over x has the value one; thus the condition of Eq. (3.5.4) holds. Assume that $f(x)$ is continuous at $x = 0$. Then,

$$\int_{-\infty}^{\infty} \delta_n(x)f(x)dx = \left(\frac{n}{2}\right) \int_{-1/n}^{1/n} f(x)dx. \qquad (3.5.8)$$

By the mean value theorem for integrals, we have

$$\int_{-1/n}^{1/n} f(x)dx = \left(\frac{2}{n}\right) f(x_0), \qquad -\left(\frac{1}{n}\right) < x_0 < \left(\frac{1}{n}\right), \qquad (3.5.9)$$

and Eq. (3.5.8) becomes

$$\int_{-\infty}^{\infty} \delta_n(x)f(x)dx = \left(\frac{n}{2}\right)\left(\frac{2}{n}\right) f(x_0) = f(x_0). \qquad (3.5.10)$$

Therefore,

$$\lim_{n \to \infty} \int_{-\infty}^{\infty} \delta_n(x)f(x)dx = f(0), \qquad (3.5.11)$$

since $x_0 \to 0$ in this limit.

We now "derive" four basic properties of the delta function:

Property 1: The delta function is even, i.e.,

$$\delta(-x) = \delta(x), \qquad (3.5.12)$$

or

$$\int_{-\infty}^{\infty} \delta(-x) f(x) dx = \int_{-\infty}^{\infty} \delta(x) f(x) dx. \qquad (3.5.13)$$

"*Proof*":

$$\int_{-\infty}^{\infty} \delta(-x) f(x) dx = -\int_{\infty}^{-\infty} \delta(y) f(-y) dy$$

$$= \int_{-\infty}^{\infty} \delta(y) f(-y) dy$$

$$= f(0). \qquad (3.5.14)$$

In the first line we transformed the variable x to $y = -x$ and rewrote this to the expression in the second line. The result in the third line follows from the definition of the delta function as stated in Eq. (3.5.3).

Property 2:

$$\int_{-\infty}^{\infty} \delta(ax) f(x) dx = \frac{1}{|a|} \cdot f(0). \qquad (3.5.15)$$

"*Proof*": First assume $a > 0$ and let $y = ax$; then

$$\int_{-\infty}^{\infty} \delta(ax) f(x) dx = \int_{-\infty}^{\infty} \delta(y) f\left(\frac{y}{a}\right) \frac{dy}{a}$$

$$= \left(\frac{1}{a}\right) \int_{-\infty}^{\infty} \delta(y) f\left(\frac{y}{a}\right) dy = \frac{f(0)}{a}. \qquad (3.5.16)$$

Next assume $a < 0$ and let $y = |a|x$; then

$$\int_{-\infty}^{\infty} \delta(-|a|x) f(x) dx = \int_{-\infty}^{\infty} \delta(|a|x) f(x) dx = \frac{f(0)}{|a|}. \qquad (3.5.17)$$

The result in Eq. (3.5.15) then follows. This can also be written as

$$\delta(ax) = \frac{1}{|a|} \cdot \delta(x). \qquad (3.5.18)$$

Property 3:

$$\int_{-\infty}^{\infty} \delta(x - a) f(x) dx = f(a). \qquad (3.5.19)$$

"*Proof*": Let $y = x - a$, then

$$\int_{-\infty}^{\infty} \delta(x - a) f(x) dx = \int_{-\infty}^{\infty} \delta(y) f(y + a) dy = f(a). \qquad (3.5.20)$$

Property 4:

$$\int_{-\infty}^{\infty} \delta(x^2 - a^2)f(x)dx = \frac{f(a) + f(-a)}{2|a|} \tag{3.5.21}$$

or

$$\delta(x^2 - a^2) = \frac{\delta(x - a) + \delta(x + a)}{2|a|} . \tag{3.5.22}$$

This result can be generalized as follows. Let $g(x)$ be a function having a finite number, N, of simple zeros and denote them by $\{\bar{x}_i : i = 1, 2, \ldots, N\}$. Then,

$$\delta[g(x)] = \sum_{i=1}^{N} \frac{\delta(x - \bar{x}_i)}{|g'(\bar{x}_i)|} . \tag{3.5.23}$$

"*Proof*": The integral on the left-side of Eq. (3.5.21) can be written as five separate integrals,

$$\int_{-\infty}^{\infty} (\ ) = \int_{-\infty}^{-a-\epsilon} (\ ) + \int_{-a-\epsilon}^{-a+\epsilon} (\ ) + \int_{-a+\epsilon}^{a-\epsilon}$$
$$+ \int_{a-\epsilon}^{a+\epsilon} (\ ) + \int_{a+\epsilon}^{\infty} , \tag{3.5.24}$$

where the first, third, and fifth integrals are zero. Now consider the second integral,

$$\int_{-a-\epsilon}^{-a+\epsilon} \delta(x^2 - a^2)f(x)dx = \int_{-a-\epsilon}^{-a+\epsilon} \delta[(x - a)(x + a)]f(x)dx$$
$$= \int_{-a-\epsilon}^{-a+\epsilon} \delta[(-2a)(x + a)]f(x)dx = \frac{f(-a)}{2|a|} . \tag{3.5.25}$$

Similarly, the fourth integral is

$$\int_{a-\epsilon}^{a+\epsilon} \delta(x^2 - a^2)f(x)dx = \int_{a-\epsilon}^{a+\epsilon} \delta[(x - a)(x + a)]f(x)dx$$
$$= \int_{a-\epsilon}^{a+\epsilon} \delta[(x - a)(2a)]f(x)dx$$
$$= \frac{f(a)}{2|a|} . \tag{3.5.26}$$

Adding all the contributions of the five integrals give the result of Eq. (3.5.21).

An additional interesting and important property of the delta function involves its derivatives. Assume that we have a delta sequence, $\delta_n(x)$, such that $\delta_n(x)$ and all its derivatives exist. For example, in Eq. (3.5.7), this holds true for the first and second examples, but not for the third one. We have, for a bounded function, $f(x)$, the result

$$\int_{-\infty}^{\infty} \delta_n'(x)f(x)dx = \delta_n(x)f(x)\Big|_{-\infty}^{\infty} - \int_{-\infty}^{\infty} \delta_n(x)f'(x)dx, \qquad (3.5.27)$$

where $\delta_n'(x) = d\delta_n/dx$ and where integration by parts was used to obtain the right-side expression. The first term on the right-side is zero. Therefore, taking the limit as $n \to \infty$, gives

$$\int_{-\infty}^{\infty} \delta'(x)f(x)dx \equiv \operatorname*{Lim}_{n\to\infty} \int_{-\infty}^{\infty} \delta_n'(x)f(x)dx$$

$$= -\operatorname*{Lim}_{n\to\infty} \int_{-\infty}^{\infty} \delta_n(x)f'(x)dx = -f'(0). \qquad (3.5.28)$$

This process can be repeated to obtain the result

$$\int_{-\infty}^{\infty} \delta^{(n)}(x)f(x)dx = (-1)^n f^{(n)}(0), \qquad (3.5.29)$$

where it is assumed that the required derivatives of $f(x)$ exist.

What forms do the delta functions take in higher dimensions? For the three dimensional case, we want

$$\int\int\int \delta^{(3)}(\vec{r})f(\vec{r})d\vec{r} = f(0) \qquad (3.5.30)$$

and

$$\int\int\int \delta^{(3)}(\vec{r} - \vec{r}_0)f(\vec{r})d\vec{r} = f(\vec{r}_0). \qquad (3.5.31)$$

For example, in Cartesian coordinates

$$\int_{-\infty}^{\infty}\int_{-\infty}^{\infty}\int_{-\infty}^{\infty} \delta^{(3)}(\vec{r} - \vec{r}_0)f(x,y,z)$$

$$= \int_{-\infty}^{\infty}\int_{-\infty}^{\infty}\int_{-\infty}^{\infty} \delta(x-x_0)\delta(y-y_0)\delta(z-z_0)f(x,y,z)dx\,dy\,dz. \qquad (3.5.32)$$

From this expression, we have

$$\delta^{(3)}(\vec{r} - \vec{r}_0) = \delta(x-x_0)\delta(y-y_0)\delta(z-z_0). \qquad (3.5.33)$$

Likewise for spherical coordinates,

$$\int\int\int \delta^{(3)}(\vec{r} - \vec{r}_0)f(r,\theta,\phi)r^2\sin\theta\,d\theta\,d\phi\,dr$$

$$= f(r_0,\theta_0,\phi_0)$$

$$= \int\int\int \delta(r-r_0)\delta(\theta-\theta_0)\delta(\phi-\phi_0)f(r,\theta,\phi)dr\,d\theta\,d\phi, \qquad (3.5.34)$$

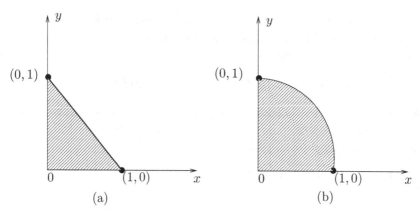

Fig. 3.6.1 Shaded areas are: (a) $x + y \leq 1$, $x > 0$, and $y > 0$. (b) $x^2 + y^2 \leq 1$, $x > 0$, $y > 0$.

and it follows that

$$\delta^{(3)}(\vec{r} - \vec{r}_0) = \left(\frac{1}{r^2 \sin\theta} \right) \delta(r - r_0)\delta(\theta - \theta_0)\delta(\phi - \phi_0). \tag{3.5.35}$$

3.6   Dirichlet Integrals

Two of the simplest cases of Dirichlet integrals are given by the following examples:

$$I = \iint\limits_{\substack{x+y\leq 1 \\ x>0, \ y>0}} dx\, dy, \tag{3.6.1}$$

$$J = \iint\limits_{\substack{x^2+y^2\leq 1 \\ x>0, \ y>0}} dx\, dy. \tag{3.6.2}$$

The integrals I and J correspond, respectively, to the areas of the shaded portions of Figure 3.6.1. I is the area of the right-triangle connecting the points $(0,0)$, $(1,0)$ and $(0,1)$; while J is the area of a quarter-circle of radius one. In both cases, the integrals can be easily calculated and give,

$$I = \int_0^1 dx \int_0^{1-x} dy = \int_0^1 (1-x)dx = \frac{1}{2}, \tag{3.6.3}$$

$$J = \iint\limits_{\substack{0\leq r\leq 1 \\ 0\leq\theta\leq\pi/2}} r\, dr\, d\theta = \int_0^{\pi/2} d\theta \int_0^1 r\, dr = \left(\frac{\pi}{2}\right)\left(\frac{1}{2}\right) = \frac{\pi}{4}. \tag{3.6.4}$$

The integral I can be generalized to the form

$$I(m,n) \equiv \iint\limits_{\substack{x+y\leq 1 \\ x>0,\, y>0}} x^m y^n dx\, dy; \qquad m,n > -1, \qquad (3.6.5)$$

where the restriction on m and n is to insure that the integrals exist. This integral can be written as

$$\begin{aligned}
I(m,n) &= \int_0^1 x^m \left[\int_0^{1-x} y^n dy \right] dx \\
&= \int_0^1 x^m \left[\frac{(1-x)^{n+1}}{n+1} \right] dx \\
&= \left(\frac{1}{n+1} \right) \int_0^1 x^m (1-x)^{n+1} dx \\
&= \left(\frac{1}{n+1} \right) B(m+1, n+2) = \frac{\Gamma(m+1)\Gamma(n+2)}{(n+1)\Gamma(m+n+3)} \\
&= \frac{\Gamma(m+1)\Gamma(n+1)}{(m+n+2)\Gamma(m+n+2)} \\
&= \frac{B(m+1, n+1)}{(m+n+2)}, \qquad (3.6.6)
\end{aligned}$$

where $B(m+1, n+1)$ is the beta function.

Now consider the $J(m,n)$ integral defined as

$$J(m,n) \equiv \iint\limits_{\substack{x^2+y^2\leq 1 \\ x>0,\, y>0}} x^n y^m dx\, dy. \qquad (3.6.7)$$

The change of variables

$$x = r\cos\theta, \qquad y = r\sin\theta, \qquad (3.6.8)$$

gives

$$\begin{aligned}
J(m,n) &= \int_0^{\pi/2} \int_0^1 (r\cos\theta)^n (r\sin\theta)^m r\, dr\, d\theta \\
&= \int_0^1 r^{n+m+1} dr \cdot \int_0^{\pi/2} (\cos\theta)^n (\sin\theta)^m d\theta \\
&= \left(\frac{1}{n+m+2} \right) \int_0^{\pi/2} (\cos\theta)^n (\sin\theta)^m d\theta. \qquad (3.6.9)
\end{aligned}$$

From Eq. (3.3.10), we have

$$\int_0^{\pi/2} (\cos\theta)^{2p-1} (\sin\theta)^{2q-1} d\theta = \left(\frac{1}{2} \right) B(p,q). \qquad (3.6.10)$$

Comparing this with the integral in Eq. (3.6.9), it follows that

$$p = \frac{n+1}{2}, \qquad q = \frac{m+1}{2}, \tag{3.6.11}$$

and

$$J(m,n) = \frac{B\left(\frac{n+1}{2}, \frac{m+1}{2}\right)}{2(m+n+2)}. \tag{3.6.12}$$

The two results for I and J can be extended to any number of variables. Consider an N-dimensional Cartesian space, $\{x_i : i = 1, 2, \ldots, N\}$. Let V_N be the volume in this space for which

$$\sum_{i=1}^{N} x_i \leq 1; \quad x_i > 0 \text{ for } i = 1, 2, \ldots, N. \tag{3.6.13}$$

Then,

$$I(m_1, m_2, \ldots, m_N) \equiv \int \cdots \int_{V_N} x_1^{m_1} x_2^{m_2} \ldots x_N^{m_N} dV_N, \tag{3.6.14}$$

where

$$dV_N = \prod_{i=1}^{N} dx_i; \qquad m_i > -1 \text{ for } i = 1, 2, \ldots, N; \tag{3.6.15}$$

is given by the expression (see Franklin in [4]),

$$I(m_1, m_2, \ldots, m_N) = \frac{\prod_{i=1}^{N} \Gamma(m_i + 1)}{\Gamma\left(\sum_{i=1}^{N} m_i + N + 1\right)}. \tag{3.6.16}$$

Similarly, let $\bar{V}_N$ be the volume for which

$$\sum_{i=1}^{N} \left(\frac{x_i}{a_i}\right)^{p_i} \leq 1; \qquad x_i > 0 \text{ for } i = 1, 2, \ldots, N; \tag{3.6.17}$$

where

$$dV_N = \prod_{i=1}^{N} dx_i; \qquad m_i > -1 \text{ for } i = 1, 2, \ldots, N; \tag{3.6.18}$$

and $p_i > 0$ for $i = 1, 2, \ldots, N$. Then, we have

$$J(m_1, m_2, \ldots, m_N) \equiv \int \cdots \int x_1^{m_1} x_2^{m_2} \ldots x_N^{m_N} dV_N$$

$$= \frac{\prod_{i=1}^{N} \left(\frac{a_i^{m_i+1}}{p_i}\right) \Gamma\left(\frac{m_i+1}{p_i}\right)}{\Gamma\left[\sum_{i=1}^{N} \left(\frac{m_i+1}{p_i}\right) + 1\right]} \tag{3.6.19}$$

3.7   Applications

We now present a number of problems illustrating the applications of the functions studied in the previous five sections.

3.7.1   Additional Properties of $\Gamma(z)$

As initially defined, the gamma function depends on the complex variable $z = x + iy$. In terms of the integral representation, we have for $\mathrm{Re}(z) > 0$,

$$\Gamma(z) = \int_0^\infty t^{z-1} e^{-t} dt = \int_0^\infty t^{(x-1)+iy} e^{-t} dt, \qquad (3.7.1)$$

and using

$$t^{iy} = e^{iy \ln t}, \qquad |t^{iy}| = 1, \qquad (3.7.2)$$

it follows that

$$|\Gamma(x + iy)| \leq \int_0^\infty |t^{(x-1)+iy}| e^{-t} dt$$

$$= \int_0^\infty t^{x-1} e^{-t} dt = \Gamma(x), \qquad (3.7.3)$$

and, we conclude that

$$|\Gamma(z)| \leq \Gamma(x), \qquad \mathrm{Re}(z) = x > 0. \qquad (3.7.4)$$

If $f(z)$ is a complex function of z, then writing $f(z)$ as

$$f(z) = u(x, y) + iv(x, y), \qquad (3.7.5)$$

it follows that the Cauchy-Riemann equations [7] must be satisfied, i.e.,

$$\frac{\partial u}{\partial x} = \frac{\partial v}{\partial y}, \qquad \frac{\partial u}{\partial y} = -\frac{\partial v}{\partial x}. \qquad (3.7.6)$$

For $f(z) = \Gamma(z)$, we have, using the relation,

$$t^z = e^{z \ln t} = e^{x \ln t + iy \ln t}$$

$$= e^{x \ln t} \left[\cos(y \ln t) + i \sin(y \ln t) \right], \qquad (3.7.7)$$

the expressions

$$u(x, y) = \int_0^\infty e^{x \ln t} \cos(y \ln t) \left(\frac{e^{-t}}{t} \right) dt, \qquad (3.7.8)$$

$$v(x, y) = \int_0^\infty e^{x \ln t} \sin(y \ln t) \left(\frac{e^{-t}}{t} \right) dt. \qquad (3.7.9)$$

An easy calculation shows that $\Gamma(z)$ does satisfy the Cauchy-Riemann equation and, consequently, $\Gamma(z)$ is a complex function of z.

Let $f(z)$ be a complex function. A *real-valued function* has the property,

$$[f(z)]^* = f(z^*), \tag{3.7.10}$$

i.e., if z is real, then $f(z)$ is real. Note that

$$z = x + iy, \qquad z^* = x - iy. \tag{3.7.11}$$

Inspection of Eq. (3.7.1) shows that $\Gamma(x)$ is a real-valued function, i.e.,

$$[\Gamma(z)]^* = \Gamma(z^*). \tag{3.7.12}$$

3.7.2   *A Definite Integral Containing Logarithms*

Consider the class of definite integrals,

$$\begin{cases} R(m,n) \equiv \int_0^1 x^m (\ln x)^n dx, \\ n = 0, 1, 2, \ldots; \text{ and } m > -1. \end{cases} \tag{3.7.13}$$

Make the following change of variables,

$$x = e^{-y}, \tag{3.7.14}$$

and obtain,

$$dx = -e^{-y} dy; \qquad x = 0 \Rightarrow y = \infty, \qquad x = 1 \Rightarrow y = 0. \tag{3.7.15}$$

Substitution of these results into the integral gives

$$\int_0^1 x^m (\ln x)^n dx = \int_\infty^0 e^{-my}(-y)^n(-e^{-y}dy)$$

$$= (-1)^n \int_0^\infty e^{-(m+1)y} y^n dy. \tag{3.7.16}$$

Let $u = (m+1)y$, then the last integral is

$$(-1)^n \int_0^\infty \left[\frac{u^n}{(m+1)^n} \right] e^{-u} \left(\frac{du}{m+1} \right)$$

$$= \left[\frac{(-1)^n}{(m+1)^{n+1}} \right] \int_0^\infty u^n e^{-u} du$$

$$= \left[\frac{(-1)^n}{(m+1)^{n+1}} \right] \Gamma(n+1). \tag{3.7.17}$$

Therefore, we conclude that

$$R(m,n) \equiv \int_0^1 x^m (\ln x)^n dx = \frac{(-1)^n n!}{(m+1)^{n+1}}. \tag{3.7.18}$$

Note that for $n = 0$, we obtain

$$R(m,n) = \int_0^1 x^m dx = \frac{1}{m+1}, \tag{3.7.19}$$

and for $m = 0$,

$$R(0, n) = \int_0^1 (\ln x)^n dx = (-1)^n n!. \tag{3.7.20}$$

Thus, a representation of the factorial function is

$$n! \equiv (-1)^n \int_0^1 (\ln x)^n dx. \tag{3.7.21}$$

3.7.3 A Class of Important Integrals

All the particular examples of the definite integrals for the exponential and Gaussian functions are special cases of the integral,

$$I(\alpha, \beta, \gamma) = \int_0^\infty t^\alpha e^{-\beta t^\gamma} dt, \tag{3.7.22}$$

where

$$\text{Re}(\alpha) > -1, \quad \text{Re}(\beta) > 0, \quad \gamma \text{ real and positive.} \tag{3.7.23}$$

To proceed, let

$$x = \beta t^\gamma \quad \text{or} \quad t = \left(\frac{x}{\beta}\right)^{1/\gamma} \tag{3.7.24}$$

and obtain

$$\begin{cases} x = 0 \Rightarrow t = 0, \quad x = \infty \Rightarrow t = \infty; \\ t^\alpha = \left(\dfrac{x}{\beta}\right)^{\alpha/\gamma}, \quad dt = \left(\dfrac{1}{\beta^{1/\gamma}}\right)\left(\dfrac{1}{\gamma}\right) x^{\frac{1-\gamma}{\gamma}} dx. \end{cases} \tag{3.7.25}$$

Substitution of these results into Eq. (3.7.22) gives

$$I(\alpha, \beta, \gamma) = \left[\frac{1}{\gamma \beta^{\frac{\alpha+1}{\gamma}}}\right] \int_0^\infty x^{\frac{\alpha+1}{\gamma}-1} e^{-x} dx, \tag{3.7.26}$$

and using the definition of the gamma function

$$I(\alpha, \beta, \gamma) = \left[\frac{1}{\gamma \beta^{\frac{\alpha+1}{\gamma}}}\right] \Gamma\left(\frac{\alpha+1}{\gamma}\right). \tag{3.7.27}$$

3.7.4 A Representation for x^{-p}

We have from the definition of the gamma function,

$$\Gamma(p) = \int_0^\infty u^{p-1} e^{-u} du, \quad \text{Re}(p) > 0. \tag{3.7.28}$$

Let $u = xt$; then, for fixed $x > 0$, we have

$$\Gamma(p) = \int_0^\infty (xt)^{p-1} e^{-xt} d(xt) = x^p \int_0^\infty t^{p-1} e^{-xt} dt, \qquad (3.7.29)$$

and

$$\frac{1}{x^p} = \left[\frac{1}{\Gamma(p)} \right] \int_0^\infty t^{p-1} e^{-xt} dt. \qquad (3.7.30)$$

3.7.5 Additional Properties of the Beta Function

The beta function satisfies the following identities:

$$B(p, q) = B(p+1, q) + B(p, q+1), \qquad (3.7.31)$$

$$B(p, q) = \left(\frac{p+q}{q} \right) B(p, q+1), \qquad (3.7.32)$$

$$B(p, q) = \left(\frac{q-1}{p} \right) B(p+1, q-1), \qquad (3.7.33)$$

$$B(p, q) B(p+q, r) = B(q, r) B(p, q+r). \qquad (3.7.34)$$

Using the definition of the beta function

$$B(p, q) = \frac{\Gamma(p)\Gamma(q)}{\Gamma(p+q)}, \qquad (3.7.35)$$

these relations can be easily checked for correctness. For example,

$$B(p, q+1) = \frac{\Gamma(p)\Gamma(q+1)}{\Gamma(p+q+1)} = \frac{q\Gamma(p)\Gamma(q)}{(p+q)\Gamma(p+q)}, \qquad (3.7.36)$$

and the relation of Eq. (3.7.32) is verified. Likewise

$$\begin{aligned}
B(p+1, q) + B(p, q+1) &= \frac{\Gamma(p+1)\Gamma(q)}{\Gamma(p+q+1)} + \frac{\Gamma(p)\Gamma(q+1)}{\Gamma(p+q+1)} \\
&= \frac{p\Gamma(p)\Gamma(q)}{(p+q)\Gamma(p+q)} + \frac{q\Gamma(p)\Gamma(q)}{(p+q)\Gamma(p+q)} \\
&= B(p, q), \qquad (3.7.37)
\end{aligned}$$

and Eq. (3.7.31) is shown to be true.

3.7.6   Fermi-Dirac Integrals

The Fermi-Dirac integrals take the form

$$J(s) = \int_0^\infty \frac{x^s}{e^x + 1}\, dx, \qquad s > 0. \tag{3.7.38}$$

They can be evaluated by first rewriting the integral as follows

$$J(s) = \sum_{n=0}^\infty (-1)^n \int_0^\infty x^s e^{-(n+1)x} dx, \tag{3.7.39}$$

where we have used

$$\frac{1}{e^x + 1} = \frac{e^{-x}}{1 + e^{-x}} = e^{-x} \sum_{n=0}^\infty (-1)^n e^{-nx} = \sum_{n=0}^\infty (-1)^n e^{-(n+1)x}, \tag{3.7.40}$$

and interchanged the summation and integration. Now let $t = (n+1)x$; under this change of variable, it follows that

$$\int_0^\infty x^s e^{-(n+1)x} dx = \left[\frac{1}{(n+1)^{s+1}} \right] \int_0^\infty t^s e^{-t} dt$$

$$= \left[\frac{1}{(n+1)^{s+1}} \right] \Gamma(s+1), \tag{3.7.41}$$

and

$$J(s) = \sum_{N=0}^\infty \left[\frac{(-1)^n}{(n+1)^{s+1}} \right] \Gamma(s+1)$$

$$= \Gamma(s+1) \sum_{n=1}^\infty \frac{(-1)^{n+1}}{n^{s+1}}$$

$$= \Gamma(s+1)\zeta(s+1;-). \tag{3.7.42}$$

Therefore the Fermi-Dirac integral is

$$J(s) \equiv \int_0^\infty \frac{x^s}{e^x + 1}\, dx = \Gamma(s+1)\zeta(s+1) \left[1 - \frac{1}{2^s} \right]. \tag{3.7.43}$$

Note that $J(s)$ is expressed as a product of both the gamma and zeta functions, along with a known factor of s.

3.7.7   An Integral Involving an Exponential

The following, rather elementary looking, integral is of interest since it occurs at some stage in many calculations,

$$I(a,b) \equiv \int_0^\infty \frac{dx}{e^{ax} + b}\,; \qquad a, b > 0. \tag{3.7.44}$$

It can be written as

$$I(a,b) = \left(\frac{-1}{ab}\right) \int_0^\infty \frac{be^{-ax}d(-ax)}{1 + be^{-ax}}$$

$$= \left(\frac{-1}{ab}\right) \ln(1 + be^{-ax})\big|_0^\infty \qquad (3.7.45)$$

or

$$I(a,b) = \frac{\ln(1 + b)}{ab}. \qquad (3.7.46)$$

Note that

$$I(a,0) = \lim_{b \to 0} \left[\frac{\ln(1 + b)}{ab}\right] = \frac{1}{a} \qquad (3.7.47)$$

and

$$I(1,1) = \ln(2). \qquad (3.7.48)$$

3.7.8   *Fermi-Dirac Integrals Containing Logarithms*

The Fermi-Dirac integral defined by Eq. (3.7.38) can be modified to include logarithmic factors, i.e.,

$$J(s,n) = \int_0^\infty \frac{x^s(\ln x)^n}{e^x + 1} dx, \qquad (3.7.49)$$

where n is a non-negative integer. This integral can be exactly evaluated from just a knowledge of $J(s)$ given by Eq. (3.7.43). To proceed, note that

$$x^s = e^{s \ln x} \qquad (3.7.50)$$

and

$$\frac{d^n}{ds^n}(x^s) = (\ln x)^n e^{s \ln x} = (\ln x)^n x^s. \qquad (3.7.51)$$

Therefore, $J(s,n)$ is

$$J(s,n) = \frac{d^n}{ds^n} J(s). \qquad (3.7.52)$$

3.7.9   *Magnetic Moment of the Electron*

The following integral occurs in the theoretical calculation of the magnetic moment of the electron,

$$M \equiv \int_0^1 [\ln(1 - x)]^2 \frac{dx}{x}. \qquad (3.7.53)$$

It can be evaluated by first making the change of variables

$$1 - x = e^{-t}, \tag{3.7.54}$$

which gives

$$\begin{cases} x = 0 \Rightarrow t = 0, & x = 1 \Rightarrow t = +\infty; \\ \dfrac{dx}{x} = \dfrac{e^{-t}dt}{1 - e^{-t}}, & [\ln(1 - x)]^2 = t^2 \end{cases} \tag{3.7.55}$$

and

$$M = \int_0^\infty \frac{t^2 e^{-t}}{1 - e^{-t}} \, dt = \int_0^\infty t^2 e^{-t} \left[\sum_{n=0}^\infty e^{-nt} \right] dt. \tag{3.7.56}$$

Now, let $x = (n + 1)t$, and obtain

$$M = \sum_{n=0}^\infty \frac{1}{(n + 1)^3} \int_0^\infty x^2 e^{-x} dx = \Gamma(3) \sum_{n=1}^\infty \frac{1}{N^3}. \tag{3.7.57}$$

Using the definition for the zeta function, we finally obtain

$$M = 2\zeta(3). \tag{3.7.58}$$

3.7.10   *Relationship Between the Theta and Delta Functions*

There are several piecewise continuous functions [8] that play significant roles in the applied mathematics relating to various areas in science and technology. Two of them are the "theta" and "sign" functions. They are defined as follows:

$$\theta(x) \equiv \begin{cases} 1, & \text{if } x > 0, \\ 0, & \text{if } x < 0; \end{cases} \tag{3.7.59}$$

$$\text{sign}(x) \equiv \begin{cases} 1, & \text{if } x > 0, \\ -1, & \text{if } x < 0. \end{cases} \tag{3.7.60}$$

Note that

$$\text{sign}(x) = \theta(x) - \theta(-x). \tag{3.7.61}$$

We will now demonstrate that the derivative of $\theta(x)$ is the delta function $\delta(x)$, i.e.,

$$\theta'(x) = \delta(x). \tag{3.7.62}$$

We do this by showing that $\theta'(x)$ has the three properties associated with a delta function as given in section 3.5. First, observe that

$$\int_{-\infty}^{\infty} \theta'(x)dx = \theta(x)\Big|_{-\infty}^{+\infty} = 1 - 0 = 1, \tag{3.7.63}$$

and thus the normalization condition holds. Second, from the definition of $\theta(x)$, it is clear that

$$\theta'(x) = 0, \qquad x \neq 0. \tag{3.7.64}$$

Finally, for a function that is continuous with a continuous derivative, we have

$$\begin{aligned}
\int_{-\epsilon}^{\epsilon} \theta'(x)f(x)dx &= \theta(x)f(x)\Big|_{-\epsilon}^{\epsilon} - \int_{-\epsilon}^{\epsilon} \theta(x)f'(x)dx \\
&= \theta(\epsilon)f(\epsilon) - \theta(-\epsilon)f(-\epsilon) - \int_{-\epsilon}^{\epsilon} \theta(x)f'(x)dx \\
&= f(\epsilon) - \int_{0}^{\epsilon} f'(x)dx \\
&= f(\epsilon) - f(\epsilon)\Big|_{0}^{\epsilon} = f(\epsilon) - [f(\epsilon) - f(0)] \\
&= f(0). \tag{3.7.65}
\end{aligned}$$

We conclude that $\theta'(x)$ satisfies the three conditions required of a delta function and consequently the result given in Eq. (3.7.62) is true. A simple calculation also gives

$$\frac{d}{dx}\,\text{sign}(x) = 2\delta(x). \tag{3.7.66}$$

3.7.11   *Evaluation of Integrals by Use of the Beta Function*

The beta function can be used to evaluate both trigonometric and rational power function definite integrals. This follows directly from the equivalent representations of $B(p,q)$:

$$B(p,q) = \frac{\Gamma(p)\Gamma(q)}{\Gamma(p+q)} \tag{3.7.67}$$

$$= \int_{0}^{1} x^{p-1}(1-x)^{q-1}dx \tag{3.7.68}$$

$$= 2\int_{0}^{\pi/2} (\cos\theta)^{2p-1}(\sin\theta)^{2q-1}d\theta, \tag{3.7.69}$$

where all these relations require, $\text{Re}(p) > 0$ and $\text{Re}(q) > 0$.

Elementary applications are

$$B\left(\frac{1}{2},\frac{1}{2}\right) = \int_0^1 \frac{dx}{\sqrt{x(1-x)}} = \frac{\Gamma\left(\frac{1}{2}\right)\Gamma\left(\frac{1}{2}\right)}{\Gamma(1)} = \pi, \qquad (3.7.70)$$

$$B\left(\frac{1}{2},\frac{3}{2}\right) = \int_0^1 \sqrt{\frac{1-x}{x}}\, dx = \frac{\Gamma\left(\frac{1}{2}\right)\Gamma\left(\frac{3}{2}\right)}{\Gamma(2)} = \frac{\pi}{2}, \qquad (3.7.71)$$

where we have used $\Gamma\left(\frac{1}{2}\right) = \sqrt{\pi}$.

The two definite integrals, given below, are equal to each other,

$$\int_0^{\pi/2} (\sin\theta)^m d\theta = \int_0^{\pi/2} (\cos\theta)^m d\theta. \qquad (3.7.72)$$

This can be demonstrated by making use of the transformation

$$\theta = \frac{\pi}{2} - \phi. \qquad (3.7.73)$$

In terms of the beta function, we have, after several rather trivial manipulations,

$$\begin{aligned}
\int_0^{\pi/2} (\sin\theta)^m d\theta &= \int_0^{\pi/2} (\cos\theta)^m d\theta \\
&= \left(\frac{1}{2}\right) B\left(\frac{m+1}{2},\frac{1}{2}\right) \\
&= \left(\frac{\sqrt{\pi}}{2}\right) \frac{\Gamma\left(\frac{m+1}{2}\right)}{\Gamma\left(\frac{m+2}{2}\right)}. \qquad (3.7.74)
\end{aligned}$$

Two cases need to be considered, m even and m odd:

$$\int_0^{\pi/2} (\cos\theta)^m d\theta = \int_0^{\pi/2} (\sin\theta)^m d\theta$$
$$= \begin{cases} \left(\dfrac{\pi}{2}\right)\dfrac{(2n-1)!!}{2^n \cdot n!}, & m = 2n, \\[3mm] \dfrac{2^n \cdot n!}{(2n+1)!!}, & m = 2n+1. \end{cases} \qquad (3.7.75)$$

3.8 Other Named Functions

A number of functions, defined by integrals, occur so frequently, that their numerical values and graphs are given in various handbooks on mathematical functions. In addition to the books by Abramowitz and Stegun, Gradshteyn and Ryzhik, and Jeffrey (listed in the bibliography to this chapter) other useful references include,

1) H. T. Davis, *Tables of Higher Mathematical Functions*, 2 Vols. (Principa Press; Bloomington, IN; 1935).

2) A. Erdélyi, *Higher Transcendental Functions, Vols. I, II, III* (Mc-Graw-Hill, New York, 1953).

3) E.W. Hobson, *Spherical and Ellipsoidal Harmonics* (Cambridge University Press, Cambridge, 1931).

4) J. Meixner and F. W. Schäfke, *Mathieusche Funktionen and Sphäroidal funktionen* (Springer-Verlag, Berlin, 1954).

5) T. Pearcey, *Table of the Fresnel Integral* (Cambridge University Press, London, 1956).

We now list a number of these "named functions" defined by integrals and refer the reader to the above indicated references for the details regarding how they arise from specific applications, their mathematical properties and tables of their numerical values. Please be aware that these functions are defined somewhat differently by various authors. Consequently, be sure that the function you are studying coincides with that defined in a particular reference book.

Cosine Integral

$$Ci(x) = \int_x^\infty \frac{\cos t}{t} \cdot dt \tag{3.8.1}$$

Sine Integral

$$Si(x) = \int_0^x \frac{\sin t}{t} \cdot dt \tag{3.8.2}$$

Logarithmic Integral

$$Li(x) = \int_0^x \frac{dt}{\ln t} \tag{3.8.3}$$

Error Function

$$\mathrm{erf}(x) = \left(\frac{2}{\sqrt{\pi}} \right) \int_0^x e^{-t^2} dt \tag{3.8.4}$$

Exponential Integral

$$Ei(x) = \int_x^\infty \frac{e^{-t}}{t} \cdot dt \tag{3.8.5}$$

Fresnel Cosine Integral

$$C(x) = \left(\sqrt{\frac{2}{\pi}} \right) \int_0^x \cos(t^2) dt \tag{3.8.6}$$

Fresnel Sine Integral

$$S(x) = \left(\sqrt{\frac{2}{\pi}}\right) \int_0^x \sin(t^2)dt \qquad (3.8.7)$$

Incomplete Gamma Function

$$\gamma(a,x) = \int_0^x t^{a-1}e^{-t}dt, \qquad \mathrm{Re}(a) > 0. \qquad (3.8.8)$$

3.9 Elliptic Integrals and Functions

Some of the simplest systems having oscillations are modeled by second-order, nonlinear ordinary differential equations. The solutions to these equations lead directly to elliptic integrals and functions. This section is divided into two parts. The first introduces several types of elliptic integrals and provides some of their elementary properties. The second part considers the so-called Jacobi elliptic function. Full accounts of both elliptic integrals and functions are presented in the books by Abramowitz and Stegun, Byrd and Friedman, and Lawden.

3.9.1 *Elliptic Integrals of the First and Second Kind*

The following nonlinear differential equation arises in the study of a pendulum,

$$\frac{d^2\theta}{dt^2} + \sin\theta = 0. \qquad (3.9.1)$$

If we define $p = d\theta/dt$, then a first-integral can be obtained by writing Eq. (3.9.1) as

$$p\frac{dp}{d\theta} + \sin\theta = 0. \qquad (3.9.2)$$

Integrating once gives

$$\frac{p^2}{2} - \cos\theta = c = \text{constant}. \qquad (3.9.3)$$

If $d\theta/dt$ is zero at $\theta = \theta_0$ then

$$c = -\cos(\theta_0) \qquad (3.9.4)$$

and

$$\frac{p^2}{2} - \cos\theta = -\cos(\theta_0). \qquad (3.9.5)$$

From this equation, we obtain

$$p = \frac{d\theta}{dt} = \pm\sqrt{2(\cos\theta - \cos\theta_0)}.$$ (3.9.6)

Selecting the positive sign gives

$$dt = \frac{d\theta}{\sqrt{2(\cos\theta - \cos\theta_0)}}.$$ (3.9.7)

Now the time it takes the pendulum to swing from $\theta = 0$ to $\theta = \theta_0$ is $T/4$ where T is the period of the oscillation. Therefore,

$$\frac{T}{4} = \left(\frac{1}{\sqrt{2}}\right)\int_0^{\theta_0} \frac{d\theta}{\sqrt{\cos\theta - \cos\theta_0}}.$$ (3.9.8)

Now let

$$k = \sin\left(\frac{\theta_0}{2}\right),$$ (3.9.9)

and use

$$\cos\theta = 1 - 2\sin^2\left(\frac{\theta}{2}\right), \qquad \cos\theta_0 = 1 - 2\sin^2\left(\frac{\theta_0}{2}\right),$$ (3.9.10)

to obtain

$$T = 2\int_0^{\theta_0} \frac{d\theta}{\sqrt{k^2 - \sin^2\left(\frac{\theta}{2}\right)}}.$$ (3.9.11)

Define the new variable ϕ to be

$$\sin\left(\frac{\theta}{2}\right) = \sin\left(\frac{\theta_0}{2}\right)\cdot\sin\phi = k\sin\phi.$$ (3.9.12)

Then,

$$\left(\frac{1}{2}\right)\cos\left(\frac{\theta}{2}\right)d\theta = k\cos\phi\,d\phi,$$ (3.9.13)

or

$$d\theta = \frac{2k\cos\phi\,d\phi}{\cos\left(\frac{\theta}{2}\right)}.$$ (3.9.14)

Now

$$\cos\left(\frac{\theta}{2}\right) = \sqrt{1 - \sin^2\left(\frac{\theta}{2}\right)} = \sqrt{1 - k^2\sin^2\phi},$$ (3.9.15)

and

$$k \cos \phi = k \sqrt{1 - \sin^2 \phi} = k \sqrt{1 - \left(\frac{1}{k^2} \right) \sin^2 \left(\frac{\theta}{2} \right)}$$

$$= \sqrt{k^2 - \sin^2 \left(\frac{\theta}{2} \right)}, \qquad (3.9.16)$$

and therefore,

$$d\theta = \frac{2 \sqrt{k^2 - \sin^2 \left(\frac{\theta}{2} \right)}}{\sqrt{1 - k^2 \sin^2 \phi}}. \qquad (3.9.17)$$

The substitution of Eq. (3.9.17) into Eq. (3.9.11) and using the fact that as θ ranges from 0 to θ_0, ϕ goes from 0 to $\pi/2$, we finally obtain

$$T = 4 \int_0^{\pi/2} \frac{d\phi}{\sqrt{1 - k^2 \sin^2 \phi}}. \qquad (3.9.18)$$

Except for the factor of four, the right-side is an example of an elliptical integral. With a change in the notation for the integration variable, the following four elliptical integrals are defined by the given expressions; for all cases, $0 \leq k^2 < 1$:

Incomplete Elliptical Integral of the First Kind

$$F(k, \phi) = \int_0^{\phi} \frac{d\theta}{\sqrt{1 - k^2 \sin^2 \theta}} \qquad (3.9.19)$$

Complete Elliptical Integral of the First Kind

$$F(k) = F \left(k, \frac{\pi}{2} \right) = \int_0^{\pi/2} \frac{d\theta}{\sqrt{1 - k^2 \sin^2 \theta}} \qquad (3.9.20)$$

Incomplete Elliptical Integral of the Second Kind

$$E(k, \phi) = \int_0^{\phi} \sqrt{1 - k^2 \sin^2 \theta} \, d\theta \qquad (3.9.21)$$

Complete Elliptical Integral of the Second Kind

$$E(k) = E \left(k, \frac{\pi}{2} \right) = \int_0^{\pi/2} \sqrt{1 - k^2 \sin^2 \theta} \, d\theta. \qquad (3.9.22)$$

Expansions in k^2 for $F(k)$ can be derived using

$$\frac{1}{\sqrt{1 - k^2 \sin^2 \theta}} = \sum_{n=0}^{\infty} \left[\frac{(2n - 1)!!}{(2n)!!} \right] (k^2)^n (\sin \theta)^{2n}, \qquad (3.9.23)$$

$$\int_0^{\pi/2} (\sin\theta)^{2n} d\theta = \frac{(2n-1)!!}{(2n)!!} \cdot \left(\frac{\pi}{2}\right). \tag{3.9.24}$$

Carrying out the required calculations gives

$$F(k) = \left(\frac{\pi}{2}\right) \left\{ 1 + \left(\frac{1}{2}\right)^2 k^2 + \left(\frac{1\cdot3}{2\cdot4}\right)^2 k^4 \right.$$

$$\left. + \cdots + \left[\frac{(2n-1)!!}{2^n n!}\right]^2 (k^2)^n + \cdots \right\}. \tag{3.9.25}$$

In a similar way, the following expression can be found for $E(k)$,

$$E(k) = \left(\frac{\pi}{2}\right) \left\{ 1 - \left(\frac{1}{2^2}\right) k^2 - \left(\frac{1\cdot3}{2\cdot4}\right)^2 \left(\frac{k^4}{3}\right) - \left(\frac{1\cdot3\cdot5}{2\cdot4\cdot6}\right) \left(\frac{k^2}{5}\right) \right.$$

$$\left. - \left[\frac{(2n-1)!!}{2^n n!}\right]^2 \left[\frac{(k^2)^n}{2n-1}\right] - \cdots \right\}. \tag{3.9.26}$$

Note that these functions have the following values for $k = 0$,

$$F(0) = \frac{\pi}{2}, \qquad E(0) = \frac{\pi}{2}. \tag{3.9.27}$$

For $k \to 1$, we have

$$F(1) = \int_0^{\pi/2} \frac{d\theta}{\sqrt{1 - \sin^2\theta}} = \int_0^{\pi/2} \frac{d\theta}{\cos\theta}$$

$$= \left(\frac{1}{2}\right) \ln\left(\frac{1+\sin\theta}{1-\sin\theta}\right) \Big|_0^{\pi/2}$$

$$= +\infty. \tag{3.9.28}$$

Also, for $k \to 1$, $E(1)$ is

$$E(1) = \int_0^{\pi/2} \sqrt{1 - \sin^2\theta}\, d\theta = \int_0^{\pi/2} \cos\theta\, d\theta = \sin\theta \Big|_0^{\pi/2} = 1. \tag{3.9.29}$$

These results show that $F(k)$ increases monotonically with k^2 over the interval $0 \le k^2 \le 1$. At $k^2 = 0$, $F(0) = \pi/2$; from this value, $F(k)$ increases and becomes unbounded at $k^2 = 1$, i.e., $F(1) = \infty$. For $E(k)$, at $k^2 = 0$, $E(0) = \pi/2$, and $E(k)$ decreases monotonically to the value, at $k^2 = 1$, of $E(1) = 1$.

3.9.2 *Jacobi Elliptic Functions*

The Jacobi elliptic functions are defined as the inverses of the elliptic integral of the first kind. Let

$$u = \int_0^\phi \frac{d\theta}{\sqrt{1 - k^2 \sin^2 \theta}}, \qquad 0 \leq k^2 \leq 1 \qquad (3.9.30)$$

and define the indicated functions as follows:

$$\begin{cases} sn(u, k) \equiv \sin \phi, \qquad cn(u, k) \equiv \cos \phi \\ dn(u, k) \equiv \sqrt{1 - k^2 \sin^2 \phi}, \\ am(u, k) \equiv \phi, \qquad \tan(u, k) \equiv \dfrac{sn(u, k)}{cn(u, k)} = \tan \phi. \end{cases} \qquad (3.9.31)$$

These functions have the associated names:

$$\begin{cases} sn(u, k) : \text{Jacobi sine function,} \\ cn(u, k) : \text{Jacobi cosine function,} \\ am(u, k) : \text{amplitude of } u. \end{cases} \qquad (3.9.32)$$

Note that for $k^2 = 0$, Eq. (3.9.30) becomes

$$u = \phi, \qquad (3.9.33)$$

and

$$\begin{cases} sn(u, 0) = \sin u, \quad cn(u, 0) = \cos u, \\ dn(u, 0) = 1, \qquad am(u, 0) = u. \end{cases} \qquad (3.9.34)$$

These results suggest that the Jacobi sine and cosine functions are generalizations of the sine/cosine trigonometric functions.

The following properties of the Jacobi elliptic functions are a direct consequence of their definitions given by Eq. (3.9.31):

$$sn(0, k) = 0, \quad cn(0, k) = 1, \quad dn(0, k) = 1, \quad am(0, k) = 0, \qquad (3.9.35)$$

$$\begin{cases} sn^2(u, k) + cn^2(u, k) = 1, \\ k^2 sn^2(u, k) + dn^2(u, k) = 1, \\ dn^2(u, k) - k^2 cn^2(u, k) = 1 - k^2 = (k')^2, \end{cases} \qquad (3.9.36)$$

$$\begin{cases} sn(-u, k) = -sn(u, k), \quad cn(-u, k) = cn(u, k), \\ dn(-u, k) = dn(u, k), \qquad am(-u, k) = -am(u, k). \end{cases} \qquad (3.9.37)$$

$$sn(u, 1) = \tanh(u), \quad cn(u, 1) = dn(u, 1) = \text{sech}(u). \qquad (3.9.38)$$

These relations indicate that the Jacobi sine and cosine elliptic functions have mathematical properties similar to those of the corresponding trigonometric functions.

For a fixed value of k, it is standard practice to not explicitly show the k in the notation for these functions; thus,

$$sn(u) \equiv sn(u, k), \qquad cn(u) = cn(u, k), \text{ etc.} \qquad (3.9.39)$$

Also, as given in Eq. (3.9.36), k' is defined by

$$(k')^2 + k^2 = 1, \qquad (3.9.40)$$

we will use

$$\begin{cases} F \equiv F(k), & F' \equiv F(k'), \\ E \equiv E(k), & E' \equiv E(k'). \end{cases} \qquad (3.9.41)$$

The various derivatives of the Jacobi elliptic functions can be easily determined. For example, from Eq. (3.9.30),

$$\frac{du}{d\phi} = \frac{1}{\sqrt{1 - k^2 \sin^2 \phi}} = \frac{1}{dn(u)}, \qquad (3.9.42)$$

and using Eq. (3.9.31), we have,

$$\frac{d}{du} sn(u) = \frac{d \sin \phi}{du} = \cos \phi \frac{d\phi}{du} = cn(u)dn(u). \qquad (3.9.43)$$

Similarly, it follows that

$$\frac{d}{du} cn(u) = \frac{d \cos \phi}{du} = -\sin \phi \frac{d\phi}{du} = -sn(u)dn(u) \qquad (3.9.44)$$

and

$$\frac{d}{du} dn(u) = -k^2 sn(u)cn(u), \qquad \frac{d}{du} am(u) = dn(u). \qquad (3.9.45)$$

The corresponding second derivatives are,

$$\begin{cases} \dfrac{d^2}{du^2} sn(u) = 2k^2 sn^3(u) - (1 + k^2)sn(u), \\ \dfrac{d^2}{du^2} cn(u) = (2k^2 - 1)cn(u) - 2k^2 cn^3(u), \\ \dfrac{d^2}{du^2} dn(u) = 2(2 - k^2) - 2dn^3(u). \end{cases} \qquad (3.9.46)$$

Inspection of Eq. (3.9.46) shows that these Jacobi elliptic functions satisfy second-order, nonlinear differential equations. For example, let

$$y(u) = cn(u), \qquad (3.9.47)$$

then

$$y'' + ay + by^3 = 0, \tag{3.9.48}$$

where

$$y'' \equiv \frac{d^2y}{du^2}, \qquad a = 1 - 2k^2, \qquad b = 2k^2. \tag{3.9.49}$$

It is a rather easy problem in the theory of nonlinear oscillations to show that all solutions to Eq. (3.9.48), for a and b given by Eq. (3.9.49) with $0 \leq k^2 \leq 1$, are periodic [6].

The following integrals can be checked by differentiation of each formula:

$$\int sn(u)du = \left(\frac{1}{k}\right)[dn(u) - kcn(u)], \tag{3.9.50}$$

$$\int \frac{du}{sn(u)} = \ln\left[\frac{sn(u)}{cn(u) + dn(u)}\right], \tag{3.9.51}$$

$$\int cn(u)du = \left(\frac{1}{k}\right)\arccos[dn(u)], \tag{3.9.52}$$

$$\int \frac{du}{cn(u)} = \left(\frac{1}{1-k^2}\right)\ln\left[\frac{(1-k^2)sn(u) + dn(u)}{cn(u)}\right], \tag{3.9.53}$$

$$\int dn(u)du = \arcsin[sn(u)] = am(u), \tag{3.9.54}$$

$$\int \frac{du}{dn(u)} = \left(\frac{1}{1-k^2}\right)\arctan\left[\frac{(1-k^2)sn(u) - cn(u)}{(1-k^2)sn(u) + cn(u)}\right]. \tag{3.9.55}$$

Our remark above on the fact that $cn(u)$ is a periodic function of u is an understatement. It turns out that the elliptic functions are doubly periodic with both a real and a purely imaginary period. These two periods are $4F = 4F(k)$ and $4iF' = 4iF(k')$, and give rise to the identities:

$$\begin{cases} sn(u \pm F) = \pm\dfrac{cn(u)}{dn(u)}, & sn(u \pm 2F) = -sn(u), \\[2mm] sn(u \pm 3F) = \mp\dfrac{cn(u)}{dn(u)}, & sn(u \pm 4F) = sn(u), \\[2mm] sn(u + iF') = \dfrac{1}{ksn(u)}, & \\[2mm] sn(u + 2mF + 2niF') = (-1)^m sn(u); \end{cases} \tag{3.9.56}$$

$$
\begin{cases}
cn(u \pm F) = \mp k' \left[\dfrac{cn(u)}{dn(u)} \right], & cn(u \pm 2F) = -cn(u), \\[2mm]
cn(u \pm 3F) = \pm k' \left[\dfrac{sn(u)}{dn(u)} \right], & cn(u \pm 4F) = cn(u), \\[2mm]
cn(u + iF') = -\left(\dfrac{i}{k} \right) \left[\dfrac{dn(u)}{sn(u)} \right] \\[2mm]
cn(u + 2mF + 2niF') = (-1)^m cn(u);
\end{cases} \tag{3.9.57}
$$

$$
\begin{cases}
dn(u \pm F) = \dfrac{k'}{dn(u)}, \\[2mm]
dn(u \pm 2F) = dn(u), \\[2mm]
dn(u + iF') = -i \left[\dfrac{cn(u)}{sn(u)} \right], \\[2mm]
dn(u + 2mF + 2niF') = (-1)^n dn(u).
\end{cases} \tag{3.9.58}
$$

Since the Jacobi elliptic functions are periodic, they have Fourier series expansions. To simplify the writing of these expressions, the following notations are used,

$$
v \equiv \left[\frac{\pi}{2F(k)} \right] u, \tag{3.9.59}
$$

$$
q(k) \equiv \exp \left[-\frac{\pi F(k')}{F(k)} \right]. \tag{3.9.60}
$$

The specific representations are given by the formulas:

$$
sn(u, k) = \left[\frac{2\pi}{kF(k)} \right] \sum_{m=0}^{\infty} \left(\frac{q^{m+1/2}}{1 - q^{2n+1}} \right) \sin[(2m + 1)v], \tag{3.9.61}
$$

$$
cn(u, k) = \left[\frac{2\pi}{kF(k)} \right] \sum_{m=0}^{\infty} \left(\frac{q^{m+1/2}}{1 + q^{2m+1}} \right) \cos[(2m + 1)v], \tag{3.9.62}
$$

$$
dn(u, k) = \frac{\pi}{2F(k)} + \left[\frac{2\pi}{F(k)} \right] \sum_{m=1}^{\infty} \left(\frac{q^m}{1 + q^{2m}} \right) \cos(2mv). \tag{3.9.63}
$$

These expansions show clearly that $sn(u, k)$ and $cn(u, k)$, when plotted versus u, resemble, respectively, the sine and cosine functions. From the definition of $dn(u, k)$, see Eq. (3.9.31), it follows that

$$
dn(u, k) > 0 \tag{3.9.64}
$$

and a plot of $dn(u, k)$ versus u looks like a curve for which oscillations take place about a constant value.

3.10 Evaluation of Integrals by Differentiation and Integration with Respect to a Parameter

A useful technique for the evaluation of certain integrals is to convert them to an equivalent differential equation that can then be solved to yield the desired integral. This procedure is illustrated with two examples.

If the integrand of an integral depends on one or more parameters, then other integrals can be generated by taking the appropriate derivative with respect to one or more of these parameters. We show the advantages of this method by way of several examples.

Consider the following definite integral,

$$I(\alpha, \beta) \equiv \int_0^\infty e^{-\alpha x^2} \cos(\beta x) dx, \tag{3.10.1}$$

where α can be complex and β real. Convergence of the integral requires $\mathrm{Re}(\alpha) > 0$. We have indicated that the value of the integral depends on the two parameters α and β. Now, what is $I(\alpha, \beta)$? To proceed, take the partial derivative of $I(\alpha, \beta)$ with respect to β and then apply the method of integration by parts to the resulting expression; doing this gives

$$\frac{\partial I}{\partial \beta} = -\int_0^\infty x e^{-\alpha x^2} \sin(\beta x) dx. \tag{3.10.2}$$

Applying integration by parts gives

$$\frac{\partial I}{\partial \beta} = \left(\frac{1}{2\alpha}\right) e^{-\alpha x^2} \cdot \sin(\beta x)\Big|_0^\infty - \left(\frac{\beta}{2\alpha}\right) \int_0^\infty e^{-\alpha x^2} \cos(\beta x) dx, \tag{3.10.3}$$

and, finally,

$$\frac{\partial I}{\partial \beta} = -\left(\frac{\beta}{2\alpha}\right) I. \tag{3.10.4}$$

This differential equation can be easily solved and we obtain

$$I(\alpha, \beta) = c(\alpha) e^{-(\beta^2/4\alpha)}, \tag{3.10.5}$$

where it is indicated that the unknown, at this point, integration "constant," $c(\alpha)$, depends on α. The value for $c(\alpha)$ can be determined by noting that from Eq. (3.10.1), we have

$$I(\alpha, 0) = \int_0^\infty e^{-\alpha x^2} dx. \tag{3.10.6}$$

Using the change of variable

$$y = \alpha x^2, \tag{3.10.7}$$

the integral becomes

$$\int_0^\infty e^{-\alpha x^2} dx = \left(\frac{1}{2\sqrt{\alpha}}\right) \int_0^\infty y^{-1/2} e^{-y} dy$$

$$= \left(\frac{1}{2\sqrt{\alpha}}\right) \Gamma\left(\frac{1}{2}\right) = \left(\frac{1}{2}\right) \sqrt{\frac{\pi}{\alpha}} = c(\alpha). \qquad (3.10.8)$$

Therefore, $I(\alpha, \beta)$ is given by the expression

$$I(\alpha, \beta) = \left(\frac{1}{2}\right) \sqrt{\frac{\pi}{\alpha}} e^{-(\beta^2/4\alpha)}. \qquad (3.10.9)$$

Observe that our success for this problem depended on the following two points: (i) taking the partial derivative with respect to β changed the cosine into a sine; (ii) the application of the integration by parts procedure, converted the sine back into a cosine and, except for an overall "constant," gave us the same integral that we wanted to evaluate. With this in mind, our conclusion is that such a method might be generally applied to integrals involving either the cosine or sine functions.

Finally, note that once we have $I(\alpha, \beta)$, other related integrals can be determined by taking the appropriate derivatives with respect to either α or β. For example,

$$\int_0^\infty x^{2n} e^{-\alpha x^2} \cos(\beta x) dx = (-1)^n \frac{\partial^n I}{\partial \alpha^n}, \qquad (3.10.10)$$

$$\int_0^\infty x^{2n+1} e^{-\alpha x^2} \sin(\beta x) dx = (-1)^{n+1} \frac{\partial^{n+1} I}{\partial \alpha^n \partial \beta}. \qquad (3.10.11)$$

For the second example, consider the integral

$$g(\lambda) \equiv \int_0^\infty \frac{e^{-z} e^{-(\lambda/z)}}{\sqrt{z}} \cdot dz. \qquad (3.10.12)$$

Setting $\lambda = 0$, gives

$$g(0) = \int_0^\infty z^{\frac{1}{2}-1} e^{-z} dz = \Gamma\left(\frac{1}{2}\right) = \sqrt{\pi}. \qquad (3.10.13)$$

Since the "nasty" term in the integrand is λ/z, let's try the following transformation of variable,

$$u = \frac{\lambda}{z}, \qquad \lambda > 0. \qquad (3.10.14)$$

Making this replacement in Eq. (3.10.12) gives

$$g(\lambda) = \sqrt{\lambda} \int_0^\infty \frac{e^{-u} e^{-(\lambda/u)}}{u^{3/2}} \cdot du. \qquad (3.10.15)$$

Since u in Eq. (3.10.15) is just an integration variable, and will not appear in the final result, we can relabel it to z, i.e., $g(\lambda)$ is

$$g(\lambda) = \sqrt{\lambda} \int_0^\infty \frac{e^{-z} e^{-(\lambda/z)}}{z^{3/2}} \cdot dz. \qquad (3.10.16)$$

Now take the derivative of $g(\lambda)$, given by Eq. (3.10.12); this gives

$$\frac{dg}{d\lambda} = \int_0^\infty \frac{e^{-z} e^{-\lambda/z} \left(-\frac{1}{z}\right)}{\sqrt{z}} \cdot dz$$

$$= -\int_0^\infty \frac{e^{-z} e^{-(\lambda/z)}}{z^{3/2}} \cdot dx \qquad (3.10.17)$$

$$= -\left(\frac{1}{\sqrt{\lambda}}\right) g, \qquad (3.10.18)$$

which is a linear, first-order, separable differential equation for $g(\lambda)$. Its solution is

$$g(\lambda) = c e^{-2\sqrt{\lambda}}, \qquad (3.10.19)$$

and the integration constant can be determined using the result in Eq. (3.10.13), i.e.,

$$c = \sqrt{\pi}. \qquad (3.10.20)$$

Therefore, the original integral is

$$g(\lambda) = \sqrt{\pi} e^{-2\sqrt{\lambda}}. \qquad (3.10.21)$$

Finally, it follows that

$$\int_0^\infty \frac{e^{-z} e^{-\lambda/z}}{z^{n+\frac{1}{2}}} \cdot dz = (-1)^n \frac{d^n g(\lambda)}{d\lambda^n}. \qquad (3.10.22)$$

Lastly, define $L(a, b)$ as the indefinite integral

$$L(a, b) \equiv \int \frac{dx}{a + bx} = \left(\frac{1}{b}\right) \ln(a + bx). \qquad (3.10.23)$$

Since

$$\frac{\partial L}{\partial a} = -\int \frac{dx}{(a + bx)^2} = \left(\frac{1}{b}\right)\left(\frac{1}{1 + bx}\right), \qquad (3.10.24)$$

we have

$$\int \frac{dx}{(a + bx)^2} = -\left(\frac{1}{b}\right)\left(\frac{1}{a + bx}\right). \qquad (3.10.25)$$

Taking the $(n-1)$th derivative of $L(a, b)$ with respect to a gives

$$\int \frac{dx}{(a + bx)^n} = -\left[\frac{1}{b(n-1)(a + bx)^{n-1}}\right]. \qquad (3.10.26)$$

Likewise,

$$\frac{\partial L}{\partial b} = -\int \frac{x\,dx}{(a+bx)^2} = -\left(\frac{1}{b^2}\right)\ln(a+bx) + \left(\frac{1}{b}\right)\left(\frac{x}{a+bx}\right) \quad (3.10.27)$$

and

$$\int \frac{x\,dx}{(a+bx)^2} = \left(\frac{1}{b^2}\right)\ln(a+bx) - \frac{x}{b(a+bx)}. \quad (3.10.28)$$

It follows that the integral

$$L(a,b,m) = \int \frac{x^m\,dx}{a+bx} \quad (3.10.29)$$

can also be determined using this procedure.

Problems

Section 3.2

3.2.1 Show that the integral in Eq. (3.2.6) converges if $\mathrm{Re}(z) > 0$.

3.2.2 Sketch $\Gamma(x)$ vs x for $-5 \le x \le 5$.

3.2.3 Use Eqs. (3.2.11) and (3.2.19) to prove that

$$\Gamma(z)\Gamma(1-z) = \frac{\pi}{\sin(\pi z)}.$$

3.2.4 From the result in the previous problem, show that $\Gamma\left(\frac{1}{2}\right) = \sqrt{\pi}$.

3.2.5 Using the recurrence relation for the gamma function and Table 3.2.1, calculate and/or estimate the following numbers:
(i) $\Gamma(4.70)$, (ii) $\Gamma(-3.05)$, (iii) $\Gamma(2.25)$.

3.2.6 Retain just the first term in the asymptotic expansion for the gamma function, i.e.,

$$\Gamma(z+1) = (2\pi z)^{1/2} z^z e^{-z} \left\{1 + O\left(\frac{1}{z}\right)\right\}$$

$$= \bar{\Gamma}(z+1)\left\{1 + O\left(\frac{1}{z}\right)\right\},$$

and compare the values of Γ and $\bar{\Gamma}$ for $z = \text{integer} = m$, $\{m : 1, 2, \ldots, 10\}$.

3.2.7 For positive integer n, the double-factorials are defined to be

$$(2n-1)!! \equiv (2n-1)(2n-3)\cdots 5\cdot 3\cdot 1.$$

Show that

$$(2n-1)!! = \frac{(2n)!}{2^n \cdot n!}.$$

3.2.8 Explain why there is no essential need to introduce $(2n)!!$.

3.2.9 Prove that

$$\Gamma\left(n + \frac{1}{2}\right) = \frac{(2n)!\sqrt{\pi}}{2^{2n} \cdot n!} \, .$$

Hint: See the book of Kahn, p. 83.

Section 3.3

3.3.1 Complete the details of the calculations in Eqs. (3.3.6), (3.3.8), and (3.3.9).

3.3.2 Show that an alternative representation for $B(p, q)$ is

$$B(p, q) = \int_0^\infty \frac{x^{p-1} dt}{(1 + x)^{p+q}} \, .$$

Hint: Use the transformation $t = x/(1 + x)$ in Eq. (3.3.1).

Section 3.4

3.4.1 Prove that the series for the Riemann zeta function converges for $\mathrm{Re}(z) > 1$.

3.4.2 Prove that $\zeta(x)$ is monotonic decreasing for $x > 1$ by calculating $d\zeta(x)/dx$.

3.4.3 Why is it legitimate to interchange the integration and summation operations for Eq. (3.4.20)?

Section 3.5

3.5.1 Prove that the first and second functions in Eq. (3.5.7) are delta sequences.

3.5.2 Show that $x\delta(x) = 0$.

3.5.3 Show that $x\delta'(x) = -\delta(x)$.

3.5.4 Construct the two-dimensional delta functions for Cartesian coordinates and plane-polar coordinates.

Section 3.6

3.6.1 Show explicitly why m and n must be greater than minus one in order for $I(m, n)$ to exist in Eq. (3.6.5).

3.6.2 Derive, in detail, the values for $I(m_1, m_2, \ldots, m_N)$ and $J(m_1, m_2, \ldots, m_N)$ as given respectively by Eqs. (3.6.16) and (3.6.19). Show that they agree with the calculated $I(m, n)$ and $J(m, n)$.

Section 3.8

3.8.1 For the exponential integral, defined by Eq. (3.8.5) calculate and/or answer the following questions:

   (i) $Ei(0)$

   (ii) $Ei(\infty)$

   (iii) What is $\frac{dEi(x)}{dx}$?

   (iv) Using the above information, sketch $Ei(x)$ versus x.

3.8.2 Carry out the same analysis for the Logarithmic integral given by Eq. (3.8.3).

Section 3.9

3.9.1 Derive from the principles of physics the pendulum equation given in Eq. (3.9.1).

3.9.2 Using the representations in Eqs. (3.9.19) to (3.9.22), show directly that $F(k,\phi)$, $F(k)$, $E(k,\phi)$, and $E(k)$ are even functions of k.

3.9.3 Derive the expansion given by Eq. (3.9.23).

3.9.4 Carry out the required expansion of the integrand for $E(k)$ and obtain the result of Eq. (3.9.26).

3.9.5 Sketch $F(k)$ vs k^2 and $E(k)$ vs k^2 for $0 \le k^2 \le 1$.

3.9.6 Show that $F(k)$ and $E(k)$ have the alternative representations:

$$F(k) = \int_0^1 \frac{dt}{\sqrt{(1-t^2)(1-k^2t^2)}},$$

$$E(k) = \int_0^1 \sqrt{\frac{1-k^2t^2}{1-t^2)}}\, dt.$$

3.9.7 Show that

$$\frac{dE}{dk} = \left(\frac{1}{k}\right)(E-F),$$

$$\frac{dF}{dk} = \left[\frac{1}{k(1-k^2)}\right]E - \left(\frac{1}{k}\right)F.$$

Section 3.10

3.10.1 Evaluate the integral

$$I_1(\alpha,\beta) = \int_0^\infty e^{-\alpha x^2}\sin(\beta x)dx.$$

3.10.2 Apply the procedure used on the integral of Eq. (3.10.1) to

$$I_2(\alpha, \beta) = \int_0^\infty e^{-\alpha x} \cos(\beta x) dx.$$

Does this method work for this case? Explain your answer.

3.10.3 Can the procedure used to determine the integrals of Eqs. (3.10.10) and (3.10.11) be used to evaluate the following two expressions

$$\int_0^\infty x^{2m+1} e^{-\alpha x^2} \cos(\beta x) dx$$

and

$$\int_0^\infty x^{2m} e^{-\alpha x^2} \sin(\beta x) dx?$$

3.10.4 Fill in the details needed to derive Eq. (3.10.15).

3.10.5 Explain why the integral of Eq. (3.10.22) exists. Hint: What happens when $z \to 0^+$ in the integrand?

3.10.6 From the basic relations

$$\int \sin(ax + b) dx = -\left(\frac{1}{a}\right) \cos(ax + b),$$

$$\int \cos(ax + b) dx = \left(\frac{1}{a}\right) \sin(ax + b),$$

derived expressions for

$$\int x^n \sin(ax + b) dx \quad \text{and} \quad \int x^n \cos(ax + b) dx.$$

3.10.7 Consider the integral

$$I(\alpha) \equiv \int_0^\infty e^{-\alpha t} \left[\frac{\sin(t)}{t}\right] dt.$$

Show that it satisfies the first-order differential equation [5]

$$\frac{dI}{d\alpha} = -\left(\frac{1}{1+\alpha^2}\right) I.$$

Use the fact that $I(\infty) = 0$, to calculate the integration constant. What is $I(0)$?

Comments and References

[1] P. A. M. Dirac, *The Principals of Quantum Mechanics*, 2nd ed. (The Clarendon Press, Oxford, 1930).

[2] An excellent and detailed discussion of generalized functions is given by Lighthill. Another good presentation on the delta function and some of its applications appear in the book by Friedman.

[3] S. M. Lea, *Mathematics for Physicists* (Thomson: Brooks/Cole; Belmont, CA; 2004).

[4] Excellent summaries of the definitions of Dirichlet integrals, how they can be evaluated, and some of their applications are given in the following references:

   (i) P. Franklin, *A Treatise on Advanced Calculus* (Wiley, New York, 1940); see problems 23, 24, and 25 on pp. 579.

   (ii) H. Jeffreys and B. S. Jeffreys, *Methods of Mathematical Physics* (Cambridge University Press, Cambridge, 1972); see section 15.08.

   (iii) P. B. Kahn, *Mathematical Methods for Scientists and Engineers* (Wiley-Interscience, New York, 1990); see section 2.2, 2.3, and 2.4.

[5] R. G. Rice and D. O. Do, *Applied Mathematics and Modeling for Chemical Engineers* (Wiley, New York, 1995).

[6] R. E. Mickens, *Nonlinear Oscillations* (Cambridge University Press, New York, 1981); see section 1.5.

[7] See the book by Speigel; chapter 13.

[8] See the books by Friedman, Kahn, and the following:

   (i) A. Erdélyi, *Operational Calculus and Generalized Functions* (Holt, Reinhart and Winston; New York, 1962).

   (ii) J. Mikusinski, *Operational Calculus* (Pergamon, New York, 1959).

Bibliography

M. Abramowitz and I. A. Stegun, *Handbook of Mathematical Functions* (Dover, New York, 1965).

G. Arken, *Mathematical Methods for Physicists*, 3rd ed. (Academic Press, New York, 1985).

F. Bowman, *Introduction to Elliptic Functions with Applications* (Dover, New York, 1961).

P. F. Byrd and M. D. Friedman, *Handbook of Elliptic Integrals for Engineers and Physicists* (Springer-Verlag, Berlin, 1954).

H. T. Davis, *Introduction to Nonlinear Differential and Integral Equations* (Dover, New York, 1962).

B. Friedman, *Principles and Techniques of Applied Mathematics* (Dover, New York, 1990).

I. S. Gradshteyn and I. M. Ryzhik, *Tables of Integrals, Series, and Products* 5th ed. (Academic Press, Boston, 1994).

A. Jeffrey, *Handbook of Mathematical Formulas and Integrals* (Academic Press, New York, 1995).

H. Jeffreys and B. S. Jeffreys, *Methods of Mathematical Physics* (Cambridge University Press, Cambridge, 1972).

P. B. Kahn, *Mathematical Methods for Scientists and Engineers* (Wiley-Interscience, New York, 1990).

D. F. Lawden, *Elliptic Functions and Applications* (Springer-Verlag, Berlin, 1989).

M. J. Lighthill, *Introduction to Fourier Series and Generalized Functions* (Cambridge University Press, London, 1958).

N. W. McLachlan, *Ordinary Nonlinear Differential Equations in Engineering and Physical Sciences*, 2nd ed. (Oxford University Press, London, 1950).

S. J. Patterson, *An Introduction to the Theory of the Riemann Zeta-Function* (Cambridge University Press, Cambridge, 1988).

A. P. Prudnikov, Yu. A. Brychkov and O. I. Marichev, *Integrals and Series, Vols. 1–4* (Gordon and Breach, New York, 1986–1992).

M. R. Spiegel, *Advanced Mathematics for Engineers and Scientists* (Schaum's Outline Series, McGraw-Hill; New York, 1999).

E. C. Titchmarsh, *The Theory of the Riemann Zeta Function* (Oxford University Press, Oxford, 1951).

Chapter 4

Qualitative Methods for Ordinary Differential Equations

4.1 Introduction

A consistent theme running throughout this text is the nonexistence of explicit exact general solutions for an arbitrary linear or nonlinear differential
equation. Even for the few instances where such expressions exist, their
functional forms may be such that the explicit forms do not provide much
useful information on the details of the solutions. Often, what is required
is a detailed understanding of the qualitative properties of the solutions
along with good analytical approximations to the solutions. The main goal
of this chapter is to introduce some techniques that allow, in part, these two
issues to be resolved. Our focus is on qualitative methods for one- and two-
dimensional dynamical systems. While these systems are restricted to just
a single and two coupled, first-order differential equations, they are general
enough to cover the mathematical modeling requirements of a broad and
interesting variety of phenomena in the natural and engineering sciences.

We begin with a study of one-dimensional autonomous systems. They
are modeled by a single first-order differential equation having the form

$$\frac{dx}{dt} = f(x). \tag{4.1.1}$$

We then define, for such equations, the concepts of fixed-points, linear
stability, and global stability. Section 4.3 shows how these techniques can
be applied to a number of particular differential equations having the form
given by Eq. (4.1.1). We also demonstrate that these general methods can,
in special cases, be also applied to first-order nonautonomous equations.

Section 4.4 discusses two-dimensional dynamical systems. Such systems
are modeled by a pair of coupled, first-order differential equations

$$\frac{dx}{dt} = f(x, y), \qquad \frac{dy}{dt} = g(x, y). \tag{4.1.2}$$

The study of these systems centers on the two-dimensional phase space, (x, y), and introduces the additional concepts of nullclines, first-integrals and symmetries. Several worked examples illustrating these concepts and their applications are presented in section 4.5.

Section 4.6, briefly introduces the concept of bifurcation. The basic idea is that the qualitative structure of trajectories in phase-space can be drastically changed when a parameter of the dynamical system varies. The concept of a Hopf-bifurcation for a nonlinear oscillating system is introduced and applied to the van der Pol equation.

4.2   One-Dimensional Systems

4.2.1   *Definition*

A one-dimensional dynamical system is characterized by a single, first-order differential equation

$$\frac{dx}{dt} = f(x), \qquad x(0) = x_0, \qquad (4.2.1)$$

where the initial condition, x_0, is specified. The function f only depends on the dependent variable x and not the time, t. This type of differential equation is called a first-order, autonomous equation. The related nonautonomous equation is

$$\frac{dx}{dt} = g(x, t), \qquad (4.2.2)$$

and, for the moment, will not be considered. We assume that $f(x)$ has the required mathematical properties such that the existence and uniqueness theorems hold. For problems in the natural and engineering sciences, this requirement is almost always satisfied.

4.2.2   *Fixed-Points*

The fixed-points are the constant solutions for Eq. (4.2.1). This means that they are determined by the solutions to the following, in general, algebraic equation

$$f(\bar{x}) = 0. \qquad (4.2.3)$$

In actual applications, the only solutions of relevance are those for which $\bar{x}$ is real, i.e., as far as the fixed-points are concerned, those having complex values do not correspond to actual states of the dynamical system. Also, it should be indicated that the number of real solutions to Eq. (4.2.3) may vary from zero to any finite integer. Clearly, purely mathematical models can be constructed such that the number of fixed-points is unbounded.

For physical systems, such as those arising in physics or mechanical engineering, the fixed-points correspond to states of equilibrium. This is the reason why only the real solutions of Eq. (4.2.3) are of interest.

4.2.3 *Sign of the Derivative*

Assume that we have a one-dimensional dynamical system for which there is one fixed-point. It can always be represented as

$$\frac{dx}{dt} = f_1(x)(x - \bar{x})^n, \tag{4.2.4}$$

where n is a non-zero, positive integer; the fixed-point is located at $x = \bar{x}$; and $f_1(x)$ has no real zeros, and, consequently, $f_1(x)$ has a definite sign. As shown in Figure 4.2.1, the solution $x(t) = \bar{x}$ divides the $t - x$ plane into two domains for n an odd positive integer:

(i) for $f_1(x) > 0$, the derivative dx/dt is positive for $x > \bar{x}$ and is negative for $x < \bar{x}$;

(ii) for $f_2(x) < 0$, the derivative dx/dt is negative for $x > \bar{x}$ and positive for $x < \bar{x}$.

Typical trajectories, i.e., $x(t)$ vs t, are shown for these two situations in Figure 4.2.2. Note that all trajectories move away from the fixed-point when $f_1(x) > 0$, while all trajectories go to the fixed-point when $f_1(x) < 0$. The first case corresponds to the fixed-point being unstable, while the second case corresponds to the fixed-point being stable. For a single fixed-point, these are the possible solution behaviors when the order of the fixed-point is odd, i.e., n in Eq. (4.2.4) is

$$n = (2m - 1), \qquad (m = 1, 2, 3, \dots). \tag{4.2.5}$$

The situation for n an even positive integer is depicted in Figure 4.2.3. This case corresponds to the fixed-point being semi-stable, i.e., trajectories on one side of the fixed-point move toward it, while those on the opposite side move away.

Since the phase space, i.e., number of dependent variables, is one dimensional, we can also represent the situations given in Figures 4.2.2 and 4.2.3 by a line where a dot indicates the fixed-point and the direction of arrows show the motion of $x(t)$ as $t \to +\infty$. For example:

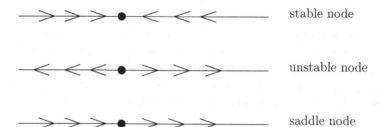

stable node

unstable node

saddle node

Note that names have been associated with each of the possible solution behaviors:

(i) If all trajectories approach the fixed-point, as $t \to +\infty$, then we call it a *stable node*.

(ii) If all trajectories move away from the fixed-point, as $t \to +\infty$, then we call it an *unstable node*.

(iii) If trajectories on one side of the fixed-point approach it as $t \to +\infty$, while trajectories on the other side move away as $t \to +\infty$, we call it a *saddle point*.

The following one-dimensional phase plots give various possibilities for the case of two-fixed points:

$$\qquad\qquad\qquad S \qquad\qquad\qquad U$$
(A) $\longrightarrow\!\!\!>\;>\!\!\bullet<\;<\!\!\bullet\;>\!\!>\!\!\longrightarrow$

$$\qquad\qquad\qquad SS \qquad\qquad\qquad S$$
(B) $\longrightarrow\!\!\!>\;>\!\!\bullet>\;>\!\!\bullet\;<\!\!<\!\!\longrightarrow$

$$\qquad\qquad\qquad SS \qquad\qquad\qquad SS$$
(C) $\longrightarrow\!\!\!>\;>\!\!\bullet>\;>\!\!\bullet\;>\!\!>\!\!\longrightarrow$

$$\qquad\qquad\qquad SS \qquad\qquad\qquad U$$
(D) $\longleftarrow\!\!\!<\;<\!\!\bullet<\;<\!\!\bullet\;>\!\!>\!\!\longrightarrow$

The letters "S", "U", and "SS" denote, respectively, stable, unstable, and semi-stable fixed-points. In more detail we have for these four cases the following situation:

Case A : a stable and an unstable node;

Case B : a saddle point and a stable node;

Case C : two saddle points;

Case D : a saddle point and an unstable node.

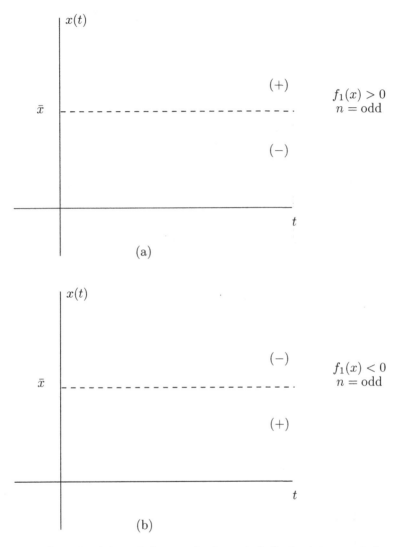

Fig. 4.2.1 One-dimensional dynamical system having a single fixed-point at $x = \bar{x}$, for $n =$ odd, positive integer.

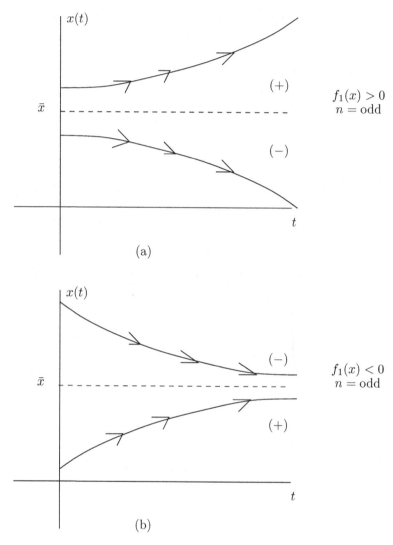

Fig. 4.2.2   Typical trajectories for a one-dimensional dynamical system having a single fixed-point at $x = \bar{x}$, for $n =$ odd, positive integer. Cases (a) and (b) correspond to the fixed-point being, respectively, unstable and stable.

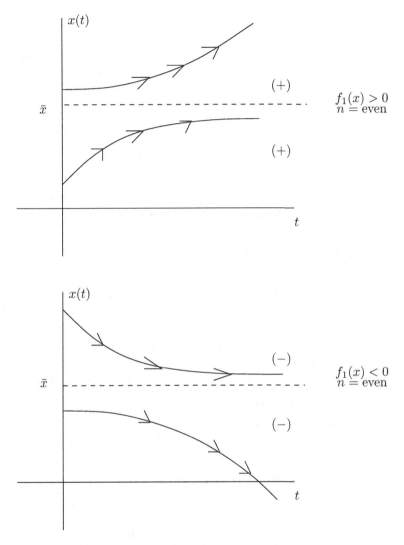

Fig. 4.2.3   Typical trajectories for a one-dimensional dynamical system having a single fixed-point at $x = \bar{x}$, but n now an even, positive integer.

It should be clear, from these diagrams, that the large time state of a one-dimensional system depends on its initial state, i.e., different initial states can lead to different (as $t \to +\infty$) final states. For example, for Case A, if we denote the two fixed-points by x_S and x_U, then for initial states, x_0, we have

$$\begin{cases} \text{if } x_0 < x_U, & \text{then } x(\infty) = x_S; \\ \text{if } x_0 > x_U, & \text{then } x(\infty) = \infty; \\ \text{if } x_0 = x_U, & \text{then } x(\infty) = x_U; \\ \text{if } x_0 = x_S, & \text{then } x(\infty) = x_S. \end{cases} \tag{4.2.6}$$

The technique discussed above can be easily extended to one-dimensional systems having m distinct fixed-points.

4.2.4 *Linear Stability*

The results of the previous section gives qualitative information on the global behavior of the solutions to a one-dimensional system. We now consider a method that is very helpful for providing more quantitative information on the behavior of solutions in a neighborhood of a fixed-point.

Consider the case where $f(x)$, as given in Eq. (4.2.1) has all simple zeros. Let $\bar{x}$ be a particular zero of $f(\bar{x}) = 0$, corresponding to the fixed-point $x(t) = \bar{x}$. For initial conditions close to the fixed-point, we have

$$x(t) = \bar{x} + \epsilon(t), \qquad |\epsilon(0)| \ll \bar{x}. \tag{4.2.7}$$

Substitution into Eq. (4.2.1) gives

$$\frac{d\epsilon}{dt} = f(\bar{x} + \epsilon). \tag{4.2.8}$$

Expanding f in a Taylor series in ϵ gives

$$\frac{d\epsilon}{dt} = f(\bar{x}) + \frac{df}{dx}\bigg|_{x=\bar{x}} \epsilon + O(\epsilon^2). \tag{4.2.9}$$

The first term on the right-side is zero because $\bar{x}$ is a fixed-point. If now only the linear term in ϵ is retained and if R is defined as

$$R \equiv \frac{df}{dx}\bigg|_{x=\bar{x}}, \tag{4.2.10}$$

then we have

$$\frac{d\epsilon}{dt} = R\epsilon, \qquad \epsilon(0) = \epsilon_0, \tag{4.2.11}$$

and its solution is

$$\epsilon(t) = \epsilon_0 e^{Rt}. \tag{4.2.12}$$

Thus, if $R > 0$, $\epsilon(t)$ increases and the neighboring trajectory moves away from the fixed-point; if $R < 0$, $\epsilon(t)$ decreases and the neighboring trajectory moves toward the fixed-point. This means that the sign of R determines the local stability of the fixed-point. In summary,

$$\begin{cases} R > 0 \Rightarrow x(t) = \bar{x}, & \text{locally unstable;} \\ R < 0 \Rightarrow x(t) = \bar{x}, & \text{locally stable.} \end{cases} \qquad (4.2.13)$$

The case,

$$R = 0, \qquad (4.2.14)$$

cannot be treated by the above linear stability analysis. However, the geometric "sign of derivative" method can be applied to determine the qualitative stability of the fixed-point.

4.3 Worked Examples

4.3.1 *Examples A*

Perhaps the simples, nontrivial first-order differential equation is

$$\frac{dx}{dt} = -x. \qquad (4.3.1)$$

Inspection shows that $f(x) = -x$ and

(i) $\bar{x} = 0$ is the only fixed-point;
(ii) the derivative has the property

$$\frac{dx}{dt} = \begin{cases} > 0, & \text{for } x < 0, \\ < 0, & \text{for } x > 0; \end{cases} \qquad (4.3.2)$$

(iii) $$R = df(0)/dx = -1. \qquad (4.3.3)$$

All of these results imply that the fixed-point, $\bar{x} = 0$, is both linearly and globally stable and that all nontrivial solutions approach zero monotonically.

Consider the differential equation

$$\frac{dx}{dt} = x. \qquad (4.3.4)$$

For this case, $f(x) = x$ and

(i) $\bar{x} = 0$ is the unique fixed-point;

(ii) the derivative has the property

$$\frac{dx}{dt} = \begin{cases} < 0, & \text{for } x < 0, \\ > 0, & \text{for } x > 0; \end{cases} \tag{4.3.5}$$

(iii) $\qquad\qquad R = df(0)/dx = 1. \tag{4.3.6}$

We conclude that the fixed-point, $\bar{x} = 0$, is both linearly and globally unstable and all nontrivial solutions become unbounded.

The fact that the respective solutions to Eq. (4.3.1) and (4.3.4) are $\exp(-x)$ and $\exp(x)$ shows the correctness of this analysis.

4.3.2 Example B

The logistic differential equation is

$$\frac{dx}{dt} = x(1 - x). \tag{4.3.7}$$

For this case, $f(x) = x(1 - x)$ and the two fixed-points are

$$\bar{x}^{(1)} = 0, \qquad \bar{x}^{(2)} = 1. \tag{4.3.8}$$

Likewise, we have $df/dx = 1 - 2x$ and

$$R_1 \equiv \frac{df(0)}{dx} = 1, \qquad R_2 \equiv \frac{df(0)}{dx} = -1. \tag{4.3.9}$$

Also, the derivative has the following properties

$$\frac{dx}{dt} = \begin{cases} < 0, & \text{for } x > 1; \\ 0, & \text{for } x = 0; \\ > 0, & \text{for } 0 < x < 1; \\ = 0, & \text{for } x = 0; \\ < 0, & \text{for } x < 0. \end{cases} \tag{4.3.10}$$

Using this information, we can draw typical solutions; these are represented in Figure 4.3.1.

Note that if $x(0) > 0$, then all solutions approach the stable fixed-point $\bar{x} = 1$, i.e.,

$$x(0) > 0 \Rightarrow \lim_{t \to \infty} x(t) = \bar{x} = 1. \tag{4.3.11}$$

For $x(0) < 0$, we can only state that, at least initially, the solutions decrease. The linear stability analysis shows that $\bar{x}^{(1)}$ and $\bar{x}^{(2)}$ are, respectively, linearly unstable and stable.

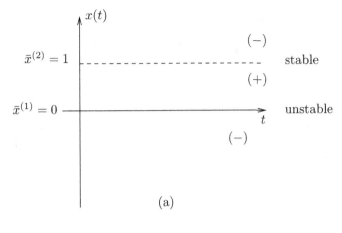

(a)

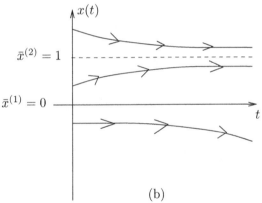

(b)

Fig. 4.3.1 The logistic equation $\dot{x} = x(1 - x)$. (a) Domains of a constant sign for the derivative. (b) Typical solutions.

4.3.3 *Example C*

The differential equation

$$\frac{dx}{dt} = x(1 - x^2), \qquad (4.3.12)$$

where $f(x) = x(1 - x^2) = x(1 - x)(1 + x)$, has three fixed-points

$$\bar{x}^{(1)} = -1, \qquad \bar{x}^{(2)} = 0, \qquad \bar{x}^{(3)} = 1. \qquad (4.3.13)$$

From the derivative $df/dx = 1 - 3x^2$, we find

$$R_1 = \frac{df(-1)}{dx} = -2, \quad R_2 = \frac{df(0)}{dx} = 1, \quad R_3 = \frac{df(1)}{dx} = -2, \qquad (4.3.14)$$

and conclude that

$$\begin{cases} \bar{x}^{(1)} = -1, & \text{linearly stable;} \\ \bar{x}^{(2)} = 0, & \text{linearly unstable;} \\ \bar{x}^{(3)} = 1, & \text{linearly stable.} \end{cases} \tag{4.3.15}$$

Figure 4.3.2 represents the domains of constant sign for the derivative and gives typical solutions. From the linear stability analysis, we conclude that $\bar{x}^{(1)}$ and $\bar{x}^{(3)}$ are stable, while $\bar{x}^{(2)}$ is unstable. Further, the following results are obtained

$$\begin{cases} x(0) > 0 \Rightarrow \operatorname*{Lim}_{t\to\infty} x(t) = \bar{x}^{(3)} = 1, \\ x(0) < 0 \Rightarrow \operatorname*{Lim}_{t\to\infty} x(t) = \bar{x}^{(1)} = -1. \end{cases} \tag{4.3.16}$$

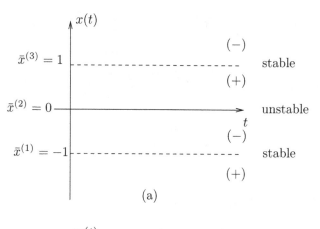

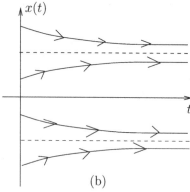

Fig. 4.3.2   The equation $\dot{x} = x(1 - x^2)$. (a) Domains of constant sign for the derivative. (b) Typical solutions.

4.3.4 Example D

Consider the differential equation

$$\frac{dx}{dt} = -\sin(\pi x). \tag{4.3.17}$$

Since $f(x) = -\sin(\pi x)$, it follows that $f(\bar{x}) = 0$ has an unlimited number of solutions; they are

$$\bar{x}_n = n, \qquad (n = 0, \pm 1, \pm 2, \dots). \tag{4.3.18}$$

Now the derivative of $f(x)$ is

$$\frac{df}{dx} = -\pi \cos(\pi x) \tag{4.3.19}$$

and

$$R_n \equiv \frac{df}{dx}\bigg|_{x=n} = -\pi \cos(\pi n) = (-1)^{n+1}. \tag{4.3.20}$$

This latter expression gives the result

$$\bar{x}^{(n)} = n = \begin{cases} \text{linearly stable} & \text{if } n = \text{even}, \\ \text{linearly unstable} & \text{if } n = \text{odd}. \end{cases} \tag{4.3.21}$$

From a study of the domains where the derivative has a definite sign, we conclude

$$\text{if } (2m - 1) < x(0) < 2m + 1, \text{ then } \lim_{t \to \infty} x(t) = \bar{x}^{(2m)} = 2m. \tag{4.3.22}$$

4.3.5 Example E

Consider a one-dimensional dynamical system modeled by the differential equation

$$\frac{dx}{dt} = t - x. \tag{4.3.23}$$

First, note that along the curve $x = t$, the derivative is zero, i.e.,

$$\text{along } x = t, \qquad \frac{dx}{dt} = 0. \tag{4.3.24}$$

Thus

$$\frac{dx}{dt} = \begin{cases} < 0, & \text{when } x > t, \\ = 0, & \text{when } x = t, \\ > 0, & \text{when } x < t. \end{cases} \tag{4.3.25}$$

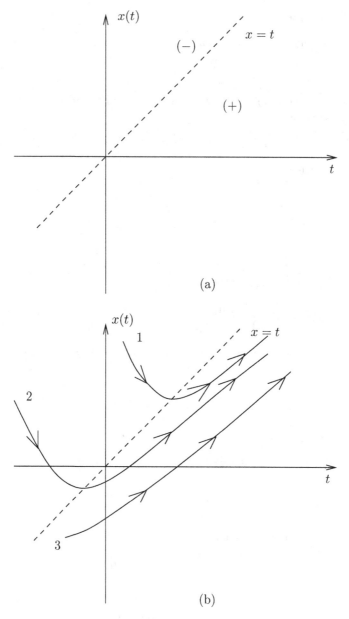

Fig. 4.3.3 The equation $\dot{x} = t - x$. (a) Domains of constant sign for the derivative. (b) Plots for three different starting points.

From this information, we can draw the behavior of the solutions $x(t)$ for various initial conditions. One conclusion that can be reached is that

$$x(t) \underset{t \text{ large}}{\longrightarrow} t. \tag{4.3.26}$$

Typical solutions are represented in Figure 4.3.3.

Equation (4.3.23) can be solved exactly and this solution is, for $x(0) = x_0$, given by

$$x(t) = (x_0 + 1)e^{-t} + t - 1. \tag{4.3.27}$$

Note that all of the qualitative properties of the solution to Eq. (4.3.23), as given in Eqs. (4.3.25) and (4.3.26) and Figure 4.3.3, are verified by the exact solution.

The procedure of this section can be extended to equations taking the form

$$\frac{dx}{dt} = f(x) \prod_{k=1}^{n} [x - \phi_k(t)], \tag{4.3.28}$$

where $f(x)$ has no real zeros and the curves

$$x = \phi_k(t), \qquad (k = 1, 2, \ldots, n) \tag{4.3.29}$$

do not intersect [1]. Elsgolts calls the equation

$$\prod_{k=1}^{n} [x - \phi_k(t)] = 0 \tag{4.3.30}$$

the degenerate equation. He also gives the conditions for which solutions to the full equation (4.3.28) approach one of the degenerate solutions, given by Eq. (4.3.29). A stability requirement is also presented.

4.4   Two-Dimensional Systems

4.4.1   *Definition*

A two-dimensional autonomous dynamical system is one defined by a pair of coupled, first-order differential equations

$$\frac{dx}{dt} = P(x, y), \qquad \frac{dy}{dt} = Q(x, y). \tag{4.4.1}$$

We assume that P and Q have a sufficient number of partial derivatives such that a unique solution exists for any given problem of interest. For many applications, P and Q are polynomials, and, as a consequence, all derivatives exist.

The phase space is the (x, y) plane for this case and the trajectories in this phase space, $y = y(x)$, are determined by the solutions to the following first-order differential equation

$$\frac{dy}{dx} = \frac{Q(x, y)}{P(x, y)}.$$ (4.4.2)

This result can be derived by noting that if $y = y(x)$, then

$$\frac{dy}{dt} = \frac{dy}{dx}\frac{dx}{dt} \Rightarrow \frac{dy}{dx} = \frac{dy/dt}{dx/dt} = \frac{Q(x, y)}{P(x, y)}.$$ (4.4.3)

The general solution to Eq. (4.4.2) is called a *first-integral* of the system given by Eq. (4.4.1).

4.4.2 *Fixed-Points*

The fixed-points are the constant solutions

$$x(t) = \bar{x}, \qquad y(t) = \bar{y}$$ (4.4.4)

to Eq. (4.4.1). Since $d\bar{x}/dt = 0$ and $d\bar{y}/dt = 0$, it follows that they are the simultaneous solutions to the equation

$$P(\bar{x}, \bar{y}) = 0, \qquad Q(\bar{x}, \bar{y}) = 0.$$ (4.4.5)

For the study of dynamical systems in the natural and engineering sciences, only the real solutions have "physical meaning." Thus, we only consider such solutions. Also, it is to be expected that Eq. (4.4.5) may have multi-fixed-points. We will present examples in the next section to illustrate this fact.

4.4.3 *Nullclines*

Nullclines are curves in the (x, y) phase space along which the solution trajectories, $y = y(x)$, have slopes of a constant value. For our purposes, only two of these nullclines are of interest. With reference to Eq. (4.4.2), make the following two definitions:

(i) The **x-nullcline** is the curve in the phase plane along which $y'(x) = dy/dx = \infty$.

(ii) The **y-nullcline** is the curve in the phase plane along which $y'(x) = 0$.

These curves are denoted, respectively, by $y_\infty(x)$ and $y_0(x)$, and can be calculated by solving the equations

$$P[x, y_\infty(x)] = 0, \qquad Q[x, y_0(x)] = 0.$$ (4.4.6)

Note that these curves are, in general, not solutions to Eq. (4.4.2). Also, it should be pointed out that the x- and y-nullclines intersect at the fixed-points of the system. It is important to be aware that the intersection of either $y_\infty(x)$ or $y_0(x)$ with itself is of no particular significance. The only intersections of importance are those where $y_\infty(x)$ and $y_0(x)$ contact each other.

The x- and y-nullclines divide the phase plane into several open domains. The boundaries of these domains are the nullclines themselves. In each open domain the "sign" of dy/dx is fixed, i.e., dy/dx is bounded and either negative or positive. The derivative, dy/dx, can only change sign by crossing from one open domain into another.

4.4.4   *First-Integrals and Symmetries*

As indicated above, the general solution to Eq. (4.4.2) is called a first-integral of Eq. (4.4.1). If such a first-integral can be obtained, then, in general, it has the structure

$$I(x,y) = C, \qquad (4.4.7)$$

where C is a constant. This means that $y(x)$ can only be obtained implicitly through the first-integral. However, even for this situation, a great deal of information can usually be determined on the properties of the solutions to the original system equations as given by Eq. (4.4.1).

A particular value for the constant C gives a curve

$$y = y(x, C), \qquad (4.4.8)$$

in the (x, y) phase plane. This curve is called a **level curve**. Different values of C give rise to different curves. The totality of allowable curves is called the level set of the system.

The first-integral, $I(x, y)$, will often be invariant under a change of coordinates x and y. If such a transformation exists, then the system is said to have a symmetry. Examples of elementary symmetries include:

(i) $x \to -x$, $y \to y$: reflection through the y-axis;
(ii) $x \to x$, $y \to -y$: reflection through the x-axis;
(iii) $x \to -x$, $y \to -y$: inversion through the origin.

The existence of first-integrals and/or the additional existence of symmetry transformations play important roles in understanding the paths of trajectories in phase space.

4.4.5   *General Features of Two-Dimensional Phase Space* [3]

In section 4.2, we were able to state quite easily the possible solution behaviors for an autonomous, one-dimensional system; this discussion included both the local and global behaviors. A similar discussion will now be done for systems defined by a two-dimensional phase space. For such systems, in general, only the following five possibilities can occur for a particular trajectory:

(i) A trajectory may approach a fixed-point as $t \to \infty$, i.e.,

$$\lim_{t\to\infty} \begin{pmatrix} x(t) \\ y(t) \end{pmatrix} = \begin{pmatrix} \bar{x} \\ \bar{y} \end{pmatrix}. \tag{4.4.9}$$

(ii) A trajectory may become unbounded and approach ∞ as $t \to +\infty$

$$\lim_{t\to\infty} \begin{pmatrix} x(t) \\ y(t) \end{pmatrix} = \begin{pmatrix} \infty \\ \infty \end{pmatrix}. \tag{4.4.10}$$

(iii) If a trajectory begins at a fixed-point, it remains there for all t.

(iv) A trajectory may be a nonintersecting closed curve.

(v) A trajectory may approach a closed curve as $t \to +\infty$ or a trajectory that begins in the neighborhood of a closed curve may move away from it as $t \to \infty$.

Note that these five possibilities describe trajectory motions for the global behaviors of a trajectory. Also, the first three cases are just extensions of possibilities that can occur for one-dimensional system. The last two cases do not have any correspondences in one-dimension. Figure 4.4.1 illustrates these five possibilities. Case (v) is shown with two possibilities: (va) represents the situation where neighboring trajectories approach a closed trajectory from both the outside and inside, while (vb) is where neighboring trajectories move away from a closed trajectory. A third possibility also exists in which trajectories approach the closed curve from one side, but move away from it on the other side.

A detailed understanding of the behavior of trajectories in phase space requires a knowledge of the local behavior for trajectories in the neighborhood of a fixed-point. The following possibilities can occur; see Figure 4.4.2.

(i) **Stable node:** All trajectories approach the fixed-point along non-spiraling curves, i.e., at $t \to \infty$, the trajectories are asymptotic to straight lines.

(ii) **Unstable node:** All trajectories leave a neighborhood of the fixed-point along non-spiraling curves.

(iii) **Stable spiral point:** All trajectories approach the fixed-point along spiral curves as $t \to +\infty$.

(iv) **Unstable spiral point:** All trajectories leave a neighborhood of the fixed-point along spiral curves at $t \to +\infty$.

(v) **Saddle point:** Trajectories initially move toward the fixed-point and then move away. Two trajectories move toward the fixed-point and stop, while two other trajectories originate at the fixed-point, but move away from it. These trajectories are called the stable and unstable manifolds of the saddle point.

(vi) **Center:** All neighboring trajectories form closed curves about a fixed-point.

4.4.6 *The Basic Procedure for Constructing Phase-Space Diagrams*

The results presented so far will, in general, permit a detailed construction of the phase space diagram for most two-dimensional systems encountered in the mathematical modeling of natural and engineering phenomena. We now outline a basic procedure for actually carrying out this construction. Section 4.5 will demonstrate the implementation of this method for a number of interesting examples.

(A) Determine the location and number of the real fixed-points, corresponding to the system

$$\frac{dx}{dt} = P(x, y), \qquad \frac{dy}{dt} = Q(x, y) \qquad (4.4.11)$$

by solving the equations

$$P(\bar{x}, \bar{y}) = 0, \qquad Q(\bar{x}, \bar{y}) = 0. \qquad (4.4.12)$$

Denote these fixed-points by

$$\{\bar{x}^{(i)}, \bar{y}^{(i)} : i = 1, 2, \ldots, I\}, \qquad (4.4.13)$$

where I is the total number of fixed-points.

(B) Calculate the x-nullcline by solving for $y_\infty(x)$ from the equation

$$P[x, y_\infty(x)] = 0. \qquad (4.4.14)$$

Note that this is a curve such that a crossing trajectory has an unbounded slope.

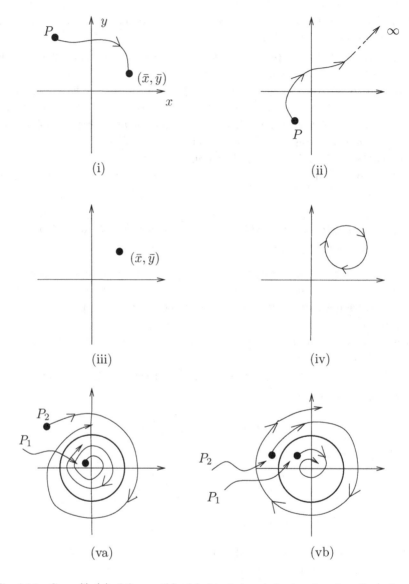

Fig. 4.4.1   Cases (i)–(v) of the possible global trajectory behaviors for a two-dimensional system. P denotes the initial starting position, while $(\bar{x}, \bar{y})$ is the location of the fixed-point.

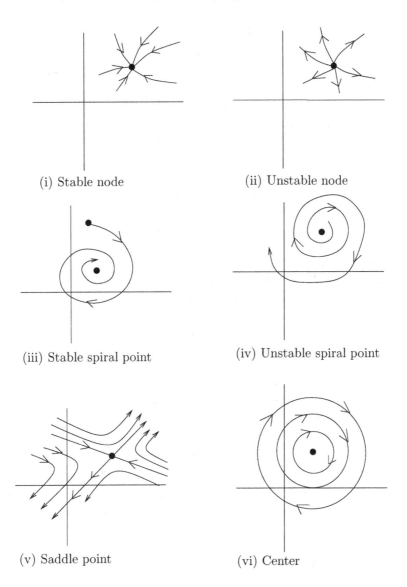

(i) Stable node   (ii) Unstable node

(iii) Stable spiral point   (iv) Unstable spiral point

(v) Saddle point   (vi) Center

Fig. 4.4.2   Possible local behaviors for two dimensional systems.

(C) Calculate the y-nullcline by solving for $y_0(x)$ from the equation

$$Q[x, y_0(x)] = 0. \qquad (4.4.15)$$

Every trajectory crossing the y-nullcline has zero slope at its intersection with this nullcline.

(D) Draw on the x-y phase plane the fixed-points and the x- and y-nullclines and check to see that they intersect at the fixed-points. Note, there should be no fixed-points that do not lie at the intersection of both an x- and y-nullcline, and vice versa.

(E) The x- and y-nullclines divide the phase plane into a number of open domains. In each of these open domains the sign of dy/dx is definite, i.e., for a given open domain y' is bounded and either positive or negative. The next step is to determine the sign of dy/dx in each open domain. This can usually be done in direct fashion by selecting a point in a given domain and evaluating $dy/dx = Q(x, y)/P(x, y)$. Whatever, the sign is for this point, all other points in this open domain have the same sign.

(F) Now start at an "appropriate point" on the phase plane and sketch carefully the trajectory passing through this point. Repeat this procedure for a number of other appropriate points until the general flow of the trajectories in the phase plane becomes obvious. If the system equations are invariant under certain coordinate transformations use the related symmetries in drawing the trajectories.

(G) The above steps will often lead to a unique geometrical structure for the trajectories in the phase plane. For the cases where there may be some ambiguities, it may be of value to examine the local behavior of trajectories in the neighborhood of the various fixed-points or at least for the one's for which some uncertainty exists as to how the nearby trajectories behavior.

In the next section, we present a brief discussion on the use of linear stability analysis to provide information on the local behavior of trajectories near fixed-points.

4.4.7   *Linear Stability*

We first examine two-dimensional linear autonomous systems

$$\begin{cases} \dfrac{dx}{dt} = ax + by, \\[2mm] \dfrac{dy}{dt} = cx + dy, \end{cases} \qquad (4.4.16)$$

where (a, b, c, d) are real constants. This form of the equations assumes that the fixed-point is at $(\bar{x}, \bar{y}) = (0, 0)$ [4]. These equations can be written in matrix form by defining

$$X = \begin{pmatrix} x \\ y \end{pmatrix}, \qquad A = \begin{pmatrix} a & b \\ c & d \end{pmatrix}. \tag{4.4.17}$$

With this notation, Eq. (4.4.16) can be written as [5].

$$\frac{dX}{dt} = AX. \tag{4.4.18}$$

Let λ_1 and λ_2 be the eigenvalues of matrix A; they are determined by the condition

$$\det(A - I\lambda) = \lambda^2 - (a + d)\lambda - (ad - bc) = 0, \tag{4.4.19}$$

where I is the 2×2 unit matrix. If $\lambda_1 \neq \lambda_2$, then associated with each of these eigenvalues is an eigenvector, V_1 and V_2, such that the general solution is

$$X(t) = c_1 V_1 e^{\lambda_1 t} + c_2 V_2 e^{\lambda_2 t}, \tag{4.4.20}$$

where c_1 and c_2 are arbitrary constants. A detailed examination of Eq. (4.4.20) leads to the following conclusions:

(i) If λ_1 *and* λ_2 are both *real and negative*, then all trajectories approach the fixed-point as $t \to +\infty$ and the fixed-point is a *stable node*.

(ii) If λ_1 *and* λ_2 are both *real and positive*, then all trajectories move away from the fixed-point as $t \to +\infty$ and the fixed-point is an *unstable node*.

(iii) If λ_1 and λ_2 are both real, but λ_1 is positive, while λ_2 is negative, then trajectories approach the fixed-point in the direction of V_2 and move away from the origin in the direction of V_1. For this case the fixed-point is a *saddle point*.

(iv) Since (a, b, c, d) are real constants, the eigenvalues λ_1 *and* λ_2 may be *complex valued*, but can only be complex conjugates of each other, i.e., $\lambda_1 = \lambda_2^*$. If

$$\operatorname{Re}\lambda_1 = \operatorname{Re}\lambda_2 \neq 0, \tag{4.4.21}$$

then the fixed-point is a *spiral point*. For $\operatorname{Re}\lambda_1 = \operatorname{Re}\lambda_2 < 0$, the trajectories spiral toward the fixed-point, while for $\operatorname{Re}\lambda_1 = \operatorname{Re}\lambda_2 > 0$, the trajectories spiral away from the fixed-point. These cases correspond, respectively, to *stable* and *unstable spiral points*.

(v) If λ_1 *and* λ_2 are *pure imaginary*, i.e.,

$$\operatorname{Re}\lambda_1 = \operatorname{Re}\lambda_2 = 0, \tag{4.4.22}$$

then the vector $X(t)$ describes a closed curve in the x-y phase space for any constants c_1 and c_2. The solution for this case is periodic.

With the above classification, the general structure of the solutions to the linear problem, as given by Eq. (4.4.16) can be easily determined.

We now consider the situation for nonlinear systems that take the form [4]

$$\begin{cases} \dfrac{dx}{dt} = ax + by + P_1(x,y), \\[2mm] \dfrac{dy}{dt} = cx + dy + Q_1(x,y), \end{cases} \tag{4.4.23}$$

where it is assumed that

$$\lim_{\substack{x \to 0 \\ y \to 0}} \left\{ \begin{array}{c} \dfrac{P_1(x,y)}{\sqrt{x^2 + y^2}} \\[3mm] \dfrac{Q_1(x,y)}{\sqrt{x^2 + y^2}} \end{array} \right\} = \begin{pmatrix} 0 \\ 0 \end{pmatrix}. \tag{4.4.24}$$

Under this assumption we can "hope" that when x and y start in a small neighborhood of the fixed-point $(0,0)$, the dynamics of the trajectory will be determined by the linear parts of Eq. (4.4.24).

However, before proceeding, we need to present a discussion on stability. For our purposes, a fixed-point of the system given by Eq. (4.4.23) is said to be stable if the starting point (x_0, y_0) is sufficiently close to the fixed-point such that $x(t)$ and $y(t)$ remain close to $(\bar{x}, \bar{y}) = (0,0)$ for all $t > 0$. This can be formulated for a general fixed-point by using a vector representation, i.e.,

$$X(t) = \begin{pmatrix} x(t) \\ y(t) \end{pmatrix}, \qquad X_0 = X(0) = \begin{pmatrix} x_0 \\ y_0 \end{pmatrix}, \qquad \bar{X} = \begin{pmatrix} \bar{x} \\ \bar{y} \end{pmatrix}, \tag{4.4.25}$$

and

$$|X_0 - \bar{X}| = \sqrt{(x_0 - \bar{x})^2 + (y_0 - \bar{y})^2}. \tag{4.4.26}$$

With this notation, the fixed-point, $\bar{X}$, is stable provided that for each $\epsilon > 0$, there exists a $\delta > 0$, such that

$$|X_0 - \bar{X}| < \delta \implies |X(t) - \bar{X}| < \epsilon, \qquad t > 0. \tag{4.4.27}$$

A fixed-point is unstable if it is not stable.

A fixed-point $\bar{X}$ is called asymptotically stable if it is stable and if every trajectory that starts sufficiently close to it approaches it as $t \to \infty$. In other words, there exists a $\delta > 0$ such that

$$|X(t) - \bar{X}| < \delta \implies \lim_{t \to \infty} X(t) = \bar{X}. \tag{4.4.28}$$

Returning to the nonlinear system given by Eq. (4.4.23), the following theorem summarizes the situation (provided that $ad - bc \neq 0$) [6].

Theorem *Let λ_1 and λ_2 be the eigenvalues associated with the matrix of the linear coefficients of the system given by Eq. (4.4.23), i.e.,*

$$\det(A - \lambda I) = \det \begin{vmatrix} a - \lambda & b \\ c & d - \lambda \end{vmatrix}$$

$$= \lambda^2 - (a + d)\lambda + (ad - bc)$$

$$= (\lambda - \lambda_1)(\lambda - \lambda_2) = 0. \tag{4.4.29}$$

Then the following results hold:

(A) *If λ_1 and λ_2 are real and equal, i.e., $\lambda_1 = \lambda_2$, the fixed-point is either a node or a spiral point. If $\lambda_1 = \lambda_2 < 0$, the fixed-point is asymptotically stable. If $\lambda_1 = \lambda_2 > 0$, the fixed-point is unstable.*

(B) *If λ_1 and λ_2 are pure imaginary, the fixed-point is either a center or a spiral point. If the fixed-point is a center, then it is stable. If the fixed-point is a spiral point, then it may be asymptotically stable or unstable.*

(C) *If λ_1 and λ_2 are real, unequal and have the same sign, then the fixed-point is a node. If $\lambda_1 > \lambda_2 > 0$, then the fixed-point is unstable. If $\lambda_1 < \lambda_2 < 0$, then the fixed-point is stable.*

(D) *If $\lambda_1 < 0 < \lambda_2$, i.e., λ_1 and λ_2 are real, unequal and have opposite sign, then the fixed-point is a saddle point.*

(E) *If λ_1 and λ_2 are complex conjugate, then the fixed-point is a spiral point. If $\mathrm{Re}\,\lambda_1 = \mathrm{Re}\,\lambda > 0$, then the spiral point is unstable. If $\mathrm{Re}\,\lambda_1 = \mathrm{Re}\,\lambda_2 < 0$, the spiral point is asymptotically stable.*

All of these results are summarized in Table 4.4.1.

There are certain difficulties that arise when applying a linear stability analysis to determine the nature of a fixed-point for a nonlinear system. First, the system may not have a linear limit. An example is

$$\frac{d^2 x}{dt^2} + x^{1/3} = 0, \tag{4.4.30}$$

which can be written as the system of equations

$$\frac{dx}{dt} = y, \qquad \frac{dy}{dt} = x^{1/3}. \tag{4.4.31}$$

Table 4.4.1   Classification of the nature of the fixed-points
for nonlinear systems

Eigenvalues	Type of Fixed-Point and Stability
$\lambda_1 > \lambda_2 > 0$	Unstable node
$\lambda_1 < \lambda_2 < 0$	Stable node
$\lambda_1 = \lambda_2 > 0$	Unstable node or spiral point
$\lambda_1 = \lambda_2 < 0$	Stable node or spiral point
$\lambda_1 < 0 < \lambda_2$	Saddle point (unstable)
$\lambda_1 = \lambda_2^* = a + bi \ (a > 0)$	Unstable spiral point
$\lambda_1 = \lambda_2^* = a + bi \ (a < 0)$	Stable spiral point
$\lambda_1 = \lambda_2^* = bi$	Stable center or stable or unstable spiral point

Second, Table 4.4.1 shows that if the linear analysis indicates that a particular fixed-point is a center, the fixed-point of the corresponding nonlinear system may or may not be a center. It can also be a stable or unstable spiral point. The implication is that centers are very fragile types of fixed-points. This can be illustrated by the following example. Consider, for simplicity, the linear system

$$\frac{dx}{dt} = y, \qquad \frac{dy}{dt} = -x, \qquad (4.4.32)$$

which can be written as

$$\frac{d}{dt}\begin{pmatrix} x \\ y \end{pmatrix} = \begin{pmatrix} 0 & 1 \\ -1 & 0 \end{pmatrix}\begin{pmatrix} x \\ y \end{pmatrix}. \qquad (4.4.33)$$

The eigenvalues of the associated matrix are

$$\lambda_1 = \lambda_2^* = i, \qquad (4.4.34)$$

and the corresponding solution is

$$\begin{cases} x(t) = c_1 \cos t + c_2 \sin t, \\ y(t) = -c_1 \sin t + c_2 \cos t, \end{cases} \qquad (4.4.35)$$

where c_1 and c_2 are arbitrary constants. Now add a small linear perturbation to the second of the differential equations in Eq. (4.4.32), i.e.,

$$\frac{dx}{dt} = y, \qquad \frac{dy}{dt} = -x - 2\epsilon y, \qquad (4.4.36)$$

where $\epsilon \neq 0$, but can be as small as we wish. The eigenvalues for this system of equations are

$$\lambda_1 = \lambda_2^* = -\epsilon + i\sqrt{1 - \epsilon^2}, \qquad |\epsilon| \ll 1. \qquad (4.4.37)$$

The solution $x(t)$ is now

$$x(t) = e^{-\epsilon t} \left[c_1 \cos\left(\sqrt{1-\epsilon^2}\, t\right) + c_2 \sin\left(\sqrt{1-\epsilon^2}\, t\right) \right], \qquad (4.4.38)$$

with a similar expression for $y(t) = dx/dt$. We now see that the center of Eq. (4.4.32) has been changed to a spiral point for the new perturbed differential equation system given by Eq. (4.4.36). Note that for $\epsilon > 0$, the spiral point is stable, while for $\epsilon < 0$, the spiral point is unstable.

In summary, indications that a fixed-point is a center for the linearized part of a system does not translate into an absolute assurance that the actual fixed-point is a center. For an actual center to be present, the neighboring trajectories must be closed curves. This means that no methods of approximation can be used to prove the existence of a genuine center. One technique that can be applied to this situation is to see if a first-integral of Eq. (4.4.23) can be explicitly constructed. The existence of a first-integral can then be used to show the existence of a center.

4.5 Worked Examples

The methods and principles presented in previous sections will now be applied to several examples. An important feature in constructing the phase space trajectories for many systems arising in the mathematical modeling of dynamical phenomena is that only a knowledge of the behavior of the trajectories in a restricted region is needed. This is a consequence of the fact that for some systems the dependent variables satisfy a positivity requirement, i.e.,

$$x(0) \geq 0, \quad y(0) \geq 0 \Longrightarrow x(t) \geq 0, \quad y(t) \geq 0. \qquad (4.5.1)$$

Particular examples include chemical reactions, where the concentrations are always non-negative, and problems involving interacting populations, where the number densities are also non-negative.

For the worked examples to follow, we will only investigate the relevant regions of phase space as defined by the particular system of interest.

4.5.1 *Example A*

Let us examine the behavior of a dynamical system described by the pair of coupled, nonlinear first-order differential equations

$$\frac{dx}{dt} = y + x(x^2 + y^2), \qquad (4.5.2)$$

$$\frac{dy}{dt} = -x + y(x^2 + y^2).$$ (4.5.3)

The only fixed-point is $(\bar{x}, \bar{y}) = (0, 0)$ and the linear approximation in a neighborhood of this fixed-point is

$$\frac{d\alpha}{dt} = \beta, \qquad \frac{d\beta}{dt} = -\alpha$$ (4.5.4)

where

$$x = \bar{x} + \alpha = \alpha, \qquad y = \bar{y} + \beta = \beta; \qquad |\alpha|, |\beta| \ll 1.$$ (4.5.5)

In matrix form, we have

$$\frac{d}{dt} \begin{pmatrix} \alpha \\ \beta \end{pmatrix} = \begin{pmatrix} 0 & 1 \\ -1 & 0 \end{pmatrix} \begin{pmatrix} \alpha \\ \beta \end{pmatrix}$$ (4.5.6)

and the eigenvalues are given by the solutions to

$$\det(A - \lambda I) = \det \begin{pmatrix} -\lambda & 1 \\ -1 & -\lambda \end{pmatrix} = \lambda^2 + 1 = 0,$$ (4.5.7)

and they are

$$\lambda_1 = \lambda_2^* = i.$$ (4.5.8)

Thus, the linear analysis suggests that the fixed-point is a center. However, this result may or may not be correct for the full nonlinear equations. In fact, for large values of x and y, we have

$$\frac{dx}{dt} \simeq r^2 x > 0, \quad \frac{dy}{dt} \simeq r^2 y > 0, \quad r^2 = x^2 + y^2,$$ (4.5.9)

and this implies that trajectories may become unbounded. If so, then $(\bar{x}, \bar{y}) = (0, 0)$ cannot be a center.

To study this issue in more detail, let us multiply Eqs. (4.5.2) and (4.5.3), respectively, by x and y, and then add these expressions. Doing this gives

$$x \frac{dx}{dt} + y \frac{dy}{dt} = (x^2 + y^2)^2,$$ (4.5.10)

or

$$\left(\frac{1}{2} \right) \frac{d}{dt} (x^2 + y^2) = (x^2 + y^2)^2.$$ (4.5.11)

Using $r^2 = x^2 + y^2$, we obtain

$$\frac{d(r^2)}{dt} = 2r^4.$$ (4.5.12)

Since, $d(r^2)/dt > 0$, for $r > 0$, it follows that all trajectories that start with $r(0) > 0$ become unbounded, i.e.,

$$x(0) \neq 0, \quad y(0) \neq 0 \Longrightarrow \lim_{t \to \infty} \begin{pmatrix} x(t) \\ y(t) \end{pmatrix} = \begin{pmatrix} \infty \\ \infty \end{pmatrix}. \tag{4.5.13}$$

This result can be directly verified because Eq. (4.5.12) can be solved exactly for $r(t)$. The solution is

$$r(t) = \frac{r_0}{\sqrt{1 - 2r_0^2 t}}, \tag{4.5.14}$$

where

$$r_0 = \sqrt{x(0)^2 + y(0)^2}. \tag{4.5.15}$$

Note that $r(t)$ becomes unbounded in a finite time, t^*, given by

$$t^* = \frac{1}{2r_0^2}. \tag{4.5.16}$$

In polar coordinates

$$x(t) = r(t)\cos\theta(t), \qquad y(t) = r(t)\sin(t), \tag{4.5.17}$$

and

$$\begin{cases} \dot{x} = \dot{r}\cos\theta - r\dot{\theta}\sin\theta, \\ \dot{y} = \dot{r}\sin\theta + r\dot{\theta}\cos\theta. \end{cases} \tag{4.5.18}$$

Substituting Eqs. (4.5.18) into Eqs. (4.5.2) and solving for $d(r^2)/dt$ and $d\theta/dt$ gives Eq. (4.5.12) and

$$\dot{\theta} = -1 \Longrightarrow \theta(t) = -t + \theta_0, \tag{4.5.19}$$

where

$$\tan\theta_0 = \frac{y(0)}{x(0)}. \tag{4.5.20}$$

Therefore,

$$x(t) = \frac{r_0 \cos(t - \theta_0)}{\sqrt{1 - 2r_0^2 t}}, \tag{4.5.21}$$

$$y(t) = -\frac{r_0 \sin(t - \theta_0)}{\sqrt{1 - 2r_0^2 t}}. \tag{4.5.22}$$

These equations correpond to an unstable spiral. Our conclusion is that while the linearized part of Eqs. (4.5.2) and (4.5.3) implies that the fixed-point $(\bar{x}, \bar{y}) = (0, 0)$ is a center, in fact, it corresponds to an unstable spiral point. Figure 4.5.1 presents these results in graphic form.

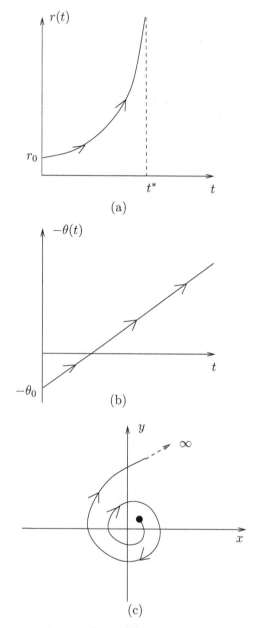

Fig. 4.5.1 Example A. (a) Plot of $r(t)$ vs t. (b) Plot of $\theta(t)$ vs t. (c) Typical phase space trajectory.

4.5.2   Example B

Consider the following system

$$\frac{dx}{dt} = x + y - x(x^2 + y^2), \tag{4.5.23}$$

$$\frac{dy}{dt} = -x + y - y(x^2 + y^2). \tag{4.5.24}$$

The only fixed-point is $(\bar{x}, \bar{y}) = (0,0)$ and the linear approximation at this fixed-point is

$$\frac{d\alpha}{dt} = \alpha + \beta, \qquad \frac{d\beta}{dt} = -\alpha + \beta, \tag{4.5.25}$$

or

$$\frac{d}{dt}\begin{pmatrix} \alpha \\ \beta \end{pmatrix} = \begin{pmatrix} 1 & 1 \\ -1 & 1 \end{pmatrix}\begin{pmatrix} \alpha \\ \beta \end{pmatrix}. \tag{4.5.26}$$

The eigenvalues are determined by the equation

$$\det\begin{pmatrix} 1-\lambda & 1 \\ -1 & 1-\lambda \end{pmatrix} = \lambda^2 - 2\lambda + 2 = 0, \tag{4.5.27}$$

and are

$$\lambda_1 = \lambda_2^* = 1 + i. \tag{4.5.28}$$

Therefore, the fixed-point is an unstable spiral point.

The analysis of the previous problem suggests that a change of variables might be useful for this pair of equations. Let

$$x(t) = r(t)\cos\theta(t), \qquad y(t) = r(t)\sin\theta(t), \tag{4.5.29}$$

then a direct calculation shows that

$$\frac{dr}{dt} = r(1 - r^2), \qquad \frac{d\theta}{dt} = -1. \tag{4.5.30}$$

The first-order, nonlinear r-equation has two fixed-points, $\bar{r}^{(1)} = 0$ and $\bar{r}^{(2)} = 1$. (The $r = -1$ solution is not considered since the only relevant values are $r \geq 0$.) Defining $R(r)$ as

$$R(r) = r(1 - r^2), \tag{4.5.31}$$

it follows

$$\frac{dR}{dr}\Big|_{r=0} = 1, \qquad \frac{dR}{dr}\Big|_{r=1} = -2, \tag{4.5.32}$$

and we find that the fixed-point at $\bar{r}^{(1)} = 0$ is unstable, while the one $\bar{r}^{(2)} = 1$ is stable. Therefore, for any initial value, $r(0) > 0$, we have

$$\lim_{t \to \infty} r(t) = 1, \qquad (4.5.33)$$

i.e., $\bar{r}^{(2)} = 1$ is globally stable. (Note that this result does not follow from the linear stability analysis, but is a consequence of the signs of the derivative in the one-dimensional phase space.)

The equation for $\theta(t)$ can be easily solved to give

$$\theta(t) = -t + \theta_0. \qquad (4.5.34)$$

Likewise, the equation for $r(t)$ can be solved if both sides are multiplied by $2r$ and the new variable $z = r^2$ is introduced; this yields

$$\frac{dz}{dt} = 2z(1 - z), \qquad (4.5.35)$$

which is the separable logistic differential equation whose solution is

$$z(t) = \frac{1}{1 + ce^{2t}}, \qquad (4.5.36)$$

where c is the constant of integration. Solving for $r(t)$ and using $r(0) = r_0$, we get

$$r(t) = \frac{r_0}{\sqrt{r_0^2 + (1 - r_0^2)e^{-2t}}}. \qquad (4.5.37)$$

The solutions to Eqs. (4.5.23) and (4.5.24) are therefore

$$x(t) = \frac{r_0 \cos(t - \theta_0)}{\sqrt{r_0^2 + (1 - r_0^2)e^{-2t}}}, \qquad (4.5.38)$$

$$y(t) = -\frac{r_0 \sin(t - \theta_0)}{\sqrt{r_0^2 + (1 - r_0^2)e^{-2t}}}. \qquad (4.5.39)$$

The fixed-point, $(\bar{x}, \bar{y}) = (0, 0)$, corresponds to $r_0 = \bar{r}^{(1)} = 0$, i.e.,

$$x(t) = \bar{x}^{(1)} = 0, \qquad y(t) = \bar{y}^{(1)} = 0. \qquad (4.5.40)$$

The analysis of Eq. (4.5.30) shows that this fixed-point is unstable, a result in agreement with the eigenvalue calculation for the system given by Eqs. (4.5.23) and (4.5.24).

The fixed-point, $\bar{r}^{(2)} = 1$, for the r differential equation, gives the periodic solution

$$x^{(p)}(t) = \cos(t - \theta_0), \quad y^{(p)}(t) = -\sin(t - \theta_0). \qquad (4.5.41)$$

In the (x, y) phase plane, this periodic solution has the representation

$$x^2 + y^2 = 1, \tag{4.5.42}$$

i.e., it is a circle. A study of the general solutions, given by Eqs. (4.5.38) and (4.5.39), shows that all trajectories in the phase space are attracted to this circle:

(i) If $r_0 > 1$, then trajectories spiral inward toward the unit circle as $t \to \infty$.

(ii) If $0 < r_0 < 1$, then trajectories spiral outward toward the unit circle as $t \to \infty$.

An *isolated, non-intersecting, closed curve* in phase space is called a *limit-cycle*. If all neighboring trajectories move toward the limit-cycle it is called a stable limit-cycle. Otherwise, the limit-cycle is called unstable. Figure 4.5.2 illustrates the behavior of limit-cycles.

The general properties of the phase plane for Eqs. (4.5.23) and (4.5.24) are shown in Figure 4.5.3.

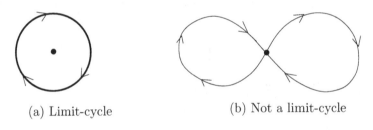

(a) Limit-cycle   (b) Not a limit-cycle

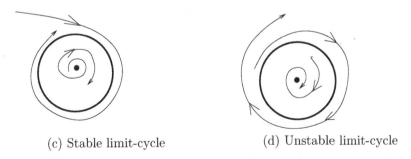

(c) Stable limit-cycle   (d) Unstable limit-cycle

Fig. 4.5.2   Limit-cycles.

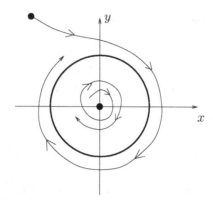

Fig. 4.5.3   General features of phase space trajectories for the system defined by
Eqs. (4.5.23) and (4.5.24).

4.5.3   *Example C* [7]

Several important dynamical systems can be modeled by differential equations having the form

$$\frac{d^2x}{dt^2} + f\left(\frac{dx}{dt}\right)x = 0, \tag{4.5.43}$$

where the function $f(z)$ has the property

$$f(0) > 0. \tag{4.5.44}$$

This equation has the system equations

$$\frac{dx}{dt} = y, \qquad \frac{dy}{dt} = -f(y)x. \tag{4.5.45}$$

Further, the trajectories in phase space are solutions to

$$\frac{dy}{dx} = -\frac{f(y)x}{y}. \tag{4.5.46}$$

Note that independent of the exact functional form for $f(y)$, Eq. (4.5.46) is invariant under the transformation

$$T_y : x \to -x. \tag{4.5.47}$$

This means that all trajectories for this system in phase space have a reflection symmetry through the y-axis. Therefore, if $\phi(x, y) = c$ is a trajectory in phase space, then $\phi(-x, y) = c$ is also a trajectory. If the two trajectories meet at some point, on the y-axis, then the two trajectories are continuations of a single trajectory.

A particular example is the equation

$$\frac{d^2x}{dt^2} + \left(1 + \frac{dx}{dt}\right)^n x = 0, \tag{4.5.48}$$

where $n = 0, 1, 2, \dots$. However, we will only examine the case $n = 1$, i.e.,

$$\frac{d^2x}{dt^2} + \left(1 + \frac{dx}{dt}\right) x = 0 \tag{4.5.49}$$

or

$$\frac{dx}{dt} = y, \qquad \frac{dy}{dt} = -(1+y)x, \tag{4.5.50}$$

and

$$\frac{dy}{dx} = -\frac{(1+y)x}{y}. \tag{4.5.51}$$

Equation (4.5.50) has a single fixed-point located at $(\bar{x}, \bar{y}) = (0,0)$. A linearization at this fixed-point gives

$$\frac{d}{dt}\begin{pmatrix} \alpha \\ \beta \end{pmatrix} = \begin{pmatrix} 0 & 1 \\ -1 & 0 \end{pmatrix}\begin{pmatrix} \alpha \\ \beta \end{pmatrix}, \tag{4.5.52}$$

and the associated eigenvalues are

$$\lambda_1 = \lambda_2^* = i, \tag{4.5.53}$$

indicating that in the linear approximation the fixed-point of Eq. (4.5.50) is a center. We will now confirm this result in two ways.

First, note that Eq. (4.5.51) is separable and can be integrated to give

$$\int \frac{y}{1+y}\, dy + \int x\, dx = 0 \tag{4.5.54}$$

and

$$H(x,y) = y - \ln(1+y) + \frac{x^2}{2} = \text{constant.} \tag{4.5.55}$$

For x and y sufficiently small, we have

$$H(x,y) = \frac{y^2}{2} + \frac{x^2}{2} + O(x^3) = \text{constant,} \tag{4.5.56}$$

and thus conclude that under these conditions the trajectories are closed and correspond to circles.

A second and, in some sense, more powerful method is to construct the trajectories in phase space using geometrical arguments based on nullclines. Equation (4.5.51) has the nullclines

$$\begin{cases} \dfrac{dy}{dx} = 0 : x = 0 \text{ or the } y\text{-axis and } y = -1; \\ \dfrac{dy}{dx} = \infty : y = 0 \text{ or the } x\text{-axis.} \end{cases} \tag{4.5.57}$$

These nullclines separate the phase plane into six distinct domains each of which the derivative, dy/dx, has a definite sign. These domains are shown in Figure 4.5.4a. Note that the x- and y-nullclines have only one intersection and that is at the location of the fixed-point $(\bar{x}, \bar{y}) = (0, 0)$. However, the y-nullcline does intersect itself at $(0, -1)$, but this point has no significance with regard to the dynamics of the system; it is not a fixed-point.

The phase plane consists of three essential different regions:

(i) Trajectories that start in the phase plane with

$$|x_0| < \infty, \qquad y_0 > -1, \qquad (4.5.58)$$

belong to closed trajectories. Thus, they correspond to periodic solutions.

(ii) $y = -1$ is an exact solution to Eq. (4.5.49) and gives

$$y(t) = -1, \qquad x(t) = -t + x_0. \qquad (4.5.59)$$

(iii) Initial trajectory conditions

$$|x_0| < \infty, \qquad y_0 < -1 \qquad (4.5.60)$$

correspond to unbounded solutions.

All three cases are illustrated in Figure 4.5.4. Figure 4.5.5 presents the geometrical proof that all trajectories originating with $y_0 > -1$ belong to closed curves and, hence, are periodic. In summary, start with some point P on the y-axis. In the first quadrant, the trajectory must have a negative slope and cross the x-axis with infinite slope. Continuing this process, the slope in the fourth quadrant is positive and crosses the y-axis with zero slope. At this point, the transformation T_y, reflection through the y-axis, is applied to obtain the last diagram. Since this is a single closed curve, it represents a periodic solution. Likewise, using similar arguments, all trajectories for $y_0 < -1$ are unbounded.

4.5.4 *Example D*

Let us now consider an elementary predator-prey model involving rabbits (x) and foxes (y). We make the following assumptions:

(i) In the absence of foxes, the rabbits reproduce at a constant rate.

(ii) If only foxes are present, then they die off at a rate proportional to their numbers.

(iii) For the situation where both foxes and rabbits are present, the rabbits rate of change is negative, while the foxes rate of change is positive. In both cases, we take the rate of change to be proportional to the product of total number of rabbits and foxes.

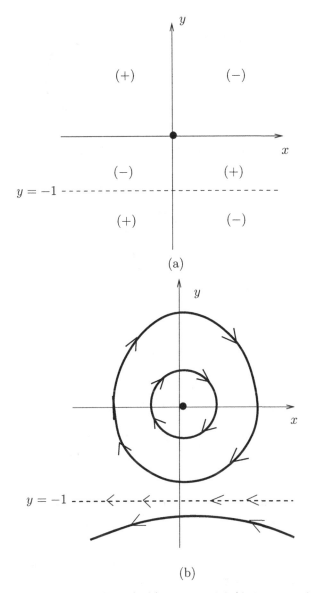

Fig. 4.5.4 Phase plane for Eq. (4.5.49). (a) The sign of dy/dx in the six domains that the phase is divided by the nullclines. (b) Typical trajectories.

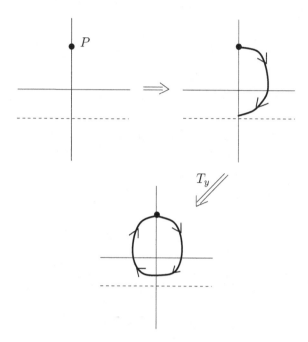

Fig. 4.5.5   Construction of trajectories for Eq. (4.5.49).

A model incorporating all these features is given by the pair of equations

$$\frac{dx}{dt} = k - r_1 xy, \qquad \frac{dy}{dt} = r_2 xy - r_3 y, \qquad (4.5.61)$$

where all the parameters are non-negative. By using dimensionless variables and setting any remaining parameters to unity, we obtain

$$\frac{dx}{dt} = 1 - xy, \qquad \frac{dy}{dt} = xy - y. \qquad (4.5.62)$$

The only fixed-point is $(\bar{x}, \bar{y}) = (1, 1)$. The behavior near this fixed-point is obtained by writing

$$x(t) = 1 + \alpha(t), \qquad y(t) = 1 + \beta(t). \qquad (4.5.63)$$

Assume that $|\alpha(0)| \ll 1$ and $|\beta(0)| \ll 1$, and substitute these expressions into Eq. (4.5.62). Neglecting nonlinear terms gives

$$\frac{d}{dt} \begin{pmatrix} \alpha \\ \beta \end{pmatrix} = \begin{pmatrix} -1 & -1 \\ 1 & 0 \end{pmatrix} \begin{pmatrix} \alpha \\ \beta \end{pmatrix}, \qquad (4.5.64)$$

which has the characteristic equation

$$\det(A - \lambda I) = \det \begin{pmatrix} -1 - \lambda & -1 \\ 1 & -\lambda \end{pmatrix} = \lambda^2 + \lambda + 1 = 0 \qquad (4.5.65)$$

for which the solutions are

$$\lambda_1 = \lambda_2^* = \frac{-1 + i\sqrt{3}}{2}. \tag{4.5.66}$$

Since the eigenvalues are complex-conjugate with negative real parts, it follows that the fixed-point is a stable spiral point.

The first-order differential equation for the trajectories is

$$\frac{dy}{dx} = \frac{y(x-1)}{1-xy}, \tag{4.5.67}$$

and the nullclines are

$$\begin{cases} \dfrac{dy}{dx} = 0 : y = 0 \text{ or the } x\text{-axis and } x = 1; \\ \dfrac{dy}{dx} = \infty : y_\infty(x) = \dfrac{1}{x}. \end{cases} \tag{4.5.68}$$

Figure 4.5.6a shows the location of the fixed-point and the two nullclines. Part b of the figure indicates that the non-negative x-axis is a solution; it corresponds to the case where $y = 0$ and thus

$$\frac{dx}{dt} = 1 \implies x(t) = x_0 + t, \qquad x_0 > 0. \tag{4.5.69}$$

Our equations are supposed to model the predator-prey dynamics of foxes and rabbits, consequently, the solutions should be non-negative. Therefore, any initial point in the first-quadrant should lead to solutions that stay in this domain. This is easy to show. First, the positive portion of the x-axis is the trajectory for the case where only rabbits exist. Second, dy/dx along the y-axis is

$$\left.\frac{dy}{dx}\right|_{x=0} = -y. \tag{4.5.70}$$

This means that at any point on the positive y-axis, trajectories always enter into the first-quadrant. The overall conclusion is that

$$x_0 > 0, \quad y_0 \geq 0 \implies x(t) > 0, \quad y(t) > 0. \tag{4.5.71}$$

In summary, this particular model has a single fixed-point that is asymptotically stable and is approached from any initial state in the first-quadrant in an oscillatory manner. The fixed-point indicates that the rabbits and foxes can coexist with each other.

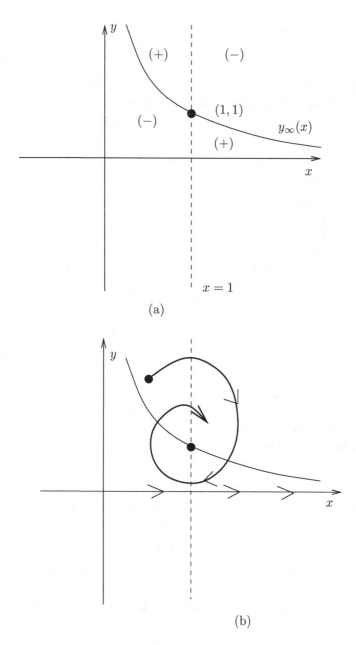

Fig. 4.5.6

4.5.5   Example E

A single-degree-of-freedom conservative system can be modeled by the following differential equation

$$\frac{d^2x}{dt^2} + g(x) = 0. \tag{4.5.72}$$

For such systems, first-integrals always exist and can be calculated by using

$$\frac{dx}{dt} = y, \quad \frac{d^2x}{dt^2} = \frac{d}{dt}\left(\frac{dx}{dt}\right) = \frac{dy}{dt} = \frac{dy}{dx}\frac{dx}{dt} = y\frac{dy}{dx}. \tag{4.5.73}$$

This means that Eq. (4.5.72) can be written as

$$y\frac{dy}{dx} + g(x) = 0. \tag{4.5.74}$$

Integrating once gives the energy function

$$E(x,y) = \frac{y^2}{2} + V(x) = \text{constant}. \tag{4.5.75}$$

Equation (4.5.72) can also be expressed in system form, i.e.,

$$\frac{dx}{dt} = y, \quad \frac{dy}{dt} = -g(x), \tag{4.5.76}$$

with the trajectories in phase-space determined by the solutions to

$$\frac{dy}{dx} = -\frac{g(x)}{y}. \tag{4.5.77}$$

The fixed-points are located at $(\bar{x}, \bar{y})$, where the $\bar{x}$ are the real zeros of

$$g(\bar{x}) = 0, \quad \{\bar{x}^{(i)} : i = 1, 2, \ldots, I\}, \tag{4.5.78}$$

and $\bar{y} = 0$. This result follows directly from Eq. (4.5.76) and shows that for conservative systems all of the fixed-points lie on the x-axis. From a physics perspective, the fixed-point $(\bar{x}, 0)$ represents a state of equilibrium of the system for which the position $\bar{x}$ is constant and the velocity is zero.

Inspection of Eq. (4.5.77) also shows that the nullclines are given by the expressions

$$\begin{cases} x\text{-nullcline}: \dfrac{dy}{dx} = \infty \text{ along the } x\text{-axis}, \\ y\text{-nullcline}: \dfrac{dy}{dx} = 0 \text{ along the curves } x = \bar{x}^{(i)}(i = 1, 2, \ldots, I). \end{cases} \tag{4.5.79}$$

A major consequence of these results is that for conservative systems, whenever trajectories cross the x-axis, they must do so with a vertical tangent.

The local stability of each fixed-point can be determined by linearizing Eq. (4.5.76) at the fixed-point. Let

$$x(t) = \bar{x}^{(i)} + \alpha(t), \qquad y(t) = 0 + \beta(t); \qquad (4.5.80)$$

then

$$\frac{d\alpha}{dt} = \beta, \qquad (4.5.81)$$

and

$$\frac{d\beta}{dt} = -g[\bar{x}^{(i)} + \alpha],$$
$$= -g[\bar{x}^{(i)}] - R_i\alpha + O(\alpha^2), \qquad (4.5.82)$$

where

$$R_i \equiv \frac{dg}{dx}\Big|_{x=\bar{x}^{(i)}}. \qquad (4.5.83)$$

Thus, the linear approximation equations at the fixed-point $(\bar{x}^{(i)}, 0)$ are

$$\frac{d\alpha}{dt} = \beta, \qquad \frac{d\beta}{dt} = -R_i\alpha, \qquad (4.5.84)$$

or

$$\frac{d}{dt}\begin{pmatrix} \alpha \\ \beta \end{pmatrix} = \begin{pmatrix} 0 & 1 \\ -R_i & 0 \end{pmatrix}\begin{pmatrix} \alpha \\ \beta \end{pmatrix}. \qquad (4.5.85)$$

The corresponding eigenvalues are obtained from

$$\det(A - \lambda I) = \det\begin{pmatrix} -\lambda & 1 \\ -R_i & -\lambda \end{pmatrix} = \lambda^2 + R_i = 0, \qquad (4.5.86)$$

and are

$$\lambda_1 = \sqrt{-R_i}, \qquad \lambda_2 = -\sqrt{-R_i}. \qquad (4.5.87)$$

Therefore, if $R_i < 0$, then $\lambda_2 < 0 < \lambda_1$, and the fixed-point is a saddle point; if $R_i > 0$, then $\lambda_1 = \lambda_2^* = -\sqrt{|R_i|}$, and the fixed-point is a center. These are the only possibilities for the types of fixed-points for conservative systems: they are either saddles or centers.

As we shall now show in several examples, particular conservative systems can possess a large number of symmetry properties.

Consider a system modeled by an odd power-law "force," i.e.,

$$g(x) = x^{2n+1}, \qquad (n = 0, 1, 2, \dots), \qquad (4.5.88)$$

and

$$\frac{dx}{dt} = y, \qquad \frac{dy}{dt} = -x^{2n+1}. \qquad (4.5.89)$$

The only fixed-point is located at $(\bar{x}, \bar{y}) = (0, 0)$. However, for this case

$$\frac{dg}{dx}\bigg|_{x=0} = (2n + 1)x^{2n}\big|_{x=0} = 0, \tag{4.5.90}$$

and our test to determine whether $(0, 0)$ is a center or saddle point fails. However, since this is a conservative system, it has the following first-integral (see Eq. (4.5.75))

$$E(x, y) = \frac{y^2}{2} + \frac{x^{2n+2}}{(2n + 2)} = C, \tag{4.5.91}$$

where the constant C is determined by the initial conditions

$$C = \frac{y_0^2}{2} + \frac{x_0^{2n+2}}{(2n + 2)} > 0, \tag{4.5.92}$$

and, as indicated, is positive if either $x_0 \neq 0$ or $y_0 \neq 0$; C can only be equal to zero at the fixed-point, i.e.,

$$E(0, 0) = C = 0. \tag{4.5.93}$$

The implicitly defined trajectory, $y = y(x, C)$, as given in Eq. (4.5.91), corresponds to a single closed curve for each $C > 0$. Consequently, all the solutions for this system are periodic.

The fact that all the curves in the phase plane for Eq. (4.5.89) are closed can also be established by a geometrical proof. We begin by writing down the first-order differential equation whose solutions give the trajectories in the phase plane; it is

$$\frac{dy}{dx} = -\frac{x^{2n+1}}{y}. \tag{4.5.94}$$

This equation is invariant under the following transformations

$$T_x : x \to x, \quad y \to -y \quad \text{(reflection through the x-axis)}, \tag{4.5.95}$$

$$T_y : x \to -x, \quad y \to y \quad \text{(reflection through the y-axis)}, \tag{4.5.96}$$

$$T_i : x \to -x, \quad y \to -y \quad \text{(inversion through the origin)}. \tag{4.5.97}$$

The two nullclines are

$$\begin{cases} x\text{-nullcline} : \dfrac{dy}{dx} = \infty \text{ along the x-axis}, \\ y\text{-nullcline} : \dfrac{dy}{dx} = 0 \text{ along the y-axis}. \end{cases} \tag{4.5.98}$$

The nullclines coincide with the coordinate axes and divide the phase plane into four domains in which the derivative, dy/dx, has a fixed sign. This is shown in Figure 4.5.7a. Let an arbitrary point be selected on the positive y-axis. The trajectory that passes through this point A will, in the first quadrant, have the form shown in Figure 4.5.7c, i.e., it starts with zero slope at A, has a negative increasing slope in the first-quadrant, and intersects the x-axis with infinite slope at B. The application of the transformation T_x gives the curve in (d). The further application of the transformation, T_y, gives the figure shown in (e) and this allows us to conclude that all the trajectories for Eq. (4.5.94) are closed. Hence, all solutions to Eq. (4.5.89) are periodic.

This type of analysis can also be applied to the equation

$$\frac{d^2x}{dt^2} + x^{1/3} = 0, \tag{4.5.99}$$

which has the system equations

$$\frac{dx}{dt} = y, \qquad \frac{dy}{dt} = -x^{1/3}, \tag{4.5.100}$$

and

$$\frac{dy}{dx} = -\frac{x^{1/3}}{y}. \tag{4.5.101}$$

Equation (4.5.100) can be generalized to the form

$$\begin{cases} \dfrac{d^2x}{dt^2} + x^{\frac{2n+1}{2m+1}} = 0, \\ (n = 0, 1, 2, \dots) \text{ and } (m = 0, 1, 2, \dots). \end{cases} \tag{4.5.102}$$

It should be noted that while the equation

$$\frac{d^2x}{dt^2} + x^2 = 0, \tag{4.5.103}$$

corresponds to a conservative system, i.e., it has the first-integral

$$E(x, y) = \frac{y^2}{2} + \frac{x^3}{3} = C, \tag{4.5.104}$$

none of its solutions are periodic.

Finally, we end with an examination of the conservative, nonlinear system

$$\frac{d^2x}{dt^2} - x + x^3 = 0, \tag{4.5.105}$$

which has the system equations

$$\frac{dx}{dt} = y, \qquad \frac{dy}{dt} = x - x^3, \tag{4.5.106}$$

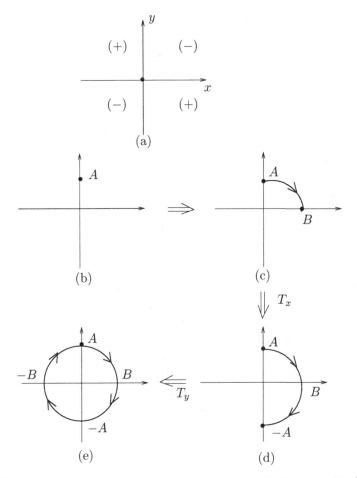

Fig. 4.5.7   Application of a geometric analysis to show that all solutions to Eq. (4.5.89) are periodic.

and

$$\frac{dy}{dx} = \frac{x(1 - x^2)}{y}.$$

(4.5.107)

This system has three fixed-points located at

$$(\bar{x}, \bar{y}) : (-1, 0), (0, 0), (1, 0),$$

(4.5.108)

and the first-integral is

$$E(x, y) = \frac{y^2}{2} - \frac{x^2}{2} + \frac{x^4}{4} = C.$$

(4.5.109)

Since $g(x) = -x + x^3$, it follows that

$$\frac{dg}{dx} = -1 + 3x^2, \qquad R_i \equiv \frac{dg}{dx}\Big|_{x=\bar{x}^{(i)}}, \qquad (4.5.110)$$

and

$$R_1 = 3, \qquad R_2 = -1, \qquad R_3 = 3. \qquad (4.5.111)$$

Since $R_1 = R_3 > 0$, the fixed-points $(-1, 0)$ and $(1, 0)$ are centers, while that at $(0, 0)$ is a saddle point.

This system has the following nullclines

$$\begin{cases} x\text{-nullcline} : \dfrac{dy}{dx} = \infty \text{ along the } x\text{-axis,} \\ y\text{-nullcline} : \dfrac{dy}{dx} = 0 \text{ along } x = -1, x = 0, , x = -1. \end{cases} \qquad (4.5.112)$$

Observe that this equation, see Eqs. (4.5.107), is invariant under the set of transformations given by Eqs. (4.5.95), (4.5.96) and (4.5.97). Figure 4.5.8 where the fixed-points are indicated by the large dots, gives a sampling of trajectories. The figure-eight shaped trajectory separates the phase plane into three regions. Inside each of the lobes, the trajectories encircle a fixed-point. For example, the left lobe contains the fixed-point $(-1, 0)$, while the right lobe contains $(1, 0)$. Note that the two lobes both begin at the saddle point $(0, 0)$ and return to it. Thus, within these lobes, the fixed-points $(-1, 0)$ and $(1, 0)$ are centers. The figure-eight trajectory is a closed *homoclinic orbit*. A simple calculation shows that the homoclinic orbit or trajectory corresponds to $C = 0$ in Eq. (4.5.109). Also, for

$$-\frac{1}{4} < C < 0, \qquad (4.5.113)$$

the periodic, closed trajectories are confined to the interior of one of the lobes. Thus, for a given value of C, in Eq. (4.5.113), there exists two periodic motions. For $C > 0$, the closed trajectories enclose both fixed-points. The general conclusion is that all solutions to Eq. (4.5.105) are periodic.

4.5.6   *Example F*

Most conservative systems interact with their environments and one of the consequences of this interaction is that the systems become dissipative, i.e., they lose energy. Under this condition, the full system, conservative system plus environmental influences, is no longer conservative. The simplest

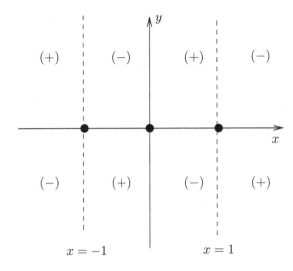

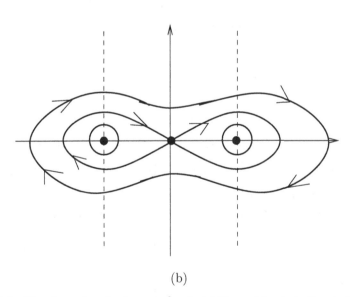

(b)

Fig. 4.5.8 The phase plane for $\ddot{x} - x + x^3 = 0$. (a) Signs of the derivatives in the eight regions whose boundaries are the nullclines. (b) Typical trajectories.

form of dissipation corresponds to adding a linear, velocity dependent term
to Eq. (4.5.72), i.e.,

$$\frac{d^2x}{dt^2} + g(x) = -2\epsilon \frac{dx}{dt}, \qquad \epsilon > 0, \qquad (4.5.114)$$

where ϵ is a positive damping coefficient. Assume that a fixed-point exists
at $(\bar{x}, \bar{y}) = (0,0)$ for the conservative system. Further, assume that ϵ is
small, i.e.,

$$0 < \epsilon \ll 1. \qquad (4.5.115)$$

Then the linearization at $(0,0)$ for Eq. (4.5.114) is

$$\frac{d}{dt}\begin{pmatrix} \alpha \\ \beta \end{pmatrix} = \begin{pmatrix} 0 & 1 \\ -R & -2\epsilon \end{pmatrix}\begin{pmatrix} \alpha \\ \beta \end{pmatrix}. \qquad (4.5.116)$$

To obtain this, we used

$$g(x) = g(0) + R\alpha + O(\alpha^2) = R\alpha, \qquad R \equiv \frac{dg}{dx}\bigg|_{x=0}, \qquad (4.5.117)$$

$$x(t) = 0 + \alpha(t), \qquad y(t) = 0 + \beta(t). \qquad (4.5.118)$$

The eigenvalues at $(0,0)$ are determined by

$$\det(A - \lambda I) = \det\begin{pmatrix} -\lambda & 1 \\ -R & -2\epsilon - \lambda \end{pmatrix} = \lambda^2 + 2\epsilon\lambda + R = 0, \qquad (4.5.119)$$

and are given by

$$\lambda_1, \lambda_2 = -\epsilon \pm \sqrt{\epsilon^2 - R}. \qquad (4.5.120)$$

If $(0,0)$ is a **center** for the conservative system, then $R > 0$ and

$$\lambda_1 = \lambda_2^* = -\epsilon + i\sqrt{R - \epsilon^2}, \qquad (4.5.121)$$

where, in addition to having $0 < \epsilon \ll 1$, we also assume that $0 < \epsilon^2 < R$. Thus, the center is transformed into a stable spiral point within the
inclusion of dissipation.

If $(0,0)$ is a saddle point, then $R < 0$, and both eigenvalues are real
with

$$\begin{cases} \lambda_1 = -\epsilon + \sqrt{|R| + \epsilon^2}, \qquad \lambda_2 = -\epsilon - \sqrt{|R| + \epsilon^2}, \\ \qquad\qquad \lambda_2 < 0 < \lambda_1. \end{cases} \qquad (4.5.122)$$

Therefore, we conclude that under the condition of dissipation, the saddle
points of the conservative system remain saddle points.

Figure 4.5.9 shows the effort of dissipation on the phase plane trajectories for the nonlinear oscillator

$$\frac{d^2x}{dt^2} + x^{1/3} = -2\epsilon \frac{dx}{dt}, \qquad \epsilon > 0. \tag{4.5.123}$$

Note that for this case the above analysis cannot be applied since $g(x) = x^{1/3}$ does not have a derivative at $x = 0$. However, the following procedure does allow us to reach the correct conclusion. To begin, observe that when $\epsilon = 0$, the resulting equation

$$\frac{d^2x}{dt^2} + x^{1/3} = 0, \tag{4.5.124}$$

has a first-integral

$$E(x,y) = \frac{y^2}{2} + \left(\frac{3}{4}\right) x^{4/3} = \text{constant}, \tag{4.5.125}$$

and from this fact we can conclude that all its trajectories in the phase plane are closed curves and thus all solutions are periodic. Also, note that the system equations are

$$\frac{dx}{dt} = y, \qquad \frac{dy}{dt} = -x^{1/3} \qquad \frac{dy}{dx} = -\frac{x^{1/3}}{y}, \tag{4.5.126}$$

and, further, the second equation is invariant under the set of transformations given by Eqs. (4.5.95), (4.5.96), and (4.5.97). This means that we can apply geometrical procedures to show that all the phase plane trajectories for Eq. (4.5.124) are closed and, therefore, periodic.

Now define the function $V(x,y)$ for the system of equations with dissipation to be

$$V(x,y) = \frac{y^2}{2} + \left(\frac{3}{4}\right) x^{4/3}, \tag{4.5.127}$$

where from Eq. (4.5.123)

$$\frac{dx}{dt} = y, \qquad \frac{dy}{dt} = -x^{1/3} - 2\epsilon y. \tag{4.5.128}$$

While $V(x,y)$ coincides with the energy function, when $\epsilon = 0$, for the full dissipative system, as given by Eq. (4.5.123), it is not related to the energy. In fact, for $\epsilon > 0$, no energy function exists. Taking the time derivative of $V(x,y)$ gives

$$\frac{dV}{dt} = y \frac{dy}{dt} + x^{1/3} \frac{dx}{dt}. \tag{4.5.129}$$

Now replace the two derivatives by the expressions in Eq. (4.5.128) to obtain

$$\frac{dV}{dt} = y(-x^{1/3} - 2\epsilon y) + x^{1/3}(y) \qquad (4.5.130)$$

or

$$\frac{dV}{dt} = -2\epsilon y^2. \qquad (4.5.131)$$

Thus, the function $V(x, y)$ will be decreasing in general unless both x and y have values corresponding to a fixed-point [8]. This means that

$$\lim_{t \to \infty} V(x(t), y(t)) = 0, \qquad (4.5.132)$$

for $x(t)$ and $y(t)$ solutions to Eq. (4.5.128) and we can conclude that

$$\lim_{t \to \infty} \begin{pmatrix} x(t) \\ y(t) \end{pmatrix} = \begin{pmatrix} 0 \\ 0 \end{pmatrix}. \qquad (4.5.133)$$

Thus, the addition of dissipation causes $(0, 0)$ to change from a center to a stable node or spiral.

4.5.7   *Example G*

A famous and important model from mathematical ecology is one derived by Lotka and Volterra for the interaction between predator and prey populations [9]. The basic assumptions of the model are listed below:

(i) The two populations consist of prey, denoted by $x(t)$, and predators, denoted by $y(t)$.

(ii) In the absence of predators, the prey population has a growth rate proportional to itself, i.e.,

$$\frac{dx}{dt} = ax, \qquad a > 0. \qquad (4.5.134)$$

(iii) Assuming the predators have no other source of food except for the prey, in the absence of the prey the predator population will decline as

$$\frac{dy}{dt} = -cy, \qquad c > 0. \qquad (4.5.135)$$

(iv) The presence of both prey and predators will have a negative effect on the prey, but a positive influence on the predators. It is assumed that these terms are proportional to the number of contacts per unit time between the predators and prey, and is taken to be $(-bxy)$ for the prey and $(+dxy)$ for the predators.

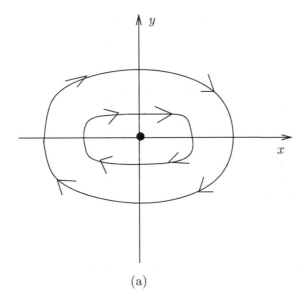

(a)

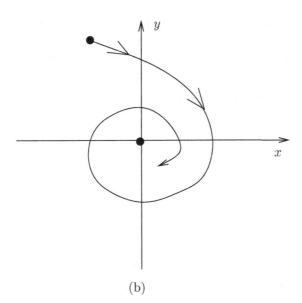

(b)

Fig. 4.5.9   The phase plane for (a) $\ddot{x} + x^{1/3} = 0$, and (b) $\ddot{x} + x^{1/3} = -2\epsilon\dot{x}$.

Putting all of these items together leads to the so-called Lotka-Volterra equations

$$\frac{dx}{dt} = ax - bxy, \qquad \frac{dy}{dt} = -cy + dxy, \qquad (4.5.136)$$

where all four parameters (a, b, c, d) are positive.

First, note that there are two fixed-points

$$(\bar{x}^{(1)}, \bar{y}^{(1)}) = (0, 0), \qquad (\bar{x}^{(2)}, \bar{y}^{(2)}) = \left(\frac{c}{d}, \frac{a}{b}\right). \qquad (4.5.137)$$

The first fixed-point corresponds to the case where no population is present and is of no direct interest. The second fixed-point is where both prey and predator can coexist. Also, note that if there are no predators, then $y(t) = 0$ and

$$\frac{dx}{dt} = ax \implies x(t) = x_0 e^{at}, \qquad (4.5.138)$$

and the prey population rapidly increases. Likewise, if predators only are present, then $x(t) = 0$, and

$$\frac{dy}{dt} = -cy \implies y(t) = y_0 e^{-ct}, \qquad (4.5.139)$$

and the predator population decreases to zero. In the (x, y) phase plane, these results show that both the x- and y-axes are trajectories of Eq. (4.5.136) and every solution $(x(t), y(t))$ that begins in the first quadrant remains there for all $t > 0$, i.e.,

$$x(0) > 0, \quad y(0) > 0 \implies x(t) > 0, \quad y(t) > 0, \quad t > 0. \qquad (4.5.140)$$

The trajectories in the phase plane are the solutions to the following first-order differential equation

$$\frac{dy}{dx} = \frac{(-c + dx)y}{(a - by)x}. \qquad (4.5.141)$$

The corresponding nullclines are

$$\begin{cases} \dfrac{dy}{dx} = 0 : \text{along the } x\text{-axis and along the curve } x = \dfrac{c}{d}; \\ \dfrac{dy}{dx} = \infty : \text{along the } y\text{-axis and along the curve } y_\infty(x) = \dfrac{a}{b}. \end{cases} \qquad (4.5.142)$$

Figure 4.5.10a illustrates the location of the nullclines and regions of constant sign for the derivative dy/dx.

The trajectory equation is separable and can be written as

$$\left(\frac{a - by}{y}\right) dy + \left(\frac{c - dx}{x}\right) dx = 0, \qquad (4.5.143)$$

with the solution

$$a\ln(y) - by + c\ln(x) - dx = p_1, \tag{4.5.144}$$

where p_1 is the integration constant. The elimination of the logarithmic expressions, by taking the exponentials of both sides of this equation gives

$$\left(\frac{y^a}{e^{by}}\right)\left(\frac{x^c}{e^{dx}}\right) = P, \tag{4.5.145}$$

where P is a constant. This equation is a first-integral for the system and determines the form of the trajectories in the (x, y) phase plane. The following two results can be shown to be true [9]:

(A) The expression given by Eq. (4.5.145) corresponds to a family of closed curves for $x_0 > 0$ and $y_0 > 0$. Note that the constant P is evaluated to be

$$P = \left(\frac{y_0^a}{e^{by_0}}\right)\left(\frac{x_0^c}{e^{dx_0}}\right). \tag{4.5.146}$$

Since all the trajectories are closed curves, in the first quadrant, it follows that all of the solutions are periodic.

(B) While we do not know the actual value of the period for these periodic solutions, a very interesting result can be found.

Let $x(t)$ and $y(t)$ be a periodic solution of Eq. (4.5.136), having period $T > 0$. Define the average values of $x(t)$ and $y(t)$ as

$$\bar{x} \equiv \left(\frac{1}{T}\right)\int_0^T x(t)dt, \qquad \bar{y} \equiv \left(\frac{1}{T}\right)\int_0^T y(t)dt. \tag{4.5.147}$$

Then, it follows that

$$\bar{x} = \frac{c}{d}, \qquad \bar{y} = \frac{a}{b}. \tag{4.5.148}$$

This means that the average values of $x(t)$ and $y(t)$ are just their values at the fixed-point.

While no simple closed form expression for the period of the Lotka-Volterra equations exists, we can obtain a value for motions close to the fixed-point by use of the linearized equations. Let

$$x(t) = \left(\frac{c}{d}\right) + \alpha(t), \qquad y(t) = \left(\frac{b}{a}\right) + \beta(t), \tag{4.5.149}$$

and substitute them into Eq. (4.5.136). If only linear terms are retained, we find

$$\frac{d\alpha}{dt} = -\left(\frac{bc}{d}\right)\beta, \qquad \frac{d\beta}{dt} = -\left(\frac{ad}{b}\right)\alpha, \tag{4.5.150}$$

or

$$\left(\frac{d^2}{dt^2} + ac\right)\begin{pmatrix} \alpha \\ \beta \end{pmatrix} = 0. \tag{4.5.151}$$

The general solution to Eq. (4.5.151) is

$$\alpha(t) = D\cos\left(\sqrt{ac}\,t + \phi\right) \tag{4.5.152}$$

where (D, ϕ) are arbitrary constants and the period is

$$T = \frac{2\pi}{\sqrt{ac}}. \tag{4.5.153}$$

Use of the first of Eq. (4.5.150) gives

$$\beta(t) = \left(\frac{d}{b}\right)\left(\sqrt{\frac{a}{c}}\right)D\sin\left(\sqrt{ac}\,t + \phi\right), \tag{4.5.154}$$

and for initial conditions in a small neighborhood of the fixed-point, the solution is

$$\begin{cases} x(t) = \left(\frac{c}{d}\right) + D\cos\left(\sqrt{ac}\,t + \phi\right), \\ y(t) = \left(\frac{a}{b}\right) + \left(\frac{d}{b}\right)\left(\sqrt{\frac{a}{c}}\right)D\sin\left(\sqrt{ac}\,t + \phi\right). \end{cases} \tag{4.5.155}$$

The corresponding trajectory equation is an ellipse

$$\frac{\left(x - \frac{c}{d}\right)^2}{D^2} + \frac{\left(y - \frac{a}{b}\right)^2}{\left(\frac{ad^2}{b^2c}\right)D^2} = 1. \tag{4.5.156}$$

4.5.8   *Example H*

A general problem that can arise is one in which a first-integral, $I(x,y)$, exists for the system of equations

$$\frac{dx}{dt} = P(x,y), \qquad \frac{dy}{dt} = Q(x,y), \tag{4.5.157}$$

and knowledge of the stability of the fixed-point is desired. We now discuss briefly several of the issues related to this problem.

Let $I(x,y)$ be a first-integral for Eq. (4.5.157). It is determined as a solution to the equation

$$\frac{dy}{dx} = \frac{Q(x,y)}{P(x,y)}, \tag{4.5.158}$$

which can be written as

$$I(x,y) = C = \text{constant}, \tag{4.5.159}$$

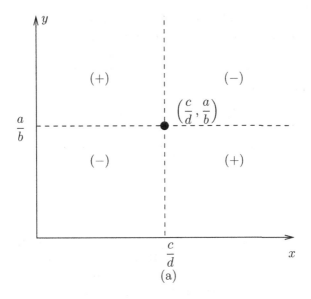

(a)

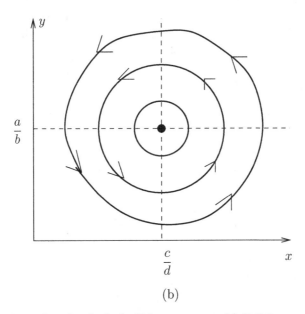

(b)

Fig. 4.5.10   Phase plane for the Lotka-Volterra system. (a) Nullclines and regions of constant sign for the derivative $dy\,dx$. (b) Schematic representation of typical trajectories.

where for the initial conditions, $x(0) = x_0$ and $y(0) = y_0$, $C = I(x_0, y_0)$.
Note that $I(x, y)$ can be called a **constant of the motion** of the system
since for any given solution $(x(t), y(t))$ its value is constant. Another way of
stating this result is to assume that $I(x, y)$ and its first partial derivatives
are continuous. Then

$$\frac{dI(x, y)}{dt} = \frac{\partial I}{\partial x} \frac{dx}{dt} + \frac{\partial I}{\partial x} \frac{dy}{dt}, \qquad (4.5.160)$$

and this implies that $I(x, y)$ is a constant of the motion for Eq. (4.5.157) if
and only if

$$P(x, y) \frac{\partial I}{\partial x} + Q(x, y) \frac{\partial I}{\partial y} = 0. \qquad (4.5.161)$$

If a **level curve** of $I(x, y)$ is defined to be the curve in the (x, y) phase
plane given by Eq. (4.5.159) for a particular value of the constant C, then it
follows that the solutions of Eq. (4.5.157) lie on the level curves of $I(x, y)$.

Let $\{\bar{x}^{(i)}, \bar{y}^{(i)} : i = 1, 2, \ldots, I\}$ be the fixed-points of the system defined
by Eq. (4.5.157). Let a first-integral, $I(x, y)$, exist for this system. The
question that can now be asked is: Can the stability of the fixed-points be
determined from a knowledge of $I(x, y)$? We will see shortly that the answer
to this question is yes. However, it should be obvious that the fixed-points
can only be centers and saddle points for this class of problems.

Second Derivative Test

Let $I(x, y)$ have continuous second partial derivatives. Let $(\bar{x}, \bar{y})$ be a fixed-
point for $I(x, y)$, i.e.,

$$\left.\frac{\partial I}{\partial x}\right|_{\substack{x=\bar{x}\\y=\bar{y}}} = 0, \qquad \left.\frac{\partial I}{\partial y}\right|_{\substack{x=\bar{x}\\y=\bar{y}}} = 0. \qquad (4.5.162)$$

Define

$$A \equiv \left.\frac{\partial^2 I}{\partial x^2}\right|_{\substack{x=\bar{x}\\y=\bar{y}}} = 0, \quad B \equiv \left.\frac{\partial^2 I}{\partial x \partial y}\right|_{\substack{x=\bar{x}\\y=\bar{y}}} = 0, \quad C = \left.\frac{\partial^2 I}{\partial y^2}\right|_{\substack{x=\bar{x}\\y=\bar{y}}} = 0. \quad (4.5.163)$$

Then:

(i) if $A > 0$ and $AC - B^2 > 0$, $(\bar{x}, \bar{y})$ is a local minimum;
(ii) if $A < 0$ and $AC - B^2 > 0$, $(\bar{x}, \bar{y})$ is a local maximum;
(iii) $AC - B^2 < 0$, then $(\bar{x}, \bar{y})$ is a saddle point;
(iv) $AC - B^2 = 0$, no conclusion can be reached.

Case (i) corresponds to the fixed-point, $(\bar{x}, \bar{y})$, surrounded by closed trajectories. This implies that all neighboring solutions oscillate about a stable center. For case (iii), the fixed-point is unstable and is a saddle point.

Let us now apply the above analysis to the Lotka-Volterra system discussed in the previous section. We write the first-integral in the form

$$I(x, y) = by - a\ln(y) + dx - c\ln(x). \qquad (4.5.164)$$

Therefore:

$$\frac{\partial I}{\partial x} = d - \frac{c}{x}, \qquad \frac{\partial I}{\partial y} = b - \frac{a}{y}, \qquad (4.5.165)$$

$$\frac{\partial^2 I}{\partial x^2} = \frac{c}{x^2}, \qquad \frac{\partial^2 I}{\partial y^2} = \frac{a}{y^2}, \qquad \frac{\partial^2 I}{\partial x \partial y} = 0. \qquad (4.5.166)$$

At the fixed-point $(\bar{x}, \bar{y}) = (c/d, a/b)$, we have

$$A = \frac{d^2}{c}, \qquad B = 0, \qquad C = \frac{b^2}{a}, \qquad (4.5.167)$$

and

$$A > 0 \quad \text{with } AC - B = \frac{b^2 d^2}{ac} > 0. \qquad (4.5.168)$$

Our conclusion is that the fixed-point is a center which is stable.

Note that while the system

$$\frac{dx}{dt} = y, \qquad \frac{dy}{dt} = -x^{1/3}, \qquad (4.5.169)$$

has a first-integral

$$I(x, y) = \frac{y^2}{2} + \left(\frac{3}{4}\right) x^{4/3}, \qquad (4.5.170)$$

the above analysis cannot be applied since the second partial derivatives do not all exist. However, the fact that the level curves of $I(x, y)$ are all closed allows us to conclude that the fixed-point at $(0, 0)$ is a (nonlinear) center.

The system

$$\frac{dx}{dt} = y, \qquad \frac{dy}{dt} = -x^3, \qquad (4.5.171)$$

has the first-integral

$$I(x, y) = \frac{y^2}{2} + \frac{x^4}{4}, \qquad (4.5.172)$$

and all its derivatives exist, i.e.,

$$\begin{cases} \dfrac{\partial I}{\partial x} = x^3, & \dfrac{\partial I}{\partial y} = y, \\[2mm] \dfrac{\partial^2 I}{\partial x^2} = 3x^2, & \dfrac{\partial^2 I}{\partial y^2} = 1, & \dfrac{\partial^2 I}{\partial x \partial y} = 0. \end{cases} \qquad (4.5.173)$$

Therefore, at the fixed-point $(\bar{x}, \bar{y}) = (0,0)$

$$A = 0, \quad B = 0, \quad C = 1 \Longrightarrow AC - B^2 = 0, \qquad (4.5.174)$$

and no conclusion can be reached regarding the fixed-point using this method. Again, because the level curves of $I(x,y)$ are closed, it can be concluded that all solutions are periodic and the fixed-point at $(0,0)$ is a center.

4.6 Bifurcations

If the differential equations modeling a particular system depend on a real parameter, λ, then it is expected that the solutions will also have a dependence on the parameter, i.e.,

$$x = x(t, \lambda), \qquad y = y(t, \lambda). \qquad (4.6.1)$$

A value λ^* of the real parameter λ such that the qualitative properties of the trajectories in phase space change their fundamental character, as λ passes through λ^*, is called a **bifurcation point**.

Figures 4.6.1, 4.6.2, and 4.6.3 illustrate this phenomena. First, Figure 4.6.1, shows the behavior of solutions to the differential equation

$$\frac{dx}{dt} = \lambda x, \qquad (4.6.2)$$

for various values of the parameter λ. For all three cases studied, the fixed-point is $\bar{x}(\lambda) = 0$. While the fixed-point does not depend on λ, its stability properties do. The three cases are:

(i) $\lambda < 0$: All solutions monotonically decrease in value to zero. The fixed-point is stable.

(ii) $\lambda = 0$: All solutions are constant, i.e.,

$$x(t) = x(0). \qquad (4.6.3)$$

These solutions have neutral stability; this means that changing the initial condition changes the solution, but only by a finite amount if the change in the initial condition was finite.

(iii) $\lambda > 0$: All solutions are monotonically increasing and become unbounded, i.e.,

$$\lim_{t \to \infty} x(t, \lambda) = \infty, \qquad \lambda > 0. \tag{4.6.4}$$

For this equation, it is clear that $\lambda^* = 0$.

The second equation presented is

$$\frac{dx}{dt} = \lambda - x^2. \tag{4.6.5}$$

There are again three cases to consider as shown in Figure 4.6.2

(i) $\lambda < 0$: All solutions monotonically decrease to negative values. There are no (real) fixed-points.

(ii) $\lambda = 0$: There is a fixed-point at $\bar{x}(0) = 0$ and it is semi-stable, i.e., trajectories with $x(0) > 0$ approach $\bar{x}(0) = 0$ from above, while trajectories with $x(0) < 1$ move away from $\bar{x}(0) = 0$.

(iii) $\lambda > 0$: For this situation, two fixed-points now appear; Abraham and Shaw [11] call this the **blue sky bifurcation**. The last diagram in Figure 4.6.2 is called a bifurcation diagram; it plots the fixed-points $\bar{x}^{(1)}(x) = \sqrt{\lambda}$ and $\bar{x}^{(2)} = -\sqrt{\lambda}$ as functions of λ. Note that $\bar{x}^{(1)}(\lambda)$ and $\bar{x}^{(2)}(\lambda)$ correspond, respectively, to the stable and unstable branches. Inspection of these diagrams shows that the bifurcation points is $\lambda^* = 0$.

The third example is the damped harmonic oscillator equation

$$\frac{d^2x}{dt^2} + 2\lambda \frac{dx}{dt} + x = 0. \tag{4.6.6}$$

A study of Figure 4.6.3 shows that the single fixed-point at $\bar{x}(\lambda) = 0$ does not depend on λ; however, its stability is a function of λ. The bifurcation point is $\lambda^* = 0$. Note that for $\lambda < 0$, the solutions are unstable, while $\lambda > 0$ corresponds to stable solutions. The transition point, $\lambda = \lambda^* = 0$, places the system in a state of neutral stability, i.e., all the solutions are periodic and the fixed-point is a center.

4.6.1 *Hopf-Bifurcations*

Bifurcation theory can generally be applied to the analysis of systems depending on parameters that change their values. Most work, to date, has centered on bifurcations depending on just a single parameter. The three books listed in the bibliography to this chapter, under the heading "bifurcations," provide good introductions to this topic on a variety of levels. In particular, the texts of Beltrami and Strogatz, and the chapter of Odell

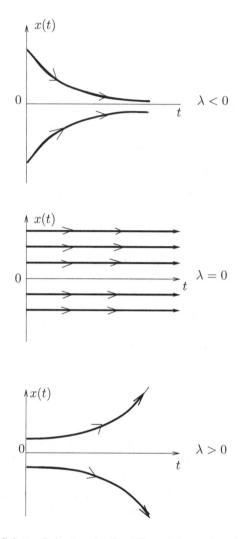

Fig. 4.6.1   Solution behaviors for the differential equations $\dot{x} = \lambda x$.

in the edited volume of Segel give applications to systems in the physical, biological, and engineering sciences.

One of the most useful applications of bifurcation theory is to systems for which limit-cycles may exist if a control parameter has values in a certain range. We now state a version of the Hopf-bifurcation theorem for planar dynamic systems and illustrate its use in the next section.

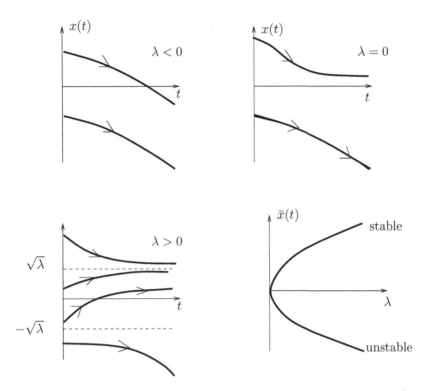

Fig. 4.6.2   Solution behaviors for the differential equation $\dot{x} = \lambda - x^2$. $\bar{x}(\lambda)$ corresponds to the fixed-points of the equation.

Consider a two-dimensional dynamical system modeled by the differential equations

$$\frac{dx}{dt} = P(x, y, \lambda), \qquad \frac{dy}{dt} = Q(x, t, \lambda), \qquad (4.6.7)$$

where λ is a control parameter. (In the following presentation, we denote the control parameter by λ and eigenvalues by the symbol "r".) Let Eq. (4.6.7) have an isolated fixed-point at $(\bar{x}(\lambda), \bar{y}(\lambda))$, where we have explicitly indicated that the fixed-point can depend on λ. For motions in a neighborhood of the fixed-point, we have

$$\begin{cases} x(t, \lambda) = \bar{x}(\lambda) + \alpha(t, \lambda), \\ y(t, \lambda) = \bar{y}(\lambda) + \beta(t, \lambda). \end{cases} \qquad (4.6.8)$$

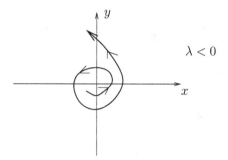

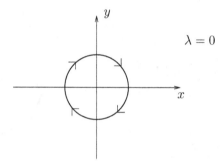

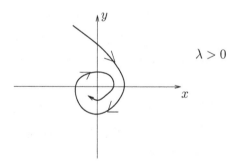

Fig. 4.6.3   Solution behaviors for the damped linear oscillator, $\ddot{x} + 2\lambda\dot{x} + x = 0$.

Substitution of Eq. (4.6.8) into Eq. (4.6.7) and retaining only the linear terms gives

$$\frac{d}{dt}\begin{pmatrix} \alpha \\ \beta \end{pmatrix} = \begin{pmatrix} \overline{\dfrac{\partial Q}{\partial x}} & \overline{\dfrac{\partial Q}{\partial y}} \\[2mm] \overline{\dfrac{\partial P}{\partial x}} & \overline{\dfrac{\partial P}{\partial y}} \end{pmatrix}\begin{pmatrix} \alpha \\ \beta \end{pmatrix}, \qquad (4.6.9)$$

where the over-bars indicate the evaluation of each term at the fixed-point. Now denote the eigenvalues of the matrix, in Eq. (4.6.9), by $r_1(\lambda)$ and $r_2(\lambda)$. Assume that the parameter λ is defined such that in the range, $|\lambda| < \delta$, where δ is a constant, the eigenvalues are complex valued functions of λ and also have a first-derivative with respect to λ, i.e.,

$$r_1(\lambda) = r_2^*(\lambda) = \alpha(\lambda) + i\beta(\lambda). \qquad (4.6.10)$$

Under these conditions, the Hopf-bifurcation theorem, for planar dynamic systems can be stated.

Hopf-Bifurcation Theorem

Let the fixed-point $(\bar{x}(\lambda), \bar{y}(\lambda))$, be asymptotically stable for $\lambda < 0$, unstable for $\lambda > 0$, and let $\alpha(0) = 0$. If

$$\frac{d\alpha(\lambda)}{d\lambda}\bigg|_{\lambda=0} > 0, \qquad \beta(0) \neq 0, \qquad (4.6.11)$$

then for all sufficiently small $|\lambda|$, an isolated closed trajectory exists for either $\lambda > 0$ or $\lambda < 0$. If $(\bar{x}(0), \bar{y}(0))$ is locally asymptotically stable, then a stable limit-cycle exists about $(\bar{x}(\lambda), \bar{y}(\lambda))$ for sufficiently small $\lambda > 0$.

The following theorem is useful for showing that if any closed trajectories exist in the phase plane, they must enclose at least one fixed-point [12].

Theorem *Let the system*

$$\frac{dx}{dt} = P(x, y), \qquad \frac{dy}{dt} = Q(x, y) \qquad (4.6.12)$$

have a closed integral curve. Then this curve must enclose fixed-points whose indices sum to +1.

Comments:

(i) Without presenting the details, the index of the various types of fixed-points are [12]

Type of Fixed-Point	Index
node	+1
saddle	−1
center	+1.

(ii) This theorem implies there is always at least one fixed-point enclosed inside any closed curve in the phase plane.

(iii) The theorem also implies that if only one fixed-point is interior to a closed curve, then it cannot be a saddle point.

Examples illustrating this theorem are shown in Figure 4.5.7, where closed trajectories surround a single center; Figure 4.5.8, where neighboring closed curves surround the two separate centers, but other closed trajectories exist and encircle the two centers and a saddle point.

4.6.2   *Two Examples*

The standard form for the van der Pol differential equation is

$$\frac{d^2z}{dt^2} + z = \lambda(1 - z^2)\frac{dz}{dt}. \tag{4.6.13}$$

However, for the application of the Hopf-bifurcation theorem another form is more useful

$$\frac{d^2x}{dt^2} + x = (\lambda - x^2)\frac{dx}{dt}. \tag{4.6.14}$$

These two differential equations can be transformed into each other. In system form, we have

$$\frac{dx}{dt} = y, \quad \frac{dy}{dt} = -x + \lambda y - x^2 y. \tag{4.6.15}$$

The only fixed-point is $(\bar{x}, \bar{y}) = (0, 0)$ and the linearization matrix at this fixed-point is

$$A = \begin{pmatrix} 0 & 1 \\ -1 & \lambda \end{pmatrix}. \tag{4.6.16}$$

The eigenvalues are determined by

$$\det(A - rI) = \det \begin{pmatrix} -r & 1 \\ -1 & \lambda - r \end{pmatrix} = r^2 - \lambda r + 1 = 0, \tag{4.6.17}$$

and

$$r_1(\lambda) = r_2^*(\lambda) = \frac{\lambda}{2} + \left(\frac{\sqrt{4 - \lambda^2}}{2}\right)i, \tag{4.6.18}$$

with

$$\alpha(\lambda) = \frac{\lambda}{2}, \quad \beta(\lambda) = \frac{\sqrt{4 - \lambda^2}}{2}, \tag{4.6.19}$$

and

$$\alpha(0) = 0, \quad \frac{d\alpha(0)}{d\lambda} = \frac{1}{2} > 0, \quad \beta(0) = 1. \tag{4.6.20}$$

Let us now consider the behavior of the fixed-point for $\lambda = 0$. For this case, we have

$$\frac{d^2x}{dt^2} + x = -x^2 \frac{dx}{dt}. \tag{4.6.21}$$

The right-side term represents positive damping and we can conclude that all solutions have the property

$$\lim_{t \to \infty} \begin{pmatrix} x(t) \\ y(t) \end{pmatrix} = \begin{pmatrix} 0 \\ 0 \end{pmatrix}, \tag{4.6.22}$$

and thus the fixed-point, $(\bar{x}(0), \bar{y}(0))$, is asymptotically stable. Based on the Hopf-bifurcation theorem, our overall conclusion is that the van der Pol equation, as represented by Eq. (4.6.14), has a stable limit-cycle for small, positive values of λ. Further, the period of the oscillations is given approximately by

$$T \approx \frac{2\pi}{\beta(0)} = 2\pi. \tag{4.6.23}$$

Chapter 9 will present a variety of approximation methods to calculate the actual amplitude and period of these oscillations.

For a second example, consider the system

$$\frac{dx}{dt} = \lambda x - \omega y - (x^2 + y^2)(x + ay), \tag{4.6.24}$$

$$\frac{dy}{dt} = \omega x + \lambda y + (x^2 + y^2)(ax - y), \tag{4.6.25}$$

where ω and a are fixed parameters, with $a > 0$, and λ will be taken as the bifurcation parameter. The only fixed-point is $(\bar{x}, \bar{y}) = (0, 0)$ and it does not depend on λ or the other two parameters. The linearization matrix at the fixed-point is

$$A = \begin{pmatrix} \lambda & -\omega \\ \omega & \lambda \end{pmatrix}, \tag{4.6.26}$$

and its eigenvalues are determined by the equation

$$(\lambda - r)^2 + \omega^2 = 0. \tag{4.6.27}$$

They are

$$r_1(\lambda) = r_2^*(\lambda) = \lambda + i\omega, \tag{4.6.28}$$

with

$$\alpha(\lambda) = \lambda, \qquad \beta(\lambda) = \omega, \qquad (4.6.29)$$

and

$$\alpha(0) = 0, \qquad \frac{d\alpha(0)}{d\lambda} = 1, \qquad \beta(0) = \omega. \qquad (4.6.30)$$

Our conclusion is that the system of equations, given by Eqs. (4.6.24) and (4.6.25), have a stable limit-cycle.

In fact, we can actually obtain more information on this system. Using a transformation to polar coordinates

$$x(t) = r(t)\cos\theta(t), \qquad y(t) = r(t)\sin\theta(t), \qquad (4.6.31)$$

these two equations reduce to

$$\frac{dr}{dt} = (\lambda - r^2)r, \qquad \frac{d\theta}{dt} = \omega + ar^2. \qquad (4.6.32)$$

The "r" equation has fixed-points at

$$\bar{r}^{(1)}(\lambda) = 0, \qquad \bar{r}^{(2)}(\lambda) = \sqrt{\lambda}, \qquad (4.6.33)$$

which are, respectively, unstable and stable. The corresponding *stationary solutions* are

$$\bar{r}^{(1)}(\lambda) = 0 : x(t,\lambda) = 0, \quad y(t,\lambda) = 0$$

$$\bar{r}^{(2)}(\lambda) = \sqrt{\lambda} : \begin{cases} x(t,\lambda) = \sqrt{\lambda}\cos\left[(\omega + a\lambda)t + \phi_0\right], \\ y(t,\lambda) = \sqrt{\lambda}\sin\left[(\omega + a\lambda)t + \phi_0\right], \end{cases} \qquad (4.6.34)$$

where ϕ_0 is an arbitrary constant.

Note that the general exact solution for $r(t)$ and $\theta(t)$, as given by Eq. (4.6.32), can be easily calculated.

Problems

Section 4.2

4.2.1 Analyze the case

$$\frac{dx}{dt} = f(x)$$

where $f(\bar{x}) = 0$ has no real zeros.

4.2.2 Draw all the possible one-dimensional phase-space solution behaviors for three distinct fixed-points.

Section 4.3

4.3.1 The logistic differential equation, as given by Eq. (4.3.7), can be solved exactly. Find this solution and use it to check on the correctness of the conclusions reached in section 4.3.2 In particular, discuss the nature of the solution if $x(0) < 0$. Plot $x(t)$ vs t for $x(0) < 0$.

4.3.2 The differential equation, given by Eq. (4.3.12), can be solved exactly. To show this, multiply both sides by $2x$ and use the new variable $z = x^2$. Carry out this calculation and verify that the discussion presented in section 4.3.3 is correct.

4.3.3 Plot the fixed-points and typical trajectories for the example in section 4.3.4 for $-5 \leq x \leq 5$.

4.3.4 Analyze the phase-space and the trajectory (x vs t) space for the differential equation

$$\frac{dx}{dt} = x^2(1 - x).$$

4.3.5 Extend the method given in section 4.3.5 to find the general qualitative properties of the solutions to the differential equations [1; 2]

(i) $\dfrac{dx}{dt} = t^2 - x,$

(ii) $\dfrac{dx}{dt} = (\sin t)^2 - 3e^x.$

Section 4.4

4.4.1 The fixed-point, $(\bar{x}, \bar{y}) = (0, 0)$, for the system of equations

$$\frac{dx}{dt} = y, \qquad \frac{dy}{dt} = 2x - y$$

is a saddle point. Show that a small linear perturbation to this equation leaves the nature of the fixed-point unchanged.

4.4.2 The system of equations

$$\frac{dx}{dt} = y, \qquad \frac{dy}{dt} = -2x - 3y$$

has a stable node for its fixed-point at $(\bar{x}, \bar{y}) = (0, 0)$. Demonstrate that under small linear perturbations, the fixed-point remains a stable node.

Section 4.5

4.5.1 Solve Eq. (4.5.12) to obtain the result of Eq. (4.5.14).

4.5.2 Construct the general phase plane diagram for the trajectories of Eq. (4.5.48) for n-even and n-odd. Is the fixed-point at $(\bar{x}, \bar{y}) = (0, 0)$ always a center?

4.5.3 Rewrite $H(x, y)$ in Eq. (4.5.55) as
$$H(x, y) = K(y) + V(x),$$
where $K(y)$ is the kinetic energy. Plot $K(y)$ vs y and compare with the case where $K(y) = y^2/2$.

4.5.4 Does the equation
$$\frac{d^2x}{dt^2} + x^{2/3} = 0$$
have periodic solutions? Explain your answer.

4.5.5 Construct the full phase plane for the Lotka-Volterra system given by Eq. (4.5.136).

4.5.6 Assume the first-integral for the Lotka-Volterra system is taken to be
$$I(x, y) = \left(\frac{y^a}{e^{by}}\right)\left(\frac{x^c}{e^{dx}}\right).$$
Apply the second derivative test, as given in section 4.5.8 and explain your result.

Section 4.6

4.6.1 Consider the following modified logistic equation
$$\frac{dx}{dt} = x(1 - x) - \mu, \qquad 0 \le \mu.$$
Determine the fixed-points and see what restrictions must be placed on μ such that the larger of the two fixed-points is non-negative.

4.6.2 Carry out the same analysis as in problem 4.6.1 for
$$\frac{dx}{dt} = x(1 - x) - \mu x.$$

4.6.3 A more complex predator-prey model is
$$\frac{dx}{dt} = rx\left(1 - \frac{x}{K}\right) - \frac{bxy}{a + x},$$
$$\frac{dy}{dt} = sy\left(1 - \frac{y}{nx}\right),$$
where (a, b, n, r, s) are positive parameters. Analyze this model and show the conditions for which a stable limit-cycle exists [13].

4.6.4 Reformulate the Hopf-bifurcation theorem for a system of three coupled, nonlinear equations. See [14] for a brief discussion of this result and an application to the "flywheel governor problem."

Comments and References

[1] L. Elsgolts, *Differential Equations and the Calculus of Variations* (MIR Publishers, Moscow, 1970). See chapter 4, section 6.

[2] R. E. Mickens, *Acta Physica Polonica* **B14**, 561 (1983).

[3] The discussion in section 4.4.5 is based on the presentation given in the text of C. M. Bender and S. A. Orszag, *Advanced Mathematical Methods for Scientists and Engineers* (McGraw-Hill, New York, 1978); see section 4.4.

[4] For these discussions, we take the fixed-point to be located at the origin, i.e., $(\bar{x}, \bar{y}) = (0, 0)$. If for a given set of equations, the fixed-point is not at the origin, i.e., $\bar{x} \neq 0$ and $\bar{y} \neq 0$, then a linear change of variables

$$x \to x' = x - \bar{x}$$
$$y \to y' = y - \bar{y}$$

will shift the fixed-point for the new variables, (x', y') to $(0, 0)$.

[5] A good introduction to the use of matrix techniques to solve systems of linear differential equations with constant coefficients is the text by M. M. Guterman and Z. H. Nitecki, *Differential Equations*, 3rd. ed. (Saunders, New York, 1992); see chapter 3.

[6] The proof of this theorem and detailed discussions of its meaning and how it can be applied are given in many places. See for example the following books:

* V. I. Arnold, *Mathematical Methods of Classical Mechanics* (Springer-Verlag, New York, 1978).

* J. Guckenheimer and P. Holmes, *Nonlinear Oscillations, Dynamical Systems, and Bifurcations of Vector Fields* (Springer-Verlag, New York, 1983).

* E. A. Jackson, *Perspectives of Nonlinear Dynamics*, Vols. 1 and 2 (Cambridge University Press, Cambridge, 1990).

* C. H. Edwards and D. E. Penney, *Differential Equations: Computing and Modeling* (Pearson Education; Upper Saddle River, NJ; 2004).

  The text by Edwards and Penney presents an excellent introduction to the concepts of stability and the classification of fixed-points. A large number of examples are given and worked out with beautiful graphics. See sections 6.1 and 6.2

[7] R. E. Mickens and D. D. Jackson, *Journal of Sound and Vibration,* **223**, 229–333 (1999).

[8] A general result for the system of equations

$$\frac{d\mathbf{x}}{dt} = \mathbf{y}, \qquad \frac{d\mathbf{y}}{dt} = -\nabla u(\mathbf{x}) - 2\epsilon\mathbf{y},$$

is given by R. E. Williamson, *Introduction to Differential Equations and Dynamical Systems* (McGraw-Hill, New York, 1997); see Example 10, pps. 375–376.

[9] See Braun, section 4.10.

[10] J. Marsden and A. Weinstein, *Calculus III,* 2nd ed. (Springer-Verlag, New York, 1985); see pp. 817.

[11] R. H. Abraham and C. D. Shaw, *Dynamics: The Geometry of Behavior. Part 4: Bifurcation Behavior* (Aerial Press; Santa Cruz, CA; 1988).

[12] See Strogatz, section 6.8.

[13] See E. Beltrami, *Mathematics for Dynamic Modeling* (Academic Press, New York, 1987); see section 6.3.

[14] See Beltrami, section 6.2. The flywheel governor problem is given and solved in section 6.4.

Bibliography

Bifurcations

E. Beltrami, *Mathematics for Dynamic Modeling* (Academic Press, Boston, 1987); see section 6.2

J. Hale and H. Koçak, *Dynamics and Bifurcation* (Springer-Verlag, New York, 1991).

G. Iooss and D. D. Joseph, *Elementary Stability and Bifurcation Theory* (Springer-Verlag, New York, 1980).

Qualitative Methods

A. A. Andronov, E. A. Leontovich, I. I. Gordon, and A. G. Maier, *Qualitative Theory of Second-Order Dynamic Systems* (Wiley, New York, 1973, Israel Program for Scientific Translations).

M. Braun, *Differential Equations and Their Applications* (Springer-Verlag, New York, 1993, 4th edition); see Chapter 4.

L. Edelstein-Keshet, *Mathematical Models in Biology* (McGraw-Hill, New York, 1988); see Chapter 5.

M. Humi and W. Miller, *Second Course in Ordinary Differential Equations for Scientists and Engineers* (Springer-Verlag, New York, 1988); see Chapter 8.

D. A. McQuarrie, *Mathematical Methods for Scientists and Engineers* (University Science Books; Sausalito, CA; 2003); see Chapter 13.

N. Minorsky, *Nonlinear Oscillations* (Van Nostrand; Princeton, NJ; 1962); see Chapters 3 and 14.

V. V. Nemytskii and V. V. Stepanov, *Qualitative Theory of Differential Equations* (Princeton University Press; Princeton, NJ; 1960); see pp. 133–134.

L. A. Segel, editor, *Mathematical Models in Molecular and Cellular Biology* (Cambridge University Press, Cambridge, 1980). See Appendix A.3: G. M. Odell, "Qualitative theory of systems of ordinary differential equations, including phase space analysis and the use of the Hopf bifurcation theorem."

S. H. Strogatz, *Nonlinear Dynamics and Chaos* (Addison-Wesley; Reading, MA; 1994); see Chapter 6.

E. T. Whittaker, *Advanced Dynamics* (Cambridge University Press, London, 1937).

Chapter 5

Difference Equations

5.1 Genesis of Difference Equations

Difference equations arise in both the modeling of discrete dynamical systems and in various purely mathematical contexts, including the numerical integration of differential equations. This chapter begins with the genesis of several difference equations and is followed by a section giving the conditions under which existence and uniqueness of solutions occur. Next the fundamental "difference" and "shift" operators are defined and their important properties derived. Sections 5.5 and 5.6 study first-order and general linear difference equations. A number of theorems are given and we show how they can be applied to determine solutions for these types of difference equations. In section 5.8, the details as to constructing general solutions to linear equations with constant coefficients are presented. Both homogeneous and inhomogeneous equations are considered, with examples to illustrate the application of the stated methods.

Nonlinear difference equations are treated in section 5.9, with worked examples in section 5.10. We only consider the cases of several special classes of nonlinear equations. They have the property of being transformed, by a change of dependent variable, into linear equations that can be solved. Finally, we give, in section 5.11, two special applications of the use of difference equations. The first is related to the Chebyshev polynomials, the second is concerned with the construction and analysis of discrete models for the logistic differential equation.

It will be noticed that there is a great similarity between the theorems of difference and differential equations [1; 2; 3] and that general relations often exist between the solutions of particular difference and difference equations [1; 2]. Our main purpose, in this chapter, is to give a general overview of what are difference equations, present some of the methods to obtain solutions for these equations, and illustrate their application. The proofs of theorems will not, in general, be provided. However, in many cases, such demonstrations can be easily found and many of the books listed in the bibliography to this chapter do give the details of the proofs [4].

References to applications in the social, natural, and engineering sciences are discussed in several of the books listed in the bibliography. In particular, see the texts by Cadzow, Goldberg, Hilderbrand, Kelley and Peterson, and Wimp. A basic, general introduction to the subject of difference equations, including partial difference equations and many applications, is my book:

R. E. Mickens, *Difference Equations: Theory and Applications*, 2nd Edition (Chapman and Hall, New York, 1990).

We now show, by means of two examples, how difference equations arise.

5.1.1   *Square-Root Iteration*

Consider the following "game" which can be played on a hand calculator. Input a real, positive number and then press the square-root key. Repeat this process until the output display either does not change or an error message appears or overflow is indicated. The question is: Which of these three possibilities occur? You should be able to give the correct answer, without any detailed mathematics. However, an analysis of this play activity is both elementary and interesting, and also leads to a difference equation.

Let x_0 be the first input number. Pressing the square-root key gives x_1 which is

$$x_1 = \sqrt{x_0}. \tag{5.1.1}$$

Pressing the square-root key again gives a number x_2, where

$$x_2 = \sqrt{x_1} = \sqrt{(\sqrt{x_1})}. \tag{5.1.2}$$

Let x_k, denote the number generated at the kth press of the key. It follows that

$$x_k = \sqrt{x_{k-1}}. \tag{5.1.3}$$

This is a first-order, nonlinear (because of the square-root operation) difference equation which describes the iteration process of inputing x_0 and continuing to press the square-root key.

5.1.2 Integral Depending on Two Parameters

Consider the following integral

$$I_k(\phi) = \int_0^\pi \frac{\cos(k\theta) - \cos(k\phi)}{\cos\theta - \cos\phi} \, d\theta, \tag{5.1.4}$$

depending on two parameters: k, which only takes on non-negative integer values, and ϕ, a real parameter. The values of $I_0(\phi)$ and $I_1(\phi)$ can be evaluated and are found to be

$$I_0(\phi) = 0, \qquad I_1(\phi) = \pi. \tag{5.1.5}$$

Since $I_k(\phi)$ is known for $k = 0$ and $k = 1$, let us attempt to determine a linear relationship among I_{k+1}, I_k, and I_{k-1}. If a relation can be found, then it can be used to recursively calculate I_k for $k \geq 2$.

Define the operator L as

$$LI_k = AI_{k+1} + BI_k + CI_{k-1}, \tag{5.1.6}$$

where (A, B, C) are independent of k, but may depend on ϕ. Now,

$$L\cos(k\theta) = A\cos(k+1)\theta + B\cos(k\theta) + C\cos(k-1)\theta$$
$$= [(A+C)\cos\theta + B]\cos(k\theta) - [(A-C)\sin\theta]\sin(k\theta). \tag{5.1.7}$$

If A, B and C are selected to be

$$A = C, \quad B = -2A\cos\phi, \quad \text{with } A = 1, \tag{5.1.8}$$

then Eq. (5.1.7) becomes

$$L\cos(k\theta) = 2(\cos\theta - \cos\phi)\cos(k\theta), \tag{5.1.9}$$

which is proportional to the denominator of the integrand in (5.1.4). Applying L to $\cos(k\phi)$ gives

$$L\cos(k\phi) = 0. \tag{5.1.10}$$

Therefore, applying L to both sides of (5.1.4) gives

$$LI_k(\phi) = \int_0^\pi \frac{L\cos(k\theta) - L\cos(k\phi)}{\cos\theta - \cos\phi} \, d\theta$$
$$= 2\int_0^\pi \cos(k\theta)d\theta = 0. \tag{5.1.11}$$

Consequently, we conclude that the integral $I_k(\phi)$ satisfies the relation

$$I_{k+1} - (2\cos\phi)I_k + I_{k-1} = 0. \tag{5.1.12}$$

A shift upward in the value by one unit in k gives

$$I_{k+2} - (2\cos\phi)I_{k+1} + I_k = 0, \tag{5.1.13}$$

or

$$I_{k+2} = (2\cos\phi)I_{k+1} - I_k = 0, \tag{5.1.14}$$

and I_k, for any positive k, can be determined recursively from a knowledge of I_0 and I_1.

Equation (5.1.14) is a linear, second-order difference equation.

5.2 Existence and Uniqueness Theorem

A sequence is a function whose domain is the set of integers. In general, we will only consider sequences whose domains are the non-negative integers. We denote the general member of a sequence by x_k and let $\{x_k\}$ represent the sequence $x_0, x_1, x_2, \ldots$.

Consider a given sequence $\{x_k\}$ and let there be a rule for determining x_k

$$x_{k+n} = F(x_{k+n-1}, x_{k+n-2}, \ldots, x_k, k), \qquad (5.2.1)$$

where F is a well-defined function of its arguments. Note that if given n successive values of x_k, then the sequence $\{x_k\}$ is uniquely defined. These given or specified values are called *initial values*.

Definition 5.2.1 An ordinary difference equation is a relation of the form given by Eq. (5.2.1).

Definition 5.2.2 The order of a difference equation is the difference between the highest and the lowest indices that appear in the equation. Note that shifting the labeling of the indices does not change the order of the equation.

Definition 5.2.3 A difference equation is linear and of order n, if it can be represented in the form

$$x_{k+n} + a_1(k)x_{k+n-1} + a_2(k)x_{k+n-2} + \cdots$$
$$+ a_{n-1}(k)x_{k+1} + a_n(k)x_k = R_k, \qquad (5.2.2)$$

where $\{a_i(k) : i = 1, 2, \ldots, n\}$ and R_k are given functions of k.

Definition 5.2.4 A difference equation is nonlinear if it is not linear.

Definition 5.2.5 A function ϕ_k is a solution to a difference equation if it reduces the equation to an identity.

The following examples illustrate the ideas introduced in the above definitions:

$$y_{k+2} + 5y_{k+1} + y_k = \cos(2k) \quad (2n\text{-order linear}), \qquad (5.2.3a)$$

$$x_{k+1} = \sqrt{x_k} \quad (1\text{st-order, nonlinear}), \qquad (5.2.3b)$$

$$x_{k+3} - x_{k-3} = 0 \Leftrightarrow x_{k+6} - x_k = 0 \quad (6\text{th-order, linear}), \qquad (5.2.3c)$$

$$x_{k+2} - 4kx_{k+1} + \left(\frac{k}{k+1}\right)x_k = e^{-k} \quad \text{(2nd-order, linear)}, \tag{5.2.3d}$$

The functions $\phi_1(k) = 1$, $\phi_2(k) = (-1)^k$ are solutions to

$$x_{k+1} - x_{k-1} = 0, \tag{5.2.4}$$

since

$$\phi_1(k+1) - \phi_1(k-1) = 1 - 1 = 0, \tag{5.2.5a}$$

$$\phi_2(k+1) - \phi_2(k-1) = (-1)^{k+1} - (-1)^{k-1} = 0. \tag{5.2.5b}$$

Likewise $\phi(k) = 3^k$ is a solution of

$$x_{k+1} - 3x_k = 0, \tag{5.2.6}$$

since

$$3^{k+1} - 3 \cdot 3^k = 0. \tag{5.2.7}$$

Let us assume that a solution has been found for Eq. (5.2.1). We still have no assurance that it is unique. On the other hand, we would also like to have the additional assurance that a particular difference equation has a solution. The following theorems give conditions which assure the existence of a unique solution.

Theorem 5.2.1 (Existence and Uniqueness) *Let*

$$x_{k+n} = F(x_{k+n-1}, x_{k+n-2}, \ldots, x_k, k) \tag{5.2.8}$$

be an nth order difference equation where F is defined for each of its arguments. This equation has a unique solution corresponding to each arbitrary selection of n initial values $x_0, x_1, x_2, \ldots, x_{n-1}$.

Proof. If the values $x_0, x_1, \ldots, x_{n-1}$ are given, then the equation with $k = 0$ uniquely specifies x_n. Once x_n is known, the equation with $k = 1$, gives x_{n+1}. Continuing in this manner, all x_k, for $k \geq n$, can be determined. ∎

5.3 The Fundamental Operators

5.3.1 The Δ Operator

The operator Δ, the difference operator, is defined as follows,

$$\Delta x_k \equiv x_{k+1} - x_k. \tag{5.3.1}$$

The Δ^2 operator is $\Delta \cdot \Delta$ and corresponds to the result

$$
\begin{aligned}
\Delta^2 x_k = \Delta(\Delta x_k) &= \Delta(x_{k+1} - x_k) \\
&= \Delta x_{k+1} - \Delta x_k = (x_{k+2} - x_{k+1}) - (x_{k+1} - x_k) \\
&= x_{k+2} - 2x_{k+1} + x_k.
\end{aligned} \tag{5.3.2}
$$

Note that Δ is a linear operator since

$$
\Delta(c_1 x_k + c_2 y_k) = c_1 \Delta x_k + c_2 \Delta y_k. \tag{5.3.3}
$$

(See problem 5.3.1.) In general, the operator Δ^n can be defined as

$$
\Delta^n x_k = \Delta(\Delta^{n-1} x_k), \tag{5.3.4}
$$

and it follows that

$$
\Delta^m \Delta^n x_k = \Delta^n \Delta^m x_k = \Delta^{m+n} x_k, \tag{5.3.5}
$$

for m and n positive integers. Note that for $m = 0$, we have

$$
\Delta^0 \Delta^n x_k = \Delta^{n+0} x_k = \Delta^n \Delta^0 x_k, \tag{5.3.6}
$$

which implies that

$$
\Delta^0 = I, \tag{5.3.7}
$$

the identity operator.

The following theorem permits the direct calculation of differences of x_k.

Theorem 5.3.1

$$
\begin{aligned}
\Delta^n x_k = x_{k+n} &- nx_{k+n-1} + \frac{n(n-1)}{2} x_{k+n-2} \\
&+ \cdots + (-1)^r \frac{n(n+1)\cdots(n-r+1)}{r!} x_{k+n-r} \\
&+ \cdots + (-1)^n x_k.
\end{aligned} \tag{5.3.8}
$$

By using the definition for the binomial coefficients

$$
\binom{n}{r} = \frac{n(n-1)\cdots(n-r+1)}{r!} = \frac{n!}{(r!)(n-r)!}, \tag{5.3.9}
$$

the result of this theorem can be rewritten to read

$$
\Delta^n x_k = \sum_{r=0}^{n} (-1)^r \binom{n}{r} x_{k+n-r}. \tag{5.3.10}
$$

Let us consider polynomial functions of the operator Δ. Let $f(s)$ be an n-degree polynomial in the variable s,

$$f(s) = a_0 s^n + a_1 s^{n-1} + \cdots + a_m, \qquad (5.3.11)$$

where the $\{a_i : i = 0, 1, \ldots, m\}$ are constants. The operator function $f(\Delta)$ is defined to be

$$\begin{aligned} f(\Delta)x_k &\equiv (a_0 \Delta^n + a_1 \Delta^{n-1} + \cdots + a_m)x_k \\ &= a_0 \Delta^n x_k + a_1 \Delta^{n-1} x_k + \cdots + a_m x_k. \end{aligned} \qquad (5.3.12)$$

Since $f(s)$ is a polynomial function of degree m, it has the factorization

$$f(s) = (s - s_1)(s - s_2) \cdots (s - s_m) = \prod_{i=1}^{m} (s - s_i), \qquad (5.3.13)$$

and we can write

$$f(\Delta)x_k = \prod_{i=1}^{m} (\Delta - s_i)x_k. \qquad (5.3.14)$$

5.3.2 *The Shift Operator, E*

The shift operator, E, is defined to be

$$Ex_k \equiv x_{k+1}. \qquad (5.3.15)$$

It follows that for positive integer p, we have

$$E^p x_k = x_{k+p}, \qquad (5.3.16)$$

with E^0 being the identity operator, i.e.,

$$E^0 x_k = x_{k+0} = x_k. \qquad (5.3.17)$$

Other properties of the shift operator include

$$E^p(c_1 x_k + c_2 y_k) = c_1 x_{k+p} + c_2 y_{k+p}, \qquad (5.3.18)$$

$$E^p E^q x_k = E^q E^p x_k = E^{p+q} x_k = x_{k+p+q}, \qquad (5.3.19)$$

$$f(E) = \prod_{i=1}^{m} (E - s_i)x_k, \qquad (5.3.20)$$

where c_1 and c_2 are constants, p and q are integers, and $f(s)$ is a polynomial function of degree m.

Let p be a positive integer; then

$$E^{-1}x_k = x_{k-1} \Rightarrow (E^{-1})^p x_k = (E^{-p})x_k = x_{k-p}. \qquad (5.3.21)$$

Using the definition of the operators Δ and E, it is easy to see that the following relationship holds

$$\Delta = E - \hat{1} \quad \text{or} \quad E = \hat{1} + \Delta, \tag{5.3.22}$$

where $\hat{1}$ is the identity operator. Therefore, if $f(s)$ and $g(s)$ are polynomial functions of s, then

$$f(\Delta) = f(E - \hat{1}), \quad g(E) = g(\hat{1} + \Delta). \tag{5.3.23}$$

This result can be used to prove the following theorem, which is the inverse problem to Theorem 5.3.1.

Theorem 5.3.2

$$x_{k+n} = x_k + n\Delta x_k + \frac{n(n-1)}{2!}\Delta^2 x_k$$
$$+ \cdots + \binom{n}{i}\Delta^i x_k + \cdots + \Delta^n x_k. \tag{5.3.24}$$

Proof. Write x_{k+n} as

$$x_{k+n} = E^n x_k = (1 + \Delta)^n x_k = \sum_{i=0}^{n} \binom{n}{i} \Delta^i x_k. \tag{5.3.25}$$

The last expression in Eq. (5.3.25) is just Eq. (5.3.24). ∎

5.3.3 *Difference of Functions*

Difference of a Product

$$\Delta(x_k y_k) = x_{k+1} y_{k+1} - x_k y_k$$
$$= x_{k+1} y_{k+1} - x_{k+1} y_k + y_k x_{k+1} - x_k y_k$$
$$= x_{k+1}(y_{k+1} - y_k) + y_k(x_{k+1} - x_k) \tag{5.3.26}$$

or

$$\Delta(x_k y_k) = x_{k+1}\Delta y_k + y_k \Delta x_k. \tag{5.3.27}$$

Leibnitz's Theorem for Differences

$$\Delta^n(x_k y_k) = x_k \Delta^n y_k + \binom{n}{1}(\Delta x_k)(\Delta^{n-1} y_{k+1})$$
$$+ \binom{n}{2}(\Delta^2 x_k)(\Delta^{n-2} y_{k+2})$$

$$+ \cdots + \binom{n}{n}(\Delta^n x_k)(y_{k+n}). \qquad (5.3.28)$$

Difference of a Quotient

$$\Delta\left(\frac{x_k}{y_k}\right) = \frac{x_{k+1}}{y_{k+1}} - \frac{x_k}{y_k} = \frac{x_{k+1}y_k - y_{k+1}x_k}{y_{k+1}y_k}$$

$$= \frac{(x_{k+1} - x_k)y_k - x_k(y_{k_1} - y_k)}{y_{k+1}y_k}, \qquad (5.3.29)$$

and

$$\Delta\left(\frac{x_k}{y_k}\right) = \frac{y_k\Delta x_k - x_k\Delta y_k}{y_{k+1}y_k}. \qquad (5.3.30)$$

Difference of a Finite Sum

Define S_k as follows

$$S_k \equiv x_1 + x_2 + \cdots + x_k; \qquad (5.3.31)$$

then

$$S_{k+1} = x_1 + x_2 + \cdots + x_k + x_{k+1} = S_k + x_{k+1} \qquad (5.3.32)$$

and

$$\Delta S_k = S_{k+1} - S_k = x_{k+1}. \qquad (5.3.33)$$

5.3.4 *The Operator* Δ^{-1}

The operator Δ^{-1} is defined to be

$$\Delta(\Delta^{-1}x_k) = x_k. \qquad (5.3.34)$$

Let $z_k = \Delta^{-1}x_k$, then

$$\Delta z_k = z_{k+1} - z_k = x_k, \qquad (5.3.35)$$

and we have

$$\begin{cases} z_{k+1} - z_k = x_k \\ z_k - z_{k-1} = x_{k-1} \\ z_{k-1} - z_{k-2} = x_{k-2} \\ \quad\vdots \qquad \vdots \qquad \vdots \\ z_2 - z_1 = x_1. \end{cases} \qquad (5.3.36)$$

Adding, respectively, the left- and right-hand sides gives

$$z_{k+1} - z_1 = x_1 + x_2 + \cdots + x_k, \qquad (5.3.37)$$

or

$$z_{k+1} = z_1 + \sum_{i=1}^{k} x_i \Rightarrow z_k = z_1 + \sum_{i=1}^{k-1} x_i. \qquad (5.3.38)$$

Substituting $z_k = \Delta^{-1} x_k$ gives, finally, the result

$$\Delta^{-1} x_k = \sum_{i=1}^{k-1} x_i + \text{constant}, \qquad (5.3.39)$$

since z_1 is an arbitrary constant.

In summary, $\Delta^{-1} x_k$ is a function whose difference is x_k. We define $\Delta^{-n} x_k$ as a function whose nth difference is x_k. Note that we have

$$\Delta^{-n} x_k = \Delta^{-1}(\Delta^{-n+1} x_k). \qquad (5.3.40)$$

Therefore,

$$\Delta^{-2} x_k = \Delta^{-1}(\Delta^{-1} x_k) = \Delta^{-1}\left(\sum_{i=1}^{k-1} x_i + c_1 \right)$$

$$= \sum_{m=1}^{k-1} \sum_{i=1}^{m-1} x_i + \Delta^{-1} c_1 + c_2, \qquad (5.3.41)$$

where c_1 and c_2 are arbitrary constants. Now

$$\Delta^{-1} c_1 = \sum_{i=1}^{k-1} c_1 = c_1 \sum_{i=1}^{k-1} 1 = c_1(k-1) = c_1 k - c_1 \qquad (5.3.42)$$

and

$$\Delta^{-1} c_1 + c_2 = c_1 k + (c_2 - c_1) = c_1 k + \bar{c}_2, \qquad (5.3.43)$$

where $\bar{c}_2 = c_2 - c_1$ is an arbitrary constant. Dropping the bar on $\bar{c}_2$ in Eq. (5.3.43), we can write Eq. (5.3.41) as

$$\Delta^{-2} x_k = \sum_{m=1}^{k-1} \sum_{i=1}^{m-1} x_i + c_1 k + c_2. \qquad (5.3.44)$$

Using an obvious notation, the above results can be generalized and the following result is obtained,

$$\Delta^{-n} x_k = \left(\sum\right)^n x_k + c_1 k^{n-1} + c_2 k^{n-2} + \cdots + c_n, \qquad (5.3.45)$$

where the c_i are n arbitrary constants.

The two operators Δ and Δ^{-1} do not commute, i.e.,

$$\Delta \Delta^{-1} \neq \Delta^{-1} \Delta. \qquad (5.3.46)$$

This is easily seen by noting that for $\Delta^{-1}\Delta$ the summation is at the end of the joint operation and, consequently, an arbitrary constant is introduced. However, for $\Delta\Delta^{-1}$ the difference operation is last and any arbitrary constant introduced by the summation operation will not appear. In summary,

$$\Delta^{-1}\Delta x_k = \Delta\Delta^{-1}x_k + c_1. \tag{5.3.47}$$

For the integral calculus, the integration by parts relation play an important role,

$$\int u\, dv = uv - \int v\, du + c, \qquad c = \text{constant.} \tag{5.3.48}$$

A similar formula exists for the difference calculus. From Eq. (5.3.27), it follows that

$$\Delta(x_k y_k) = x_{k+1}\Delta y_k + y_k\Delta x_k, \tag{5.3.49}$$

or

$$y_k\Delta x_k = \Delta(x_k y_k) - x_{k+1}\Delta y_k. \tag{5.3.50}$$

Applying Δ^{-1} to Eq. (5.3.50) gives

$$\Delta^{-1}(y_k\Delta x_k) = \Delta^{-1}\Delta(x_k y_k) - \Delta^{-1}(x_{k+1}\Delta y_k), \tag{5.3.51}$$

which on evaluation is

$$\sum_{i=1}^{k-1} y_i\Delta x_i = x_k y_k - \sum_{i=1}^{k-1} x_{i+1}\Delta y_i + c_1. \tag{5.3.52}$$

This last relation is the discrete analogue to Eq. (5.3.48).

Again, in correspondence with continuous variable calculus, the following is the fundamental theorem of the sum calculus.

Theorem 5.3.3 *If* $\Delta F_k = f_k$, *and* (a, b) *are integers with* $a \leq b$, *then*

$$\sum_{k=a}^{b} f_k = F(b+1) - F(a). \tag{5.3.53}$$

As a final result, we derive the so-called Abel transformation,

$$\sum_{k=1}^{n} x_k y_k = x_{n+1}\sum_{k=1}^{n} y_k - \sum_{k=1}^{n}\left(\Delta x_k \sum_{i-1}^{k} y_i\right). \tag{5.3.54}$$

We proceed by using the relation for summation by parts, given by Eq. (5.3.52) and the fundamental theorem of the sum calculus, to obtain the result

$$\sum_{k=1}^{n} f_k\Delta g_k = f_{n+1}g_{n+1} - f_1 g_1 - \sum_{k=1}^{n} g_{n+1}\Delta f_k, \tag{5.3.55}$$

where f_k and g_k are functions of k. Now take

$$f_k = x_k \qquad \Delta g_k = y_k. \tag{5.3.56}$$

It follows that

$$\sum_{k=1}^{n-1} \Delta g_k = g_n - g_1 = \sum_{k=1}^{n-1} y_k \tag{5.3.57}$$

and

$$g_n = g_1 + \sum_{k=1}^{n-1} y_k. \tag{5.3.58}$$

Substitution of these results into Eq. (5.3.55) gives

$$\sum_{k=1}^{n} x_k y_k = x_{k+1} \left(g_1 + \sum_{k=1}^{n} y_k \right) - x_1 g_1$$

$$- \sum_{k=1}^{n} \left[\Delta x_k \left(g_1 + \sum_{i=1}^{k} y_i \right) \right]$$

$$= x_{n+1} \sum_{k=1}^{n} y_k - \sum_{k=1}^{n} \left(\Delta x_k \sum_{i=1}^{k} y_i \right)$$

$$+ x_{n+1} g_1 - x_1 g_1 - \sum_{k=1}^{n} g_i \Delta x_k. \tag{5.3.59}$$

However, the following is true

$$x_{n+1} g_1 - x_1 g_1 - \sum_{k=1}^{n} g_1 \Delta x_k = x_{n+1} g_1 - x_1 g_1 - g_1 \sum_{k=1}^{n} (x_{k+1} - x_k) = 0.$$
$$\tag{5.3.60}$$

Using this in Eq. (5.3.59) gives the result of Eq. (5.3.54), the Abel transformation.

5.4   Worked Problems Based on Section 5.3

5.4.1   *Examples A*

(i) Let $x_k = c$ where c is a constant; then

$$\begin{cases} \Delta c = c - c = 0, \\ Ec = c. \end{cases} \tag{5.4.1}$$

(ii) Let $x_k = (-1)^k k$, then

$$\Delta[(-1)^k k] = (-1)^{k+1}(k+1) - (-1)^k = -(-1)^k(2k+1). \qquad (5.4.2)$$

(iii) Let $x_k = k^n$, where n is a positive integer; then

$$\Delta k^n = (k+1)^n - k^n = nk^{n-1} + \binom{n}{2}k^{n-2} + \cdots + \binom{n}{n-1}k + 1. \quad (5.4.3)$$

(iv) Let $x_k = a^k$, then

$$\Delta a^k = a^{k+1} - a^k = (a-1)a^k. \qquad (5.4.4)$$

(v) Let $x_k = \cos(ak)$, where a is a constant; then

$$\begin{aligned}
\Delta \cos(ak) &= \cos(ak + a) - \cos(ak) \\
&= (\cos a)\cos(ak) - (\sin a)\sin(ak) - \cos(ak) \\
&= (\cos a - 1)\cos(ak) - (\sin a)\sin(ak).
\end{aligned} \qquad (5.4.5)$$

5.4.2 Example B

What is $\Delta^n(k^2 v_k)$? We can apply the Leibnitz's theorem for differences, Eq. (5.3.28), with

$$x_k = k^2, \qquad y_k = v_k, \qquad (5.4.6)$$

to obtain

$$\begin{aligned}
\Delta^n(k^2 v_k) &= k^2 \Delta^n v_k + \binom{n}{1}(\Delta k^2)(\Delta^{n-1} v_{k+1}) \\
&\quad + \binom{n}{2}(\Delta^2 k^2)(\Delta^{n-2} v_{k+2}),
\end{aligned} \qquad (5.4.7)$$

where it should be clear that all the other terms are zero since

$$\Delta^m k^2 = 0, \qquad m > 2. \qquad (5.4.8)$$

Therefore,

$$\Delta^n(k^2 v_k) = k^2 \Delta^n v_k + (2k+1)\Delta^{n-1} v_k + n(n-1)\Delta^{n-2} v_{k+2}. \quad (5.4.9)$$

As an application, let $v_k = a^k$, then

$$\begin{aligned}
\Delta^n(k^2 a^k) &= [(a-1)^2 k^2 + na(a-1)(2k+1) \\
&\quad + a^2 n(n-1)](a-1)^{n-2} a^k.
\end{aligned} \qquad (5.4.10)$$

5.4.3   *Example C*

Consider a function $f(x)$ such that all the derivatives exist. Define $E_a f(x)$ to be

$$E_a f(x) \equiv f(x + a). \qquad (5.4.11)$$

Now define the derivative operator, D, as

$$Df(x) \equiv \frac{df(x)}{dx}. \qquad (5.4.12)$$

We now show that

$$e^{aD} f(x) = E_a f(x), \qquad (5.4.13)$$

where the operator e^{aD} is defined to be

$$e^{aD} = \sum_{i=0}^{\infty} \left(\frac{1}{i!}\right) a^i D^i. \qquad (5.4.14)$$

The expansion of $f(x + a)$ is

$$f(x + a) = \sum_{i=0}^{\infty} \left(\frac{1}{i!}\right) a^i \frac{d^i f(x)}{dx^i}. \qquad (5.4.15)$$

Note that we have expanded in a for fixed x. In terms of D, the right-side of Eq. (5.4.14) can be rewritten to the form

$$
\begin{aligned}
f(x + a) &= \sum_{i=0}^{\infty} \left(\frac{1}{i!}\right) a^i (D^i f(x)) \\
&= \left[\sum_{i=0}^{\infty} \left(\frac{1}{i!}\right) a^i D^i\right] f(x) \\
&= e^{aD} f(x).
\end{aligned}
\qquad (5.4.16)
$$

Applying the result given by Eq. (5.4.11) to the term on the left-side of Eq. (5.4.16), we obtain Eq. (5.4.13).

5.4.4   *Examples D*

We show below the use of the operator Δ^{-1} to calculate the anti-differences of several functions.

   (i) Consider

$$x_k = \Delta^{-1} c, \qquad c = \text{constant}. \qquad (5.4.17)$$

From Eq. (5.3.39), we have

$$x_k = c\Delta^{-1}(1) = c\sum_{i=1}^{k-1} 1 = ck + c_1, \qquad (5.4.18)$$

where c_1 is an arbitrary constant.

(ii) Calculate $\Delta^{-1}a^k$. We proceed as follows. First, we have

$$\Delta a^k = (a-1)a^k, \qquad (5.4.19)$$

and applying Δ^{-1} gives

$$\Delta^{-1}\Delta a^k = (a-1)\Delta^{-1}a^k \qquad (5.4.20)$$

or

$$a^k = (a-1)\Delta^{-1}a^k + c_1 \qquad (5.4.21)$$

Dividing by $(a-1)$ and defining $c_2 = c_1/(1-a)$, we get

$$\Delta^{-1}a^k = \frac{a^k}{(a-1)} + c_2, \qquad a \neq 1. \qquad (5.4.22)$$

(iii) Define the *factorial function*, $k^{(n)}$, as follows

$$k^{(N)} \equiv k(k-1)(k-2)\cdots(k-n-1) = \frac{k!}{(k-n)!}. \qquad (5.4.23)$$

So, for example,

$$\begin{cases} k^{(0)} = 1, \\ k^{(1)} = k, \\ k^{(2)} = k^2 - k, \\ k^{(3)} = k^3 - 3k^2 + 2k, \\ k^{(4)} = k^4 - 6k^3 + 11k^2 - 6k, \\ k^{(5)} = k^5 - 10k^4 + 35k^3 - 50k^2 + 24k, \\ \text{etc.} \end{cases} \qquad (5.4.24)$$

These relations can be inverted to give the powers of k in terms of the factorial functions. Applying this to the expression in Eq. (5.4.24) gives:

$$\begin{cases} 1 = k^{(0)}, \\ k = k^{(1)}, \\ k^2 = k^{(2)} + k^{(1)}, \\ k^3 = k^{(3)} + 3k^{(2)} + k^{(1)}, \\ k^4 = k^{(4)} + 7k^{(3)} + 6k^{(2)} + k^{(1)}, \\ k^5 = k^{(5)} + 15k^{(4)} + 25k^{(3)} + 10k^{(2)} + k^{(1)}, \\ \text{etc.} \end{cases} \qquad (5.4.25)$$

The relations given in Eqs. (5.4.24) and (5.4.25) can be generalized to the expressions,

$$\begin{cases} k^{(n)} = \sum_{i=1}^{n} s_i^n k^i, \\ k^n = \sum_{i=1}^{n} S_i^n k^{(i)}, \end{cases} \tag{5.4.26}$$

where the coefficients are called

$$\begin{cases} s_i^n = \text{Stirling numbers of the first-kind}, \\ S_i^n = \text{Stirling numbers of the second-kind}. \end{cases}$$

See problem 5.4.1.

From the definition of $k^{(n)}$, it follows that

$$k^{(k)} = k!, \tag{5.4.27}$$

$$k^{(n)} = 0, \qquad n > k, \tag{5.4.28}$$

$$k^{(n)}(k - n)^{(j)} = k^{(n+j)}. \tag{5.4.29}$$

Let $n = -j > 0$, then from the last equation,

$$k^{(-n)}(k + n)^{(n)} = k^{(-n+n)} = k^{(0)} = 1, \tag{5.4.30}$$

and

$$k^{(-n)} = \frac{1}{(k + n)^{(n)}} = \frac{1}{(k + n)(k + n - 1)\cdots(k + 1)}. \tag{5.4.31}$$

From the continuous calculus, we have

$$\frac{d}{dx}(x^n) = nx^n, \qquad \frac{d^2}{dx^2}(x^n) = n(n - 1)x^{n-1}, \text{ etc.} \tag{5.4.32}$$

Likewise, for the discrete calculus, the following holds

$$\Delta k^n = nk^{n-1} + \binom{n}{2}k^{n-2} + \cdots + 1. \tag{5.4.33}$$

We now show that the use of the factorial polynomials gives a result similar to Eq. (5.4.32), namely,

$$\Delta k^{(n)} = nk^{(n-1)}. \tag{5.4.34}$$

A consequence of Eq. (5.4.34) is that

$$\Delta^2 k^{(n)} = \Delta(\Delta k^{(n)}) = \Delta[nk^{(n-1)}] = n(n-1)k^{(n-2)}, \qquad (5.4.35)$$

$$\Delta^3 k^{(n)} = \Delta(\Delta^2 k^{(n)}) = n(n-1)(n-2)k^{(n-3)} \qquad (5.4.36)$$

$$\vdots \qquad \vdots \qquad \qquad \vdots$$

$$\Delta^n k^{(n)} = n!, \qquad (5.4.37)$$

$$\Delta^m k^{(n)} = 0, \qquad m > n. \qquad (5.4.38)$$

To prove the result in Eq. (5.4.34), three cases need to be considered: $n > 0$, $n < 0$, and $n = 0$. For the last case, it follows that

$$\Delta k^{(0)} = \Delta 1 = 0. \qquad (5.4.39)$$

For $n > 0$, we have

$$\Delta k^{(n)} = (k+1)^{(n)} - k^{(n)}$$
$$= (k+1)k(k-1)\cdots(k-n+2) - k(k-1)\cdots(k-n+2)(k-n+1)$$
$$= [(k+1) - (k-n+1)]k(k-1)\cdots(k-n+2)$$
$$= nk^{(n-1)}. \qquad (5.4.40)$$

For $n < 0$, then from Eq. (5.4.31)

$$\Delta k^{(n)} = \frac{1}{(k+2)(k+3)\cdots(k-n)(k-n+1)}$$
$$\qquad - \frac{1}{(k+1)(k+2)\cdots(k-n-1)(k-n)}$$
$$= \frac{(k+1) - (k-n+1)}{(k+1)(k+2)\cdots(k-n)(k-n+1)}$$
$$= nk^{(n-1)}. \qquad (5.4.41)$$

Therefore, Eq. (5.4.34) is correct for all integer values of k.

5.5 First-Order Linear Difference Equations

The general form of a first-order linear difference equation is

$$x_{k+1} - p_k x_k = q_k, \qquad (5.5.1)$$

where p_k and q_k are specified functions of k. The case for which $q_k \equiv 0$, i.e.,

$$x_{k+1} - p_k x_k = 0, \qquad (5.5.2)$$

is called the homogeneous equation, while Eq. (5.5.1), for $q_k \not\equiv 0$, is the inhomogeneous equation. The general solution to Eq. (5.5.1) consists of the sum of the solution to the homogeneous Eq. (5.5.2) and the particular solution to the inhomogeneous Eq. (5.5.1). We now derive a procedure to give the "formal" solution to Eq. (5.5.1).

First consider the homogeneous Eq. (5.5.2). If x_0 is given, then

$$
\begin{cases}
x_1 = p_0 x_0, \\
x_2 = p_1 x_2 = p_1 p_0 x_0, \\
x_3 = p_2 x_2 = p_2 p_1 p_0 x_0, \\
\vdots \quad \vdots \\
x_k = p_{k-1} x_{k-1} = p_{k-1} p_{k-2} \cdots p_1 p_0 x_0,
\end{cases}
\tag{5.5.3}
$$

where the last relation can be rewritten to the form

$$
x_k = x_0 \prod_{i=0}^{k-1} p_i.
\tag{5.5.4}
$$

Since x_0 can take any arbitrary value, Eq. (5.5.4) is the general solution to the homogeneous Eq. (5.5.2).

We begin the study of the inhomogeneous equation by dividing it by $\prod_{i=0}^{k} p_i$ to obtain

$$
\left(x_{k+1} \Big/ \prod_{i=0}^{k} p_i \right) - \left(x_k \Big/ \prod_{i=0}^{k-1} p_i \right) = q_k \Big/ \prod_{i=0}^{k} p_i.
\tag{5.5.5}
$$

This expression can be written as

$$
\Delta \left(x_k \Big/ \prod_{i=0}^{k-1} p_i \right) = q_k \Big/ \prod_{i=0}^{k} p_i,
\tag{5.5.6}
$$

with the particular solution

$$
x_k \Big/ \prod_{i=0}^{k-1} p_i = \Delta^{-1} \left(q_k \Big/ \prod_{i=0}^{k} p_i \right),
\tag{5.5.7}
$$

or

$$
x_k = \left(\prod_{i=0}^{k-1} p_i \right) \sum_{i=0}^{k-1} \left(q_i \Big/ \prod_{r=0}^{i} p_r \right).
\tag{5.5.8}
$$

The general solution to Eq. (5.5.1) is given by the sum of the expressions in Eqs. (5.5.4) and (5.5.8),

$$x_k = A \prod_{i=0}^{k-1} p_i + \left(\prod_{i=0}^{k-1} p_i \right) \sum_{i=0}^{k-1} \left(q_i \Big/ \prod_{r=0}^{i} p_r \right). \tag{5.5.9}$$

We now illustrate the use of Eqs. (5.5.4) and Eq. (5.5.9) by using them to solve several linear, first-order difference equation.

5.5.1 Example A

Consider the inhomogeneous equation

$$x_{k+1} - b x_k = a, \qquad x_0 = A = \text{given}, \tag{5.5.10}$$

where (a, b) are constants and $b \neq 1$. For this case,

$$p_k = b, \qquad q_k = a. \tag{5.5.11}$$

Therefore

$$\prod_{i=0}^{k-1} p_i = \prod_{i=0}^{k-1} b = b^k, \tag{5.5.12}$$

and

$$\sum_{i=0}^{k-1} \left(q_i \Big/ \prod_{r=1}^{i} p_r \right) = \sum_{i=0}^{k-1} \left(\frac{a}{b^i} \right) = a \sum_{i=0}^{k-1} b^{-i}. \tag{5.5.13}$$

Using the fact that

$$\sum_{i=0}^{k-1} z^i = \frac{1 - z^k}{1 - z}, \tag{5.5.14}$$

and letting $z = b^{-1}$, we obtain

$$b^k \left(a \sum_{i=0}^{k-1} b^{-i} \right) = \left(\frac{a}{b-1} \right) b^k - \frac{a}{b-1}, \tag{5.5.15}$$

and the general solution to Eq. (5.5.10) is

$$x_k = C b^k - \left(\frac{a}{b-1} \right), \qquad b \neq 1, \tag{5.5.16}$$

where all the constants have been absorbed into the now arbitrary constant C.

Note that for $b = 1$,

$$x_{k+1} - x_k = a, \tag{5.5.17}$$

with $p_k = 1$ and $q_k = a$, and we have

$$\prod_{i=0}^{k-1} p_k = \prod_{i=0}^{k-1} 1 = 1, \tag{5.5.18}$$

$$\sum_{i=0}^{k-1} \left(q_i \Big/ \prod_{r=0}^{i} p_r \right) = -\sum_{i=0}^{k-1} a/(1) = a \sum_{i=0}^{k-1} = ak, \tag{5.5.19}$$

and the general solution is

$$x_k = A + ak, \qquad \text{for } b = 1. \tag{5.5.20}$$

5.5.2 *Example B*

Consider the following homogeneous equation

$$x_{k+1} - e^{2k} x_k = 0, \tag{5.5.21}$$

corresponding to

$$p_k = e^{2k}. \tag{5.5.22}$$

Therefore, the solution is given by the expression

$$x_k = A \prod_{i=0}^{k-1} p_i = A \prod_{i=0}^{k-1} e^{2i} = A \exp\left(2 \sum_{i=0}^{k-1} i \right). \tag{5.5.23}$$

However, it can easily be shown that

$$\sum_{i=0}^{k-1} i = \frac{k(k-1)}{2}, \tag{5.5.24}$$

with the consequence that Eq. (5.5.23) takes the form

$$x_k = A \exp[k(k-1)]. \tag{5.5.25}$$

5.5.3 *Examples C*

The equation

$$(k+1)x_{k+1} - k x_k = 0 \tag{5.5.26}$$

can be transformed to

$$z_{k+1} - z_k = 0, \tag{5.5.27}$$

if we define $z_k = k x_k$. The solution to this equation is

$$z_k = A \Rightarrow x_k = \frac{A}{k}, \tag{5.5.28}$$

where A is an arbitrary constant.

Note that writing Eq. (5.5.26) as

$$x_{k+1} - \left(\frac{k}{k+1}\right) x_k = 0, \tag{5.5.29}$$

identifies p_k as $p_k = k/(k+1)$. Therefore,

$$x_k = A \prod_{i=1}^{k-1} p_i = A \prod_{i=1}^{k-1} \left(\frac{i}{i+1}\right). \tag{5.5.30}$$

Now, it is easy to show that

$$\prod_{i=1}^{k-1} i = (k-1)! \quad \text{and} \quad \prod_{i=1}^{k-1}(i+1) = k!, \tag{5.5.31}$$

and, as a consequence,

$$x_k = A \frac{(k-1)!}{k!} = \frac{A}{k}. \tag{5.5.32}$$

Now consider an inhomogeneous form of Eq. (5.5.26),

$$(k+1)x_{k+1} - kx_k = k, \tag{5.5.33}$$

which can be written as

$$\Delta(kx_k) = k. \tag{5.5.34}$$

Applying the operator Δ^{-1}, we obtain

$$kx_k = \Delta^{-1}k = \sum_{i=1}^{k-1} i = \frac{k(k-1)}{2} + A, \tag{5.5.35}$$

where A is an arbitrary constant. Therefore, the general solution to Eq. (5.5.32) is

$$x_k = \left(\frac{k-1}{2}\right) + \frac{A}{k}. \tag{5.5.36}$$

Also, observe that Eq. (5.5.33) can also be written as

$$x_{k+1} - \left(\frac{k}{k+1}\right) x_k = \frac{k}{k+1}, \tag{5.5.37}$$

for which

$$p_k = q_k = \frac{k}{k+1}, \tag{5.5.38}$$

$$\prod_{i=1}^{k-1} p_i = \prod_{i=1}^{k-1} \left(\frac{i}{i+1}\right) = \frac{1}{k}, \quad \prod_{i=1}^{k-1} q_i = \frac{1}{k}, \tag{5.5.39}$$

and

$$\sum_{i=1}^{k-1} \left(q_i \Big/ \prod_{r=1}^{i} p_r \right) = \sum_{i=1}^{k-1} \left(\frac{i}{i+1} \right) (i+1)$$

$$= \sum_{i=1}^{k-1} i = \frac{k(k-1)}{2}. \qquad (5.5.40)$$

Substitution of these expressions into Eq. (5.5.9) gives the result of (5.5.36).

5.6   General Linear Difference Equations

5.6.1   *General Properties*

Let the function $\{a_i(k) : i = 1, 2, \ldots, n\}$ and R_k be defined over a set of integers

$$k_1 \leq k \leq k_2, \qquad (5.6.1)$$

where (k_1, k_2) can be finite or unbounded. The equation

$$x_{k+n} + a_1(k)x_{k+n-1} + \cdots + a_n x_n = R_k \qquad (5.6.2)$$

is said to be a linear, nth order difference equation, provided

$$a_n(k) \neq 0, \qquad (5.6.3)$$

for any k of Eq. (5.6.1).

   Equation (5.6.2) is said to be homogeneous if R_k is identically zero for k in $k_1 \leq k \leq k_2$. If $R_k \not\equiv 0$, then the equation is called inhomogeneous.

   The following three theorems are immediate consequences of the above definitions.

Theorem 5.6.1 *Let c be an arbitrary constant. If x_k is a solution of the homogeneous equation*

$$x_{k+n} + a_1(k)x_{k+n-1} + \cdots + a_n x_k = 0, \qquad (5.6.4)$$

then cx_k is also a solution.

Theorem 5.6.2 *Let c_1 and c_2 be arbitrary constants. Let $x_k^{(1)}$ and $x_k^{(2)}$ be solutions of the homogeneous Eq. (5.6.4); then*

$$x_k = c_1 x_k^{(1)} + c_2 x_k^{(2)}, \qquad (5.6.5)$$

is also a solution.

The result from this theorem is called the *principle of superposition*.

Theorem 5.6.3 *Let $x_k^{(1)}$ be a solution to the homogeneous Eq. (5.6.4) and let X_k be a solution to the inhomogeneous Eq. (5.6.2); then*

$$x_k = x_k^{(1)} + X_k, \qquad (5.6.6)$$

is a solution to Eq. (5.6.2).

Theorem 5.6.4 (Existence and uniqueness) *There exists one, and only one, solution to Eq. (5.6.2) for which*

$$x_k = A_0, \quad x_{k+1} = A_1, \ldots, y_{k+n-1} = A_{n-1}, \qquad (5.6.7)$$

where the $(A_i : i = 0, 1, \ldots, n - 1)$ are n arbitrary constants and k to $k + n - 1$, lies in the interval $[k_1, k_2]$.

The proofs of these four theorems are direct and are assigned to the problems.

5.6.2 *Linearly Independent Functions*

A set of n functions $\{f_i(k) : i = 1, 2, \ldots, n\}$ is said to be linearly dependent over the interval $k_1 \leq k \leq k_2$, if there exists a set of n constants $\{c_i : i = 1, 2, \ldots, n\}$, not all zero, such that

$$c_1 f_1(k) + c_2 f_2(k) + \cdots + c_n f_n(k) = 0. \qquad (5.6.8)$$

If the set of functions is not linearly dependent, then the set is said to be linearly independent.

How can we determine if a set of functions is linearly independent or dependent. The introduction of the so-called Casorati determinant will provide aid in this quest.

The Casorati determinant or Casoratian of n functions $\{f_i(k) : i = 1, 2, \ldots, n\}$ is defined as

$$C(k) \equiv \begin{vmatrix} f_1(k) & f_2(k) & \cdots & f_n(k) \\ f_1(k+1) & f_2(k+1) & \cdots & f_n(k+1) \\ \vdots & \vdots & & \\ f_1(k+n-1) & f_2(k+n-1) & \cdots & f_n(k+n-1) \end{vmatrix}. \qquad (5.6.9)$$

In the following two theorems and for the remainder of this chapter, whenever the term "for all k" is used, we mean the set of k values, $k_1 \leq k \leq k_2$, for which the relevant functions are defined.

Theorem 5.6.5 *Let the N functions, $\{f_i(k) : i = 1, 2, \ldots, n\}$, be linearly dependent functions; then their Casoratian is zero for all k.*

Theorem 5.6.6 *Let the n functions, $\{f_i(k) : i = 1, 2, \ldots, n\}$, have their Casoratian equal to zero for all k; then they are linearly dependent.*

Note that Theorems 5.6.5 and 5.6.6 show that if the Casoratian for n functions is non-zero for all k, then these functions are linearly independent.

5.6.3 Fundamental Theorems for Homogeneous Equations

The following four theorems apply to the nth order, linear difference equation

$$x_{k+1} + a_1(k)x_{k+n-1} + \cdots + a_n x_k = 0. \qquad (5.6.10)$$

Their proofs are found in R. E. Mickens, *Difference Equations: Theory and Applications*, 2nd ed. (Chapman and Hall, New York, 1990).

Theorem 5.6.7 *Let the n functions, $\{a_i(k) : i = 1, 2, \ldots, n\}$, be defined for all k, with $a_n(k)$ nonzero for all k; then Eq. (5.6.10) has n linearly independent solutions, $\{x_i(k) : i = 1, 2, \ldots, n\}$.*

Definition A fundamental set of solutions of Eq. (5.6.10) is any n functions, $\{x_i(k) : i = 1, 2, \ldots, n\}$, which are solutions of Eq. (5.6.10) and whose Casoratian, $C(k)$, is nonzero for all values of k.

Theorem 5.6.8 *Every fundamental set of solutions of Eq. (5.6.10) is linearly independent.*

Theorem 5.6.9 *An nth order homogeneous linear difference equation has n and only n linearly independent solutions.*

Theorem 5.6.10 *The general solution of Eq. (5.6.10) is given by*

$$x_k = c_1 x_1(k) + c_2 x_2(k) + \cdots + c_n x_n(k), \qquad (5.6.11)$$

where $\{c_i : i = 1, 2, \ldots, n\}$ are n arbitrary constants, and the n functions, $\{x_i(k) : i = 1, 2, \ldots, n\}$, are a fundamental set of solutions.

It can be shown that the Casoratian satisfies the first-order, linear difference equation

$$C(k + 1) = (-1)^n a_n(k) C(k). \qquad (5.6.12)$$

Therefore, the Casoratian is essentially defined in terms of the coefficients $a_n(k)$ appearing in Eq. (5.6.10). Since

$$a_n(k) \neq 0, \qquad k_1 \leq k \leq k_2, \tag{5.6.13}$$

it follows that for starting at some fixed value of $k = \bar{k}$, in this interval, the Casoratian is never equal to zero and it can be calculated by solving Eq. (5.6.12). Doing this gives

$$C(k) = (-1)^{nk} C(\bar{k}) \prod_{i=\bar{k}}^{k-1} a_n(i). \tag{5.6.14}$$

5.6.4 *Inhomogeneous Equations*

We now present a technique for calculating a particular solution to the inhomogeneous equation

$$x_{k+n} + a_1(k)x_{k+n-1} + \cdots + a_n(k)x_k = R_k, \tag{5.6.15}$$

provided we know the general solution to the homogeneous equation

$$x_{k+n} + a_1(k)x_{k+n-1} + \cdots + a_n(k)x_k = 0. \tag{5.6.16}$$

This technique is called the method of variation of constants and is similar to the one developed for differential equations [5].

We begin by assuming the n linearly independent solutions are already known for the homogeneous equation. They will be denoted as $x_i(k)$, for $1 \leq i \leq n$. We wish to determine functions $C_i(k)$, $1 \leq i \leq n$, such that the particular solution X_k of the inhomogeneous equation takes the form

$$X_k = C_1(k)x_1(k) + C_2(k)x_2(k) + \cdots + C_n(k)x_n(k). \tag{5.6.17}$$

We have

$$X_{k+1} = \sum_{i=1}^{n} C_i(k+1)x_i(k+1) \tag{5.6.18}$$

and

$$C_i(k+1) = C_i(k) + \Delta C_i(k). \tag{5.6.19}$$

If we now require

$$\sum_{i=1}^{n} x_i(k+1)\Delta C_i(k) = 0, \tag{5.6.20}$$

then

$$X_{k+1} = \sum_{i=1}^{n} C_i(k)x_i(k+1). \tag{5.6.21}$$

Likewise, we have from Eq. (5.6.21)

$$X_{k+2} = \sum_{i=1}^{n} C_i(k+1)x_i(k+2), \tag{5.6.22}$$

and this is equal to

$$X_{k+2} = \sum_{i=1}^{n} C_i(k)x_i(k+2), \tag{5.6.23}$$

if we set

$$\sum_{i=1}^{n} x_i(k+2)\Delta C_i(k) = 0. \tag{5.6.24}$$

Continuing this procedure gives

$$X_{k+n-1} = \sum_{i=1}^{n} C_i(k)x_i(k+n-1) \tag{5.6.25}$$

$$\sum_{i=1}^{n} x_i(k+n-1)\Delta C_i(k) = 0, \tag{5.6.26}$$

and

$$X_{k+n} = \sum_{i=1}^{n} C_i(k)x_i(k+n) + \sum_{i=1}^{n} x_i(k+n)\Delta C_i(k). \tag{5.6.27}$$

Now the substitution of Eqs. (5.6.21), (5.6.23), and (5.6.27) into the inhomogeneous equation (5.6.15) gives

$$\sum_{i=1}^{n} x_i(k+n)\Delta C_i(k) = R_k. \tag{5.6.28}$$

Note that Eqs. (5.6.20), (5.6.24)–(5.6.26), and (5.6.28) are a set of n linear equations for the $\Delta C_i(k)$, $1 \leq i \leq n$. Consequently, they can be solved to give

$$\Delta C_i(k) = \frac{f_i(k)}{C(k+1)}, \qquad 1 \leq i \leq n, \tag{5.6.29}$$

where $f_i(k)$ are known functions given in terms of the $x_i(k)$, $1 \leq i \leq n$, and R_k; and $C(k+1)$ is the Casoratian of the linearly independent set of solutions to the homogeneous equation. Its value is given by Eq. (5.6.14). Note that

$$C(k+1) \neq 0, \tag{5.6.30}$$

and this implies that when Eq. (5.6.29) is solved for the $C_i(k)$, the solutions are defined for all k.

It should be indicated that since the particular solution should not contain terms with arbitrary constants or terms appearing in the homogeneous solutions, all such terms can be dropped on their first appearance.

5.7 Worked Problems

5.7.1 *Example A*

The second-order linear, homogeneous difference equation

$$x_{k+2} - x_{k+1} - 2x_k = 0, \tag{5.7.1}$$

has the two solutions

$$x_1(k) = (-1)^k, \qquad x_2(k) = 2^k, \tag{5.7.2}$$

as can be shown by direct substitution into the difference equation, i.e.,

$$\begin{cases} x_1(k) = (-1)^k : (-1)^{k+2} - (-1)^{k+1} - 2(-1)^k = 0, \\ x_2(k) = 2^k : 2^{k+2} - 2^{k+1} - 2 \cdot 2^k = 0. \end{cases} \tag{5.7.3}$$

Let C_1 and C_2 be arbitrary constants. Then

$$x_k = C_1(-1)^k + C_2 2^k, \tag{5.7.4}$$

is also a solution since

$$[C_1(-1)^{k+2} + C_2 2^{k+2}] - [C_1(-1)^{k+1} + C_2 2^{k+1}] - 2[C_1(-1)^k + C_2 2^k]$$
$$= (-1)^k[C_1 + C_1 - 2C_1] + 2^k[4C_2 - 2C_2 - 2C_2]$$
$$= 0. \tag{5.7.5}$$

The Casoratian for $x_1(k) = (-1)^k$ and $x_2(k) = 2^k$ is

$$C(k) = \begin{vmatrix} (-1)^k & 2^k \\ (-1)^{k+1} & 2^{k+1} \end{vmatrix} = (-1)^k 2^k \begin{vmatrix} 1 & 1 \\ -1 & 2 \end{vmatrix}$$
$$= (-1)^k 2^k \cdot 3 \neq 0. \tag{5.7.6}$$

This result confirms the fact that $x_1(k)$ and $x_2(k)$ are linearly independent.

The Casoratian can also be calculated using Eq. (5.6.14). For this problem, $n = 2$ and $a_2(k) = -2$ and

$$C(k) = (-1)^{2k} A \prod_{i=0}^{k-1} (-2) = (-1)^k 2^k A, \tag{5.7.7}$$

where A is a constant.

5.7.2   Example B

The two functions

$$x_1(k) = 2^k, \qquad x_2(k) = 3^k, \qquad (5.7.8)$$

can be shown by direct substitution to be solutions to the second-order, homogeneous equation

$$x_{k+2} - 5x_{k+1} + 6x_k = 0. \qquad (5.7.9)$$

By inspection or by calculating the Casoratian, it follows that $x_1(k)$ and $x_2(k)$ are linearly independent, i.e.,

$$C(k) = \begin{vmatrix} 2^k & 3^k \\ 2^{k+1} & 3^{k+1} \end{vmatrix} = 2^k \cdot 3^k \begin{vmatrix} 1 & 1 \\ 2 & 3 \end{vmatrix} = 6^k \neq 0. \qquad (5.7.10)$$

Consider now the inhomogeneous equation

$$x_{k+2} - 5x_{k+1} + 6x_k = k^2. \qquad (5.7.11)$$

Again, direct substitution of

$$X_k = \left(\frac{1}{2}\right)(k^2 + 3k + 5), \qquad (5.7.12)$$

into Eq. (5.7.11) shows that it is a particular solution to the equation. Therefore, the general solution to Eq. (5.7.11) is given by the sum of the expressions in Eqs. (5.7.8) and (5.7.12), i.e.,

$$x_k = C_1 \cdot 2^k + C_2 \cdot 3^k + \left(\frac{1}{2}\right)(k^2 + 3k + 5). \qquad (5.7.13)$$

5.7.3   Example C

Let ϕ be a parameter such that $\sin\phi \neq 0$, and let R_k be a given function of k. Our task is to determine a general solution to the inhomogeneous equation

$$x_{k+1} - (2\cos\phi)x_k + x_{k-1} = R_k. \qquad (5.7.14)$$

Two linearly independent solutions to the homogeneous equation

$$x_{k+1} - (2\cos\phi)x_k + x_{k-1} = 0, \qquad (5.7.15)$$

are

$$x_1(k) = \cos(\phi k), \qquad x_2(k) = \sin(\phi k). \qquad (5.7.16)$$

The unknown functions, $C_1(k)$ and $C_2(k)$, in the expression for the particular solution, satisfy the following relations,

$$\begin{cases} [\cos\phi(k+1)]\Delta C_1(k) + [\sin\phi(k+1)]\Delta C_2(k) = 0, \\ [\cos\phi(k+2)]\Delta C_1(k) + [\sin\phi(k+2)]\Delta C_2(k) = R_{k+1}. \end{cases} \tag{5.7.17}$$

These two equations are linear in $\Delta C_1(k)$ and $\Delta C_2(k)$ and can be solved to obtain the expressions

$$\begin{cases} \Delta C_1(k) = -\dfrac{R_{k+1}\sin\phi(k+1)}{\sin\phi}, \\ \Delta C_2(k) = \dfrac{R_{k+1}\cos\phi(k+1)}{\sin\phi}. \end{cases} \tag{5.7.18}$$

These two equations can be solved by applying the Δ^{-1} operator to both sides of the equations. Carrying out this operation gives

$$\begin{cases} C_1(k) = -\displaystyle\sum_{i=1}^{k} \dfrac{R_i\sin(\phi i)}{\sin\phi}, \\ C_2(k) = \displaystyle\sum_{i=1}^{k} \dfrac{R_i\cos(\phi i)}{\sin\phi}. \end{cases} \tag{5.7.19}$$

Therefore, the particular solution to the inhomogeneous equation is

$$X_k = -\frac{\cos(\phi k)}{\sin\phi}\sum_{i=1}^{k} R_i\sin(\phi i) + \frac{\sin(\phi k)}{\sin\phi}\sum_{i=1}^{k} R_i\cos(\phi i)$$

$$= \left(\frac{1}{\sin\phi}\right)\sum_{i=1}^{k} R_i\sin\phi(k-i), \tag{5.7.20}$$

where the following trigonometric substitution was used to obtain the last expression

$$\sin(a-b) = \sin a\cos b - \cos a\sin b. \tag{5.7.21}$$

Finally, the general solution to Eq. (5.7.14) is given by a linear combination of the two solutions to the homogeneous equation along with X_k from Eq. (5.7.21), i.e.,

$$x_k = A\cos(\phi k) + B\sin(\phi k) + \left(\frac{1}{\sin\phi}\right)\sum_{i=1}^{k} R_k\sin\phi(k-i), \tag{5.7.22}$$

where A and B are arbitrary constants.

5.7.4 Example D

The homogeneous second-order difference equation

$$x_{k+2} - 3x_{k+1} + 2x_k = 0, \tag{5.7.23}$$

has the solutions

$$x_1(k) = 1, \qquad x_2(k) = 2^k. \tag{5.7.24}$$

Its general solution is therefore

$$x_k = A + B \cdot 2^k, \tag{5.7.25}$$

where A and B are arbitrary constants.

Now consider the inhomogeneous equation

$$x_{k+2} - 3x_{k+1} + 2x_k = 4^k + 3k^2, \tag{5.7.26}$$

where the inhomogeneous term is

$$R_k = 4^k + 3k^2. \tag{5.7.27}$$

The particular solution can be written as

$$X_k = C_1(k)2^k + C_2(k), \tag{5.7.28}$$

where the functions $C_1(k)$ and $C_2(k)$ are solutions to the equations

$$\begin{cases} 2^{k+1}\Delta C_1(k) + \Delta C_2(k) = 0, \\ 2^{k+2}\Delta C_1(k) + \Delta C_2(k) = 4^k + 3k^2. \end{cases} \tag{5.7.29}$$

Solving for $\Delta C_1(k)$ and $\Delta C_2(k)$ gives

$$\begin{cases} \Delta C_1(k) = 2^{k-1} + 3 \cdot \left(\dfrac{1}{2}\right)^{k+1} k^2, \\ \Delta C_2(k) = -4^k - 3k^2. \end{cases} \tag{5.7.30}$$

Solutions for both $C_1(k)$ and $C_2(k)$ can be obtained by applying the Δ^{-1} operator to both sides of the two expressions given in Eq. (5.7.30). Doing this gives

$$\begin{aligned}
C_1(k) &= \bar{A} + \left(\frac{1}{2}\right)\sum_{i=0}^{k-1} 2^i + \left(\frac{3}{2}\right)\sum_{i=0}^{k-1}\left(\frac{1}{2}\right)^i \cdot i^2 \\
&= \bar{A} - \left(\frac{1}{2}\right)(1 - 2^k) - 6\left[\left(\frac{k^2}{2} + k + \frac{3}{2}\right)\left(\frac{1}{2}\right)^k - \frac{3}{2}\right] \\
&= A_1 + \left(\frac{1}{2}\right)2^k - (3k^2 + 6k + 9)\left(\frac{1}{2}\right)^k
\end{aligned} \tag{5.7.31}$$

and

$$C_2(k) = \bar{B} - \sum_{i=0}^{k-1} 4^i - 3 \sum_{i=0}^{k-1} i^2$$

$$= B_1 - \left(\frac{1}{3}\right) 4^k - \left[k^3 - \left(\frac{3}{2}\right) k^2 + \frac{k}{2} \right], \qquad (5.7.32)$$

where $\bar{A}$ and $\bar{B}$ are arbitrary constants, and

$$A_1 = \bar{A} - \frac{1}{2} + 9, \qquad B_1 = \bar{B} + \frac{1}{3}. \qquad (5.7.33)$$

Since the particular solution should not contain terms having arbitrary constants, it follows that $C_1(k)$ and $C_2(k)$ are

$$C_1(k) = \left(\frac{1}{2}\right) 2^k - (3k^2 + 6k + 9) \left(\frac{1}{2}\right)^k, \qquad (5.7.34a)$$

$$C_2(k) = -\left(\frac{1}{3}\right) 4^k - \left[k^3 - \left(\frac{3}{2}\right) k^2 + \left(\frac{1}{2}\right) k \right]. \qquad (5.7.34b)$$

Substituting $C_1(k)$ and $C_2(k)$ into Eq. (5.7.28) gives

$$X_k = C_1(k) 2^k + C_2(k)$$

$$= \left[\left(\frac{1}{2}\right) 2^k - (3k^2 + 6k + 9) \left(\frac{1}{2}\right)^k \right] 2^k$$

$$- \left\{ \left(\frac{1}{3}\right) 4^k + \left[k^3 - \left(\frac{3}{2}\right) k^2 + \left(\frac{1}{2}\right) k \right] \right\}$$

$$= \left(\frac{1}{2}\right) 4^k - (3k^2 + 6k + 9) - \left(\frac{1}{3}\right) 4^k$$

$$- \left[k^3 - \left(\frac{3}{2}\right) k^2 + \left(\frac{1}{2}\right) k \right]$$

$$= \left(\frac{1}{6}\right) 4^k - k^3 - \left(\frac{3}{2}\right) k^2 - \left(\frac{13}{2}\right) k - 9. \qquad (5.7.35)$$

The constant, (-9), can be dropped since its inclusion gives a term already appearing in the homogeneous solution. Thus, the general solution to Eq. (5.7.26) is

$$x_k = A + B \cdot 2^k + \left(\frac{1}{6}\right) 4^k - k^3 - \left(\frac{3}{2}\right) k^2 - \left(\frac{13}{2}\right) k. \qquad (5.7.36)$$

5.8   Linear Difference Equations with Constant Coefficients

In this section, we study nth order, linear difference equations with constant coefficients. It will be shown that the homogeneous equation can always be solved exactly and the solutions expressed in terms of elementary functions. As pointed out earlier in this chapter, the theory of these equations match closely the corresponding theory for nth order, linear differential equations with constant coefficients. We also treat briefly the case of inhomogeneous equations for a special class of R_k.

5.8.1   *Homogeneous Equations*

Consider the following nth order, linear homogeneous difference equation with constant coefficients

$$x_{k+n} + a_1 x_{k+n-1} + a_2 x_{k+n-2} + \cdots + a_n x_k = 0, \qquad (5.8.1)$$

where $(a_i : i = 1, 2, \ldots, n)$ are given constants. Using the shift operator, E, Eq. (5.8.1) can be written as

$$f(E)x_k = 0, \qquad (5.8.2)$$

where $f(E)$ is the operator function

$$f(E) = E^n + a_1 E^{n-1} a_2 E^{n-2} + \cdots + a_{n-1} E + a_n. \qquad (5.8.3)$$

Definition   The characteristic equation associated with Eq. (5.8.1) is

$$f(r) = r^n + a_1 r^{n-1} + a_2 r^{n-2} + \cdots + a_{n-1} r + a_n = 0. \qquad (5.8.4)$$

Since $f(r)$ is an nth degree polynomial, it has n roots, $(r_i : i = 1, 2, \ldots, n)$ and can be written in factored form, i.e.,

$$f(r) = \prod_{i=1}^{n} (r - r_i) = (r - r_1)(r - r_2) \cdots (r - r_n) = 0. \qquad (5.8.5)$$

Theorem 5.8.1 *Let r_i be any solution of the characteristic equation, then*

$$x_k = (r_i)^k \qquad (5.8.6)$$

is a solution to the homogeneous Eq. (5.8.1).

Theorem 5.8.2 *Assume the n roots of the characteristic equation are distinct; then a fundamental set of solutions is*

$$x_i(k) = (r_i)^k, \qquad i = 1, 2, 3, \ldots, n. \qquad (5.8.7)$$

An immediate consequence of this theorem is that the general solution of the homogeneous equation (5.8.1) is

$$x_k = \sum_{i=1}^{n} C_i (r_i)^k, \tag{5.8.8}$$

where the C_i are n arbitrary constants. The next theorem deals with the case of multiple roots.

Theorem 5.8.3 *Let the roots of the characteristic equation*

$$r^n + a_1 r^{n-1} + \cdots + a_{n-1} r + a_n = 0, \tag{5.8.9}$$

be r_i with multiplicity $(m_i : i = 1, 2, \ldots, \ell)$, where $m_i \geq 1$, $\ell \leq n$, and

$$m_1 + m_2 + \cdots + m_\ell = n. \tag{5.8.10}$$

Then the general solution of

$$x_{k+n} + a_1 x_{k+n-1} + \cdots + a_n x_{k+1} + a_n x_k = 0, \tag{5.8.11}$$

is

$$\begin{aligned}
x_k = (r_1)^k & \left[A_1^{(1)} + A_2^{(1)} k + \cdots + A_{m_1}^{(1)} k^{m_1 - 1} \right] \\
& + (r_2)^k \left[A_1^{(2)} + A_2^{(2)} k + \cdots + A_{m_2}^{(2)} k^{m_2 - 1} \right] + \cdots \\
& + (r_{m_\ell})^k \left[A_1^{(\ell)} + A_2^{(\ell)} k + \cdots + A_{m_\ell}^{(\ell)} k^{m_\ell - 1} \right],
\end{aligned} \tag{5.8.12}$$

where the A's are arbitrary constants. (Note that there are exactly n coefficients, i.e., different A's.)

5.8.2 *Inhomogeneous Equations*

We now turn to the issue of how particular solutions are to be determined for an nth order, linear, inhomogeneous difference equations having constant coefficients, i.e., the equation

$$x_{k+n} + a_1 x_{k+n-1} + \cdots + a_{n-1} x_{k+1} + a_n x_k = R_k. \tag{5.8.13}$$

The procedure to be presented works only for R_k being a linear combination of terms having the form

$$a^k, \quad e^{bk}, \quad \sin(ck), \quad \cos(ck), \quad k^\ell, \tag{5.8.14}$$

or products of such terms,

$$a^k \sin(ck), \quad k^\ell e^{bk}, \quad a^k k^\ell \cos(ck), \quad \text{etc.} \tag{5.8.15}$$

In Eq. (5.8.14) (a, b, c) are real constants and ℓ is a non-negative integer. Note that if we allow for the possibility that the constant a can be complex valued, then each of the terms in Eqs. (5.8.14) and (5.8.15) are but special cases of $k^\ell a^k$.

To proceed, several definitions are needed.

Definition A family of a term R_k is the set of all functions of which R_k and $E^m R_k$, for $m = 1, 2, 3 \ldots$, are linear combinations.

Definition A finite family is a family that contains only a finite number of functions.

An example of a function having a finite family is $R_k = e^{ak}$. For this case

$$E^m a^k = a^m a^k, \qquad (5.8.16)$$

where the family consists of just one member, namely a^k.

As a second example, let $R_k = k^\ell$, where $\ell \geq 0$, is an integer. Now,

$$E^m k^\ell = (k + m)^\ell, \qquad (5.8.17)$$

and it should be clear that the family of k^ℓ is the set of functions $(1, k, k^2, \ldots, k^\ell)$.

We can also check that the family of $\cos(ck)$ or $\sin(ck)$ is $[\cos(ck), \sin(ck)]$.

For the case where R_k is a product, the family consists of all possible products of the distinct members of the individual term families. To illustrate this, consider $R_k = k^\ell a^k$. We have

$$\text{Family}(k^\ell) = \{1, k, \ldots, k^\ell\}, \quad \text{Family}(a^k) = \{a^k\}, \qquad (5.8.18)$$

and the family of $k^\ell a^k$ is

$$\text{Family}(k^\ell a^k) = \{a^k, ka^k, \ldots, k^\ell a^k\}. \qquad (5.8.19)$$

Similarly, we have for $R_k = k^\ell \cos(ck)$,

$$\text{Family}(k^\ell \cos(ck)) = \{\, \cos(ck), k\cos(ck), \ldots, k^\ell \cos(ck),$$
$$\sin(ck), k\sin(ck), \ldots, k^\ell \sin(ck)\}. \qquad (5.8.20)$$

With the above definitions, we can now provide a summary of the procedure needed to determine particular solutions to an n-th order, linear, inhomogeneous difference equation having constant coefficients:

(i) Construct the family of R_k. If it does not have a finite number of members, then stop. Some other procedure must be used to obtain the particular solution.

(ii) If the family of R_k contains a finite number of members and if the family contains no terms of the homogeneous solution, then write the particular solution X_k as a linear combination of the members of that family. Substitution of X_k into the inhomogeneous equation (5.8.13) and setting the coefficients of the linearly independent terms to zero will allow for a complete determination of all the constants of combination.

(iii) If the family of R_k contains terms of the homogeneous solution, then multiply each member of the family by the smallest integral power of k for which all such terms are removed. The particular solution X_k can then be written as a linear combination of the members of this modified family. Again, determine the constants of combination such that the inhomogeneous difference equation is identically satisfied.

The next section provides worked examples illustrating the application of this technique to several problems.

5.8.3 *Worked Examples*

Example A

The second-order equation

$$x_{k+2} - x_k = 0, \tag{5.8.21}$$

has the characteristic equation

$$r^2 - 1 = (r - 1)(r + 1) = 0, \tag{5.8.22}$$

with roots $r_1 = 1$ and $r_2 = -1$. Therefore, two linearly independent or fundamental solutions are

$$x_1(k) = 1, \qquad x_2(k) = (-1)^k. \tag{5.8.23}$$

For C_1 and C_2 arbitrary constants, the general solution is

$$x_k = C_1 + C_2(-1)^k. \tag{5.8.24}$$

Example B

Consider the following third-order difference equation

$$x_{k+3} + x_{k+2} - x_{k+1} - x_k = 0. \tag{5.8.25}$$

The characteristic equation with its factorization is
$$r^3 + r^2 - r - 1 = (r-1)(r+1)^2 = 0. \tag{5.8.26}$$
Therefore,
$$r_1 = 1, \qquad r_2 = r_3 = -1, \tag{5.8.27}$$
where the second root, $r = -1$, has multiplicity two. For this case, the fundamental solutions are
$$x_1(k) = 1, \qquad x_2(k) = (-1)^k, \qquad x_3(k) = (-1)^k k, \tag{5.8.28}$$
and the general solution to Eq. (5.8.25) is
$$x_k = C_1 + (C_2 + C_3 k)(-1)^k. \tag{5.8.29}$$

Example C

The second-order difference equation
$$x_{k+2} - 2x_{k+1} + 2x_k = 0, \tag{5.8.30}$$
has the characteristic equation
$$r^2 - 2r + 2 = 0, \tag{5.8.31}$$
with the solutions
$$r_1 = 1 + i, \quad r_2 = 1 - i, \quad i \equiv \sqrt{-1}. \tag{5.8.32}$$
We can rewrite the two complex roots to the forms
$$r_1 = r_2^* = \sqrt{2}e^{\pi i/4}. \tag{5.8.33}$$
Therefore, the two fundamental roots are
$$x_1(k) = (2)^{k/2} e^{\pi i k/4}, \qquad x_2(k) = (2)^{k/2} e^{-\pi i k/4}. \tag{5.8.34}$$
If we use
$$e^{\pm i\theta} = \cos\theta \pm i \sin\theta, \tag{5.8.35}$$
and note that $\cos\theta$ and $\sin\theta$ are linearly independent, then another set of fundamental solutions can be formed,
$$\bar{x}_1(k) = (2)^{k/2} \cos\left(\frac{\pi 4}{4}\right), \qquad \bar{x}_2(k) = (2)^{k/2} \sin\left(\frac{\pi 4}{4}\right). \tag{5.8.36}$$
Therefore, the general solution can be expressed in two equivalent ways, namely,
$$x_k = \left[C_1 \cos\left(\frac{\pi k}{4}\right) + C_2 \sin\left(\frac{\pi k}{4}\right)\right] (2)^{k/2} \tag{5.8.37}$$
or
$$x_k = \left[Ce^{i\pi k/4} + C^* e^{-i\pi k/4}\right] (2)^{k/2}, \tag{5.8.38}$$
where C_1 and C_2 are arbitrary real constants, and C is an arbitrary complex constant.

Example D

The characteristic equation for the difference equation

$$x_{k+1} = (2\cos\phi)x_k + x_{k-1} = 0, \tag{5.8.39}$$

where ϕ is a real parameter is

$$r^2 - (2\cos\phi)r + 1 = 0. \tag{5.8.40}$$

The two solutions to Eq. (5.8.40) are

$$r_{1,2} = \left(\frac{1}{2}\right)\left[2\cos\phi \pm \sqrt{4\cos^2\phi - 4}\right]$$

$$= \cos\phi \pm i\sin\phi = e^{\pm i\phi}. \tag{5.8.41}$$

Therefore, a fundamental set of solutions are

$$x_1(k) = \cos(\phi k), \qquad x_2(k) = \sin(\phi k), \tag{5.8.42}$$

and the general solution to Eq. (5.8.39) is

$$x_k = C_1\cos(\phi k) + C_2\sin(\phi k), \tag{5.8.43}$$

where C_1 and C_2 are arbitrary real constants.

Example E

The second-order difference equation

$$x_{k+2} - 5x_{k+1} + 6x_k = 2 + 4k \tag{5.8.44}$$

has $R_k = 2 + 4k$. The characteristic equation for the homogeneous part of Eq. (5.8.44) is

$$r^2 - 5r + 6 = (r - 3)(r - 2) = 0, \tag{5.8.45}$$

and, consequently, the homogeneous solution is

$$x_k^{(H)} = C_1 \cdot 3^k + C_2 \cdot 2^k. \tag{5.8.46}$$

Examination of R_k allows us to determine its family of functions, i.e.,

$$\text{Family}(2 + 4k) = \{1, k\}. \tag{5.8.47}$$

Since neither member occurs in $x_k^{(H)}$, the particular solution is

$$X_k = A + Bk, \tag{5.8.48}$$

where A and B are to be determined. Substitution of Eq. (5.8.48) into Eq. (5.8.44) gives

$$[A + B(k + 2)] - 5[A + B(k + 1)] + 6[A + Bk] = 2 + 4k. \tag{5.8.49}$$

Setting the coefficients of the terms k^0 and k equal to zero gives

$$2A - 3B = 2, \quad 2B = 4 \quad \text{or} \quad A = 4, \quad B = 2, \tag{5.8.50}$$

and the particular solution is

$$X_k = 4 + 2k. \tag{5.8.51}$$

Finally, we can write the general solution to Eq. (5.8.44) as

$$x_k = C_1 e^k + C_2 \cdot 2^k + 4 + 2k. \tag{5.8.52}$$

Example F

Consider the third-order difference equation

$$x_{k+3} - 7x_{k+2} + 16x_{k+1} - 12x_k = 2^k \cdot k. \tag{5.8.53}$$

The characteristic equation for the homogeneous part of the equation is

$$r^3 - 7r^2 + 16r - 12 = (r - 2)^2(r - 3) = 0, \tag{5.8.54}$$

and the homogeneous solution is

$$x_k^{(H)} = (C_1 + C_2 k)2^k + C_3 \cdot 3^k. \tag{5.8.55}$$

The family of $R_k = 2^k \cdot k$ is $[2^k, k \cdot 2^k]$. However, note that both members appear in the homogeneous solution. Therefore, we must multiply the family of R_k by k^2 to obtain a new family that does not contain any function appearing in $x_k^{(H)}$. Consequently, the particular solution takes the form

$$X_k = (Ak^2 + Bk^3)2^k, \tag{5.8.56}$$

where A and B are to be determined. Substitution of X_k into Eq. (5.8.53) and simplifying gives

$$2^k(-8A + 24B) + 2^k \cdot k(-24B) = 2^k \cdot k, \tag{5.8.57}$$

or

$$8A - 24B = 0, \quad -24B = 1 \quad \text{or} \quad A = -\left(\frac{1}{8}\right), \quad B = -\left(\frac{1}{24}\right). \tag{5.8.58}$$

Therefore the particular solution is

$$X_k = -\left(\frac{1}{24}\right)(3 + k)k^2 \cdot 2^k, \tag{5.8.59}$$

and the general solution to Eq. (5.8.53) is

$$x_k = (C_1 + C_2 k)2^k + C_3 \cdot 3^k - \left(\frac{1}{24}\right)(3 + k)k^2 \cdot 2^k. \tag{5.8.60}$$

5.9 Nonlinear Difference Equations

The situation for nonlinear difference equations is similar to that of nonlinear differential equations, i.e., no general methods exist for the construction of their solutions. In fact, from a structural viewpoint, the classes of difference equations that can be exactly solved correspond to the same classes of differential equations. The main purpose of this section is to introduce several special types of nonlinear difference equations and indicate how they may be solved.

5.9.1 *Homogeneous Equations*

A difference equation that can be expressed in the form

$$f\left(\frac{x_{k+1}}{x_k}, k\right) = 0, \tag{5.9.1}$$

is defined to be homogeneous equation. For the case where f is a polynomial function of x_{k+1}/x_k, then it can be written as

$$\prod_{i=1}^{n} [z_k - A_i(k)] = 0 \tag{5.9.2}$$

if it is of n-degree; the $A_i(k)$ are known functions of k; and z_k is taken to be

$$z_k = \frac{x_{k+1}}{x_k}. \tag{5.9.3}$$

It should be clear that the solutions to each of the linear equations

$$z_k - A_i(k) = 0 \Rightarrow x_{k+1} - A_i(k)x_k = 0, \tag{5.9.4}$$

gives a solution to Eq. (5.9.1).

5.9.2 *Riccati Equations*

Riccati equations are first-order, nonlinear difference equations taking the form

$$P(k)x_{k+1}x_k + Q(k)x_{k+1} + R(k)x_k = 0, \tag{5.9.5}$$

where $P(k)$, $Q(k)$, and $R(k)$ are given functions of k. Note that the substitution

$$y_k = \frac{1}{x_k}, \tag{5.9.6}$$

gives the first-order, linear, inhomogeneous equation

$$R(k)y_{k+1} + Q(k)y_k + P(k) = 0, \tag{5.9.7}$$

and this equation can be solved using methods presented in section 5.5.

A generalization of Eq. (5.9.5) is

$$P(k)x_{k+1}x_k + Q(k)x_{k+1}R(k)x_k = S(k). \tag{5.9.8}$$

Dividing by $P(k)$ and shifting the index by one unit gives

$$x_k x_{k-1} + A(k)x_k + B(k)x_{k-1} = C(k), \tag{5.9.9}$$

where

$$A(k) = \frac{Q(k-1)}{P(k-1)}, \quad B(k) = \frac{R(k-1)}{P(k-1)}, \quad C(k) = \frac{S(k-1)}{P(k-1)}. \tag{5.9.10}$$

We will now demonstrate that the transformation

$$x_k = \frac{y_k - B(k)y_{k+1}}{y_{k+1}}, \tag{5.9.11}$$

reduces Eq. (5.9.9) to a linear, second-order homogeneous equation.

The substitution of Eq. (5.9.11) into Eq. (5.9.9) gives, when simplified, the expression

$$y_k \{y_{k-1} + [A(k) - B(k-1)]y_k - [A(k)B(k) + C(k)]y_{k+1}\} = 0, \tag{5.9.12}$$

which for $y_k \neq 0$ is

$$[A(k)B(k) + C(k)]y_{k+1} - [A(k) - B(k-1)]y_k - y_{k-1} = 0. \tag{5.9.13}$$

Note that while Eq. (5.9.9) is first-order, Eq. (5.9.13) is second-order. Thus, we expect the general solutions to, respectively, have one and two arbitrary constants. We might expect the solution y_k to lead to x_k having two arbitrary constants through the transformation of Eq. (5.9.11). However, this is not the case as will be shown below.

Let the general solution to Eq. (5.9.13) be written as

$$y_k = D_1 v_k = D_2 w_k, \tag{5.9.14}$$

where v_k and w_k are any two linearly independent solutions, and D_1 and D_2 are two arbitrary constants. (We don't need to actually know the explicit forms for v_k and w_k for the arguments that follow.) The substitution of Eq. (5.9.14) into Eq. (5.9.11) gives

$$x_k = \frac{[v_k - B(k)v_{k+1}]D_1 + [w_k - B(k)w_{k+1}]D_2}{D_1 v_{k+1} + D_2 w_{k+1}} \tag{5.9.15}$$

$$= \frac{[v_k - B(k)v_{k+1}] + [w_k - B(k)w_{k+1}]\bar{D}}{v_{k+1} + \bar{D}w_{k+1}} \tag{5.9.16}$$

and an inspection of this relation shows that x_k depends not on D_1 and D_2, but only on the combination

$$\bar{D} = \frac{D_2}{D_1}. \tag{5.9.17}$$

Consequently, the correct number, i.e., one, of integration constants exist for x_k and y_k.

5.9.3 *Clairaut's Equation*

The Clairaut difference equation is a first-order, nonlinear equation having the structure

$$x_k = k\Delta x_k + f(\Delta x_k), \tag{5.9.18}$$

where f is a nonlinear function. If we define y_k to be

$$y_k = \Delta x_k, \tag{5.9.19}$$

then

$$x_k = ky_k + f(y_k), \tag{5.9.20}$$

and

$$\Delta x_k = y_k = (k+1)y_{k+1} - ky_k + f(y_{k+1}) - f(y_k). \tag{5.9.21}$$

Using the fact that

$$(k+1)y_{k+1} = (k+1)\Delta y_k + (k+1)y_k, \tag{5.9.22}$$

it follows that

$$(k+1)\Delta y_k + f(y_k + \Delta y_k) - f(y_k) = 0, \tag{5.9.23}$$

and either

$$\Delta y_k = 0 \quad \text{or} \quad (k+1) + \frac{f(y_k + \Delta y_k) - f(y_k)}{\Delta y_k} = 0. \tag{5.9.24}$$

The first possibility gives

$$y_k = c = \text{constant}, \tag{5.9.25}$$

and, from Eq. (5.9.20), the solution

$$x_k = ck + f(c). \tag{5.9.26}$$

The second possibility may also provide a second solution.

5.9.4   *Miscellaneous Forms*

The following special class of nonlinear difference equations can always be transformed into linear, nth order, inhomogeneous difference equations:

$$(x_{k+n})^{\gamma_1}(x_{k+n-1})^{\gamma_2}\cdots(x_k)^{\gamma_{n+1}} = f(k). \tag{5.9.27}$$

To obtain the required equation, take the logarithm of this equation

$$\gamma_1\ln(x_{k+n}) + \gamma_2\ln(x_{k+n-1}) + \cdots + \gamma_{n+1}\ln(x_k) = \ln[f(k)], \tag{5.9.28}$$

and define

$$y_k = \ln(x_k), \qquad g(k) = \ln[f(k)]. \tag{5.9.29}$$

Therefore, y_k satisfies the equation

$$\gamma_1 y_{k+n} + \gamma_2 y_{k+n-1} + \cdots + \gamma_{n+1}y_k = g(k), \tag{5.9.30}$$

an equation that can be solved by use of methods given earlier in the chapter. Denote the n linear independent solutions of Eq. (5.9.30) as

$$\{\bar{y}_i(k) : i = 1, 2, \ldots, n\}, \tag{5.9.31}$$

then the general solution is

$$y_k = \sum_{i=1}^{n} C_i\bar{y}_i(k), \tag{5.9.32}$$

where the C_i are n arbitrary constants. The solution x_k is then

$$x_k = \exp\left[\sum_{i=1}^{n} C_i\bar{y}_i(k)\right]. \tag{5.9.33}$$

Finally, it should be indicated that if a solution to the difference equation

$$F(x_{k+n}, x_{k+n-1}, \ldots, x_k, k) = 0 \tag{5.9.34}$$

is known, then a solution of the difference equation

$$F[h(y_{k+n}), h(y_{k+n-1}), \ldots, h(y_k), k] = 0, \tag{5.9.35}$$

where h is a given function, is also known, and given by

$$x_k = h(y_k). \tag{5.9.36}$$

Equation (5.9.27) is a special case of Eq. (5.9.35).

5.10 Worked Examples of Nonlinear Equations

5.10.1 *Example A*

The homogeneous equation

$$(x_{k+1})^2 + (3 - k)x_{k+1}x_k - 3k(x_k)^2 = 0, \qquad (5.10.1)$$

can be factored as

$$(x_{k+1} + 3x_k)(x_{k+1} - kx_k) = 0. \qquad (5.10.2)$$

Therefore,

$$x_{k+1} + 3x_k = 0 \quad \text{or} \quad x_{k+1} - kx_k = 0, \qquad (5.10.3)$$

and the solution to Eq. (5.10.2) is

$$x_k = c_1(-3)^k \quad \text{or} \quad x_k = c_2(k - 1)!, \qquad (5.10.4)$$

where c_1 and c_2 are arbitrary constants.

5.10.2 *Example B*

The nonlinear difference equation

$$x_{k+1}x_k + ax_{k+1} + bx_k = c, \qquad (5.10.5)$$

with (a, b, c) known constants, can be transformed into the second-order, linear equation

$$(ab + c)y_{k+1} - (a - b)y_k - y_{k-1} = 0, \qquad (5.10.6)$$

by use of the substitution

$$x_k = \frac{y_k}{y_{k+1}} - b. \qquad (5.10.7)$$

The characteristic equation for Eq. (5.10.6) is

$$(ab + c)r^2 - (a - b)r - 1 = 0. \qquad (5.10.8)$$

If the two solutions to Eq. (5.10.8) are denoted as r_1 and r_2, then

$$y_k = D_1(r_1)^k + D_2(r_2)^k, \qquad (5.10.9)$$

where D_1 and D_2 are arbitrary constants. Substitution of Eq. (5.10.9) into Eq. (5.10.7) gives, after some algebra, the result

$$x_k = \frac{(1 - br_1) + (1 - br_2)\bar{D}(r_2/r_1)^k}{r_1 + \bar{D}(r_2/r_1)^{k+1}}, \qquad (5.10.10)$$

where $\bar{D} = D_2/D_1$.

5.10.3   *Example C*

Consider the Clairaut equation

$$x_k = k\Delta x_k \neq (\Delta x_k)^2, \tag{5.10.11}$$

and let $y_k = \Delta x_k$; then

$$x_k = k y_k + y_k^2. \tag{5.10.12}$$

Applying the operator Δ gives

$$(k+1)\Delta y_k + 2y_k\Delta y_k + (\Delta y_k)^2 = 0, \tag{5.10.13}$$

and either

$$\Delta y_k = 0 \Rightarrow x_k = ck + c^2, \tag{5.10.14}$$

or, we have

$$(k+1) + 2y_k + \Delta y_k = 0, \tag{5.10.15}$$

which has the solution

$$y_k = c_1(-1)^k - \frac{k}{2} - \frac{1}{4}, \tag{5.10.16}$$

where c_1 is an arbitrary constant. Substitution into Eq. (5.10.12) gives the following second solution of Eq. (5.10.11),

$$x_k = \left[c_1(-1)^k - \frac{1}{4} \right]^2 - \frac{k^2}{4}. \tag{5.10.17}$$

5.10.4   *Example D*

The second-order, nonlinear difference equation

$$x_{k+2}x_k = (x_{k+1})^2 \tag{5.10.18}$$

can be transformed into the following second-order, linear equation

$$y_{k+2} - 2y_{k+1} + y_k = 0, \tag{5.10.19}$$

by means of the transformation

$$y_k = \ln(x_k). \tag{5.10.20}$$

The solution of Eq. (5.10.19) is easily found to be

$$y_k = c_1 + c_2 k, \tag{5.10.21}$$

where c_1 and c_2 are arbitrary constants. Thus, using Eq. (5.10.20), the general solution to Eq. (5.10.18) is

$$x_k = c_3 e^{c_2 k}, \qquad c_3 = e^{c_1} > 0. \tag{5.10.22}$$

5.10.5   *Example E*

Consider the equation

$$(x_{k+2})^2 - 4(x_{k+1})^2 + 3(x_k)^2 = k, \qquad (5.10.23)$$

and make the substitution

$$y_k = (x_k)^2 \qquad (5.10.24)$$

to obtain

$$y_{k+2} - 4y_{k+1} + 3y_k = k. \qquad (5.10.25)$$

The general solution to this equation is

$$y_k = c_1 + c_2(3)^k - \frac{k^2}{4}, \qquad (5.10.26)$$

and, consequently,

$$(x_k)^2 = c_1 + c_2(3)^k - \frac{k^2}{4}. \qquad (5.10.27)$$

Therefore, Eq. (5.10.23) has the two solutions

$$x_k^{(1)} = + \left[c_1 + c_2(3)^k - \frac{k^2}{4} \right]^{1/2}, \qquad (5.10.28)$$

or

$$x_k^{(2)} = - \left[c_1 + c_2(3)^k - \frac{k^2}{4} \right]^{1/2}. \qquad (5.10.29)$$

5.11   Two Applications

5.11.1   *Chebyshev Polynomials*

Special classes of polynomials are of value for the study of certain issues that arise in both applied and pure mathematics. An important example of such a class of polynomial functions is the Chebyshev polynomials [6] denoted by the symbol $T_k(x)$. They are defined by the recurrence formula

$$T_{k+2} - xT_{k+1} + \left(\frac{1}{4} \right) T_k = 0, \qquad (5.11.1)$$

where x is taken to be in the interval

$$|x| \leq 1, \qquad (5.11.2)$$

and

$$T_0 = 2, \qquad T_1 = x. \qquad (5.11.3)$$

We now investigate some of the properties of these functions.

Using the starting values given in Eq. (5.11.3), it is easy to calculate the other Chebyshev polynomials; for example

$$\begin{cases} T_2(x) = x^2 - \dfrac{1}{2}, \\[2mm] T_3(x) = x^3 - \dfrac{3x}{4}, \\[2mm] T_4(x) = x^4 - x^2 + \dfrac{1}{8}. \end{cases} \qquad (5.11.4)$$

Proceeding in this manner, the $T_k(x)$ for any finite positive integer can be obtained. However, this procedure is very laborious and it would be much better to have a compact expression giving $T_k(x)$ explicitly in terms of x and k. This can be easily done since Eq. (5.11.1) is a second-order, linear difference equation, with constant (in k) coefficients.

To proceed, note that the characteristic equation for Eq. (5.11.1) is

$$r^2 - xr + \frac{1}{4} = 0, \qquad (5.11.5)$$

and its two solutions are

$$2r_{1,2} = x \pm \sqrt{x^2 - 1}. \qquad (5.11.6)$$

Therefore, the kth Chebyshev polynomial is given by

$$T_k(x) = \left(\frac{1}{2^k}\right)\left[A(r_1)^k + B(r_2)^k\right] \qquad (5.11.7)$$

where the constants A and B are determined by the requirement given on T_0 and T_1, i.e.,

$$\begin{cases} A + B = 2, \\ r_1 A + r_2 B = 2x. \end{cases} \qquad (5.11.8)$$

The solution for A and B is

$$A = B = 1, \qquad (5.11.9)$$

and, consequently,

$$T_k(x) = \left(\frac{1}{2^k}\right)\left[\left(x + \sqrt{x^2 - 1}\right)^k + \left(x - \sqrt{x^2 - 1}\right)^k\right]. \qquad (5.11.10)$$

Let us now examine Eq. (5.11.10) in more detail. Since $|x| \leq 1$, we have $x^2 \leq 1$, and

$$\sqrt{x^2 - 1} = i\left(\sqrt{1 - x^2}\right), \qquad i = \sqrt{-1}. \qquad (5.11.11)$$

Therefore,

$$x \pm \sqrt{x^2 - 1} = x \pm i\sqrt{1 - x^2} = e^{\pm i\phi}, \tag{5.11.12}$$

where

$$\tan \phi(x) = \frac{\sqrt{1 - x^2}}{x}, \tag{5.11.13}$$

and

$$\left(x + \sqrt{x^2 - 1}\right)^k + \left(x - \sqrt{x^2 - 1}\right)^k = e^{ik\phi} + e^{-ik\phi} = 2\cos(k\phi). \tag{5.11.14}$$

This last result means that $T_k(x)$ has the representation

$$T_k(x) = \frac{\cos[k\phi(x)]}{2^{k-1}}. \tag{5.11.15}$$

It follows from Eq. (5.11.12) that

$$\cos \phi = x \quad \text{or} \quad \phi = \cos^{-1} x. \tag{5.11.16}$$

Therefore, $T_k(x)$ can be written as

$$T_k(x) = \frac{\cos(k \cos^{-1} x)}{2^{k-1}}, \qquad |x| \le 1. \tag{5.11.17}$$

Note that while the results given in Eqs. (5.11.10) and (5.11.17) are mathematically equivalent, in general, the form expressed by Eq. (5.11.10) would be the one to use for an explicit calculation of $T_k(x)$.

As an elementary application of the Chebyshev polynomials, we show how to expand the function

$$f(x) = 2x^4 - 3x^2 + x + 7 \tag{5.11.18}$$

in terms of these polynomials. Using Eqs. (5.11.3) and (5.11.4), the first five Chebyshev polynomials can be inverted to express $(1, x, x^2, x^3, x^4)$ as follows:

$$\begin{cases} 1 = \dfrac{T_0(x)}{2}, \\ x = T_1(x), \\ x^2 = T_2(x) + \frac{1}{2}, \\ x^3 = T_3(x) + \left(\frac{3}{4}\right) T_1(x), \\ x^4 = T_4(x) - T_2(x) - \frac{3}{8}. \end{cases} \tag{5.11.19}$$

Substituting these results into $f(x)$ gives

$$f(x) = \left(\frac{5}{4}\right) T_0(x) + T_1(x) - T_2(x) + 2T_4(x). \tag{5.11.20}$$

It can be shown [6], under suitable conditions, that a function $h(x)$, defined on $-1 \leq x \leq 1$, can be expanded in terms of a convergent series of the $T_k(x)$, i.e.,

$$h(x) = \sum_{n=0}^{\infty} a_n T_n(x). \qquad (5.11.21)$$

5.11.2   A Discrete Logistic Equation

The logistic differential equation is

$$\frac{dx}{dt} = x(1 - x), \qquad x(0) = x_0 \geq 0. \qquad (5.11.22)$$

This equation is separable and with the use of partial fractions, an exact solution can be calculated; it is

$$x(t) = \frac{x_0}{x_0 + (1 - x_0)e^{-t}}. \qquad (5.11.23)$$

The differential equation has two fixed-points or equilibrium solutions; they are

$$\bar{x}^{(1)} = 0, \qquad \bar{x}^{(2)} = 1. \qquad (5.11.24)$$

Examination of Eq. (5.11.22) shows that all solutions, with initial values $x_0 > 0$, approach $x = 1$ monotonically. The reason for this conclusion is based on the observation that if $0 < x < 1$, then dx/dt is positive, while if $x > 1$, the derivative is negative. Since $x = 1$ is a fixed-point, all solutions with $x_0 > 0$ must approach it monotonically. Therefore, $x = \bar{x}^{(1)} = 0$ is unstable and $x = \bar{x}^{(2)}$ is stable. A study of the explicit solution, as given in Eq. (5.11.23) also shows this to be correct.

Consider the following discrete model for the logistic equation [3]

$$\frac{x_{k+1} - x_k}{h} = x_k(1 - x_{k+1}), \qquad x_0 = \text{given}, \qquad (5.11.25)$$

where we have used

$$\begin{cases} t \to t_k = hk, \qquad h = \Delta t, \quad k = \text{integer}, \\ x(t) \to x(t_k) \simeq x_k. \end{cases} \qquad (5.11.26)$$

Note that the derivative is approximated by a forward-Euler representation

$$\frac{dx}{dt} \to \frac{x(t_k + h) - x(t_k)}{h} \qquad (5.11.27)$$

the linear term by

$$x \to x_k, \qquad (5.11.28)$$

and the quadratic, nonlinear term

$$x^2 \to x_k x_{k+1}. \qquad (5.11.29)$$

Our main reason for introducing this particular discrete model for the logistic equation is to show that its solution has the same qualitative properties as the exact solution.

Equation (5.11.25) can be solved exactly. Let

$$x_k = \frac{1}{y_k}, \qquad (5.11.30)$$

then y_k satisfies the following first-order, linear, inhomogeneous equation

$$y_{k+1} - \left(\frac{1}{1+h} \right) y_k = \frac{h}{1+h}, \qquad (5.11.31)$$

and has the solution

$$y_k = 1 + A(1+h)^{-k}. \qquad (5.11.32)$$

Using the initial condition, $y_0 = 1/x_0$, then A is

$$A = \frac{1 - x_0}{x_0}, \qquad (5.11.33)$$

and x_k can be written as

$$x_k = \frac{x_0}{x_0 + (1 - x_0)(1+h)^{-k}}. \qquad (5.11.34)$$

An examination of Eq. (5.11.34) shows that its solution has the same qualitative properties as the exact solution to the logistic equation given by Eq. (5.11.23). Note that the exponential function in the exact solution is approximated by the term $(1 + h)^{-k}$ in the solution to the discrete model of the logistic equation.

For a more detailed discussion of the construction of discrete models for differential equations and difficulties that can arise see Mickens [3].

Problems

Section 5.3

5.3.1 Prove that Δ is a linear operator. See Eq. (5.3.3). Also, show that

$$E(x_k y_k) = (Ex_k)(Ey_k),$$
$$E(x_k)^n = (Ex_k)^n,$$
$$Ef(x_k) = f(Ex_k).$$

5.3.2 Prove Theorem 5.3.1 (see Eq. (5.3.8)). Hint: Use mathematical induction.

5.3.3 Prove Leibnitz's theorem for differences, i.e., Eq. (5.3.28).

5.3.4 Derive the following result

$$\Delta^{-3}x_k = \sum_{\ell=1}^{k-1}\sum_{m=1}^{\ell-1}\sum_{i=1}^{m-1} x_i + c_1 k^2 + c_2 k + c_3.$$

5.3.5 Write out an explicit representation for $(\sum)^n$ which appears in (5.3.45).

5.3.6 Is the operator Δ^{-1} linear? Prove your answer.

5.3.7 Prove the fundamental theorem of the sum calculus given by (5.3.53).

5.3.8 Let $x_k = a^k$ and $y_k = 1 + k + 3^k$. Calculate the following expressions:

(a) $\Delta(x_k y_k)$,   (b) $\Delta^2(x_k y_k)$,

(c) $\Delta\left(\dfrac{x_k}{y_k}\right)$,   (d) $\Delta^n x_k$,

(e) $\Delta^n y_k$.

5.3.9 Use Leibnitz's theorem to evaluate the expressions:

(a) $\Delta^n(k^3 a^k)$ (b) $\Delta^n(k^3 a^{-k})$

(b) $\Delta^n[(-1)^k a^k]$ (d) $\Delta^n[a^k \cdot b^k]$

Section 5.4

5.4.1 Derive the following recursion relations for the two types of Stirling numbers

$$s_i^{n+1} = s_{i-1}^n - n s_i^n,$$
$$S_i^{n+1} = S_{i-1}^n - i S_i^n,$$

Note that these are partial difference equations since two discrete variables i and n appear.

5.4.2 Prove the results given in Eqs. (5.4.27), (5.4.28), and (5.4.29).

5.4.3 Show that the first difference of $(a + bk)^{(n)}$, where a and b are constants, is

$$\Delta(a + bk)^n = bn(a + bk)^{(n-1)}.$$

Also, prove that

(i) $\Delta^m(a + bk)^{(n)} = b^m n(n - 1)\cdots(n - m + 1)(a + bk)^{(n-m)}$, $0 < m < n$,

(ii) $\Delta^n(a + bk)^{(n)} = b^n n!$,

(iii) $\Delta^p(a + bk)^{(n)} = 0$, $p > n$.

Section 5.5

5.5.1 Use the sum calculus to prove the result given by Eq. (5.5.14).

5.5.2 Find the general solutions to the following equations:

(a) $x_{k+1} - x_k = e^{-k}$,

(b) $x_{k+1} + x_k = e^{-k}$,

(c) $x_{k+1} - \left(\frac{k+1}{k}\right) x_k = (k+1)k$,

(d) $x_{k+1} - ax_k = \cos(bk)$.

5.5.3 Why must the lower limit for i in Eq. (5.5.30) be taken as one rather than zero?

5.5.4 Obtain the solution to the equation

$$x_{k+1} - \left[\frac{a(k - \alpha_1)(k - \alpha_2) \cdots (k - \alpha_n)}{(k - \beta_1)(k - \beta_2) \cdots (k - \beta_m)}\right] x_k = 0,$$

where the $(\alpha_i : i = 1, 2, \ldots, n)$ and $(\beta_j : j = 1, 2, \ldots, m)$ are constants, and a is a constant.

Section 5.6

5.6.1 Construct the proofs for Theorems 5.6.1, 5.6.2, 5.6.3, and 5.6.4.

5.6.2 Prove Theorems 5.6.5 and 5.6.6.

5.6.3 Using Eq. (5.6.13), show that for n even, the Casoratian has a definite sign for all k.

Section 5.7

5.7.1 Determine which of the following sets of functions are linearly dependent and which are linearly independent:

(a) $\{1, (-1)^k, k\}$

(b) $\{a^k, ka^k, k^2 a^k\}$,

(c) $\{k, k^2, k^5\}$,

(d) $\{1, a^k, b^k\}$, $a \neq b$,

(e) $\{e^{ik}, e^{-ik}, \cos k\}$,

(f) $\{2^k, 3 \cdot 2^k, k\}$,

(g) $\{1, (-1)^k, k, (-1)^k k\}$

5.7.2 Determine particular solutions to

$$x_{k+1} - (2\cos\phi)x_k + x_{k-1} = R_k, \qquad \sin\phi = 0,$$

for the following R_k:

(a) $R_k = R = $ constant,

(b) $R_k = e^{-bk}$, $b > 0$,

(c) $R_k = \cos(ak)$,

(d) $R_k = e^{-bk}\cos(ak)$.

5.7.3 Calculate a particular solution to the equation

$$x_{k+2} - 2x_{k+1} + x_k = 1 + 3e^{-k},$$

given that the homogeneous equation has the solutions $x_1(k) = 1$, $x_2(k) = k$.

Section 5.8

5.8.1 Prove Theorem 5.8.1.

5.8.2 Prove Theorem 5.8.2.

5.8.3 Prove Theorem 5.8.3.

5.8.4 Show that a linear relationship holds between the real constants (c_1, c_2) in Eq. (5.8.37) and the real and imaginary parts of the complex valued constant c in Eq. (5.8.38).

5.8.5 Express the solution to Eq. (5.8.39) as a pair of complex-conjugated pair of functions. Calculate the general solutions to the following equations:

(a) $x_{k+2} - 6x_{k+1} + 8x_k = 2 + 3k^2 - 5 \cdot 3^k$,

(b) $x_{k+2} - x_{k-1} = 4^k \cdot k$,

(c) $x_{k+2} - (2\cos\phi)x_k + x_{k-1} = 2\cos(\phi k)$

(d) $x_{k+2} - 2x_{k+1} + x_k = 5 + k^3$,

(e) $x_{k+4} + x_k = 1 + (-1)^k$.

Section 5.9

5.9.1 Assume that $A_j(k) = A_\ell(k)$, where

$$i \leq i < j \leq n,$$

in Eq. (5.9.2). Does the analysis of how solutions to the homogeneous Eq. (5.9.1) require any changes?

5.9.2 Derive Eq. (5.9.13) from Eqs. (5.9.9) and (5.9.11).

5.9.3 Show that the transformation

$$x_k = \frac{y_{k+1} - A(k+1)y_k}{y_k}$$

reduces

$$x_k x_{k-1} + A(k)x_k + B(k)x_{k-1} = C(k)$$

to the second-order, linear, homogeneous equation

$$y_{k+1} - [A(k+1) - B(k)]y_k - [A(k)B(k) + C(k)]y_{k-1} = 0.$$

Section 5.10

5.10.1 Does the following nonlinear equation have a solution

$$(x_{k+1})^2 + (x_k)^2 = 0?$$

5.10.2 Solve the following Riccati-type difference equations. All constants are assumed to be positive.

(a) $x_{k+1} = \frac{1}{x_k}$,
(b) $x_{k+1} = \lambda x_k(1 - x_{k+1})$,
(c) $x_{k+1} = \frac{\lambda x_k}{1 + b x_k} - h$.

Plot, for each equation, x_k vs k.

5.10.3 Consider the following question: Is every exactly solvable nonlinear difference equation reducible to a linear difference equation?

5.10.4 Are the following two nonlinear, first-order difference equations equivalent

(a) $\sqrt{x_{k+1}} = x_k$,
(b) $x_{k+1} = (x_k)^2$?

5.10.5 Construct solutions to

$$x_{k+1} = k\Delta x_k + (\Delta x_k)^3.$$

Section 5.11

5.11.1 Use $T_k(x)$, as expressed in Eq. (5.11.10), to calculate $T_2(x)$, $T_3(x)$, and $T_4(x)$. Compare these values with those given in Eq. (5.11.4).

5.11.2 Calculate using either Eqs. (5.11.10) or (5.11.15), the following values: $T_k(1)$, $T_k(-1)$, and $T_k(0)$.

5.11.3 Derive a differential equation satisfied by the Chebyshev polynomials.

5.11.4 Apply the transformation given in Eq. (5.11.29) to Eq. (5.11.24) and show that the result is Eq. (5.11.30). Solve this equation to get x_k as a function of x_0 and k.

5.11.5 Show that for $0 < h \ll 1$, $k = $ fixed, the following relation holds,

$$\frac{1}{(1+h)^k} = e^{-hk}[1 + O(h^2 k)] = e^{-t_k}[1 + O(ht_k)].$$

5.11.6 Prove that for any $h > 0$, the solutions of the discrete model, given by Eq. (5.11.25), have the same qualitative properties as the exact solution to the logistic differential equation.

5.11.7 The following is a second discrete model for the logistic equation

$$\frac{x_{k+1} - x_k}{h} = x_k(1 - x_k).$$

Show that this equation can be transformed to

$$y_{k+1} = \lambda y_k(1 - y_k).$$

Plot y_k vs k for values of λ equal to 1.01, 2.500, 3.200, 3.500, 3.566, 3.840, and 4.000. For each value of λ, select several values of x_0 and study the various behaviors. What conclusions can you make regarding the suitability of this scheme to provide accurate numerical solutions for the logistic equation.

5.11.8 Analyze the following discrete model for the logistic equation

$$\frac{x_{k+1} - x_k}{\phi(h)} = x_k(1 - x_k),$$

where

$$\phi(h) = 1 - e^{-h}.$$

Take h to be 0.01, 0.1, 1.0, 10.0, 100.0.

Comments and References

[1] R. B. Potts, *American Mathematical Monthly* **89**, 402 (1982).

[2] R. E. Mickens, *Numerical Methods for Partial Differential Equations* **5**, 313 (1989).

[3] R. E. Mickens, *Nonstandard Finite Difference Models of Differential Equations* (World Scientific, Singapore, 1994).

[4] See refs. [Brand (1966), Elaydi (1995), Fort (1948), Miller (1960, 1968), Kelley (1991)] in the Bibliography to this chapter.

[5] S. L. Ross, *Introduction to Ordinary Differential Equations*, 4th ed. (Wiley; Hoboken, NJ; 1989).

[6] N. N. Lebedev, *Special Functions and Their Applications* (Dover, New York, 1972).

Bibliography

P. M. Batchelder, *An Introduction to Linear Difference Equations* (Harvard University Press, Cambridge, 1927).

G. Boole, *Calculus of Finite Differences*, 4th ed. (Chelsea, New York, 1958).

L. Brand, *Differential and Difference Equations* (Wiley, New York, 1966).

J. A. Cadzow, *Discrete-Time Systems* (Prentice-Hall; Englewood Cliffs, NJ; 1973)

F. Chorlton, *Differential and Difference Equations* (Van Nostrand, London, 1965).

S. N. Elaydi, *An Introduction to Difference Equations* (Springer, New York, 1995).

T. Fort, *Finite Differences and Difference Equations in the Real Domain* (Clarendon Press, Oxford, 1948).

S. Goldberg, *Introduction to Difference Equations* (Wiley, New York, 1958).

F. B. Hilderbrand, *Methods of Applied Mathematics*, 2nd ed. (Prentice-Hall; Englewood Cliffs, NJ; 1956).

F. B. Hilderbrand, *Finite-Difference Equations and Simulations* (Prentice-Hall; Englewood Cliffs, NJ; 1968).

C. Jordan, *Calculus of Finite Differences*, 3rd ed. (Chelsea, New York, 1965).

H. Levy and F. Lessman, *Finite Difference Equations* (Macmillan, New York, 1961).

K. S. Miller, *An Introduction to the Calculus of Finite Differences and Difference Equations* (Holt, New York, 1960).

K. S. Miller, *Linear Difference Equations* (W. A. Benjamin, New York, 1968).

L. M. Milne-Thomson, *The Calculus of Finite Differences* (Macmillan, London, 1960).

W. G. Kelley and A. C. Peterson, *Difference Equations: An Introduction with Applications* (Academic Press, Boston, 1991).

C. H. Richardson, *An Introduction to the Calculus of Finite Differences* (Van Nostrand, New York, 1954).

M. R. Spiegel, *Calculus of Finite Differences and Difference Equations* (McGraw-Hill, New York, 1971).

J. Wimp, *Computation and Recurrence Relations* (Pitman; Marshfield, MA; 1984).

Chapter 6

Sturm-Liouville Problems

6.1 Introduction

Mathematical descriptions of systems often are based on prescribed values of dependent variables on the boundaries of the system. The vibrations of a clamped string is an example of such a system. For this case, no string motions occur at the clamped or endpoints. A second example is that of an object placed in a uniform fluid flow; we expect the flow velocity to be constant at large distances from the object. The purpose of this chapter is to consider some of the mathematical properties of systems modeled by second-order differential equations for which the solutions take particular values at their boundaries. A large class of systems are characterized mathematically as being Sturm-Liouville problems. We introduce the various issues related to this situation by first examining in detail the vibrating string. Next, we state, without proof, several important theorems which form the background knowledge needed to understand Sturm-Liouville problems. We show that the differential equations and associated boundary conditions for the special (polynomial) functions are Sturm-Liouville problems and indicate how Green's functions can be calculated. The chapter ends with a discussion of how the asymptotic behavior of certain types of second-order, linear differential equations can be determined. The method is illustrated by application to Bessel functions.

As stated above, we do not provide proofs of any of the stated theorems and related results. However, in many instances such proofs are straightforward to obtain. The books listed in the bibliography to this chapter do contain these proofs and other materials extending the discussion of Sturm-Liouville problems and the asymptotics of solutions to differential equations.

6.2   The Vibrating String

Consider an elastic string of length L. Its equation of motion is

$$\frac{d^2u}{dx^2} + \lambda u = 0, \tag{6.2.1}$$

where x denotes the space coordinate, $0 \leq x \leq L$, and $u(\bar{x})$ is the displacement of the string at $x = \bar{x}$. The parameter λ is *a priori* not specified, but will take in values depending on what are selected for the magnitudes of u and its first-derivative at $x = 0$ and $x = L$. Three possibilities exist for these boundary values:

(i) $x(0) = 0$, $x(L) = 0$; $\tag{6.2.2}$
(ii) $x(0) = 0$, $x'(L) = 0$; $\tag{6.2.3}$
(iii) $x'(0) = 0$, $x'(L) = 0$. $\tag{6.2.4}$

Case (i) corresponds to the string being clamped at both ends. Case (ii) is the situation for which the string is clamped at $x = 0$, but allowed to be free at $x = L$. Finally, in case (iii) both ends are allowed to be free.

Our task is to calculate the solutions to Eq. (6.2.1) and determine what values of λ are possible. Note that each possible λ will have an associated solution u_λ. Consequently, the second part of the task will be to also find the u_λ.

6.2.1   *Fixed Ends*

The general solution to Eq. (6.2.1) is

$$u(x) = A\sin\left(\sqrt{\lambda}x\right) + B\cos\left(\sqrt{\lambda}x\right). \tag{6.2.5}$$

For fixed ends, we have

$$u(0) = 0, \qquad u(L) = 0, \tag{6.2.6}$$

with

$$\begin{cases} u(0) = 0 \implies B = 0; \\ u(L) = 0 \implies A\sin\left(\sqrt{\lambda}L\right) = 0. \end{cases} \tag{6.2.7}$$

The latter requirement gives

$$\sqrt{\lambda}L = n\pi, \qquad (n = 1, 2, 3, \dots). \tag{6.2.8}$$

Since L is the fixed length of the string, then λ can only take on a discrete set of values given by

$$\lambda_n = \left(\frac{n\pi}{L}\right)^2, \qquad (n = 1, 2, 3 \dots). \tag{6.2.9}$$

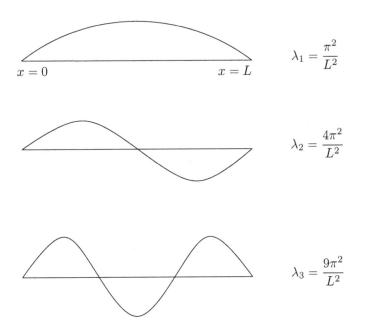

$$\lambda_1 = \frac{\pi^2}{L^2}$$

$$\lambda_2 = \frac{4\pi^2}{L^2}$$

$$\lambda_3 = \frac{9\pi^2}{L^2}$$

Fig. 6.2.1 Plots of $u_1(x)$, $u_2(x)$ and $u_3(x)$ for the vibrating string with both ends fixed.

If ω_0 is defined as

$$\omega_0 = \left(\frac{\pi}{L}\right), \tag{6.2.10}$$

then

$$\sqrt{\lambda_n} = n\omega_0. \tag{6.2.11}$$

The corresponding function $u_n(x)$, associated with λ_n, is

$$u_n(x) = A_n \sin\left(\frac{n\pi x}{L}\right) = A_n \sin(n\omega_0 x), \tag{6.2.12}$$

where A_n is a constant that can be selected in some manner suitable to the particular requirements of the system being studied.

Figure 6.2.1 presents $u_1(x)$, $u_2(x)$ and $u_3(x)$.

6.2.2   One Fixed and One Free Ends

The general solution, for this case, is still given by Eq. (6.2.5), but now the restrictions of Eq. (6.2.3) must be satisfied. We obtain

$$\begin{cases} u(0) = 0 \implies B = 0; \\ u'(L) = 0 \implies \left. \frac{du(x)}{dx} \right|_{x=L} = -\sqrt{\lambda}A\cos\left(\sqrt{\lambda}L\right) = 0, \end{cases} \tag{6.2.13}$$

and

$$\cos\left(\sqrt{\lambda}L\right) = 0 \;\Rightarrow\; \sqrt{\lambda}L = (2n-1)\left(\frac{\pi}{2}\right),\qquad(6.2.14)$$

and, therefore,

$$\lambda_n = \left[\frac{(2n-1)\pi}{2L}\right]^2\qquad(6.2.15)$$

or

$$\sqrt{\lambda_n} = (2n-1)\left(\frac{\omega_0}{2}\right).\qquad(6.2.16)$$

The associated solutions are

$$u_n(x) = A_n \sin\left[\frac{(2n-1)\pi x}{2L}\right] = A_n \sin\left[\left(n-\frac{1}{2}\right)\omega_0 x\right].\qquad(6.2.17)$$

Figure 6.2.2 presents $u_1(x)$, $u_2(x)$, and $u_3(x)$.

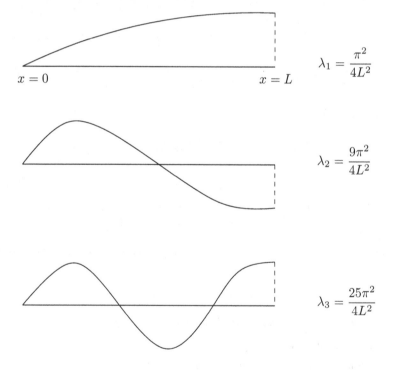

Fig. 6.2.2   Plots of $u_1(x)$, $u_2(x)$ and $u_3(x)$ for the vibrating string with one end fixed and the other free.

6.2.3 Both Ends Free

For this case, we obtain

$$u'(x) = \sqrt{\lambda}A\cos\left(\sqrt{\lambda}x\right) - \sqrt{\lambda}B\sin\left(\sqrt{\lambda}x\right), \qquad (6.2.18)$$

and

$$\begin{cases} u'(0) = 0 \implies A = 0, \\ u'(L) = 0 \implies B\sin\left(\sqrt{\lambda}L\right) = 0, \end{cases} \qquad (6.2.19)$$

and, therefore

$$\lambda_n = \left(\frac{n\pi}{L}\right)^2 = n^2\omega_0^2, \qquad (6.2.20)$$

with $u_n(x)$ given by

$$u_n(x) = B_n\cos\left(\frac{n\pi x}{L}\right) = B_n\cos(n\omega_0 x). \qquad (6.2.21)$$

Figure 6.2.3 gives $u_1(x)$, $u_2(x)$ and $u_3(x)$.

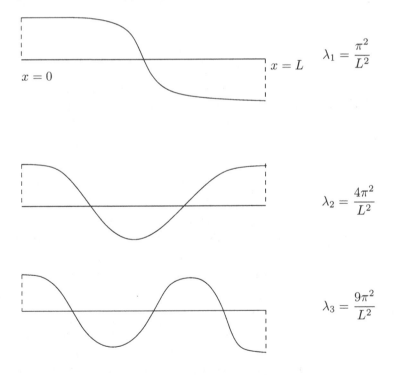

Fig. 6.2.3 Plots of $u_1(x)$, $u_2(x)$ and $u_3(x)$ for the vibrating string with both ends free.

6.2.4   *Discussion*

Let us introduce the notation

$$\begin{cases} ff : u(0) = 0, & u(L) = 0; \\ fF : u(0) = 0, & u'(L) = 0; \\ FF : u'(0) = 0, & u'(L) = 0. \end{cases} \tag{6.2.22}$$

Thus $\{u_n^{ff}(x), \lambda_n^{ff}\}$, $\{u_n^{fF}(x), \lambda_n^{fF}\}$, and $\{u_n^{FF}(x), \lambda_n^{FF}\}$ correspond, respectively to cases (i), (ii) and (iii), with $n = 1, 2, 3 \ldots$. Inspection of these solutions and the plots in Figures 6.2.1 to 6.2.3 direct us to the following conclusions:

(i) The boundary conditions play a major role in determining both the values of the λ_n and their associated functions u_n. Thus, for a given set of boundary conditions there are an infinite number of characteristic values λ_n and characteristic functions $u_n(x)$.

(ii) There is a smallest characteristic value, λ_1, and they are ordered in an increasing, unbounded sequence, i.e.,

$$\lambda_1 < \lambda_2 < \lambda_3 < \cdots < \lambda_n < \cdots, \tag{6.2.23}$$

with

$$\operatorname*{Lim}_{n \to \infty} \lambda_n = \infty. \tag{6.2.24}$$

(iii) Not including possible zeros on the boundary, the characteristic function, $u_n(x)$, has a finite number of zeros whose value depends on n. In general, the number of zeros of the characteristic functions satisfies a linear relation, i.e., let $cz(n)$ be the number of zeros associated with $u_n(x)$, then for the vibrating string

$$\begin{cases} cz(n)^{ff} = n - 1, \\ cz(n)^{fF} = n, \\ cz(n)^{FF} = n - 1. \end{cases} \tag{6.2.25}$$

(iv) The $\sqrt{\lambda_n}$ are related to the frequencies of oscillations of the elastic string. Thus, our analysis shows that different ways of clamping the string lead to different physical modes of oscillation. Note also that while cases (i) and (ii) have exactly the same "spectrum" of frequencies, i.e.,

$$\sqrt{\lambda_n} = n\omega_0, \qquad (n = 1, 2, 3, \ldots), \tag{6.2.26}$$

their characteristic shape functions differ.

In the next two sections, we present results that generalize the results obtained for the elastic string problem. We show that this system is just a particular example of the class of equations associated with the so-called Sturm-Liouville problems.

6.3 Sturm Separation and Comparison Theorem [1]

Consider the linear, second-order differential equation

$$y'' + a_1(x)y' + a_2(x)y = 0, \qquad (6.3.1)$$

where $a_1(x)$ and $a_2(x)$ are defined and continuous on some interval, $a \le x \le b$. Let $y_1(x)$ and $y_2(x)$ be two linearly independent solutions. Under these conditions we know that every solution to Eq. (6.3.1) can be expressed as a linear combination of $y_1(x)$ and $y_2(x)$, i.e.,

$$y(x) = c_1 y_1(x) + c_2 y_2(x), \qquad (6.3.2)$$

where (c_1, c_2) are arbitrary constants. The following theorem gives useful information on the relative locations of the zeros of $y_1(x)$ and $y_2(x)$.

Theorem 6.3.1 *Let $y_1(x)$ and $y_2(x)$ be linearly independent solutions of Eq. (6.3.1); then $y_1(x)$ must have a zero between any two consecutive zeros of $y_2(x)$ and, likewise, $y_2(x)$ must have a zero between any two consecutive zeros of $y_1(x)$.*

Another way of stating the conclusions of the theorem is to say that the zeros of $y_1(x)$ and $y_2(x)$ alternate. This result is called the Sturm separation theorem.

Another very useful result is given in the next theorem. This theorem is called the Sturm Comparison Theorem. The main value of the theorem is that it allows knowledge of the solution behaviors of one differential equation to restrict the possible solution behaviors of related second differential equations, provided certain conditions hold.

Theorem 6.3.2 *Let $f(x)$ and $g(x)$ be nontrivial solutions, respectively, of the second-order, linear differential equations*

$$u'' + p(x)u = 0, \qquad v'' + q(x)v = 0, \qquad (6.3.3)$$

where $p(x)$ and $q(x)$ are defined on some interval, $a \le x \le b$, and they have the property

$$p(x) \ge q(x), \qquad a \le x \le b. \qquad (6.3.4)$$

Then $f(x)$ has at least one zero between any two zeros of $g(x)$, unless $p(x) \equiv q(x)$ and, $f(x)$ and $g(x)$ are linearly dependent.

Note that one of the consequences of this theorem is that if $q(x) \leq 0$, then no solution of

$$v'' + q(x)v = 0, \tag{6.3.5}$$

can have more than one zero. (Of course, we exclude the trivial $v = 0$, solution.) The proof of this statement is straightforward and we give it. In Eq. (6.3.3), let

$$p(x) = 0, \qquad q(x) \leq 0. \tag{6.3.6}$$

Then, a nontrivial solution for $u(x)$ is $u(x) = 1$. Applying Theorem 6.3.2, we conclude that $u(x) = 1$ must have at least one zero between any two zeros of $v(x)$. But the solution $u(x) = 1$ has no zeros and it follows that $v(x)$ has at most one zero.

An additional result is that if $q(x) = \lambda^2 > 0$, where λ is constant, and if $p(x)$ is such that

$$p(x) \geq q(x) = \lambda^2, \tag{6.3.7}$$

then every solution of

$$u'' + p(x)u = 0, \tag{6.3.8}$$

must have a zero between any two consecutive zeros of the solutions to

$$v'' + \lambda^2 v = 0 \tag{6.3.9}$$

which are

$$v(x) = A \sin\left[\lambda(x - x_0)\right], \tag{6.3.10}$$

where (A, x_0) are arbitrary constants. Since the zeros of $v(x)$, as given by the function in Eq. (6.3.10), appear at intervals of π/λ, it follows that $u(x)$ must have at least one zero in every interval in x of length π/λ.

We now illustrate the application of these two theorems by presenting several examples.

6.3.1   *Example A*

A corollary to Theorem 6.3.2 is that if the equation

$$y''(x) + p(x)y(x) = 0, \tag{6.3.11}$$

has all its nontrivial solutions oscillatory, then

$$u''(x) + P(x)u(x) = 0, \tag{6.3.12}$$

where

$$P(x) \geq p(x), \qquad (6.3.13)$$

has all oscillatory solutions.

To make this concrete, consider the following Euler-Cauchy differential equation

$$x^2 y'' + \left(\frac{\lambda^2}{4}\right) y = 0. \qquad (6.3.14)$$

The substitution $x = \exp(t)$ gives

$$\frac{d^2 y}{dt^2} - \frac{dy}{dt} + \left(\frac{\lambda^2}{4}\right) y = 0. \qquad (6.3.15)$$

The characteristic equation is

$$r^2 - r + \left(\frac{\lambda^2}{4}\right) = 0, \qquad (6.3.16)$$

and its two solutions are

$$r_{\pm} = \left(\frac{1}{2}\right) \left[1 \pm \sqrt{1 - \lambda^2}\right]. \qquad (6.3.17)$$

Since for $\lambda > 1$, r_+ and r_- are complex conjugates, then the solution to Eq. (6.3.15) is

$$y(t) = e^{t/2} \left[A \cos\left(\sqrt{\lambda^2 - 1}\, t\right) + B \sin\left(\sqrt{\lambda^2 - 1}\, t\right)\right], \qquad (6.3.18)$$

where A and B are arbitrary integration constants. Replacing t by $t = \ln x$, we obtain

$$y(t) = \sqrt{x} \left\{A \cos\left[\left(\sqrt{\lambda_1^2 - 1}\right) \ln x\right] + B \sin\left[\left(\sqrt{\lambda^2 - 1}\right) \ln x\right]\right\}. \qquad (6.3.19)$$

Thus, we conclude that the solutions to Eq. (6.3.14) are oscillatory. Rewriting this differential equation to the form

$$\frac{d^2 y}{dx^2} + \left(\frac{1}{4x^2}\right) y = 0, \qquad (6.3.20)$$

we conclude, based on Theorem 6.3.2, that if

$$q(x) > \frac{1}{4x^2}, \qquad (6.3.21)$$

then all its solutions oscillate.

6.3.2   *Example B*

The linear, second-order differential equation

$$y''(x) + y(x) = 0, \tag{6.3.22}$$

has the following two linearly independent solutions

$$y_1(x) = \sin x, \qquad y_2(x) = \cos x. \tag{6.3.23}$$

The zeros of $y_1(x)$ and $y_2(x)$ occur, respectively, at

$$\begin{cases} x_n^{(1)} = \pm n\pi, \qquad x_n^{(2)} = \pm \left(\frac{2n+1}{2} \right) \pi, \\ \quad n = 0, 1, 2, \ldots. \end{cases} \tag{6.3.24}$$

It clearly follows that between any two consecutive zeros of $y_1(x)$, there lies a single zero of $y_2(x)$ and vice versa. This result is consistent with Theorem 6.3.1.

6.3.3   *Example C*

The Airy differential equation is

$$y''(x) + xy(x) = 0. \tag{6.3.25}$$

Note that since

$$x > 1, \qquad \text{for } x > 1, \tag{6.3.26}$$

and all to the solutions to

$$y''(x) + y(x) = 0 \tag{6.3.27}$$

are oscillatory, it follows that all solutions to the Airy equation are oscillatory for $x > 1$.

On the other hand, we have $x < 0$ for negative values of x and a non-trivial solution to

$$y''(x) = 0, \tag{6.3.28}$$

is $y(x) = 1$. From these results we can conclude that the Airy differential equations have no oscillatory solutions for $x < 0$.

6.4   Sturm-Liouville Problems [2]

6.4.1   *Fundamental Definition*

Definition 1   Let real functions $\{p(x), q(x), r(x)\}$ exist, on the interval $a \leq x \leq b$, such that $p(x)$ has a continuous first derivative and has the property, $p(x) > 0$; let $q(x)$ and $r(x)$ be continuous with $r(x) > 0$; and let λ be a parameter not dependent on x. Let real constants (A_1, A_2, B_1, B_2) exist such that the following two conditions hold for $\{y(a), y(b), y'(a), y'(b)\}$,

$$\begin{cases} A_1 y(a) + A_2 y'(a) = 0, \\ B_1 y(b) + B_2 y'(b) = 0, \end{cases} \tag{6.4.1}$$

where A_1 and A_2, and B_1 and B_2, are not both equal to zero. Then the second-order, linear differential equation

$$\frac{d}{dx}\left[p(x)\frac{dy}{dx}\right] + [q(x) + \lambda r(x)]y = 0 \tag{6.4.2}$$

along with the boundary-value conditions of Eq. (6.4.1) define the Sturm-Liouville Problem.

Definition 2   The values of λ for which Eq. (6.4.2) has nontrivial solutions are called *eigenvalues* of the problem. The associated solutions are called the *eigenfunctions* of the problem.

6.4.2   *Properties of Eigenvalues and Eigenfunctions*

The eigenvalues and eigenfunctions of Sturm-Liouville Problems have the following properties:

(a) For a given Sturm-Liouville Problem, there exists an infinite number of eigenvalues and they form an unbounded monotonic increasing sequence, i.e.,

$$\lambda_1 < \lambda_2 < \lambda_3 < \cdots < \lambda_n < \cdots, \tag{6.4.3}$$

with

$$\lim_{n \to \infty} \lambda_n = \infty. \tag{6.4.4}$$

(b) Associated with each eigenvalue, λ_n, there exists a unique eigenfunction $\phi_n(x)$, defined up to an overall multiplicative constant.

(c) The eigenfunction $\phi_n(x)$, associated with the eigenvalue λ_n, has exactly $(n-1)$ zeros in the open interval $a < x < b$.

Some clarification is needed here. The problem of the vibrating string, with specified boundary conditions, is a Sturm-Liouville Problem with

$$p(x) = 1, \qquad q(x) = 0, \qquad r(x) = 1. \tag{6.4.5}$$

The boundary conditions for cases (i) and (ii) gave eigenfunctions that satisfied the statement (c) above. However, the boundary condition for case (iii) had eigenfunctions for which $\phi_n(x)$ has n zeros. What all this really means is that we need to be careful when applying (c). In any case, the general situation is that the smallest eigenvalue either has no zeros or at most one zero, and in either case the number of zeros for the eigenfunctions increase by one as we go from $\phi_n(x)$ to $\phi_{n+1}(x)$.

6.4.3   *Orthogonality of Eigenfunctions*

Definition 3 Let $f(x)$ and $g(x)$ be defined on the interval, $a \le x \le b$. Let there exist a function $w(x) > 0$, defined on the interval, $a \le x \le b$, such that

$$\int_a^b f(x)g(x)w(x)dx = 0. \tag{6.4.6}$$

Then $f(x)$ and $g(x)$ are said to be *orthogonal* on the interval, $a \le x \le b$, with respect to the *weight function $w(x)$*.

Definition 4 Let $\{\psi_n(x) : n = 1, 2, 3, \dots\}$ be an infinite set of functions defined on the interval, $a \le x \le b$. Let

$$\int_a^b \psi_m(x)\psi_n(x)w(x)dx = 0, \qquad m \ne n. \tag{6.4.7}$$

Then the set $\{\psi_n(x) : n = 1, 2, 3, \dots\}$ is called an orthogonal system on the interval, $a \le x \le b$, with respect to the weight function $w(x)$.

Theorem 6.4.1 *Consider the Sturm-Liouville Problem with a corresponding infinite set of eigenvalues and eigenfunctions,* $\{\lambda_n, \phi_n(x) : n = 1, 2, 3, \dots\}$. *Then*

$$\int_a^b \phi_m(x)\phi_n(x)r(x)dx = 0, \qquad m \ne n, \tag{6.4.8}$$

i.e., the eigenfunctions $\phi_m(x)$ and $\phi_n(x)$ are orthogonal with respect to the weight function $w(x) = r(x)$ on the interval, $a \le x \le b$.

6.4.4 *Expansion of Functions*

Definition 5 Let $f(x)$ be defined on the interval, $a \le x \le b$, such that the following integral exists,

$$\int_a^b [f(x)]^2 w(x) dx = 1. \tag{6.4.9}$$

Then $f(x)$ is said to be normalized with respect to the weight function $w(x)$ on this interval.

Definition 6 Let $\{\psi_n(x) : n = 1, 2, 3, \dots\}$ be an infinite set of functions defined on the interval, $a \le x \le b$. Let

$$\int_a^b \psi_m(x)\psi_n(x)w(x) dx = \delta_{mn}. \tag{6.4.10}$$

Then this set is said to be an orthonormal system with respect to the weight function, $w(x)$ on the interval $a \le x \le b$.

Theorem 6.4.2 *Consider the Sturm-Liouville Problem with a corresponding infinite set of eigenvalues and orthonormal eigenfunctions* $\{\lambda_n, \phi_n(x) : n = 1, 2, 3, \dots\}$. *Let $f(x)$ be continuous on the interval, $a \le x \le b$, have a piecewise continuous first derivative on this interval. Let $f(x)$ and all the $\phi_n(x)$ satisfy the same boundary conditions at $x = a$ and $x = b$. Then the series*

$$\sum_{n=1}^{\infty} c_n \phi_n(x), \tag{6.4.11}$$

where

$$c_n = \int_a^b f(x)\phi_n(x)r(x) dx, \qquad (n = 1, 2, 3, \dots), \tag{6.4.12}$$

converges uniformly and absolutely to $f(x)$ on the interval $a \le x \le b$.

This theorem informs us that if we are given a function defined on the interval, $a \le x \le b$, having the above specified properties, then it can be represented as an infinite series expansion in terms of an orthonormal system. Therefore, we can write

$$f(x) = \sum_{n=1}^{\infty} c_n \phi_n(x). \tag{6.4.13}$$

6.4.5   The Completeness Relation

We now demonstrate that a relationship holds between the orthonormal set $\{\lambda_n, \phi_n(x) : n = 1, 2, 3 \ldots\}$ and the Dirac delta function. To show this, substitute the expression for c_n, calculated from Eq. (6.4.12) into Eq. (6.4.13); doing this gives

$$f(x) = \sum_{n=1}^{\infty} \left[\int_a^b f(x')\phi_n(x')r(x')dx' \right] \phi_n(x). \qquad (6.4.14)$$

Interchanging the summation and integration, we obtain

$$f(x) = \int_a^b f(x') \left[\sum_{n=1}^{\infty} \phi_n(x)\phi_n(x')r(x') \right] dx'. \qquad (6.4.15)$$

Based on the discussion of the Dirac delta function, in Chapter 3, we conclude that

$$\sum_{n=1}^{\infty} \phi_n(x)\phi_n(x')r(x') = \delta(x - x'). \qquad (6.4.16)$$

This formula is called the *completeness relation*.

6.5   Applications

6.5.1   The Special Functions

Chapter 7 will be concerned with the so-called special functions. However, we would like to point out that the polynomial special functions are the solutions to Sturm-Liouville Problems. The tables list several of the important properties of these functions:

   (i) Table 6.5.1 – The function name, symbol, interval of definition and weight function;

   (ii) Table 6.5.2 – The differential equation for these functions in the standard format;

   (iii) Table 6.5.3 – The differential equation for the functions written in the Sturm-Liouville representation;

   (iv) Table 6.5.4 – The identification of function $p(x)$, $q(x)$ and $r(x)$ from their Sturm-Liouville representation;

   (v) Table 6.5.5 – Normalizations for the polynomial special functions.

A word of caution: There exists in published tables and handbooks a variety of normalization conditions for the polynomial special functions. This situation is a consequence of the fact that these tables and handbooks give orthogonal representations for these functions, but they are not in general orthonormal.

Table 6.5.1 Polynomial Special Functions

Function Name	Symbol	Interval	Weight Function
Legendre	$P_n(x)$	$-1 \le x \le 1$	1
Associated Legendre	$P_n^m(x)$	$-1 \le x \le 1$	1
Chebyshev	$T_n(x)$	$-1 \le x \le 1$	$\sqrt{1-x^2}$
Hermite	$H_n(x)$	$-\infty < x < \infty$	e^{-x^2}
Laguerre	$L_n(x)$	$0 \le x < \infty$	e^{-x}
Associated Laguerre	$L_n^{(\alpha)}(x)$	$0 \le x < \infty$	$x^\alpha e^{-x}$

Table 6.5.2 Differential Equations for the Polynomial Special Functions

Symbol*	Differential Equations
$P_n^m(x)$	$(1-x^2)y'' - 2xy' + \left[n(n+1) - \frac{m^2}{1-x^2} \right] y = 0$
$T_n(x)$	$(1-x^2)y'' - xy' + n^2 y = 0$
$H_n(x)$	$y'' - 2xy' + 2ny = 0$
$L_n^{(\alpha)}(x)$	$xy'' + (\alpha + 1 - x)y' + ny = 0$

*Note that $P_n(x) = P_n^0(x)$ and $L_n(x) = L_n^{(0)}(x)$.

Table 6.5.3 Sturm-Liouville Problem Representation of the Differential Equations for the Polynomial Special Functions

Symbol	Differential Equation
$P_n^m(x)$	$[(1-x^2)y']' + \left[n(n+1) - \frac{m^2}{1-x^2} \right] y = 0$
$T_n(x)$	$\left[\sqrt{1-x^2} y' \right]' + \lambda(1-x^2)^{-1/2} y = 0$
$H_n(x)$	$[e^{-x^2} y']' + 2\lambda e^{-x^2} y = 0$
$L_n^{(\alpha)}(x)$	$[x^{\alpha+1} e^{-x} y']' + \lambda x^\alpha e^{-x} y = 0$

Table 6.5.4 Sturm-Liouville $p(x)$, $q(x)$, and $r(x)$ Functions for the Polynomial Special Functions

Symbol	$p(x)$	$q(x)$	$r(x)$	λ
$P_n^m(x)$	$1 - x^2$	$-\frac{m^2}{1-x^2}$	1	$n(n+1)$
$T_n(x)$	$\sqrt{1-x^2}$	0	$(1-x^2)^{-1/2}$	n^2
$H_n(x)$	e^{-x^2}	0	e^{-x^2}	n
$L_n^{(\alpha)}(x)$	$x^{\alpha+1} e^{-x}$	0	$x^\alpha e^{-x^2}$	n

Table 6.5.5 Normalizations for the Polynomial Special Functions

Symbol	C_n*
$P_n^m(x)$	$\left(\frac{2}{2n+1}\right)\frac{(n+m)!}{(n-m)!}$
$T_n(x)$	π for $n = 0$; $\pi/2$ for $n = 1, 2, \ldots$
$H_n(x)$	$\sqrt{\pi}2^n \cdot n!$
$L_n^{(\alpha)}(x)$	$\frac{\Gamma(\alpha+n+1)}{n!}$

$*C_n \equiv \int_a^b [\phi_n(x)]^2 r(x)dx.$

6.5.2 *Fourier Expansion of $f(x) = x(1 - x)$*

A Fourier series representation of the function $f(x)$ defined on the interval, $0 \leq x \leq 1$, can be obtained by first calculating the eigenfunctions to the following Sturm-Liouville Problem:

$$\begin{cases} y'' + \lambda y = 0, \\ y(0) = 0, \quad y(1) = 0. \end{cases} \tag{6.5.1}$$

The general solution is

$$y(x) = A\cos\left(\sqrt{\lambda}x\right) + B\sin\left(\sqrt{\lambda}x\right). \tag{6.5.2}$$

Requiring the boundary conditions to be satisfied gives

$$\begin{cases} y(0) = 0 \Rightarrow A = 0, \\ y(1) = B\sin\left(\sqrt{\lambda}\right) \Rightarrow \sqrt{\lambda} = n\pi, \end{cases} \tag{6.5.3}$$

and the eigenvalues and normalized eigenfunctions are

$$\lambda_n = n^2\pi, \qquad \phi_n(x) = \sqrt{2}\sin(n\pi x), \qquad (n = 1, 2, 3\ldots). \tag{6.5.4}$$

The expansion of $f(x) = x(1 - x)$ is given by

$$f(x) = x(1 - x) = \sqrt{2}\sum_{n=1}^{\infty} c_n \sin(n\pi x), \tag{6.5.5}$$

where the coefficients c_n are to be calculated from the expression

$$c_n = \sqrt{2}\int_0^1 x(1 - x)\sin(n\pi x)dx = \left(\frac{2^{3/2}}{\pi^3}\right)\left[\frac{1 + (-1)^{n+1}}{n^3}\right]. \tag{6.5.6}$$

Now

$$1 + (-1)^{n+1} = \begin{cases} 2; & n = 1, 3, 5, \ldots, \\ 0; & n = 2, 4, 6, \ldots, \end{cases} \tag{6.5.7}$$

and, therefore,

$$c_{2m-1} = \frac{2^{5/2}}{\pi^3 (2m-1)^3}, \qquad (m = 1, 2, 3, \ldots); \tag{6.5.8}$$

thus, $f(x) = x(1-x)$ has the series representation

$$x(1-x) = \left(\frac{2^{5/2}}{\pi^3} \right) \sum_{m=1}^{\infty} \frac{\sin(2m-1)\pi x}{(2m-1)^3}. \tag{6.5.9}$$

6.5.3 Representation of $\delta(x)$ in Terms of Cosine Functions

On the interval $-L \leq x \leq L$, the following infinite set of functions form an orthonormal system:

$$\begin{cases} \phi_1(x) = \dfrac{1}{\sqrt{2L}}, \\[2mm] \phi_{2n}(x) = \left(\dfrac{1}{\sqrt{L}} \right) \cos \left(\dfrac{n\pi x}{L} \right), & (n = 1, 2, 3, \ldots), \\[2mm] \phi_{2n+1}(x) = \left(\dfrac{1}{\sqrt{L}} \right) \sin \left(\dfrac{n\pi x}{L} \right), & (n = 1, 2, 3, \ldots). \end{cases} \tag{6.5.10}$$

The weight function is $r(x) = 1$ for this system. According to the completion relation, as given by Eq. (6.4.16), the Dirac delta function has the following representation in terms of these functions

$$\delta(x - x') = \left(\frac{1}{2L} \right) + \left(\frac{1}{L} \right) \sum_{n=1}^{\infty} \left[\cos \left(\frac{n\pi x}{L} \right) \cos \left(\frac{n\pi x'}{L} \right) \right.$$

$$\left. + \sin \left(\frac{n\pi x}{L} \right) \sin \left(\frac{n\pi x'}{L} \right) \right]. \tag{6.5.11}$$

6.5.4 Reality of the Eigenvalues

We now prove that the eigenvalues for a Sturm-Liouville Problem can only be real. Note that the solutions to a Sturm-Liouville Problem may be complex as the following equation shows,

$$\begin{cases} y'' + \lambda y = 0, & \lambda > 0; \\ y(x) = c e^{i\sqrt{\lambda} x} + c^* e^{-i\sqrt{\lambda} x}, \end{cases} \tag{6.5.12}$$

where c is an arbitrary complex number. What we will now do is to show that $\lambda = \lambda^*$ and this implies that λ is real.

To begin, assume that $p(x)$, $q(x)$, $r(x)$, as well as the constants (A_1, A_2, B_1, B_2) appearing in the boundary conditions are all real. Therefore, taking the complex conjugate of

$$\frac{d}{dx}\left[p(x)\frac{dy}{dx}\right] + [q(x) + \lambda r(x)]y = 0, \qquad (6.5.13)$$

gives

$$\frac{d}{dx}\left[p(x)\frac{dy*}{dx}\right] + [q(x) + \lambda^* r(x)]y = 0. \qquad (6.5.14)$$

Now multiply Eq. (6.5.13) by y^* and Eq. (6.5.14) by y, and subtract the resulting expressions. Doing this gives the following result

$$\frac{d}{dx}\left\{p(x)\left[y\frac{dy^*}{dx} - y^*\frac{dy}{dx}\right]\right\} = (\lambda - \lambda^*)r(x)yy^*. \qquad (6.5.15)$$

If Eq. (6.5.15) is integrated in x from a to b, then we obtain

$$p(x)\left[y\frac{dy^*}{dx} - y^*\frac{dy}{dx}\right]\Bigg|_{x=a}^{x=b} = (\lambda - \lambda^*)\int_a^b r(x)|y|^2 dx. \qquad (6.5.16)$$

To evaluate the left-side of this equation, make use of the two sets of boundary value conditions, i.e.,

$$\begin{cases} A_1 y(a) + A_2 y'(a) = 0, & B_1 y(b) + B_2 y'(b) = 0, \\ A_1 y^*(a) + A_2 y^{*\prime}(a) = 0, & B_1 y^*(b) + B_2 y^{*\prime}(b) = 0, \end{cases} \qquad (6.5.17)$$

where the second relation is just the complex conjugate of the first one. Using Eq. (6.5.17) in the left-side of Eq. (6.5.16) gives the result that it is zero. This means that

$$(\lambda - \lambda^*)\int_a^b r(x)|y|^2 dx = 0. \qquad (6.5.18)$$

Since $r(x) > 0$ and $|y|^2 > 0$, it follows that $(\lambda - \lambda^*)$ is equal to zero, i.e.,

$$\lambda = \lambda^*. \qquad (6.5.19)$$

Thus, our conclusion is that λ is real.

6.5.5   *A Boundary Value Problem* [3]

An interesting Sturm-Liouville Problem is the following one:

$$\begin{cases} y'' + k^2 y = 0 \\ y(0) = 0, \quad 5y(1) + y'(1) = 0, \end{cases} \tag{6.5.20}$$

where the interval is $0 \leq x \leq 1$ and $k^2 = \lambda$. Note that the boundary condition at $x = 1$ involves both $y(1)$ and its derivative $y'(1)$.

The general solution to the differential equation is

$$y(x) = A \cos(kx) + B \sin(kx). \tag{6.5.21}$$

Applying the boundary conditions gives

$$\begin{cases} y(0) = 0 \;\Rightarrow\; A = 0, \\ 5y(1) + y'(1) = 0 \;\Rightarrow\; 5 \sin k + k \sin k = 0. \end{cases} \tag{6.5.22}$$

Unlike other problems that we have examined for which explicit formulas were obtained for the eigenvalues, the eigenvalues here are determined by the equation

$$\tan k_n = - \left(\frac{k_n}{5} \right), \tag{6.5.23}$$

which cannot be solved explicitly for the k_n. However, there are an infinite number of distinct values k_n and these will give us the eigenvalues by means of the relation

$$\lambda_n = k_n^2. \tag{6.5.24}$$

To each k_n, there corresponds an eigenfunction $\phi_n(x)$ given by

$$\phi_n(x) = c_n \sin(k_n x), \qquad (n = 1, 2, 3, \dots), \tag{6.5.25}$$

where c_n is a normalization constant.

Note that since this is a Sturm-Liouville Problem we are assumed that these eigenvalues and eigenfunctions exist, even if, at the moment, the exact values for the k_n are not known.

The normalization constant, c_n, can be calculated from

$$\int_0^1 [\phi_n(x)]^2 dx = 1, \tag{6.5.26}$$

where we have used the fact that the weight function is $r(x) = 1$. Therefore,

$$\int_0^1 [\sin(k_n x)]^2 dx = \left(\frac{1}{2} \right) \int_0^1 [1 - \cos(2k_n x)] \, dx$$

$$= \left(\frac{1}{2}\right)\left[x - \frac{\sin(2k_n x)}{2k_n}\right]\Bigg|_{x=0}^{x=1}$$

$$= \left(\frac{1}{2}\right)\left[1 - \frac{\sin(2k_n x)}{2k_n}\right]$$

$$= \frac{2k_n - \sin(2k_n)}{4k_n} \tag{6.5.27}$$

and it follows that the normalized wave functions are

$$\phi_n(x) = \left[\frac{4k_n}{2k_n - \sin(2k_n)}\right]^{1/2} \sin(k_n x), \qquad (n = 1, 2, 3, \dots). \tag{6.5.28}$$

We now show that these eigenfunctions are orthogonal, i.e.,

$$\int_0^1 \sin(k_n x) \sin(k_m x) dx = 0, \qquad m \neq n. \tag{6.5.29}$$

We start by noting the following two trigonometric and integral relations

$$\sin(k_n x) \sin(k_m x) = \left(\frac{1}{2}\right)\cos(k_n - k_m x) - \left(\frac{1}{2}\right)\cos(k_n + k_m)x, \tag{6.5.30}$$

$$\int_0^1 \cos(k_n - k_m)x \, dx = \frac{\sin(k_n - k_m)x}{(k_n - k_m)}\Bigg|_{x=0}^{x=1} = \frac{\sin(k_n - k_m)}{(k_n - k_m)}$$

$$= \frac{\sin k_n \cos k_m - \cos k_n \sin k_m}{(k_n - k_m)}, \tag{6.5.31}$$

$$\int_0^1 \cos(k_n + k_m)x \, dx = \frac{\sin k_n \cos k_m + \cos k_n \sin k_m}{(k_n + k_m)}. \tag{6.5.32}$$

Therefore,

$$2\int_0^1 \sin(k_n x)(\sin(k_m x)dx = \frac{\sin k_n \cos k_m - \cos k_n \sin k_m}{(k_n - k_m)}$$

$$- \frac{\sin k_n \cos k_m + \cos k_n \sin k_m}{(k_n + k_m)}. \tag{6.5.33}$$

If on the right-side of Eq. (6.5.33) we use

$$\sin k_n = -\frac{k_n \cos k_n}{5}, \tag{6.5.34}$$

then after some algebraic manipulations, we find that the right-side is zero. This means that

$$\int_0^1 \sin(k_n x) \sin(k_m x) dx = 0, \tag{6.5.35}$$

and it follows that the $\{\phi_n(x) : n = 1, 2, 3, \dots\}$, as given by Eq. (6.5.28) are orthogonal.

6.6 Green's Functions

Suppose we want to solve the inhomogeneous differential equation

$$Ly(x) = -f(x), \qquad a \le x \le b, \qquad (6.6.1)$$

satisfying the boundary conditions

$$\begin{cases} A_1 y(a) + A_2 y'(a) = 0, \\ B_1 y(b) + B_2 y'(b) = 0, \end{cases} \qquad (6.6.2)$$

where L is the Sturm-Liouville operator

$$L \equiv \frac{d}{dx}\left[p(x)\frac{d}{dx}\right] + [q(x) + \lambda r(x)]. \qquad (6.6.3)$$

Note that λ is just a constant, *a priori* given by the nature of the problem, and is assumed not to be an eigenvalue of the homogeneous problem, $Ly(x) = 0$, having the same boundary conditions as given by Eq. (6.6.2).

The function $-f(x)$ can be considered a distributed source in the variable x. One way to proceed in the construction of the solution to Eqs. (6.6.1) with the conditions of Eq. (6.6.2) is to study the following equation for a "point" source

$$LG(x, x') = -\delta(x - x'), \qquad (6.6.4)$$

where $\delta(x - x')$ is the Dirac delta function and $G(x, x')$, which depends on both x and x', is called the **Green's function**. The Green's function is to also satisfy the boundary conditions required for $y(x)$, i.e.,

$$\begin{cases} A_1 G(a, x') + A_2 G'(a, x') = 0, \\ B_1 G(b, x') + B_2 G'(b, x') = 0. \end{cases} \qquad (6.6.5)$$

The required solution $y(x)$ can be expressed in terms of $G(x, x')$ by means of the formula

$$y(x) = \int_a^b G(x, x') f(x') dx'. \qquad (6.6.6)$$

It is easy to check that this is the correct solution. Applying the operator L to both sides of Eq. (6.6.6) gives

$$Ly(x) = \int_a^b [LG(x, x')] f(x') dx'$$
$$= \int_a^b [-\delta(x - x')] f(x') dx' = -f(x), \qquad (6.6.7)$$

and this is just the original differential equation. However, what is needed is an explicit representation for $G(x, x')$. We now demonstrate how this function can be constructed.

To begin, consider the set of eigenvalues and eigenfunctions, $\{\lambda_n, \phi_n(x) : n = 1, 2, 3, \ldots\}$, associated with the following Sturm-Liouville Problem

$$
\begin{cases}
\dfrac{d}{dx}\left[p(x)\dfrac{d\phi_n}{dx}\right] + [q(x) + \lambda_n r(x)]\phi_n = 0, \\[2mm]
A_1\phi_n(a) + A_2\phi_n'(a) = 0, \\[2mm]
B_1\phi_n(b) + B_2\phi_n'(b) = 0, \\[2mm]
n = 1, 2, 3, \ldots .
\end{cases} \tag{6.6.8}
$$

Now since this set of functions is complete, we can expand the Green's function in terms of them and obtain the representation

$$
G(x, x') = \sum_{n=1}^{\infty} g_n(x')\phi_n(x), \tag{6.6.9}
$$

where the $g_n(x')$ are to be be determined. Applying the operator L to both sides of Eq. (6.6.9) gives

$$
\begin{aligned}
LG(x, x') &= \sum_{n=1}^{\infty} g_n(x')L\phi_n(x) \\
&= \sum_{n=1}^{\infty} g_n(x')(\lambda - \lambda_n)r(x)\phi_n(x) \\
&= -\delta(x - x'). \tag{6.6.10}
\end{aligned}
$$

The second line on the right-side of Eq. (6.6.10) was obtained by use of the following calculation

$$
\begin{aligned}
L\phi_n(x) &= \frac{d}{dx}\left[p(x)\frac{d\phi_n}{dx}\right] + q(x)\phi_n + \lambda r(x)\phi_n \\
&= \frac{d}{dx}\left[p(x)\frac{d\phi_n}{dx}\right] + [q(x) + \lambda_n r(x)]\phi_n + (\lambda - \lambda_n)r(x)\phi_n \\
&= (\lambda - \lambda_n)r(x)\phi_n . \tag{6.6.11}
\end{aligned}
$$

Now multiply both sides of Eq. (6.6.10) by $\phi_m(x)$ and integrate over x to obtain

$$
\sum_{n=1}^{\infty} g_n(x')(\lambda - \lambda_n)\int_a^b \phi_n(x)\phi_m(x)r(x)dx = -\phi_m(x'). \tag{6.6.12}
$$

If the eigenfunctions are orthonormal, then

$$\int_a^b \phi_n(x)\phi_m(x)r(x)dx = \delta_{nm}, \tag{6.6.13}$$

and Eq. (6.6.12) becomes

$$g_n(x')(\lambda - \lambda_n) = -\phi_m(x') \tag{6.6.14}$$

or

$$g_n(x') = \frac{\phi_m(x')}{\lambda_n - \lambda}. \tag{6.6.15}$$

Substituting this result into Eq. (6.6.9) gives the final expression for the Green's function

$$G(x, x') = \sum_{n=1}^{\infty} \frac{\phi_n(x')\phi_n(x)}{\lambda_n - \lambda}, \tag{6.6.16}$$

where, as previously stated, the parameter λ does not have a value equal to any of the eigenvalues of the associated Sturm-Liouville Problem as given by Eq. (6.6.8).

While we have written the original differential equation in a form for which the left-side appears as a Sturm-Liouville type differential equation, this is not a restriction at all. It can be shown [4] that any second-order, linear, inhomogeneous differential equation

$$\frac{d^2y}{dx^2} + a_1(x)\frac{dy}{dx} + a_2(x)y = h(x), \qquad a \leq x \leq b, \tag{6.6.17}$$

can be transformed to the form

$$\frac{d}{dx}\left[p(x)\frac{dy}{dx}\right] + s(x)y = -f(x), \qquad a \leq x \leq b, \tag{6.6.18}$$

where $\{p(x), s(x), f(x)\}$ are determined by $\{a_1(x), a_2(x), h(x)\}$.

The minus sign on the right-side, for example, of Eqs. (6.6.1) and (6.6.18), are there for historical reasons, i.e., in many of the original studies on certain inhomogeneous, linear partial differential equations and the construction of their associated Green's functions, a minus sign was displayed in front of the source terms.

There is also an alternative method for constructing Green's functions based on the use of two fundamental solutions to the homogeneous problem for Eq. (6.6.1). The texts listed in the bibliography to this section provide clear and detailed discussions as to how this is to be done as well as many applications. An excellent introduction to Green's functions for both ordinary and partial differential equations is the textbook by E. A. Kraut, *Fundamentals of Mathematical Physics* (McGraw-Hill, New York, 1967).

6.7 Worked Examples for Green Functions

6.7.1 $y''(x) = -f(x)$; $y(0) = y(L) = 0$

Consider the following boundary value problem

$$y''(x) = -f(x), \qquad y(0) = 0, \quad y(L) = 0. \qquad (6.7.1)$$

The associated Green's function is determined by the equations

$$G''(x, x') = -\delta(x - x'), \qquad G(0, x') = G(L, x') = 0. \qquad (6.7.2)$$

The orthogonal set of functions, $\{\sin\left(\frac{n\pi x}{L}\right) : n = 1, 2, 3, \dots\}$, satisfy the boundary conditions and can be used to represent $G(x, x')$ in the following manner

$$G(x, x') = \sum_{n=1}^{\infty} g_n(x') \sin\left(\frac{n\pi x}{L}\right). \qquad (6.7.3)$$

Also, we have

$$\delta(x - x') = \sum_{n=1}^{\infty} h_n(x') \sin\left(\frac{n\pi x}{L}\right), \qquad (6.7.4)$$

where

$$h_n(x') = \left(\frac{2}{L}\right) \int_0^L \delta(x - x') \sin\left(\frac{n\pi x}{L}\right) dx = \left(\frac{2}{L}\right) \sin\left(\frac{n\pi x'}{L}\right), \quad (6.7.5)$$

and

$$\delta(x - x') = \left(\frac{2}{L}\right) \sum_{n=1}^{\infty} \sin\left(\frac{n\pi x'}{L}\right) \sin\left(\frac{n\pi x}{L}\right). \qquad (6.7.6)$$

Therefore, substituting Eqs. (6.7.3) and (6.7.6) into Eq. (6.7.2) and equating the corresponding coefficients of $\sin(n\pi x/L)$ gives

$$\left(\frac{n^2\pi^2}{L^2}\right) h_n(x) = \left(\frac{2}{L}\right) \sin\left(\frac{n\pi x'}{L}\right). \qquad (6.7.7)$$

Placing this result in Eq. (6.7.3) gives the required Green's function for Eq. (6.7.1)

$$G(x, x') = \left(\frac{2L}{\pi^2}\right) \sum_{n=1}^{\infty} \left(\frac{1}{n^2}\right) \sin\left(\frac{n\pi x'}{L}\right) \sin\left(\frac{n\pi x}{L}\right). \qquad (6.7.8)$$

The solution, $y(x)$, to the original boundary value problem is given by

$$y(x) = \int_0^L G(x, x') f(x') dx'$$

$$= \left(\frac{2L}{\pi^2}\right) \sum_{n=1}^{\infty} \left(\frac{1}{n^2}\right) \sin\left(\frac{n\pi x}{L}\right) \int_0^L f(x') \sin\left(\frac{n\pi x'}{L}\right) dx'$$

$$= \left(\frac{L^2}{\pi^2}\right) \sum_{n=1}^{\infty} \left(\frac{f_n}{n^2}\right) \sin\left(\frac{n\pi x}{L}\right), \qquad (6.7.9)$$

where f_n are the Fourier sine coefficients of $f(x)$.

6.7.2 $y''(x) = -f(x) : y(0) = 0, y'(L) = 0$

The infinite set of complete orthonormal eigenfunctions $\{A_n \sin[(2n - 1)\pi x/L] : n = 1, 2, 3, \ldots\}$, where $A_n = \sqrt{2/L}$ satisfy the boundary conditions $y(0) = 0$ and $y'(L) = 0$, and thus can be used to represent the Green's function, i.e.,

$$G(x, x') = \left(\frac{8L}{\pi^2}\right) \sum_{n=1}^{\infty} \frac{\sin\left[\frac{(2n-1)\pi x'}{2L}\right] \sin\left[\frac{(2n-1)\pi x}{2L}\right]}{(2n-1)^2}. \qquad (6.7.10)$$

Therefore, the solution $y(x)$ is

$$y(x) = \int_0^L G(x, x') f(x') dx'$$

$$= \left(\frac{8L}{\pi^2}\right) \sum_{n=1}^{\infty} \left[\frac{\sin\left[\frac{(2n-1)\pi x}{2L}\right]}{(2n-1)^2}\right] \int_0^L f(x') \sin\left[\frac{(2n-1)\pi x'}{2L}\right] dx'$$

$$= \left(\frac{4L^2}{\pi^2}\right) \sum_{n=1}^{\infty} \left[\frac{f_n}{(2n-1)^2}\right] \sin\left[\frac{(2n-1)\pi x}{2L}\right], \qquad (6.7.11)$$

where

$$f_n = \left(\frac{2}{L}\right) \int_0^L f(x') \sin\left[\frac{(2n-1)\pi x'}{2L}\right] dx'. \qquad (6.7.12)$$

6.7.3 $y''(x) + k^2 y(x) = -f(x) : y(0) = y(L) = 0$

The orthonormal eigenfunctions satisfying this set of boundary conditions are

$$\begin{cases} \phi_n(x) = \left(\sqrt{\frac{2}{L}}\right) \sin\left(\frac{n\pi x}{L}\right), \quad \lambda_n = \left(\frac{n\pi}{L}\right)^2, \\ n = 1, 2, 3, \ldots. \end{cases} \qquad (6.7.13)$$

Therefore, the Green's function, based on Eq. (6.6.16) is,

$$G(x,x') = \left(\frac{2}{L}\right) \sum_{n=1}^{\infty} \frac{\sin(n\pi x'/L)\sin(n\pi x/L)}{\left(\frac{n\pi}{L}\right)^2 - k^2}, \tag{6.7.14}$$

provided $k^2 \neq \lambda_n$ for $n = 1, 2, 3, \ldots$.

6.7.4 $y''(x) + y(x) = x : y(0) = y(1) = 0$

This boundary value problem is a special case of the problem in section 6.7.3 where

$$L = 1, \quad k = 1, \quad f(x) = -x. \tag{6.7.15}$$

The corresponding Green's function is

$$G(x,x') = 2\sum_{n=1}^{\infty} \frac{\sin(n\pi x')\sin(n\pi x)}{n^2\pi^2 - 1}, \tag{6.7.16}$$

and the solution $y(x)$ is

$$
\begin{aligned}
y(x) &= -\int_0^1 G(x,x')x'dx' \\
&= -2\sum_{n=1}^{\infty} \left[\frac{\sin(n\pi x)}{n^2\pi^2 - 1}\right] \int_0^1 x'\sin(n\pi x')dx' \\
&= 2\sum_{n=1}^{\infty} \left[\sin\frac{(n\pi x)}{n^2\pi^2 - 1}\right] \left[\frac{(-1)^n}{n\pi}\right] \\
&= \left(\frac{2}{\pi}\right)\sum_{n=1}^{\infty} \frac{(-1)^n \sin(n\pi x)}{n[n^2\pi^2 - 1]}.
\end{aligned} \tag{6.7.17}
$$

It can be shown that this solution is equal to the following expression

$$y(x) = x - \frac{\sin(x)}{\sin(1)}. \tag{6.7.18}$$

6.8 Asymptotic Behavior of Solutions to Differential Equations

Often what is needed to fully understand a system modeled by differential equations is not the explicit solution, $y(x)$, valid for all values of x, but the behavior as x becomes large. The scattering of waves in quantum mechanics [7] and acoustics [8] are particular examples of this situation. It turns out that often the "asymptotic" behavior of the solutions can be determined without first obtaining exact solutions. The purpose of this section is to introduce a method for constructing the asymptotic solutions that works for many of the differential equations appearing in the natural and engineering sciences. However, before demonstrating the procedure, we give background information on how the first-derivative term can be eliminated in a second-order, linear differential equation. We also discuss the Liouville-Green transformation. This transformation allows us to change the structure of the original differential equation to one that is more suitable for the asymptotic analysis.

6.8.1   *Elimination of First-Derivative Terms*

The general second-order, linear, inhomogeneous differential equation is

$$\frac{d^2y}{dx^2} + p(x)\frac{dy}{dx} + q(x)y = f(x). \tag{6.8.1}$$

We now show that it is possible to introduce a transformation of the dependent variable such that the new equation does not contain a first-derivative term. To do this define $u(x)$ as

$$u(x) = y(x)\exp\left[\left(\frac{1}{2}\right)\int^x p(x')dx'\right]. \tag{6.8.2}$$

Therefore,

$$y(x) = u(x)e^{-I(x)}, \tag{6.8.3}$$

where

$$I(x) = \left(\frac{1}{2}\right)\int^x p(x')dx'. \tag{6.8.4}$$

Note that

$$I'(x) = \frac{p(x)}{2}, \qquad I''(x) = \left(\frac{1}{2}\right)p'(x), \tag{6.8.5}$$

and

$$\begin{cases} y'(x) = [u' - I'u]e^{-I}, \\ y''(x) = \{u'' - 2I'u' + [(I')^2 - I'']u\}e^{-I}. \end{cases} \tag{6.8.6}$$

Substitution of Eqs. (6.8.3), (6.8.5) and (6.8.6) into Eq. (6.8.1) gives, after some algebraic manipulation the following result

$$u''(x) + Q(x)u(x) = F(x), \tag{6.8.7}$$

where

$$Q(x) = q(x) - \left(\frac{1}{2}\right)p'(x) - \frac{p(x)^2}{4}, \tag{6.8.8}$$

$$F(x) = f(x)e^{I(x)}. \tag{6.8.9}$$

To illustrate the method consider the weakly damped harmonic oscillator differential equation

$$y'' + 2\epsilon y' + y = 0, \tag{6.8.10}$$

where $0 < \epsilon \ll 1$. Here we have

$$p(x) = 2\epsilon, \quad q(x) = 1, \quad f(x) = 0. \tag{6.8.11}$$

Therefore

$$Q(x) = 1 - \epsilon^2, \qquad F(x) = 0, \tag{6.8.12}$$

and the transformed differential equation is

$$u'' + (1 - \epsilon^2)u = 0, \tag{6.8.13}$$

which has the solution

$$u(x) = c_1 \cos\left(\sqrt{1 - \epsilon^2}\, x\right) + c_2 \sin\left(\sqrt{1 - \epsilon^2}\, x\right), \tag{6.8.14}$$

where c_1 and c_2 are arbitrary constants. Since

$$I(x) = \left(\frac{1}{2}\right)\int^x p(x')dx' = \left(\frac{1}{2}\right)\int^x 2\epsilon\, dx' = \epsilon x, \tag{6.8.15}$$

we have, in terms of the original dependent variable $y(x)$, the result

$$\begin{aligned} y(x) &= e^{-I(x)}u(x) \\ &= e^{-\epsilon x}c_1 \cos\left(\sqrt{1 - \epsilon^2}\, x\right) + c_2 \sin\left(\sqrt{1 - \epsilon^2}\, x\right). \end{aligned} \tag{6.8.16}$$

A second example is the differential equation

$$y'' + \left(\frac{1}{x}\right)y' + \left(1 - \frac{n^2}{x^2}\right)y = 0, \tag{6.8.17}$$

for which

$$p(x) = \frac{1}{x}, \qquad q(x) = \left(1 - \frac{n^2}{x^2}\right), \quad f(x) = 0. \tag{6.8.18}$$

Now

$$u(x) = y\exp\left[\left(\frac{1}{2}\right)\int^x \frac{dx'}{x'}\right] = \sqrt{x}\, y, \tag{6.8.19}$$

and $u(x)$ satisfies the equation

$$u''(x) + \left[1 + \frac{1 - 4n^2}{4x^2}\right] u(x) = 0. \tag{6.8.20}$$

Note that if $n = \pm(1/2)$, then Eq. (6.8.20) has the exact solution

$$u(x) = c_1 \cos x + c_2 \sin x, \tag{6.8.21}$$

and the exact solution to

$$y'' + \left(\frac{1}{x}\right) y' + \left(1 - \frac{1}{4x^2}\right) y = 0, \tag{6.8.22}$$

is

$$y(x) = \frac{c_1 \sin x + c_2 \cos x}{\sqrt{x}}. \tag{6.8.23}$$

If x is large and has values such that

$$\frac{1 - 4n^2}{4x^2} \ll 1, \tag{6.8.24}$$

then it seems reasonable that the exact solution to Eq. (6.8.20) should be "close" to the function given by Eq. (6.8.21) for all values of n. If this is correct, then the asymptotic behavior of the solution to Eq. (6.8.17) should be dominated by the expression in Eq. (6.8.23), i.e., for x large, the solution to the original differential equation has the property

$$y(x) \xrightarrow[x \text{ large}]{} \frac{c_1 \cos x + c_2 \sin x}{\sqrt{x}}. \tag{6.8.25}$$

The above discussion indicates that there may be some advantages in eliminating the derivative term in a second-order, linear differential equation. It may give a transformed equation for which at least the leading term of the asymptotic solution can be readily calculated.

6.8.2 *The Liouville-Green Transformation*

Suppose we start with a differential equation for which no first-order derivative term occurs. An example is the Airy equation

$$y''(x) + xy(x) = 0. \tag{6.8.26}$$

If we wish to determine its asymptotic form, then it's not clear how to proceed. The method to be presented, called the Liouville-Green transformation allows us to effectively deal with this situation.

Consider the following homogeneous differential equation

$$\frac{d^2y}{x^2} + p(x)\frac{dy}{dx} + q(x)y = 0. \tag{6.8.27}$$

The transformation given in the previous section allows for the elimination of the first-derivative term and carrying out this procedure gives

$$\frac{d^2v}{dx^2} + Q(x)v = 0, \tag{6.8.28}$$

where

$$v(x) = y(x)\exp\left[\left(\frac{1}{2}\right)\int^x p(x')dx'\right], \tag{6.8.29}$$

$$Q(x) = q(x) - \left(\frac{1}{2}\right)p'(x) - \left(\frac{1}{4}\right)p(x)^2. \tag{6.8.30}$$

Note that the Airy differential equation is of this form with $Q(x) = x$.

We now introduce a new independent variable s, defined as

$$s = \int^x \sqrt{Q(x')}\,dx' = s(x). \tag{6.8.31}$$

Our task now is to replace x by s in Eq. (6.8.28). However, to do this, we must first calculate the first- and second-derivatives of $Q(x(s))$ with respect to s; they are

$$\frac{dv}{dx} = \frac{dv}{ds}\frac{ds}{dx} = \sqrt{Q(x)}\frac{dv}{ds}, \tag{6.8.32}$$

$$\frac{d^2v}{dx^2} = Q(x)\frac{d^2v}{d^2s} + \left(\frac{1}{2\sqrt{Q(x)}}\right)\left(\frac{dQ}{dx}\right)\frac{dv}{ds}. \tag{6.8.33}$$

From Eq. (6.8.31), $s = s(x)$ can be inverted to give $x = x(s)$ and therefore $Q(x) = Q(x(s))$. Thus, all of the expressions on the right-sides of Eqs. (6.8.32) and (6.8.33) are functions of s. The substitution of these expressions for the derivatives into Eq. (6.8.28) gives finally

$$\frac{d^2v}{ds^2} + \left[\left(\frac{1}{2}\right)\left(\frac{dQ}{dx}\right)\left(\frac{1}{Q^{3/2}}\right)\right]\frac{dv}{ds} + v = 0. \tag{6.8.34}$$

Now the first-derivative term of this equation can be eliminated by the change of variable

$$w(s) = v(s)\exp\left[\left(\frac{1}{2}\right)\int^s\left(\frac{dQ}{dx}\right)\left(\frac{1}{2Q^{3/2}}\right)ds'\right]. \tag{6.8.35}$$

Using

$$ds' = \frac{ds'}{dx'} dx' = \sqrt{Q(x')} \, dx',\qquad(6.8.36)$$

Eq. (6.8.35) can be rewritten as

$$w(s) = v(s) \exp\left(\frac{1}{4}\right) \int^x \left(\frac{dQ}{dx'}\right)\left(\frac{1}{Q}\right) dx' = [Q(x)]^{1/4} v(s),\qquad(6.8.37)$$

where $Q(x)$ can be written in terms of s by use of the relation $x = x(s)$ obtained from Eq. (6.8.31). Putting all these items together, Eq. (6.8.34) takes the form

$$\frac{d^2 w}{ds^2} + \left[1 - \left(\frac{1}{2}\right)\frac{dh}{ds} - \left(\frac{1}{4}\right)h^2\right] w = 0,\qquad(6.8.38)$$

where

$$h(s) = \left(\frac{1}{2}\right)\left(\frac{dQ}{dx}\right)\left(\frac{1}{Q^{3/2}}\right).\qquad(6.8.39)$$

Why have we gone through such a complex set of calculational manipulations to obtain the result of Eqs. (6.8.38) and (6.8.39)? Since our purpose is to determine the behavior of $y(x)$ as $x \to \infty$, it may be that Eq. (6.8.38) has a structure such that $h(s)$ and its derivative both go to zero as $x \to \infty$, i.e.,

$$\lim_{s \to \infty} \begin{pmatrix} h(s) \\ h'(s) \end{pmatrix} = \begin{pmatrix} 0 \\ 0 \end{pmatrix}.\qquad(6.8.40)$$

If this is true, then the leading term to the asymptotic behavior, as determined by Eq. (6.8.38), will be

$$w(s) = A\cos(s) + B\sin(s),\qquad(6.8.41)$$

where A and B are arbitrary integration constants. Reversing the transforms from $y \to v \to w$, then provides the leading term to the asymptotic behavior of $y(x)$.

The examples of the next section will be used to illustrate this technique.

6.9 Worked Examples

6.9.1 *The Airy Equation*

The Airy differential equation is

$$\frac{d^2 y}{dx^2} + xy = 0,\qquad(6.9.1)$$

and contains no first-derivative term. Therefore, we can make the variable change $y \to v$ and obtain

$$\frac{d^2v}{ds^2} + xv = 0, \tag{6.9.2}$$

which corresponds to Eq. (6.8.28) with

$$Q(x) = x, \tag{6.9.3}$$

$$s = \int^x \sqrt{x'}\, dx' = \left(\frac{2}{3}\right) x^{3/2}. \tag{6.9.4}$$

Written in terms of the s variable, Eq. (6.9.2) becomes (see Eq. (6.8.34))

$$\frac{d^2v}{ds^2} + \left(\frac{1}{3s}\right)\frac{dv}{ds} + v = 0. \tag{6.9.5}$$

The transformed dependent variable $w(s)$ is, from Eq. (6.8.37), given by

$$w(s) = |x|^{1/4} v(s), \tag{6.9.6}$$

and satisfies the equation

$$\frac{d^2w}{ds^2} + \left(1 + \frac{1}{18s^2}\right) w = 0; \tag{6.9.7}$$

see Eqs. (6.8.38) and (6.8.39) where $h(s)$ is calculated to be

$$h(s) = \frac{1}{3s^2}. \tag{6.9.8}$$

Therefore, for $s \gg 1$, we have the following approximation for $w(s)$

$$w(s) \simeq A\cos(s) + B\sin(s), \tag{6.9.9}$$

since under this condition the differential equation, as given by Eq. (6.9.6) becomes

$$\frac{d^2w}{ds^2} + w = O\left(\frac{1}{x^2}\right). \tag{6.9.10}$$

Since $v(s) = w(s)/|x|^{1/4}$ and $s = \left(\frac{2}{3}\right) x^{3/2}$, we have

$$v(x) \xrightarrow[x \gg 1]{} \left(\frac{1}{|x|^{1/4}}\right) \left\{ A\cos\left[\left(\frac{2}{3}\right) x^{3/2}\right] + B\sin\left[\left(\frac{2}{3}\right) x^{3/2}\right] \right\}. \tag{6.9.11}$$

This is the leading term in the asymptotic expansion for the solution to the Airy differential equation given by Eq. (6.9.2).

6.9.2 The Bessel Equation

The following second-order, linear, homogeneous differential equation is called the Bessel equation of order n

$$x^2 y'' + xy' + (x^2 - n^2)y = 0. \tag{6.9.12}$$

We have previously studied this equation, see section 6.8.1, and found that under the transformation

$$u(x) = \sqrt{x}\, y(x), \tag{6.9.13}$$

the differential equation for $u(x)$ is

$$u'' + \left(1 + \frac{1}{x^2}\right) u = 0, \tag{6.9.14}$$

where we have selected for study the case of $n = 0$; see Eqs. (6.8.18) to (6.8.20). Under the assumption that $x \gg 1$, we obtained the approximation

$$u(x) \simeq A \sin x + B \cos x. \tag{6.9.15}$$

How can a better approximation to Eq. (6.9.14) be determined? The answer is provided by assuming that Eq. (6.9.14) has an asymptotic expansion in $(1/x)$ of the form

$$
\begin{aligned}
u(x) = (A \sin x) & \left[1 + \frac{a_1}{x} + \frac{a_2}{x^2} + \cdots + \frac{a_n}{x^n} + O\left(\frac{1}{x^{n+1}}\right)\right] \\
+ (B \cos x) & \left[1 + \frac{b_1}{x} + \frac{b_2}{x^2} + \cdots + \frac{b_n}{x^n} + O\left(\frac{1}{x^{n+1}}\right)\right], \quad (6.9.16)
\end{aligned}
$$

where the constants $(a_1, \ldots, a_n)$ and $(b_1, \ldots, b_n)$ are calculated by substituting Eq. (6.9.16) into the differential equation, given by Eq. (6.9.14), and equating to zero the coefficient of x^{-m}, where $m = 1, 2, 3, \ldots, n$. For $m = 2$, we have

$$
\begin{aligned}
v(x) = (A \sin x) & \left[1 + \frac{a_1}{x} + \frac{a_2}{x^2}\right] \\
+ (B \cos x) & \left[1 + \frac{b_1}{x} + \frac{b_2}{x^2}\right] + O\left(\frac{1}{x^3}\right)
\end{aligned} \tag{6.9.17}
$$

$$
\begin{aligned}
\frac{dv(x)}{dx} = & (A \cos x)\left[1 + \frac{a_1}{x} + \frac{a_2}{x^2}\right] + (A \sin x)\left[-\frac{a_1}{x^2}\right] \\
& - (B \cos x)\left[1 + \frac{b_1}{x} + \frac{b_2}{x^2}\right] + (B \cos x)\left[-\frac{b_1}{x^2}\right] \\
& + O\left(\frac{1}{x^3}\right)
\end{aligned} \tag{6.9.18}
$$

$$= (\cos x) \left[A + \frac{a_1 A}{x} + \frac{Aa_2 - Bb_1}{x^2} \right]$$

$$- (\sin x) \left[B + \frac{Bb_1}{x} + \frac{Aa_1 + Bb_2}{x^2} \right] + O\left(\frac{1}{x^3}\right) \qquad (6.9.19)$$

$$\frac{d^2 v(x)}{dx^2} = -(\sin x) \left[A + \frac{a_1}{x} + \frac{Aa_2 - Bb_1}{x^2} \right] + (\cos x) \left(-\frac{a_1 A}{x^2} \right)$$

$$- (\cos x) \left[B + \frac{Bb_1}{x} + \frac{Aa_1 + Bb_2}{x^2} \right] - (\sin x) \left(-\frac{Bb_1}{x^2} \right)$$

$$+ O\left(\frac{1}{x^3}\right), \qquad (6.9.20)$$

and

$$\frac{d^2 v(x)}{dx^2} = (\sin x) \left[-A - \frac{a_1 A}{x} + \frac{2Bb_1 - Aa_2}{x^2} \right]$$

$$- (\cos x) \left[B + \frac{Bb_1}{x} + \frac{2Aa_1 + Bb_2}{x^2} \right] + O\left(\frac{1}{x^3}\right). \qquad (6.9.21)$$

Also,

$$\left(1 + \frac{1}{4x^2}\right) v = \left(1 + \frac{1}{4x^2}\right) \left\{ (A \sin x) \left[1 + \frac{a_1}{x} + \frac{a_2}{x^2} \right] \right.$$

$$\left. + (B \cos x) \left[1 + \frac{b_1}{x} + \frac{b_2}{x^2} \right] \right\} + O\left(\frac{1}{x^3}\right)$$

$$= (A \sin x) \left[1 + \frac{a_1}{x} + \frac{a_2}{x^2} \right] + (A \sin x) \left(\frac{1}{4x^2} \right)$$

$$+ (B \cos x) \left[1 + \frac{b_1}{x} + \frac{b_2}{x^2} \right] + (B \cos x) \left(\frac{1}{4x^2} \right) + O\left(\frac{1}{x^3}\right)$$

$$= (\sin x) \left[A + \frac{a_1 A}{x} + \frac{A\left(a_2 + \frac{1}{4}\right)}{x^2} \right]$$

$$+ (\cos x) \left[B + \frac{b_1 B}{x} + \frac{B\left(b_2 + \frac{1}{4}\right)}{x^2} \right]$$

$$+ O\left(\frac{1}{x^3}\right). \qquad (6.9.22)$$

The coefficients of the various terms, obtained by substituting Eqs. (6.9.21) and (6.9.22) into Eq. (6.9.14) are:

$$(\sin x) : -A + A, \qquad (6.9.23a)$$

$$\frac{\sin x}{x} : -a_1 A + a_1 A, \tag{6.9.23b}$$

$$\frac{\sin x}{x^2} : 2Bb_1 - Aa_2 + A\left(a_2 + \frac{1}{4}\right), \tag{6.9.23c}$$

$$\cos x : -B + B, \tag{6.9.23d}$$

$$\frac{\cos x}{x} : -Bb_1 + b_1 B, \tag{6.9.23e}$$

$$\frac{\cos x}{x^2} : -2Aa_1 - Bb_2 + B\left(b_2 + \frac{1}{4}\right). \tag{6.9.23f}$$

Setting these coefficients to zeros gives the following nontrivial relations

$$2Bb_1 + \frac{A}{4} = 0, \qquad -2Aa_1 + \frac{B}{4} = 0, \tag{6.9.24}$$

and

$$a_1\left(\frac{A}{B}\right) = -b_1\left(\frac{B}{A}\right) = \frac{1}{8}. \tag{6.9.25}$$

or

$$a_1 A = \frac{B}{8}, \qquad b_1 B = -\frac{A}{8}. \tag{6.9.26}$$

Note that to this level of calculation (a_2, b_2) cannot be determined.
Inserting the expressions of Eq. (6.9.26) into Eq. (6.9.17) gives

$$v(x) = A\left[\sin x - \frac{\cos x}{8x}\right] + B\left[\cos x + \frac{\sin x}{8x}\right] + O\left(\frac{1}{x^2}\right). \tag{6.9.27}$$

This process can be continued to calculate all the coefficients $(a_1, a_2, \ldots, a_n)$ and $(b_1, b_2, \ldots, b_n)$. However, as we have seen, to obtain explicit values for these coefficients, we must include terms up to x^{-n-1}, i.e., the asymptotic representation for $v(x)$ to be used is

$$
\begin{aligned}
v(x) = {}& (A\sin x)\left[1 + \frac{a_1}{x} + \frac{a_2}{x^2} + \cdots + \frac{a_n}{x^n} + \frac{a_{n+1}}{x^{n+1}} + O\left(\frac{1}{x^{n+2}}\right)\right] \\
& + (B\cos x)\left[1 + \frac{b_1}{x} + \frac{b_2}{x^2} + \cdots + \frac{b_n}{x^n} + \frac{b_{n+1}}{x^{n+1}} + O\left(\frac{1}{x^{n+2}}\right)\right].
\end{aligned}
\tag{6.9.28}
$$

6.9.3 A General Expansion Procedure

A number of methods exist for constructing the asymptotic behavior of special types of second-order, linear differential equations. The texts listed in the bibliography provide excellent summaries of almost all the known methods and, in many cases, provide proofs that these solutions are in fact asymptotic representations.

Consider the following special class of differential equations

$$y''(x) + f(x)y = 0, \tag{6.9.29}$$

where $f(x)$ has the form (or at least is asymptotic to this form)

$$f(x) = a_0 + \frac{a_1}{x} + \frac{a_2}{x^2} + \cdots + \frac{a_n}{x^n} + O\left(\frac{1}{x^{n+1}}\right), \tag{6.9.30}$$

and the coefficients $(a_0, a_1, a_2, \dots)$ are known. (Note that by a rescaling of the differential equation, a_0 can be made to equal one. For the remainder of this discussion we take $a_0 = 1$.) Then it can be shown that the asymptotic solutions to Eq. (6.9.29) can be expressed as

$$y(x) = e^{\lambda x} x^\sigma \left[1 + \frac{\alpha_1}{x} + \frac{\alpha_2}{x^2} + \cdots + \frac{\alpha_n}{x^n} + O\left(\frac{1}{x^{n+1}}\right)\right], \tag{6.9.31}$$

where the parameters $(\lambda, \sigma, \alpha_i)$ are to be determined in terms of the coefficients $(a_1, a_2, \dots, a_n)$.

The substitution of Eq. (6.9.32) into Eq. (6.9.29), along with the cancelling of an overall factor of $x^\sigma \exp(\lambda x)$; and the collecting together of all terms in like powers of x^{-m}, for $m = 1, 2, 3, \dots, n$, gives the following relations:

$$\lambda^{(1)} = i, \qquad \lambda^{(2)} = -i; \tag{6.9.32}$$

$$\sigma^{(1)} = i\left(\frac{a_1}{2}\right), \qquad \sigma^{(2)} = -i\left(\frac{a_1}{2}\right); \tag{6.9.33}$$

$$\begin{cases} \alpha_1^{(1)} = -\left(\frac{i}{2}\right)\left[-i\left(\frac{a_1}{2}\right) - \left(\frac{a_1^2}{4}\right) + a_2\right], \\[2mm] \alpha_1^{(2)} = \left(\frac{i}{2}\right)\left[i\left(\frac{a_1}{2}\right) - \left(\frac{a_1^2}{4}\right) + a_2\right], \end{cases} \tag{6.9.34}$$

and where the general alpha coefficient values are given by the following recurrence relation

$$\alpha_{n+1} = \left[\frac{1}{2\lambda(n+1)}\right]\left\{\left[n(n+1) - (2n+1)\sigma + \sigma^2 + a_2\right]\alpha_n\right.$$
$$\left. + a_3\alpha_{n-1} + \cdots + a_{n+2}\right\}, \qquad (n = 0, 1, 2, \dots). \tag{6.9.35}$$

Note that each parameter $(\lambda, \sigma, \alpha_i)$ has two values. Thus, in using Eq. (6.9.35) we must be consistent in which value we select. However, this (maybe) minor problem, in general, does not arise. Since the exponential function appears in the assumed form for the asymptotic solution, it should be clear that for oscillatory solutions λ will be pure imaginary and this has the consequence that the other parameter (σ, α_i) will also be complex. For differential equations modeling physical systems, the coefficients $(a_1, a_2, \dots)$ will be real. Therefore, if we denote the two separate linearly independent solution by

$$y^{(1)}(x) = e^{\lambda^{(1)} x} x^{\sigma^{(1)}} \left[1 + \frac{\alpha_1^{(1)}}{x} + \cdots + \frac{\alpha_n^{(1)}}{x^n} + O\left(\frac{1}{x^{n+1}}\right) \right],$$

$$y^{(2)}(x) = e^{\lambda^{(2)} x} x^{\sigma^{(2)}} \left[1 + \frac{\alpha_1^{(2)}}{x} + \cdots + \frac{\alpha_n^{(2)}}{x^n} + O\left(\frac{1}{x^{n+1}}\right) \right], \qquad (6.9.36)$$

then

$$[y^{(1)}(x)]^* = y^{(2)}(x), \qquad (6.9.37)$$

and

$$y_1(x) = \operatorname{Re} y^{(1)}(x), \qquad y_2(x) = \operatorname{Im} y^{(1)}(x), \qquad (6.9.38)$$

are also linearly independent solutions.

An excellent discussion of this method of expansion, along with extensions to other forms for $f(x)$, is given in the text of Murray.

Problems

Section 6.2

6.2.1 Think of and discuss an alternative physical oscillating system that can be modeled by the three cases presented in this section.

6.2.2 For the case of one end fixed, the other end free, work out $\{\lambda_n, u_n(x) : n = 1, 2, 3, \dots\}$ for the boundary conditions

$$u(0) = 0, \qquad u'(L) = a.$$

Plot $u_1(x)$, $u_2(x)$ and $u_3(x)$, and discuss physically what your solutions mean.

6.2.3 Analyze the case

$$u(0) = 0, \qquad u(L) = b > 0.$$

Section 6.3

6.3.1 Consider the differential equation

$$\frac{d^2y}{dx^2} + \left[1 + \left(\frac{1 - 4n^2}{4x^2}\right)\right] y = 0.$$

Show that for all real n and for x sufficiently large, that the solutions $y_n(x)$ oscillate.

6.3.2 Prove that if $p(x)$ is continuous and $p(x) > 0$ for $x > 0$ and, further λ is a real number, then the equation

$$\frac{d^2y}{dx^2} + [p(x) + \lambda^2]y = 0,$$

has solutions with an infinite number of zeros.

6.3.3 Discuss and plot the qualitative behavior of the following two differential equations:

(i) $y'' + x^2 y = 0$,

(ii) $xy'' + \lambda y' = 0$, $\lambda = $ real.

Section 6.4

6.4.1 Consider a Sturm-Liouville type problem, but with the boundary conditions, as given by Eq. (6.4.1) changed to

$$\begin{cases} A_1 y(a) + A_2 y'(a) = A_3 \\ B_1 y(b) + B_2 y'(b) = B_3 \end{cases}$$

where $(A_1, A_2, A_3, B_1, B_2, B_3)$ are given real constants with not both A_1 and A_2 zero, and not both B_1 and B_2 zero. Is this a valid generalization of the Sturm-Liouville Problem? Select an elementary example of such a system to study.

6.4.2 Discuss why the weight function $w(x)$ should be of one sign over the interval, $a \le x \le b$. Examine a case where $w(x)$ is taken to have a zero at $x = x_0$, $a < x_0 < b$.

6.4.3 Write down the corresponding form of the completeness relation, given in Eq. (6.4.16), for the case where the ϕ_n are orthogonal, but not normalized.

Section 6.5

6.5.1 How "good" an approximation is $f(x) = x(1 - x)$ to $\sin(\pi x)$ over the interval $0 \le x \le 1$? See section 6.5.2.

6.5.2 Is it possible to construct a representation for the Dirac delta function in terms of only the sine or cosine functions?

6.5.3 Using essentially the same method as given in section 6.5.4 to prove the reality of the eigenvalues, show that the eigenfunctions belonging to two different eigenvalues are orthogonal with respect to $r(x)$ on the interval $a \leq x \leq b$.

6.5.4 To obtain good evidence that Eq. (6.5.23) has an infinite number of solutions for k_n, plot $y_1(x) = \tan x$ and $y_2(x) = -\left(\frac{x}{5}\right)$ on the same graph. Give arguments to show that for large x, the zeros of

$$\tan x + \frac{x}{5} = 0$$

are separated by (approximately) π. Calculate numerically the first ten zeros.

Section 6.6

6.6.1 Prove that $G(x, x') = G(x', x)$.

6.6.2 Rederive all of the results presented in section 6.6 with the $-f(x)$ replaced by $f(x)$ in Eq. (6.6.1).

6.6.3 Assume that $\{\lambda_n, \phi_n(x) : n = 1, 2, 3, \ldots\}$ are not normalized. How does this change the result presented in Eq. (6.6.16) for the Green's function?

6.6.4 Given $a_1(x)$, $a_2(x)$ and $h(x)$ appearing in Eq. (6.6.17), derive expressions for the functions $p(x)$, $s(x)$ and $f(x)$ in Eq. (6.6.18).

Section 6.7

6.7.1 Solve the following boundary value problem by both direct integration of the differential equation and by use of the Green's function:

$$y''(x) = x, \qquad y(0) = y(1) = 0.$$

6.7.2 Show that

$$G(x, x') = \left[\frac{1}{\sin(1)}\right] \begin{cases} -\sin(x')\sin(x-1), & 0 \leq x' < x, \\ -\sin(x')\sin(x'-1), & x < x' \leq 1, \end{cases}$$

is a Green's function for the boundary value problem

$$y''(x) + y(x) = -f(x), \qquad y(0) = y(1) = 0.$$

6.7.3 Solve the problem of section 6.7.4, i.e.,

$$y''(x) + y(x) = x, \qquad y(0) = y(1) = 0,$$

using the Green's function given in Problem 6.7.3.

6.7.4 Consider the boundary value problem

$$y''(x) = \sin x, \qquad y(0) = y(1) = 0.$$

Calculate the solution using a Green's function and compare it with the solution obtained by application of the method of undetermined coefficients.

Section 6.8

6.8.1 Calculate the first- and second-derivative expressions given by Eqs. (6.8.32) and (6.8.33).

6.8.2 In Eq. (6.8.39), what restriction must be placed on $Q(x)$ such that $h(s)$ and $h'(s)$ both go to zero as $s \to \infty$?

Section 6.9

6.9.1 Can the next term in the asymptotic solution for the Airy equation be calculated using the expansion form given in section 6.9.2 for the Bessel equation?

6.9.2 Show that the next terms in the asymptotic expansion for the solution of Bessel's equations are

$$-\frac{9A \sin x}{2(8x)^2} \quad \text{and} \quad -\frac{9B \cos x}{2(8x)^2}.$$

Note that this calculation suggests that the expansion variable is $(8x)^{-1}$ rather than x^{-1}.

6.9.3 Apply the Liouville-Green transformation to the differential equation

$$\frac{d^2 v}{dx^2} + x^2 v = 0$$

and determine the first two terms in its asymptotic expansion.

6.9.4 Consider the following version of the Airy differential equation

$$\frac{d^2 v}{dx^2} - xv = 0.$$

Determine its asymptotic behavior for $x \gg 1$?

6.9.5 The Coulomb equation arises in the scattering of two elementary particles [5]. One form of this differential equation is

$$y''(x) + \left[k^2 + \frac{k_1}{x} + \frac{k_2}{x^2} \right] y(x) = 0,$$

where the constants (k, k_1, k_2) are specified in advance. Use the method given in Problem 6.9.3 to obtain the first two terms in its asymptotic expansion.

6.9.6 Find the first two terms in the asymptotic solution $(x \to \infty)$ of the equation

$$y''(x) + \left(\frac{x^2}{1 + x^2} \right) y(x) = 0.$$

6.9.7 In the asymptotic expansion for $f(x)$, given by Eq. (6.9.30), let $a_0 < 0$. Carry out the construction of the asymptotic solution for this case.

Comments and References

[1] Proofs of all of the stated results given in this section are given in the books by Birkhoff and Rota, Coddington and Levinson, and Ross.

[2] We follow closely, in this section, the presentation of Ross as given in sections 12.1 to 12.3 of his textbook.

[3] This problem is taken from the textbook by D. A. McQuarrie, *Mathematical Methods for Scientists and Engineers* (University Science Books; Sausalito, CA; 2003).

[4] E. A. Kraut, *Fundamentals of Mathematical Physics* (McGraw-Hill, New York, 1967); see section 6.22.

[5] For example see:
  * P. M. Morse and H. Feshback, *Methods of Theoretical Physics*, Vols. I and II (McGraw-Hill, New York, 1953).
  * P. R. Wallace, *Mathematical Analysis of Physical Problems* (Dover, New York, 1984).

[6] See the textbook of McQuarrie in [3] above; pps. 708–709.

[7] J. L. Powell and B. Crasemann, *Quantum Mechanics* (Addison-Wesley; Reading, MA; 1961).

[8] P. M. Moore, *Vibration and Sound* (McGraw-Hill, New York, 1936).

Bibliography

Asymptotics for Differential Equations

R. Bellman, *Stability Theory of Differential Equations* (Dover, New York, 1953).

R. Bellman, *Perturbation Techniques in Mathematics, Engineering and Physics* (Holt, Rinehart and Winston; New York, 1966).

C. Bender and S. A. Orszag, *Advanced Mathematical Methods for Scientists and Engineers* (McGraw-Hill, New York, 1978).

N. G. De Bruign, *Asymptotic Methods in Analysis* (North-Holland, Amsterdam, 1958).

A. Erdelyi, *Asymptotic Expansion* (Dover, New York, 1956).

H. Jeffreys and B. Jeffreys, *Methods of Mathematical Physics*, 3rd Ed. (Cambridge University Press, Cambridge, 1966).

P. B. Kahn, *Mathematical Methods for Scientists and Engineers: Linear and Nonlinear Systems* (Wiley-Interscience, New York, 1990).

J. D. Murray, *Asymptotic Analysis* (Springer-Verlag, New York, 1984).

W. Wasow, *Asymptotic Expansions for Ordinary Differential Equations* (Interscience, New York, 1965).

Green's Functions

G. B. Arfken and H. J. Weber, *Mathematical Methods for Physicists*, 4th ed. (Academic Press, New York, 1995).

P. K. Chattopadhyay, *Mathematical Physics* (Wiley, New York, 1990).

D. G. Duffy, *Green's Functions with Applications* (Chapman and Hall/CRC; New York, 2001).

S. M. Lea, *Mathematics for Physicists* (Brooks/Cole; Belmont, CA; 2004).

I. Stakgold, *Green's Functions and Boundary Value Problems* (Wiley-Interscience, New York, 1979).

E. C. Tichmarsh, *Eigenfunction Expansions Associated with Second Order Differential Equations*, 2nd ed. Vol. I (Oxford University Press, London, 1962).

Sturm-Liouville Problems

G. Birkhoff and G.-C. Rota, *Ordinary Differential Equations* (Ginn and Company, Boston, 1962).

E. A. Coddington and N. Levinson, *Theory of Ordinary Differential Equations* (McGraw-Hill, New York, 1955).

S. L. Ross, *Differential Equations* (Blaisdell; Waltham, MA; 1964).

H. Sagan, *Boundary and Eigenvalue Problems in Mathematical Physics* (Wiley, New York, 1961).

Chapter 7

Special Functions and Their Properties

7.1 Introduction

The main purpose of this chapter is to introduce the classical orthogonal polynomials and provide a concise summary of their mathematical properties. These functions are solutions to certain second-order differential equations that repeatedly occur in the mathematical modeling of a wide range of dynamical systems in the natural and engineering sciences. Our interest in studying them comes from their usefulness in applications and for their ability to provide mathematical representations of a broad class of functions, i.e., they provide sets of basis functions in terms of which other functions can be expanded. In general, they satisfy second-order differential equations that belong to the class of Sturm-Liouville problems.

In the materials to follow, we state a number of relevant facts on these functions and their associated properties, but do not give the proofs. Several excellent texts already exist and it is very hard to see how their presentations can be done better. Also, our philosophy, as clearly stated in the preface, is to provide a textbook that introduces certain concepts for the main purpose of seeing how they can be applied to problems of importance in the various natural and engineering sciences; the proofs of related mathematical statements are not stressed nor actually needed for the required analysis. However, when possible, references to readily available papers and/or books are given. This is certainly the case for the topics covered in this chapter.

Three excellent references, covering essentially all the topics in this chapter, except for the last section on applications, are listed below:

1) H. S. Wilf, *Mathematics for the Physical Sciences* (Dover, New York, 1978).

2) C. W. Wong, *Introduction to Mathematical Physics, Methods and Concepts* (Oxford University Press, New York, 1991).

3) S. Hassani, *Mathematical Physics, A Modern Introduction to Its Foundations* (Springer-Verlag, New York, 1999). They collectively provide proofs for essentially all of the mathematical statements made in this chapter.

To illustrate how the "special functions" arise, we consider a system modeled by the wave equation in spherical coordinates [1]. The general wave equation is

$$\nabla^2 \phi = \left(\frac{1}{c^2}\right) \frac{\partial^2 \phi}{\partial t^2}, \qquad \phi = \phi(\mathbf{r}, t). \tag{7.1.1}$$

For spherical coordinates, we have

$$\left(\frac{1}{r^2}\right) \frac{\partial}{\partial r}\left(r^2 \frac{\partial \phi}{\partial r}\right) + \left(\frac{1}{r^2 \sin\theta}\right) \frac{\partial}{\partial \theta}\left(\sin\theta \frac{\partial \phi}{\partial \theta}\right)$$
$$+ \left(\frac{1}{r^2 \sin^2\theta}\right) \frac{\partial^2 \phi}{\partial \psi^2} = \left(\frac{1}{c^2}\right) \frac{\partial^2 \phi}{\partial t^2}, \tag{7.1.2}$$

where $\mathbf{r} = (r, \theta, \psi)$ and c is the constant speed of propagation of the wave. A particular solution can be found by applying the method of separation of variables, i.e., assume ϕ can be written as,

$$\phi(r, \theta, \psi, t) = R(r)\Theta(\theta)\Psi(\psi)T(t). \tag{7.1.3}$$

Substitution of this form into Eq. (7.1.2) gives the following four ordinary differential equations,

$$\frac{d^2 R}{dr^2} + \left(\frac{2}{r}\right) \frac{dR}{dr} + \left(p^2 - \frac{\lambda}{r^2}\right) R = 0, \tag{7.1.4}$$

$$\left(\frac{1}{\sin\theta}\right) \frac{d}{d\theta}\left(\sin\theta \frac{d\Theta}{d\theta}\right) + \left[\lambda - \frac{m^2}{\sin^2\theta}\right] \Theta = 0. \tag{7.1.5}$$

$$\frac{d^2 \Psi}{d\psi^2} + m^2 \Psi = 0, \tag{7.1.6}$$

$$\frac{d^2 T}{dt^2} + c^2 p^2 T = 0, \tag{7.1.7}$$

where the (λ, m, p) are constants of separation. Eqs. (7.1.6) and (7.1.7) have well known solutions expressible in terms of the sine and cosine functions. Equations (7.1.4) and (7.1.5), respectively, are the Bessel and associated Legendre's differential equations. The most interesting feature of these particular differential equations is that they show up repeatedly in the mathematical modeling of classical systems. Consequently, to understand such systems, an understanding of the solutions to these differential equations must be obtained.

The next section gives an overview of a number of topics related to the so-called classical orthogonal polynomials. This is followed by separate sections detailing the major mathematical properties of the Legendre, Hermite, Chebyschev and Laguerre polynomials. Section 7.7 discusses the Bessel functions. While it is not one of the classical orthogonal polynomials, it certainly plays a role in the sciences as large as any of the polynomial functions. Finally, in the last section, we show how these functions can be used to solve problems.

7.2 Classical Orthogonal Polynomials

The classical orthogonal polynomials share a number of common features that can be characterized by particular formulas. In the following, we state these properties and present the related mathematical formulations of the associated concepts. We will denote a sequence of classical orthogonal polynomials (COP) by $\{\phi_n(x) : n = 0, 1, 2, \dots\}$.

7.2.1 *Differential Equation and Interval of Definition*

The COP satisfy second-order, linear differential equations having the form

$$P_2(x)\phi'' + P_1(x)\phi' + Q(n)\phi = 0, \qquad (7.2.1)$$

where $P_1(x)$ and $P_2(x)$ are, respectively, linear and quadratic functions of x, and $Q(n)$ is, in general, a quadratic function of n.

The interval in x, $[a, b]$, over which a particular COP is defined, can be finite, semi-infinite, or infinite. The standard procedure is to take these three possibilities to be,

$$\begin{cases} \text{finite interval:} & -1 \leq x \leq 1, \\ \text{semi-finite interval:} & 0 \leq x \leq \infty, \\ \text{finite interval:} & -\infty < x < \infty. \end{cases} \qquad (7.2.2)$$

7.2.2   Weight Functions and Rodrique's Formulas

The sequence of COP can be explicitly calculated by means of a so-called Rodrique's formula. This relationship has the structure,

$$\phi_n(x) = \left[\frac{1}{K_n w(x)} \right] \frac{d^n}{dx^n} \left\{ w(x)[g(x)]^n \right\}, \qquad (7.2.3)$$

where $w(x)$ is the "weight function" and $g(x)$ is a quadratic function of x. The weight function, $w(x)$, is defined on $[a, b]$ and satisfies the three conditions,

(i) $w(x) \geq 0$, $a \leq x \leq b$;

(ii) $\int_a^b x^n w(x) dx$ exists for $(n : 0, 1, 2 \ldots)$;

(iii) $w(x)$ is infinitely differentiable on $[a, b]$.

The K_n is a given normalization constant. Its particular functional dependence on n is determined by the historical fact that the various COP are normalized differently.

7.2.3   Orthogonality Relations

The COP are solutions to Sturm-Liouville problems and $w(x)$ is the weight function for the associated differential equation and other related properties. In particular, the COP satisfy the following orthogonality relation

$$\int_a^b \phi_i(x)\phi_j(x)w(x)dx = N_i \delta_{ij}, \qquad (7.2.4)$$

where N_i is a given function of i for each particular sequence of COP.

7.2.4   Generating Function

One very nice result for each of the COP is that they can be generated by repeated differentiation of a single function of two variables. This generating function, $G(x, t)$, when expanded as a Taylor series in t, takes the form,

$$G(x, t) = \sum_{n=0}^{\infty} a_n \phi_n(x) t^n, \qquad (7.2.5)$$

where the constant coefficients, $\{a_n : n = 0, 1, 2, \ldots\}$, are selected such that the $\phi_n(x)$ have certain desired properties, usually related to their normalization. Note that if $G(x, t)$, on the left-side, is known in closed form, then

$\phi_n(x)$ is given by a relationship involving the nth partial derivative of the generating function with respect to t, i.e.,

$$\phi_n(x) = \left[\frac{1}{(a_n)(n!)}\right] \frac{\partial^n G(x,t)}{\partial t^n}\bigg|_{t=0}. \qquad (7.2.6)$$

A knowledge of the generating function is very important since it can be used to derive many of the interesting properties of the COP, including various relationships between them and their derivatives. However, in general, the actual calculation of $G(x,t)$ can be quite difficult [2].

7.2.5 *Recurrence Relations*

The sequence of functions defining a given COP satisfy a three term recurrence relation having the form,

$$\phi_{n+1}(x) = (A_n x + B_n)\phi_n(x) + C_n \phi_{n-1}(x), \qquad (7.2.7)$$

where (A_n, B_n, C_n) are given functions of n. An advantage of having such a relation is that given $\phi_0(x)$ and $\phi_1(x)$ all the remaining $\phi_n(x)$, $n \geq 2$, can be explicitly calculated in terms of them.

7.2.6 *Differential Recurrences*

The first derivatives of a particular COP satisfy a recurrence relationship involving only the COP. This differential recurrence relation takes the form

$$g(x)\phi'_n(x) = (A n x + S_n)\phi_n(x) + R_n \phi_{n-1}(x), \qquad (7.2.8)$$

where $g(x)$ is the same function as appears in Eq. (7.2.3), A is a given constant, and (S_n, R_n) are specified functions of n. A consequence of Eq. (7.2.8) is that first derivatives of $\phi_n(x)$ can be expressed in terms of $\phi_n(x)$ and $\phi_{n-1}(x)$.

7.2.7 *Special Values*

For $\{\phi_n : n = 0, 1, 2, \dots\}$ defined on a finite interval, which can always be taken as $[-1, 1]$, the values of $\phi_n(x)$ at the end-points and at zero are generally of interest. Almost always there exist patterns in the values. Also, a knowledge of the special values can often lead to great simplifications in calculations involving the COP.

7.2.8 *Zeros of COP*

The $\{\phi_n(x) : n = 0, 1, 2, 3, \dots\}$ have important properties related to their zeros. A partial listing of some are given below:

(i) $\phi_n(x)$ has exactly n zeros, i.e., roots of the equation

$$\phi_n(\bar{x}) = 0. \tag{7.2.9}$$

(ii) These zeros are real and simple.
(iii) All are contained in the interval of definition.

The next four sections list the essential properties of four different classical orthogonal polynomials. These are the polynomial functions associated with the names of Legendre, Hermite, Chebyschev, and Laguerre. The reader should be aware that in various texts and handbooks, the formulas for special features of these equations may involve different overall constants than the ones we use. Also, the name Chebyschev appears in several spellings; for example, it is often written as Tschebycheff.

7.3 Legendre Polynomials: $P_n(x)$

Differential Equation

$$(1 - x^2)\frac{d^2\phi}{dx^2} - 2x\frac{d\phi}{dx} + n(n+1)\phi = 0, \qquad n = 0, 1, 2, \dots \tag{7.3.1}$$

Interval and Weight Function

$$\text{Interval:} \qquad -1 \leq x \leq 1, \tag{7.3.2}$$

$$\text{Weight Function:} \quad w(x) = 1. \tag{7.3.3}$$

Generating Function

$$\frac{1}{\sqrt{1 - 2xt + t^2}} = \sum_{n=0}^{\infty} P_n(x)t^k. \tag{7.3.4}$$

Rodrique's Formula

$$P_n(x) = \left(\frac{1}{2^n n!}\right)\frac{d^n}{dx^n}\left[(x^2 - 1)^n\right], \qquad (n = 0, 1, 2, \dots). \tag{7.3.5}$$

Orthogonality Condition

$$\int_{-1}^{1} P_m(x)P_n(x)dx = \left(\frac{2}{2n+1}\right)\delta_{nm}, \qquad (n = 0, 1, 2, \ldots). \qquad (7.3.6)$$

Recurrence Relations

$$(n+1)P_{n+1}(x) = (2n+1)xP_n(x) - nP_{n-1}(x),$$
$$(x^2 - 1)P'_n(x) = nxP_n(x) - nP_{n-1}(x)$$
$$= \left[\frac{n(n+1)}{2n+1}\right][P_{n+1}(x) - P_{n-1}(x)], \qquad (n = 0, 1, 2, \ldots).$$
$$(7.3.7)$$

List of $P_0(x)$ to $P_5(x)$

$$P_0(x) = 1,$$
$$P_1(x) = x,$$
$$P_2(x) = \left(\frac{1}{2}\right)[3x^2 - 1],$$
$$P_3(x) = \left(\frac{1}{2}\right)[5x^3 - 3x],$$
$$P_4(x) = \left(\frac{1}{8}\right)[35x^4 - 30x^2 + 3],$$
$$P_5(x) = \left(\frac{1}{8}\right)[63x^5 - 70x^3 + 15x]. \qquad (7.3.8)$$

Special Properties and Values

For $n = 0, 1, 2, \ldots$, we have
$$P_n(-x) = (-1)^n P_n(x),$$
$$P_n(1) = 1, \qquad P_n(-1) = (-1)^n,$$
$$P_{2n+1}(0) = 0, \qquad P_{2n}(0) = (-1)^n \left[\frac{(2n)!}{2^{2n}(n!)^2}\right]. \qquad (7.3.9)$$

7.4 Hermite Polynomials: $H_n(x)$

Differential Equation

$$\frac{d^2\phi}{dx^2} - 2x\frac{d\phi}{dx} + 2n\phi = 0, \qquad (n = 0, 1, 2, \ldots). \qquad (7.4.1)$$

Interval and Weight Function

$$\text{Interval:} \qquad -\infty < x < \infty, \qquad (7.4.2)$$

$$\text{Weight Function:} \quad w(x) = e^{-x^2}. \qquad (7.4.3)$$

Generating Function

$$\exp(-t^2 + 2xt) = \sum_{n=0}^{\infty} H_n(x) \left(\frac{t^n}{n!} \right). \qquad (7.4.4)$$

Rodrique's Formula

$$H_n(x) = (-1)^n e^{x^2} \frac{d^n}{dx^n} \left[e^{-x^2} \right], \qquad (n = 0, 1, 2, \dots). \qquad (7.4.5)$$

Orthogonality Condition

$$\int_{-\infty}^{\infty} e^{-x^2} H_m(x) H_n(x) dx = \sqrt{\pi}\, 2^n\, n!\, \delta_{nm}, \qquad (n = 0, 1, 2, \dots). \quad (7.4.6)$$

Recurrence Relations

$$H_{n+1}(x) = 2x H_n(x) - 2n H_{n-1}(x),$$
$$H_n'(x) = 2n H_{n-1}(x). \qquad (7.4.7)$$

List of $H_0(x)$ to $H_5(x)$

$$H_0(x) = 1,$$
$$H_1(x) = 2x,$$
$$H_2(x) = 4x^2 - 2,$$
$$H_3(x) = 8x^3 - 12x,$$
$$H_4(x) = 16x^4 - 48x^2 + 12,$$
$$H_5(x) = 32x^5 - 160x^3 + 120x. \qquad (7.4.8)$$

Special Properties and Values

For $n = 0, 1, 2, \ldots$, we have

$$H_n(-x) = (-1)^n H_n(x),$$

$$H_{2n+1}(0) = 0, \qquad H_{2n}(0) = (-1)^n \left[\frac{(2n!)}{n!} \right]. \tag{7.4.9}$$

7.5 Chebyschev Polynomials: $T_n(x)$ and $U_n(x)$

Differential Equation

There are two types of Chebyshev polynomials denoted by $T_n(x)$ and $U_n(x)$. They are both solutions to the differential equation,

$$(1 - x^2)\frac{d^2\phi}{dx^2} - x\frac{d\phi}{dx} + n^2\phi = 0, \qquad (n = 0, 1, 2, \ldots). \tag{7.5.1}$$

The functions $\sqrt{1 - x^2}\, U_n(x)$ and $T_n(x)$ are two linearly independent solutions of Eq. (7.5.1).

Interval and Weight Function

$$\text{Interval:} \qquad -1 \le x \le 1, \tag{7.5.2}$$

$$\text{Weight Function:} \quad 1/\sqrt{1 - x^2} = w(x). \tag{7.5.3}$$

Generating Function

$$\frac{1 - t^2}{1 - 2xt + t^2} = T_0(x) + 2\sum_{n=1}^{\infty} T_n(x)t^n, \tag{7.5.4}$$

$$\frac{1}{1 - 2xt + t^2} = \sum_{n=0}^{\infty} U_n(x)t^n. \tag{7.5.5}$$

Rodrique's Formula

For $n = 0, 1, 2, \ldots$, we have,

$$T_n(x) = \left[\frac{(-1)^n \sqrt{\pi(1 - x^2)}}{2^n \Gamma\left(n + \frac{1}{2}\right)} \right] \frac{d^n}{dx^n} \left[(1 - x^2)^{n - \frac{1}{2}} \right], \tag{7.5.6}$$

$$U_n(x) = \left[\frac{(-1)^n (n + 1)\sqrt{\pi/(1 - x^2)}}{2^{n+1} \Gamma\left(n + \frac{3}{2}\right)} \right] \frac{d^n}{dx^n} \left[(1 - x^2)^{n + \frac{1}{2}} \right]. \tag{7.5.7}$$

Orthogonality Condition

$$\int_{-1}^{1} T_m(x)T_n(x)(1-x^2)^{-1/2}dx = \begin{cases} 0, & m \neq n, \\ \frac{\pi}{2}, & m = n \neq 0, \\ \pi, & m = n = 0, \end{cases} \tag{7.5.8}$$

$$\int_{-1}^{1} U_m(x)U_n(x)(1-x^2)^{-1/2}dx = \left(\frac{\pi}{8}\right)\delta_{nm}. \tag{7.5.9}$$

Recurrence Relations

$$T_{n+1}(x) = 2xT_n(x) - T_{n-1}(x)$$
$$U_{n+1}(x) = 2xU_n(x) - U_{n-1}(x)$$
$$T_n(x) = U_n(x) - xU_{n-1}(x)$$
$$(1-x^2)U_n(x) = xT_{n+1} - T_{n+2}(x)$$
$$(1-x^2)T_n'(x) = -nxT_n(x) + nT_{n-1}(x). \tag{7.5.10}$$

List of $T_n(x)$ to $U_n(x)$ for $n = 0$ to 5

$$T_0(x) = 1 \qquad\qquad U_0(x) = 1$$
$$T_1(x) = x \qquad\qquad U_1(x) = 2x$$
$$T_2(x) = 2x^2 - 1 \qquad\qquad U_2(x) = 4x^2 - 1$$
$$T_3(x) = 4x^3 - 3x \qquad\qquad U_3(x) = 8x^3 - 4x$$
$$T_4(x) = 8x^4 - 8x^2 + 1 \qquad\qquad U_4(x) = 16x^4 - 12x^2 + 1$$
$$T_5(x) = 16x^5 - 20x^3 + 5x \qquad U_5(x) = 32x^5 - 32x^3 + 6x. \tag{7.5.11}$$

Special Properties and Values

$$T_n(-x) = (-1)^n T_n(x) \qquad U_n(-x) = (-1)^n U_n(x)$$
$$T_{2n}(0) = (-1)^n \qquad\qquad T_{2n+1}(0) = 0$$
$$T_n(1) = 1 \qquad\qquad T_n(-1) = (-1)^n$$
$$U_{2n}(0) = (-1)^n \qquad\qquad U_{2n+1}(0) = 0. \tag{7.5.12}$$

7.6 Laguerre Polynomials: $L_n(x)$

Differential Equations

$$x\frac{d^2\phi}{dx^2} + (1-x)\frac{d\phi}{dx} + ny = 0, \qquad (n = 0, 1, 2\ldots, 0). \qquad (7.6.1)$$

Interval and Weight Function

$$\text{Interval:} \qquad 0 \le x \le \infty, \qquad (7.6.2)$$
$$\text{Weight Function:} \quad w(x) = e^{-x}. \qquad (7.6.3)$$

Generating Functions

$$\left(\frac{1}{1-t}\right)\exp\left(\frac{xt}{t-1}\right) = \sum_{n=0}^{\infty} L_n(k)t^n. \qquad (7.6.4)$$

Rodrique's Formula

For $n = 0, 1, 2, \ldots$, we have,

$$L_n(x) = \left(\frac{e^x}{n!}\right)\frac{d^n}{dx^n}\left[x^n e^{-x}\right]. \qquad (7.6.5)$$

Orthogonality Condition

$$\int_0^{\infty} e^{-x}L_m(x)L_n(x)dx = \delta_{nm}. \qquad (7.6.6)$$

Recurrence Relations

$$(n+1)L_{n+1}(x) = (2n+1-x)L_n(x) - nL_{n-1}(x),$$
$$xL_n'(x) = nL_n(x) - nL_{n-1}(x). \qquad (7.6.7)$$

List of $L_0(x)$ to $L_5(x)$

$$L_0(x) = 1$$
$$L_1(x) = 1 - x$$
$$L_2(x) = \left(\frac{1}{2!}\right)\left[2 - 4x + x^2\right]$$

$$L_3(x) = \left(\frac{1}{3!}\right)\left[6 - 18x + 9x^2 - x^3\right]$$

$$L_4(x) = \left(\frac{1}{4!}\right)\left[24 - 96x + 72x^2 - 16x^3 + x^4\right]$$

$$L_5(x) = \left(\frac{1}{5!}\right)\left[120 - 600x + 600x^2 - 200x^3 + 25x^4 - x^5\right]. \qquad (7.6.8)$$

7.7 Legendre Functions of the Second Kind and Associated Legendre Functions

The Legendre differential equation (7.3.1) has a second linearly independent solution that is a nonpolynomial function of x. These functions are denoted by $Q_n(x)$ and are named Legendre functions of the second kind and order n. These functions are unbounded at $x = \pm 1$ and thus are defined over the interval $-1 < x < 1$.

The $Q_n(x)$ satisfy the same recurrence relations as the Legendre functions of the first kind, $P_n(x)$. However, the Rodrique's formula is given by the following expression

$$Q_n(x) = \left(\frac{1}{2}\right) P_n(x) \cdot \ln\left(\frac{1+x}{1-x}\right) - W_{n-1}(x), \qquad (7.7.1)$$

where

$$W_{n-1}(x) = \left[\frac{2n-1}{1 \cdot n}\right] P_{n-1}(x) + \left[\frac{2n-5}{3(n-1)}\right] P_{n-3}(x)$$

$$+ \left[\frac{2n-9}{5(n-2)}\right] P_{n-5}(x) + \cdots$$

$$= \sum_{m=1}^{n} \left(\frac{1}{m}\right) P_{m-1}(x) P_{n-m}(x), \qquad (7.7.2)$$

with $W_{-1}(x) = 0$.

The integrals of products of $Q_n(x)$ with $Q_m(x)$ and $P_m(x)$ are rather complex and we refer the reader to the handbook by Abramowitz and Stegun for these evaluations.

The first six $Q_n(x)$ are:

$$Q_0(x) = \left(\frac{1}{2}\right)\ln\left(\frac{1+x}{1-x}\right)$$

$$Q_1(x) = \left(\frac{x}{2}\right)\ln\left(\frac{1+x}{1-x}\right) - 1$$

$$Q_2(x) = \left(\frac{1}{4}\right)(3x^2 - 1)\ln\left(\frac{1+x}{1-x}\right) - \left(\frac{3}{2}\right)x$$

$$Q_3(x) = \left(\frac{1}{4}\right)(5x^3 - 3x)\ln\left(\frac{1+x}{1-x}\right) - \left(\frac{5}{2}\right)x^2 + \frac{2}{3}$$

$$Q_4(x) = \left(\frac{1}{16}\right)(35x^4 - 30x^2 + 3)\ln\left(\frac{1+x}{1-x}\right) - \left(\frac{35}{8}\right)x^3 + \left(\frac{55}{24}\right)x$$

$$Q_5(x) = \left(\frac{1}{16}\right)(63x^5 - 70x^3 + 15x)\ln\left(\frac{1+x}{1-x}\right)$$

$$- \left(\frac{63}{8}\right)x^4 + \left(\frac{49}{8}\right)x^2 - \frac{8}{15}. \tag{7.7.3}$$

The Legendre associated differential equation is

$$(1 - x^2)\frac{d^2\phi}{dx^2} - 2x\frac{d\phi}{dt} + \left[n(n+1) - \frac{m^2}{1-x^2}\right]\phi = 0. \tag{7.7.4}$$

Note that for $m = 0$, this equation reduces to the usual Legendre differential equation. In the following, we only consider values of m and n that are non-negative integers. The two linearly solutions to Eq. (7.7.4) are denoted by $P_n^m(x)$ and $Q_n^m(x)$. Both functions can be determined by differentiation of the first and second Legendre functions. The required relations are,

$$P_n^m(x) = (1 - x^2)^{m/2}\frac{d^m}{dx^m}[P_n(x)], \tag{7.7.5}$$

$$Q_n^m(x) = (1 - x^2)^{m/2}\frac{d^m}{dx^m}[Q_n(x)]. \tag{7.7.6}$$

The $P_n^m(x)$ and $Q_n^m(x)$ satisfy the same recurrence relations. Several of them are listed below:

$$\begin{cases} P_n^{m+1}(x) = \left[\frac{1}{\sqrt{x^2-1}}\right]\left[(n-m)xP_n^m(x) - (n+m)P_{n-1}^m(x)\right], \\[2mm] (n-m+1)P_{n+1}^m(x) = (2n+1)xP_n^m(x) - (n+m)P_{n-1}^m(z), \\[2mm] P_{n+1}^m(x) = P_{n-1}^m(x) + (2n+1)(x^2-1)^{1/2}P_n^{m-1}(x), \\[2mm] (x^2-1)\frac{dP_n^m(x)}{dx} = nxP_n^m(x) - (m+n)P_{n-1}^m(x). \end{cases} \tag{7.7.7}$$

The associated Legendre functions satisfy a number of integral relations over the interval $-1 < x < 1$. Several of the simpler ones are given below:

$$\begin{cases} \displaystyle\int_{-1}^{1} P_n^m(x)P_\ell^m(x)dx = \left(\frac{1}{n+\frac{1}{2}}\right)\left[\frac{(n+m)!}{(n-m)!}\right]\delta_{n\ell}, \\[12pt] \displaystyle\int_{-1}^{1} \frac{P_n^m(x)P_\ell^m(x)}{1-x^2}\,dx = \left[\frac{(n+m)!}{m(n-m)!}\right]\delta_{n\ell}, \\[12pt] \displaystyle\int_{-1}^{1} Q_n^m(x)P_\ell^m(x)dx = (-1)^m\left[\frac{1-(-1)^{\ell+n}(n+m)!}{(\ell-n)(\ell+n+1)m(n-m)!}\right]. \end{cases} \tag{7.7.8}$$

Finally, we present several formulas giving special values for the associated Legendre functions and their first derivatives:

$$\begin{cases} P_n^m(0) = \left(\dfrac{2^m}{\sqrt{\pi}}\right)\left[\dfrac{\cos[\pi(n-m)/2]\Gamma\left(\frac{m+n+1}{2}\right)}{\Gamma\left(\frac{n-m+2}{2}\right)}\right], \\[16pt] Q_n^m(0) = [-\sqrt{\pi}\,2^{m-1}]\left[\dfrac{\sin[\pi(n+m)/2]\Gamma\left(\frac{m+n+1}{2}\right)}{\Gamma\left(\frac{n-m+2}{2}\right)}\right], \end{cases} \tag{7.7.9}$$

and

$$\begin{cases} \dfrac{dP_n^m(x)}{dx}\bigg|_{x=0} = \left(\dfrac{2^{m+1}}{\sqrt{\pi}}\right)\left[\dfrac{\sin[\pi(n+m)/2]\Gamma\left(\frac{m+n+2}{2}\right)}{\Gamma\left(\frac{n-m+1}{2}\right)}\right], \\[16pt] \dfrac{dQ_n^m(x)}{dx}\bigg|_{x=0} = (2^m\sqrt{\pi})\left[\dfrac{\cos[\pi(m+n)/2]\Gamma\left(\frac{m+n+2}{2}\right)}{\Gamma\left(\frac{n-m+1}{2}\right)}\right]. \end{cases} \tag{7.7.10}$$

7.8   Bessel Functions

The general form of the Bessel differential equation is

$$x^2\frac{d^2\phi}{dx^2} + x\frac{d\phi}{dx} + (\lambda^2 x^2 - \nu^2)\phi = 0, \tag{7.8.1}$$

where (λ, ν) are parameters. The transformation of variable $\lambda x \to x$, gives the standard form

$$x^2\frac{d^2\phi}{dx^2} + x\frac{d\phi}{dx} + (x^2 - \nu^2)\phi = 0. \tag{7.8.2}$$

The two linearly independent solutions of Eq. (7.8.2) are denoted by $J_\nu(x)$ and $Y_\nu(x)$. For the case where ν is not an integer, both $J_\nu(x)$ and $J_{-\nu}(x)$ are linearly independently solutions and $Y_\nu(x)$ can be taken as

$$Y_\nu(x) = \frac{J_\nu(x)\cos(\nu\pi) - J_{-\nu}(x)}{\sin(\nu\pi)}, \qquad \nu \neq \text{integer.} \tag{7.8.3}$$

For integer ν, we have

$$J_{-n}(x) = (-1)^n J_n(x) \tag{7.8.4}$$

and $Y_n(x)$ is

$$Y_n(x) = \lim_{\nu \to n} Y_\nu(x). \tag{7.8.5}$$

The functions $J_\nu(x)$ and $Y_\nu(x)$ are called, respectively, Bessel functions of the first and second kind of order ν.

The $J_\nu(x)$ and Y_ν, for general ν, have the following **series representations**:

$$J_\nu(x) = x^\nu \sum_{k=0}^{\infty} \frac{(-1)^k x^{2k}}{2^{2k+\nu} k! \Gamma(\nu + k + 1)}, \tag{7.8.6}$$

$$Y_\nu(x) = \left[\frac{1}{\sin(\nu\pi)} \right] \left\{ \cos(\nu\pi) \left(\frac{x}{2} \right)^\nu \sum_{k=0}^{\infty} \frac{(-1)^k x^{2k}}{2^{2k} k! \Gamma(\nu + k + 1)} \right.$$
$$\left. - \left(\frac{x}{2} \right)^{-\nu} \sum_{k=0}^{\infty} \frac{(-1)^k x^{2k}}{2^{2k} k! \Gamma(k - \nu + 1)} \right\}; \tag{7.8.7}$$

if $\nu = n = $ integer, then

$$Y_n(x) = -\left(\frac{1}{\pi} \right) \left(\frac{2}{x} \right)^n \sum_{k=0}^{n-1} \frac{(n-k-1)!}{k!} \left(\frac{x^2}{4} \right)^k + \left[\left(\frac{2}{\pi} \right) \ln \left(\frac{x}{2} \right) \right] J_n(x)$$
$$- \left(\frac{1}{\pi} \right) \left(\frac{x}{2} \right)^n \sum_{k=0}^{\infty} (-1)^k \left[\frac{\psi(k+1) + \psi(n+k+1)}{k!(n+k)!} \right] \left(\frac{x^2}{4} \right)^k, \tag{7.8.8}$$

where $\psi(x)$ is the digamma function

$$\psi(x) = \frac{d}{dx} \left[\ln \Gamma(x) \right] = \frac{\Gamma'(x)}{\Gamma(x)}, \tag{7.8.9}$$

and is given by

$$\psi(x) = \sum_{n=1}^{\infty} \left[\frac{1}{n} - \frac{1}{x-1+n} \right] - \gamma \tag{7.8.10}$$

with γ being the Euler-Mascheroni constant,

$$\gamma = 0.577215\ldots. \tag{7.8.11}$$

A **Rodrique type formula** for the Bessel $J_n(x)$ functions is

$$J_n(x) = (-1)^n x^n \left(\frac{1}{x} \frac{d}{dx} \right)^n J_0(x). \tag{7.8.12}$$

The corresponding **generating function** is

$$\exp\left[\left(\frac{x}{2}\right)\left(t - \frac{1}{t}\right)\right] = \sum_{n=-\infty}^{+\infty} J_n(x)t^n, \qquad (7.8.13)$$

where it should be noted that the summation is over both positive and negative integers.

The two Bessel functions have the following **limiting values** when $x \to 0$ and $x \to \infty$:

$$\begin{cases} J_0(0) = 1, & J_n(0) = 0 & (n = 1, 2, 3, \ldots); \\ J_0'(0) = 0, & J_1'(0) = \frac{1}{2}, & J_n'(0) = 0 & (n = 2, 3, 4, \ldots); \\ \operatorname*{Lim}_{x\to\infty} J_n(x) = 0 & (n = 0, 1, 2 \ldots); \end{cases} \qquad (7.8.14)$$

$$\operatorname*{Lim}_{x\to 0} Y_n(x) = -\infty, \qquad \operatorname*{Lim}_{x\to\infty} Y_n(x) = 0 \qquad (n = 0, 1, 2 \ldots). \qquad (7.8.15)$$

For large x, the following **asymptotic representations** hold

$$J_\nu(x) = \sqrt{\frac{2}{\pi x}}\left[\cos\left(x - \frac{\nu\pi}{2} - \frac{\pi}{4}\right) + O\left(\frac{1}{x}\right)\right], \qquad (7.8.16)$$

$$Y_\nu(x) = \sqrt{\frac{2}{\pi x}}\left[\sin\left(x - \frac{\nu\pi}{2} - \frac{\pi}{4}\right) + O\left(\frac{1}{x}\right)\right]. \qquad (7.8.17)$$

Both $J_\nu(x)$ and $Y_\nu(x)$ satisfy the same *recurrence and differentiable relations*; several of them are listed below for $J_\nu(x)$:

$$2\nu J_\nu(x) = x[J_{\nu+1}(x) + J_{\nu-1}(x)], \qquad (7.8.18)$$

$$2\frac{d}{dx}[J_\nu(x)] = J_{\nu-1}(x) - J_{\nu+1}(x), \qquad (7.8.19)$$

$$\left(\frac{1}{x}\frac{d}{dx}\right)^m [x^\nu J_\nu(x)] = x^{\nu-m} J_{\nu-m}(x), \qquad (7.8.20)$$

$$\left(\frac{1}{x}\frac{d}{dx}\right)^m [x^{-\nu} J_\nu(x)] = (-1)^m x^{-\nu-m} J_{\nu+m}(x). \qquad (7.8.21)$$

Also, for integer n, we have

$$J_{-n}(x) = (-1)^n J_n(x), \qquad Y_{-n}(x) = (-1)^n Y_n(x). \qquad (7.8.22)$$

The $J_n(x)$ Bessel function has a number of both interesting and important integral representations; they include the following forms:

$$J_n(x) = \begin{cases} \left(\dfrac{1}{\pi}\right) \displaystyle\int_0^\pi \cos[x\cos\theta - n\theta]d\theta, \\[2ex] \left[\dfrac{2}{\sqrt{\pi}\Gamma\left(n+\frac{1}{2}\right)}\right] \left(\dfrac{x}{2}\right)^n \displaystyle\int_0^{\pi/2} \cos(x\sin\theta)(\cos\theta)^{2n}d\theta, \\[2ex] \left[\dfrac{2}{\sqrt{\pi}\Gamma\left(n+\frac{1}{2}\right)}\right] \left(\dfrac{x}{2}\right)^n \displaystyle\int_0^{\pi/2} \cos(x\sin\theta)(\sin\theta)^{2n}d\theta. \end{cases} \qquad (7.8.23)$$

There are several other functions that are related to the Bessel functions. The **Hankel functions** of the first and second kinds are defined as

$$H_n^{(1)}(x) = J_n(x) + iY_n(x), \qquad H_n^{(2)}(x) = J_n(x) - iY_n(x). \qquad (7.8.24)$$

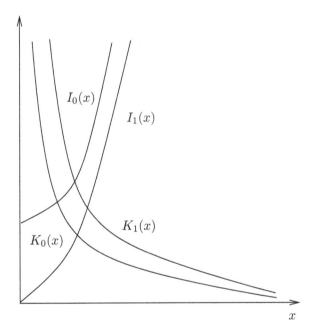

Fig. 7.8.1   Plots of $I_0(x)$, $I_1(x)$, $K_0(x)$, and $K_1(x)$ vs x.

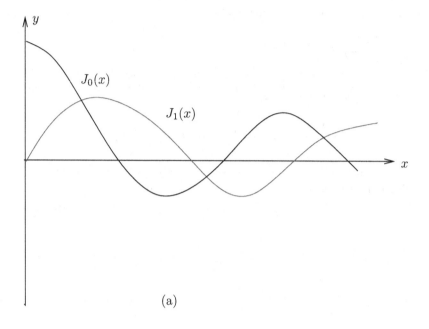

(a)

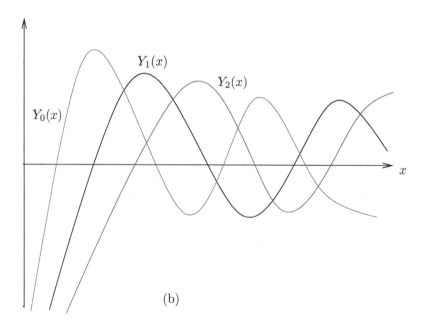

(b)

Fig. 7.8.2   Plots of $J_0(x)$, $J_1(x)$, $Y_0(x)$, and $Y_1(x)$.

Note that for real x, these two functions are complex conjugates of each other, i.e.,

$$\left[H_n^{(1)}(x)\right]^* = H_n^{(2)}(x), \qquad (7.8.25)$$

with the real part equal to $J_n(x)$ and the imaginary part equal to $Y_n(x)$.

The modified Bessel functions, $I_\nu(x)$ and $K_\nu(x)$, are two linearly independent solutions of the differential equation

$$x^2 \frac{d^2\phi}{dx^2} + x \frac{d\phi}{dx} - (x^2 + \nu^2)\phi = 0. \qquad (7.8.26)$$

They are defined as

$$I_\nu(x) = i^{-\nu} J_\nu(ix), \qquad (7.8.27a)$$

with

$$I_{-n}(x) = I_n(x), \qquad n = \text{integer}; \qquad (7.8.27b)$$

and

$$K_\nu(x) = \left(\frac{\pi}{2}\right) \left[\frac{I_{-\nu}(x) - I_\nu(x)}{\sin(\nu\pi)}\right], \qquad \nu \neq \text{integer}, \qquad (7.8.28)$$

$$K_n(x) = \operatorname*{Lim}_{\nu \to n} \left(\frac{\pi}{2}\right) \left[\frac{I_{-\nu}(x) - I_\nu(x)}{\sin(\nu\pi)}\right], \qquad (7.8.29)$$

for $n = \text{integer}$.

Representative plots of $J_n(x)$, $Y_n(x)$, $I_n(x)$, and $K_n(x)$ are given in Figures 7.8.1 and 7.8.2.

7.9 Some Proofs and Worked Problems

This section presents a variety of problems and some proofs of results stated earlier in this chapter. A large number of additional worked problems, proofs, and discussion of related issues are also given in the following four books:

1. P. K. Chattopadhyay, *Mathematical Physics* (Wiley, New York, 1990).

2. J. Irving and N. Mullineux, *Mathematics in Physics and Engineering* (Academic Press, New York, 1959).

3. M. R. Spiegel, *Advanced Mathematics for Engineers and Scientists* (Schaum's Outline Series, McGraw-Hill, New York, 1999).

4. P. R. Wallace, *Mathematical Analysis of Physical Problems* (Dover, New York, 1984).

7.9.1 *Proof of Rodriques Formula for $P_n(x)$*

Rodriques formula for the Legendre polynomials is

$$P_n(x) = \left(\frac{1}{2^n n!}\right) \frac{d^n}{dx^n} (x^2 - 1)^n. \tag{7.9.1}$$

To prove this relation, define the function $F(x, n)$ as follows

$$F(x, n) = (x^2 - 1)^n, \qquad (n = 0, 1, 2, \ldots). \tag{7.9.2}$$

Note that

$$(x^2 - 1)F' = 2nxF. \tag{7.9.3}$$

Let now Eq. (7.9.3) be differentiated $(n + 1)$ times. The application of the Leibnitz relation gives

$$(x^2 - 1)\frac{d^{n+1}F}{dx^{n+2}} + [2(n+1)x]\frac{d^{n+1}F}{dx^{n+1}} + n(n+1)\frac{d^n F}{dx^n}$$
$$= 2nx\frac{d^{n+1}F}{dx^{n+1}} + 2n(n+1)\frac{d^n F}{dx^n}, \tag{7.9.4}$$

and this can be rewritten as

$$(1 - x^2)\frac{d^2}{dx^2}\left(\frac{d^n F}{dx^n}\right) - (2x)\frac{d}{dx}\left(\frac{d^n F}{dx^n}\right) + n(n+1)\left(\frac{d^n F}{dx^n}\right) = 0. \tag{7.9.5}$$

However, inspection of this last equation shows that it is the Legendre differential equation for $F^{(n)}(x, n)$. From Eq. (7.9.2), it follows that $F(x, n)$ is a polynomial of degree $2n$; therefore, $F^{(n)}(x, n)$ is a polynomial of nth degree. This implies that $F^{(n)}(x, n)$ must be proportional to $P_n(x)$ since $P_n(x)$ is also a polynomial of degree n, i.e., $F^{(n)}(x, n) = cP_n(x)$, where $c = $ constant. This constant can be determined by another application of the Leibniz relation for the nth derivative of a product, i.e.,

$$\frac{d^n F}{dx^n} = \frac{d^n}{dx^n}\left[(x+1)^n(x-1)^n\right]$$
$$= \sum_{k=0}^{n}\binom{n}{k}\left[\frac{d^{n-k}}{dx^{n-k}}(x+1)^n\right]\left[\frac{d^k}{dx^k}(x-1)^n\right], \tag{7.9.6}$$

where

$$\binom{n}{k} = \frac{n!}{(n-k)!k!}. \tag{7.9.7}$$

On the right-side of Eq. (7.9.6), the term with $k = n$ does not depend on the factor $(x - 1)$. Therefore, if we evaluate $F^{(n)}(x, n)$ at $x = 1$, the following is obtained

$$\frac{d^n F}{dx^n}\bigg|_{x=1} = \binom{n}{n}(x + 1)^n n!\bigg|_{x=1} = \frac{(n!)^2(2^n)}{n!} = 2^n \cdot n!. \tag{7.9.8}$$

Consequently, we conclude that

$$P_n(x) = \left(\frac{1}{2^n n!}\right)\frac{d^n}{dx^n}[(x^2 - 1)^n], \tag{7.9.9}$$

and this is the Rodrique formula for the Legendre polynomials.

7.9.2 *Integrals of x^m with $P_n(x)$*

We now study the following integral

$$\int_{-1}^{1} x^m P_n(x)dx. \tag{7.9.10}$$

Replacing $P_n(x)$ by its Rodriques formula representation

$$\int_{-1}^{1} x^m P_n(x)dx = \left(\frac{1}{2^n n!}\right)\int_{-1}^{1} x^m \frac{d^n}{dx^n}[(x^2 - 1)^n]dx \tag{7.9.11}$$

and integrating by parts gives

$$\int_{-1}^{1} x^m P_n(x)dx = \left(\frac{1}{2^n n!}\right)\left\{x^m \frac{d^{n-1}}{dx^{n-1}}(x^2 - 1)^n\bigg|_{-1}^{1}\right.$$

$$\left. - m\int_{-1}^{1} x^{m-1}\frac{d^{n-1}}{dx^{n-1}}[(x^2 - 1)^n]\,dx\right\}$$

$$= -\left(\frac{m}{2^n n!}\right)\int_{-1}^{1} x^{m-1}\frac{d^{n-1}}{dx^{n-1}}[(x^2 - 1)^n]\,dx. \tag{7.9.12}$$

This last relation follows from the fact that all terms in the first expression on the right-side of Eq. (7.9.12) contain a factor of $(x^2 - 1)$ and vanish at $x = \pm 1$. If we continue this process n times, then at the nth step x^m must be differentiated n times. For $m < n$, it follows that the result must be zero, i.e.,

$$\int_{-1}^{1} x^m P_n(x)dx = 0, \qquad m < n. \tag{7.9.13}$$

If $m = n$, then

$$\int_{-1}^{1} x^n P_n(x)dx = \left(\frac{1}{2^n n!}\right)[(-1)^n n(n-1)\cdots 2\cdot 1]\int_{-1}^{1}(x^2-1)^n dx$$

$$= \left[\frac{(-1)^n n!}{2^n n!}\right]\int_{-1}^{1}(x^2-1)^n dx. \tag{7.9.14}$$

The integral on the right-side of Eq. (7.9.14) has the value

$$\int_{-1}^{1}(x^2-1)^n dx = (-1)^n\left[\frac{(n!)^2\cdot 2^{2n+1}}{(2n+1)!}\right]. \tag{7.9.15}$$

Therefore,

$$\int_{-1}^{1} x^n P_n(x)dx = \frac{(n!)^2\cdot 2^{n+1}}{(2n+1)!}. \tag{7.9.16}$$

7.9.3   Expansion of $\delta(x)$ in Terms of $P_n(x)$

We now derive an expansion for the Dirac delta function in terms of the Legendre polynomials. Let $\delta(x)$ be written as

$$\delta(x) = \sum_{n=0}^{\infty} a_n P_n(x), \tag{7.9.17}$$

where

$$a_n = \left(\frac{2n+1}{2}\right)\int_{-1}^{1} P_n(x)\delta(x)dx = \left(\frac{2n+1}{2}\right)P_n(0). \tag{7.9.18}$$

Since $P_{2k+1}(x)$ is an odd function for k a positive integer, it follows that $P_{2k+1}(0) = 0$. However, using the result in Eq. (7.3.9), we obtain

$$a_{2k} = \left(\frac{4k+1}{2}\right)\left[\frac{(-1)^k(2k)!}{2^{2k}(k!)^2}\right] \tag{7.9.19}$$

and

$$\delta(x) = \sum_{n=0}^{\infty}(-1)^n\left(\frac{4n+1}{2}\right)\left[\frac{(2n)!}{2^{2n}(n!)^2}\right]P_{2n}(x). \tag{7.9.20}$$

Note that since $\delta(x)$ is an even "function," i.e.,

$$\delta(-x) = \delta(x), \tag{7.9.21}$$

it is consistent to have only even degree Legendre polynomials in its representation.

What about $\delta(x - x')$? It follows from Eq. (7.9.21) that

$$\delta(x - x') = \delta(x' - x). \tag{7.9.22}$$

A nice symmetric way to formulate the Legendre representation is to write Eq. (7.9.20) as

$$\delta(x) = \sum_{n=0}^{\infty} \left(\frac{2n+1}{2} \right) P_n(0) P_n(x). \tag{7.9.23}$$

It follows that the symmetric extension of the last equation is

$$\delta(x - x') = \sum_{n=0}^{\infty} \left(\frac{2n+1}{2} \right) P_n(x') P_n(x). \tag{7.9.24}$$

7.9.4 *Laplace's Equation in Spherical Coordinates*

The general Laplace equation is a second-order partial differential equation. In spherical coordinates, (r, θ, ψ), it has the representation

$$\frac{\partial}{\partial r} \left(r^2 \frac{\partial U}{\partial r} \right) + \left(\frac{1}{\sin \theta} \right) \frac{\partial}{\partial \theta} \left(\sin \theta \frac{\partial U}{\partial \theta} \right) + \left(\frac{1}{\sin^2 \theta} \right) \frac{\partial^2 U}{\partial \psi^2} = 0, \tag{7.9.25}$$

where $U = U(r, \theta, \psi)$.

For the case of full spherical symmetry, U depends only on r and it satisfies the equation

$$\frac{\partial}{\partial r} \left(r^2 \frac{\partial U}{\partial r} \right) = 0. \tag{7.9.26}$$

Integrating once gives

$$r^2 \frac{\partial U}{\partial r} = \text{constant}. \tag{7.9.27}$$

A second integration leads to

$$U(r) = \frac{A}{r} + B, \tag{7.9.28}$$

where A and B are arbitrary constants.

For axial symmetry, U depends on r and θ, i.e., it is independent of ψ. This gives the equation

$$\frac{\partial}{\partial r} \left(r^2 \frac{\partial U}{\partial r} \right) + \left(\frac{1}{\sin \theta} \right) \frac{\partial}{\partial \theta} \left(\sin \theta \frac{\partial U}{\partial \theta} \right) = 0. \tag{7.9.29}$$

Assuming a solution of the form

$$U(r, \theta) = R(r) \Theta(\theta), \tag{7.9.30}$$

and substituting this result in Eq. (7.9.29), we obtain

$$\left(\frac{1}{R} \right) \frac{d}{dr} \left(r^2 \frac{dR}{dr} \right) = - \left(\frac{1}{\Theta \sin \theta} \right) \frac{d}{d\theta} \left(\sin \theta \frac{d\Theta}{d\theta} \right). \tag{7.9.31}$$

Since the left-side depends only on r, while the right-side depends just on θ, it follows that each must be equal to a constant which we take as λ. Therefore, the second-order differential equations for $R(r)$ and $\Theta(\theta)$ are

$$r^2 \frac{d^2 R}{dr^2} + 2r \frac{dR}{dr} - \lambda R = 0, \tag{7.9.32}$$

$$\left(\frac{1}{\sin\theta}\right) \frac{d}{d\theta}\left(\sin\theta \frac{d\Theta}{d\theta}\right) + \lambda\Theta = 0. \tag{7.9.33}$$

If we let $x = \cos\theta$, then

$$\frac{d}{d\theta} = -(\sin\theta)\frac{d}{dx}, \tag{7.9.34}$$

and Eq. (7.9.33) becomes

$$(1 - x^2)\frac{d^2\Theta}{dx^2} - 2x\frac{d\Theta}{dx} + \lambda\Theta = 0, \tag{7.9.35}$$

which is Legendre's equation. If $\lambda = n(n+1)$, where n is a non-negative integer, then Eq. (7.9.35) has two linearly independent solutions,

$$\Theta_n(x) = c_1 P_n(x) + c_2 Q_n(x), \tag{7.9.36}$$

where c_1 and c_2 are arbitrary constants.

Setting $\lambda = n(n+1)$ in Eq. (7.9.32) gives

$$r^2 \frac{d^2 R}{dr^2} + 2r \frac{dR}{dr} - n(n+1)R = 0, \tag{7.9.37}$$

which is a Cauchy-Euler equation. A complete solution can be determined by making the transformation of variables

$$r = e^s, \tag{7.9.38}$$

and obtaining

$$\frac{d^2 R}{ds^2} + \frac{dr}{ds} - n(n+1)R = 0. \tag{7.9.39}$$

This is a second-order linear equation with constant coefficients and its general solution is

$$R_n(s) = D_1 e^{ns} + D_2 e^{-(n+1)s}, \tag{7.9.40}$$

or in terms of r

$$R_n(r) = D_1 r^n + D_2/r^{n+1}, \tag{7.9.41}$$

where D_1 and D_2 are arbitrary constants.

The most general solution to Eq. (7.9.29) is therefore given by

$$U(r,\theta) = \sum_{n=0}^{\infty} R_n(r)\Theta_n(\theta)$$

$$= \sum_{n=0}^{\infty}\left(A_n r^n + \frac{B_n}{r^{n+1}}\right)[C_n P_n(\cos\theta) + D_n Q_n(\cos\theta)], \tag{7.9.42}$$

where the (A, B, C, D) are arbitrary constants. There dependence on n is determined by the particular problem studied. In the next section, we apply these results and show how the constants can be calculated.

7.9.5 Sphere in a Uniform Flow

Consider a sphere of radius a placed in a uniform flow of fluid. The fluid is assumed to have an undisturbed velocity flow that is constant and has the magnitude c. The direction of the flow is taken to be in the direction of the z-axis. The velocity potential U satisfies the Laplace equation

$$\nabla^2 U = 0, \tag{7.9.43}$$

subject to the boundary conditions

$$\begin{cases} U = \text{finite at all places in the fluid;} \\ U \to cz \text{ as } r \to \infty. \end{cases} \tag{7.9.44}$$

For this problem we can use spherical coordinates. The variable z can be replaced by

$$z = r \cos\theta = r P_1(\cos\theta), \tag{7.9.45}$$

and the variable ψ will not appear because of the symmetry of the problem, i.e., $U(r, \theta, \psi) \to U(r, \theta)$.

To proceed, we need to add the further condition

$$\frac{\partial U}{\partial r} = 0, \qquad r = a. \tag{7.9.46}$$

This is the requirement that the fluid flow velocity is zero on the surface of the sphere.

The general form of the solution has already been calculated and is given by Eq. (7.9.42). For our problem the D_n coefficients are all zero since the $Q_n(\cos\theta)$ become unbounded for $\theta = 0$ or π and U is assumed bounded. Therefore,

$$U(r, \theta) = \sum_{n=0}^{\infty} \left[A_n r^n + \frac{B_n}{r^{n+1}} \right] P_n(\cos\theta). \tag{7.9.47}$$

The second condition of Eq. (7.9.44) can be written, using the result of Eq. (7.9.45), as

$$U \to cz = cr P_1(\cos\theta), \qquad r \to \infty. \tag{7.9.48}$$

This implies that in Eq. (7.9.47), we have

$$A_0 = \text{arbitrary}, \qquad A_1 = c; \qquad A_n = 0, \quad n \geq 2. \tag{7.9.49}$$

Therefore, U takes the form

$$U(r, \theta) = A_0 + crP_1(\cos\theta) + \sum_{n=0}^{\infty} \left(\frac{B_n}{r^{n+1}} \right) P_n(\cos\theta). \qquad (7.9.50)$$

The requirement of Eq. (7.9.46) gives

$$\frac{\partial U}{\partial r}\bigg|_{r=a} = cP_1(\cos\theta) - \sum_{n=0}^{\infty} \left[\frac{(n+1)B_n}{a^{n+2}} \right] P_n(\cos\theta) = 0. \qquad (7.9.51)$$

Since the $\{P_n(x) : n = 0, 1, 2, \ldots\}$ form an orthogonal set of functions, it follows that

$$B_0 = 0, \qquad B_1 = \frac{ca^3}{2}; \qquad B_n = 0, \qquad n \geq 2. \qquad (7.9.52)$$

Using these results, we have

$$U(r, \theta) = A_0 + c \left[r + \frac{a^3}{2r^2} \right] \cos\theta. \qquad (7.9.53)$$

The velocity $\mathbf{v}(r, \theta)$ at any point (r, θ, ψ) is given by

$$\mathbf{v}(r, \theta) = -\nabla U(r, \theta). \qquad (7.9.54)$$

In terms of the various velocity components, we have

$$\begin{cases} v_r = -\dfrac{\partial U}{\partial r} = -c \left[1 - \dfrac{a^3}{r^3} \right], \\[2mm] v_\theta = -\left(\dfrac{1}{r} \right) \dfrac{\partial U}{\partial \theta} = c \left[1 + \dfrac{a^3}{2r^3} \right] \sin\theta, \\[2mm] v_\psi = -\left(\dfrac{1}{r\sin\theta} \right) \dfrac{\partial U}{\partial \psi} = 0. \end{cases} \qquad (7.9.55)$$

Note that A_0 does not influence the values for the velocity or its components. This is a general feature for any quantity whose value is determined by calculating the gradient of a potential function. Physically, this means that only potential differences can be measured or have influence on the behavior of a system. The absolute value of the potential is meaningless.

7.9.6 *The Harmonic Oscillator*

The harmonic oscillator provides an extremely useful first approximation for many systems in the natural and engineering sciences [3; 4]. For microscopic systems the relevant equation for a particle of mass m subject to a linear potential

$$V(x) = \left(\frac{1}{2} \right) kx^2 = \left(\frac{1}{2} \right) m\omega^2 x^2, \qquad (7.9.56)$$

is

$$-\left(\frac{\hbar^2}{2m}\right)\frac{d^2\psi}{dx^2} + \left(\frac{1}{2}\right)m\omega^2x^2\psi = E\psi. \tag{7.9.57}$$

In these equations, $\omega^2 = k/m$ and is the square of the classical angular frequency; $\hbar = h/2\pi$, where h is the constant associated with Max Planck; and E is the energy of the particle having mass m. The required solutions must satisfy the boundary conditions

$$\psi(-\infty) = 0, \qquad \psi(\infty) = 0. \tag{7.9.58}$$

To determine this solution, it is convenient to rewrite the equation in terms of the dimensionless variable

$$y = \sqrt{\frac{m\omega}{\hbar}}\, x. \tag{7.9.59}$$

Doing this gives

$$\frac{d^2\psi}{dy^2} = (y^2 - K)\psi, \tag{7.9.60}$$

where

$$K = \frac{2E}{\hbar\omega}. \tag{7.9.61}$$

Our task is to calculate the allowed values for K and the associated "wave-functions," $\psi(x, K)$.

Note that for large y, Eq. (7.9.60) can be approximated by the equation

$$\frac{d^2\psi}{dy^2} \approx y^2\psi, \tag{7.9.62}$$

and the approximation to the solution of this equation takes the form

$$\psi(y) \approx c_1 e^{-y^2/2} + c_2 e^{y^2/2}, \tag{7.9.63}$$

where (c_1, c_2) are arbitrary constants. Let's check this result by calculating the first and second derivatives of $\psi(y)$ as given in Eq. (7.9.63):

$$\begin{cases} \psi'(y) \simeq -c_1 y e^{-y^2/2} + c_2 y e^{y^2/2}, \\ \psi''(y) \simeq c_1 y^2 e^{-y^2/2} + c_2 y^2 e^{y^2/2} \\ \qquad -c_1 y^{-y^2/2} + c_2 e^{y^2/2}. \end{cases} \tag{7.9.64}$$

Since the last two terms in the expression for ψ'' are small compared to the first two terms, we conclude that for large values of y, Eq. (7.9.63) is a valid approximation to the actual solution.

To proceed, we assume the exact solution to be

$$\psi(y) = g(y)e^{-x^2/2}. \tag{7.9.65}$$

The reason why the $\exp(x^2/2)$ term was not included is the fact that such a term does not satisfy the boundary conditions given by Eq. (7.9.58). If this $\psi(y)$ is now substituted into Eq. (7.9.60), the following equation is found for $g(y)$

$$\frac{d^2 g}{dy^2} - 2y\frac{dg}{dy} + (K - 1)g = 0. \tag{7.9.66}$$

However, this is just the differential equation leading to the Hermite polynomials; see section 7.4. If we make the identification

$$K - 1 = 2n; \qquad n = 0, 1, 2, \ldots, \tag{7.9.67}$$

then

$$\begin{cases} g(y) \to H_n(y), \\ K \to K_n = 2n + 1. \end{cases} \tag{7.9.68}$$

From Eq. (7.9.61), it follows that the allowable values for E_n are

$$E_n = \hbar\omega\left(n + \frac{1}{2}\right). \tag{7.9.69}$$

We conclude that the microscopic harmonic oscillator has its energy level quantized, i.e., unlike the classical oscillator, for which all positive energies are possible, the quantum oscillator has only the discrete values given by Eq. (7.9.69). Associated with these of these levels is a wavefunction

$$\psi_n(y) = e^{-y^2/2}H_n(y). \tag{7.9.70}$$

7.9.7 *Matrix Elements of the Electric Dipole*

The harmonic oscillator can emit or absorb electromagnetic radiation if the so-called dipole matrix has nonzero elements. This matrix is defined as

$$M_{mn} \equiv \int_{-\infty}^{\infty} e^{-x^2} x H_m(x) H_n(x) dx. \tag{7.9.71}$$

Let us replace $x H_m(x)$ by

$$x H_m = \left(\frac{1}{2}\right) H_{m+1} + m H_{m-1}. \tag{7.9.72}$$

This result is a consequence of the recurrence relation given in Eq. (7.4.7). We obtain

$$M_{nm} = \int_{-\infty}^{\infty} e^{-x^2} \left[\left(\frac{1}{2} \right) H_{m+1}(x) + m H_{m-1}(x) \right] H_n(x) dx$$

$$= \sqrt{\pi} 2^{n-1} n! \delta_{m+1,n} + \sqrt{\pi} 2^n (n+1)! \delta_{m-1,n}$$

$$= \begin{cases} \sqrt{\pi} 2^{n-1} n!, & \text{if } m = n - 1, \\ \sqrt{\pi} 2^n (n+1)!, & \text{if } m = n + 1. \end{cases} \tag{7.9.73}$$

This last set of relations informs us that dipole transitions can only take place between neighboring levels, i.e., the quantum harmonic oscillator can change its energy by amount ΔE given by

$$\Delta E = \begin{cases} +\hbar \omega, & \text{if } m = n + 1, \\ -\hbar \omega, & \text{if } m = n - 1. \end{cases} \tag{7.9.74}$$

The $\pm \hbar \omega$ changes in energy correspond to energy absorption and emission by the harmonic oscillator.

7.9.8 *General Solution Hermite's Differential Equation for* $n = 0$

Our discussion of the Hermite differential equation in section 7.4 only considered the polynomial solutions. However, for each value of integer n, there also must be a second linearly independent solution. We can demonstrate that this second solution may be quite complex in form.

The Hermite differential equation for $n = 0$ is

$$y'' - 2xy' = 0. \tag{7.9.75}$$

This equation can be solved by letting $z = y'$; this gives

$$z' - 2xz = 0, \tag{7.9.76}$$

with the solution

$$z = \frac{dy}{dx} = c_2 e^{x^2}, \tag{7.9.77}$$

where c_2 is an arbitrary integration constant. Likewise, solving for $y(x)$ gives

$$y(x) = c_1 + c_2 \int_0^x e^{t^2} dt, \tag{7.9.78}$$

where c_1 is a second arbitrary constant. Since, $H_0(x) = 1$, we can write $y(x)$ as

$$y(x) = c_1 H_0(x) + c_2 \int_0^x e^{t^2} dt. \tag{7.9.79}$$

Therefore, Eq. (7.9.79) is the general solution to the Hermite differential equation for $n = 0$.

7.9.9   *Three Dimensional Harmonic Oscillator*

The Schrödinger equation for the harmonic oscillator in three dimensions is

$$\nabla^2\psi + (\lambda - r^2)\psi = 0, \tag{7.9.80}$$

where the equation has been put in a dimensionless form. In Cartesian coordinates (x, y, z), the substitution

$$\psi(x, y, z) = X(x)Y(y)Z(z), \tag{7.9.81}$$

gives

$$\frac{d^2X}{dx^2} + (\lambda_x - x^2)X = 0, \tag{7.9.82}$$

$$\frac{d^2Y}{dy^2} + (\lambda_y - y^2)Y = 0, \tag{7.9.83}$$

$$\frac{d^2Z}{dz^2} + (\lambda_z - z^2)Z = 0, \tag{7.9.84}$$

where

$$\lambda_x + \lambda_y + \lambda_z = \lambda. \tag{7.9.85}$$

This set of individual, uncoupled differential equations have already been solved in section 7.9.6 in terms of the Hermite polynomials. Using this result, we obtain

$$\psi_{n_x,n_y,n_z}(x, y, z) = C(n_x, n_y, n_z)e^{-r^2/2}H_{n_x}(x)H_{n_y}(y)H_{n_z}(z), \tag{7.9.86}$$

where

$$\lambda_x = 2n_x + 1, \quad \lambda_n = 2n_y + 1, \quad \lambda_z = 2n_z + 1, \tag{7.9.87}$$

and (n_x, n_y, n_z) are non-negative integers, r^2 is

$$r^2 = x^2 + y^2 + z^2, \tag{7.9.88}$$

and $C(n_x, n_y, n_z)$ is

$$C(n_x, n_y, n_z) = \left[2^n(n_x)!(n_y)!(n_z)!\pi^{3/2}\right]^{-1/2}, \tag{7.9.89}$$

where

$$n = n_x + n_y + n_z \tag{7.9.90}$$

$$\lambda \to \lambda_{n_x,n_y,n_z} = 2(n_x + n_y + n_z) + 3 = 2n + 3. \tag{7.9.91}$$

For spherical coordinates, the solution to Eq. (7.9.80) can be expressed as

$$\psi(r, \theta, \phi) = R(r)\Theta(\theta)\Phi(\phi). \tag{7.9.92}$$

Substitution into Eq. (7.9.80) gives the following equation for the radial function $R(r)$

$$\left(\frac{1}{r^2}\right)\frac{d}{dr}\left(r^2\frac{dR}{dr}\right) + \left[\lambda - r^2 - \frac{\ell(\ell+1)}{r^2}\right]R = 0, \tag{7.9.93}$$

where ℓ is a non-negative integer and this is related to the fact that the $\theta - \phi$ dependence is given in terms of the so-called spherical harmonic functions, i.e.,

$$\Theta(\theta)\Phi(\phi) \to Y_\ell^m(\theta, \phi) = C_\ell^m e^{im\phi}P_\ell^m(\cos\theta), \tag{7.9.94}$$

where C_ℓ^m is a normalization constant depending on m and ℓ, the $P_\ell^m(\cos\theta)$ functions are solutions to the associated Legendre differential equations, and the integers (ℓ, m) satisfy the restrictions

$$\ell = 0, 1, 2, \ldots; \qquad |m| \le \ell. \tag{7.9.95}$$

Inspection of Eq. (7.9.86) suggests that it may be of value to write $R(r)$ as

$$R(r) = e^{-r^2/2}\phi(r). \tag{7.9.96}$$

Substitution of Eq. (7.9.96) into Eq. (7.9.93) gives the following equation for $\phi(r)$

$$\phi'' + \left(\frac{2}{r} - 2r\right)\phi' + \left[\lambda - 3 - \frac{\ell(\ell+1)}{r^2}\right]\phi = 0. \tag{7.9.97}$$

The function $\phi(r)$ should have the property that

$$\phi(0) = \text{finite}. \tag{7.9.98}$$

This means that Eq. (7.9.97) should be checked as to whether it has such solutions. For r small, the solutions are determined by the equation

$$\phi'' + \left(\frac{2}{r}\right)\phi' - \left[\frac{\ell(\ell+1)}{r^2}\right]\phi \simeq 0 \tag{7.9.99}$$

or

$$r^2\phi'' + 2r\phi' - \ell(\ell+1)\phi \simeq 0, \tag{7.9.100}$$

with the solutions

$$\phi_1(r) = r^\ell, \qquad \phi_2(r) = r^{-(\ell+1)}. \tag{7.9.101}$$

[Eq. (7.9.100) is a Cauchy-Euler equation and can be solved using standard methods.] Examination shows that since $\ell = 0, 1, 2, \ldots$, only $\phi_1(r)$ gives a bounded solution at $r = 0$. Therefore, we can define a new function $v(r)$ related to $\phi(r)$ by

$$\phi(r) = r^\ell v(r). \tag{7.9.102}$$

Continuing with the substitution of this expression into Eq. (7.9.97), we obtain for $v(r)$ the differential equation

$$r^2 v'' + 2r[(\ell + 1) - r^2]v' + [2(n - \ell)r^2]v = 0, \tag{7.9.103}$$

where λ has been replaced by

$$\lambda - 3 = 2n. \tag{7.9.104}$$

The final change of variable

$$r \to t = r^2, \tag{7.9.105}$$

with

$$\frac{dv}{dr} = \frac{dv}{dt}\frac{dt}{dr} = 2r\frac{dv}{dt} \tag{7.9.106}$$

$$\frac{d^2v}{dr^2} = \frac{d}{dr}\left(\frac{dv}{dr}\right) = \frac{d}{dr}\left(2r\frac{dv}{dt}\right) = 2r\frac{d}{dt}\left(2r\frac{dv}{dt}\right)$$

$$= 2r\left[2\frac{dr}{dt}\frac{dv}{dt} + 2r\frac{d^2v}{dt^2}\right] = 2\frac{dv}{dt} + 4t\frac{d^2v}{dt^2}, \tag{7.9.107}$$

gives

$$t\frac{d^2v}{dt^2} + \left[\left(\ell + \frac{3}{2}\right) - t\right]\frac{dv}{dt} + \left(\frac{n - \ell}{2}\right)v = 0, \tag{7.9.108}$$

which can be rewritten as

$$t\frac{d^2v}{dt^2} + (\alpha + 1 - t)\frac{dv}{dt} + kv = 0, \tag{7.9.109}$$

with

$$\alpha = \frac{2\ell + 1}{2}, \qquad k = \frac{n - \ell}{2}. \tag{7.9.110}$$

Equation (7.9.109) is the differential equation for the associated Laguerre functions $L_k^\alpha(t)$. They can be generated by use of the following formula

$$L_k^\alpha(t) = \left[\frac{(-1)^k}{k!}\right]e^t\frac{d^k}{dt^k}\left[t^{k+\alpha}e^{-t}\right]. \tag{7.9.111}$$

Putting all these calculations together, we finally obtain an expression for the radial wave function given by

$$R(r) \to R_k^\alpha(r) = r^\ell e^{-r^2/2}L_k^\alpha(r^2). \tag{7.9.112}$$

7.9.10 Proof that $J_{-n}(x) = (-1)^n J_n(x)$

The series expansion for the Bessel function of the first kind having order n is

$$J_n(x) = \sum_{k=0}^{\infty} \frac{(-1)^k \left(\frac{x}{2}\right)^{n+2k}}{(k!)\Gamma(n+k+1)}. \tag{7.9.113}$$

Let $n \to (-n)$ and obtain

$$J_{-n}(x) = \sum_{k=0}^{\infty} \frac{(-1)^k \left(\frac{x}{2}\right)^{-n+2k}}{(k!)\Gamma(-n+k+1)}$$

$$= \sum_{k=0}^{n-1} \frac{(-1)^k \left(\frac{x}{2}\right)^{-n+2k}}{(k!)\Gamma(-n+k+1)} + \sum_{k=n}^{\infty} \frac{(-1)^k \left(\frac{x}{2}\right)^{-n+2k}}{(k!)\Gamma(-n+r+1)}. \tag{7.9.114}$$

In the first term, $\Gamma(-n+k+1)$ is infinite for $k = 0, 1, 2, \ldots, n-1$; consequently, this sum is zero. For the second term, replace k by $n+r$, and obtain

$$\sum_{r=0}^{\infty} \frac{(-1)^{n+r} \left(\frac{x}{2}\right)^{n+2r}}{(n+r)!\Gamma(r+1)} = (-1)^n \sum_{r=0}^{\infty} \frac{(-1)^r \left(\frac{x}{2}\right)^{n+2r}}{\Gamma(n+r+1)r!}$$

$$= (-1)^n J_n(x). \tag{7.9.115}$$

Therefore we have shown that

$$J_{-n}(x) = (-1)^n J_n(x). \tag{7.9.116}$$

7.9.11 Calculation of $J_n(x)$ from $J_0(x)$

We will start with the recurrence relation

$$J_{k-1} - J_{k+1} = 2J_k'(x), \tag{7.9.117}$$

for the Bessel function of first kind of order k. For $k = 0$, we have

$$J_{-1}(x) - J_1(x) = 2J_0'(x), \tag{7.9.118}$$

and on using Eq. (7.9.116)

$$J_1(x) = -J_0'(x). \tag{7.9.119}$$

Note that this equation can also be written as

$$J_1(x) = (-1)^1 x^1 \left(\frac{1}{x}\frac{d}{dx}\right) J_0(x). \tag{7.9.120}$$

We want to now use mathematical induction to prove that

$$J_n(x) = (-1)^n x^n \left(\frac{1}{x}\frac{d}{dx}\right)^n J_0(x). \qquad (7.9.121)$$

To proceed, we clearly have Eq. (7.9.121) true for $n = 1$. Now assume that it is true for $n = k$. Then for $n = k + 1$, it follows that

$$(-1)^{k+1} x^{k+1} \left(\frac{1}{x}\frac{d}{dx}\right)^{k+1} J_0(x) = (-1)^{k+1} x^{k+1} \cdot \left(\frac{1}{x}\frac{d}{dx}\right)\left[\left(\frac{1}{x}\frac{d}{dx}\right)^k J_0(x)\right]$$

$$= (-1)^{k+1} x^k \frac{d}{dx}\left[(-1)^k \left(\frac{1}{x^k}\right) J_k(m)\right]$$

$$= -\frac{dJ_k(x)}{dx} + \left(\frac{k}{x}\right) J_k. \qquad (7.9.122)$$

From the recurrence relations, see Eqs. (7.8.18) and (7.8.19),

$$J_{k+1}(x) = \left(\frac{k}{x}\right) J_k(x) - \frac{dJ_k(x)}{dx}, \qquad (7.9.123)$$

and can thus conclude

$$(-1)^{k+1} x^{k+1} \left(\frac{1}{x}\frac{d}{dx}\right)^{k+1} J_0(x) = J_{k+1}(x). \qquad (7.9.124)$$

Therefore, by induction, we see that the result in Eq. (7.9.121) is true for all n.

7.9.12   *An Integral Representation for $J_n(x)$*

To determine an integral representation for the Bessel functions of first kind, we start with the generating function

$$\exp\left\{\left(\frac{x}{2}\right)\left[t - \frac{1}{t}\right]\right\} = \sum_{n=-\infty}^{\infty} J_n(x)t^n. \qquad (7.9.125)$$

The change of variable

$$t = e^{i\theta}, \qquad (7.9.126)$$

gives

$$\exp[ix\sin\theta] = \sum_{n=-\infty}^{\infty} J_n(x)e^{in\theta}. \qquad (7.9.127)$$

To obtain the result on the left-side, the following calculation was done,

$$\begin{cases} t - \dfrac{1}{t} = e^{i\theta} - e^{-i\theta} = 2i\sin\theta, \\ \left(\dfrac{x}{2}\right)\left[t - \dfrac{1}{t}\right] = \left(\dfrac{x}{2}\right)(2i\sin\theta) = ix\sin\theta. \end{cases} \qquad (7.9.128)$$

Taking the real and imaginary parts of Eq. (7.9.127) provides the following two expressions,

$$\cos(x \sin \theta) = \sum_{n=-\infty}^{\infty} J_n(x) \cos(n\theta)$$

$$= J_0(x) + 2 \sum_{k=1}^{\infty} J_{2k}(x) \cos(2k\theta), \qquad (7.9.129)$$

$$\sin(x \sin \theta) = 2 \sum_{k=1}^{\infty} J_{2k-1}(x) \sin(2k-1)\theta. \qquad (7.9.130)$$

Now multiply Eq. (7.9.129) by $\cos(m\theta)$, where $m \neq 0$, and integrate from 0 to π; doing this gives

$$\int_0^{\pi} \cos(x \sin \theta) \cos(m\theta) d\theta)$$

$$= 2 \sum_{k=1}^{\infty} J_{2k}(x) \int_0^{\pi} \cos(2k\theta) \cos(m\theta) d\theta. \qquad (7.9.131)$$

Since

$$\int_{-\pi}^{\pi} \cos(2k\theta) \cos(m\theta) d\theta = 2 \int_0^{\pi} \cos(2k\theta)(\cos m\theta) d\theta$$

$$= \pi \delta_{2k,m} \quad (k, m \neq 0), \qquad (7.9.132)$$

we obtain

$$\left(\frac{1}{\pi}\right) \int_0^{\pi} \cos(x \sin \theta) \cos(2k\theta) d\theta = J_{2k}(x). \qquad (7.9.133)$$

Using the same techniques, it follows that

$$\left(\frac{1}{\pi}\right) \int_0^{\pi} \sin(x \sin \theta) \sin(2k-1)\theta \, d\theta = J_{2k-1}(x). \qquad (7.9.134)$$

Therefore for arbitrary k, even or odd, we can combine Eqs. (7.9.133) and (7.9.134), in the single expression

$$J_k(x) = \left(\frac{1}{\pi}\right) \int_0^{\pi} [\cos(x \sin \theta) \cos(k\theta) + \sin(x \sin \theta) \sin(k\theta] \, d\theta. \qquad (7.9.135)$$

Using

$$\cos(\theta_1 - \theta_2) = \cos \theta_1 \cos \theta_2 + \sin \theta_2 \sin \theta_2, \qquad (7.9.136)$$

with

$$\theta_1 = x \sin \theta, \qquad \theta_2 = k\theta, \qquad (7.9.137)$$

we finally obtain

$$J_k(x) = \left(\frac{1}{x}\right) \int_0^{\pi} \cos(k\theta - x \sin \theta) d\theta. \qquad (7.9.138)$$

7.9.13   Expansion of Cosine in Terms of $J_n(x)$

We show that the cosine can be represented as an expansion in the Bessel functions. Since $\cos x$ is an even function, it follows that in the expansion only even orders of Bessel functions will appear.

Again, we start from the generating function

$$\exp\left[\left(\frac{x}{2}\right)\left(t - \frac{1}{t}\right)\right] = \sum_{n=-\infty}^{\infty} J_n(x)t^n, \tag{7.9.139}$$

and replace t by $t = i = \sqrt{-1}$. Doing this gives

$$t - \frac{1}{t} = i - \frac{1}{i} = 2i, \tag{7.9.140}$$

and

$$e^{ix} = \sum_{n=-\infty}^{\infty} J_n(x)i^n. \tag{7.9.141}$$

Taking the real part of each side

$$\cos x = J_0(x) + \left[i^2 J_2(x) + i^{-2} J_{-2}(x)\right] \\ + \left[i^4 J_4(x) + i^{-4} J_{-4}(x)\right] + \cdots \tag{7.9.142}$$

using

$$i^{2n} J_{2n}(x) + i^{-2n} J_{-2n}(x) = i^{2n} J_{2n}(x) + i^{-2n}(-1)^{2n} J_{2n}(x), \tag{7.9.143}$$

and

$$i^{2n} = (i^2)^n = (-1)^n, \tag{7.9.144}$$

we obtain

$$\cos x = J_0(x) + 2\sum_{n=1}^{\infty} (-1)^n J_{2n}(x). \tag{7.9.145}$$

7.9.14   Derivation of a Recurrence Relation from the Generating Function

We now derive the following recurrence relation

$$J_{n-1}(x) + J_{n+1}(x) = \left(\frac{2n}{x}\right) J_n(x). \tag{7.9.146}$$

To begin, differentiate the generating function, given by Eq. (7.9.139), to obtain

$$\exp\left[\left(\frac{x}{2}\right)\left(t - \frac{1}{t}\right)\right]\left[\left(\frac{x}{2}\right)\left(1 + \frac{1}{t^2}\right)\right] = \sum_{n=-\infty}^{\infty} n J_n(x)t^{n-1} \tag{7.9.147}$$

and

$$\sum_{n=-\infty}^{\infty} \left(\frac{x}{2}\right)\left(1 + \frac{1}{t^2}\right) J_n(x)t^n = \sum_{n=-\infty}^{\infty} n J_n(x)t^{n-1}, \qquad (7.9.148)$$

where the last equation can be rewritten as

$$\left(\frac{x}{2}\right) \sum_{n=-\infty}^{\infty} J_n(x)t^n + \left(\frac{x}{2}\right) \sum_{n=-\infty}^{\infty} J_n(x)t^{n-2}$$

$$= \sum_{n=-\infty}^{\infty} n J_n(x)t^{n-1}. \qquad (7.9.149)$$

Shifting indices gives

$$\left(\frac{x}{2}\right) \sum_{n=-\infty}^{\infty} J_n(x)t^n + \left(\frac{x}{2}\right) \sum_{n=-\infty}^{\infty} J_{n+2}(x)t^n$$

$$= \sum_{n=-\infty}^{\infty} (n+1) J_{n+1}(x)t^n. \qquad (7.9.150)$$

Comparing coefficients of t^n, we find

$$\left(\frac{x}{2}\right) [J_n(x) + J_{n+2}(x)] = (n+1) J_{n+1}(x), \qquad (7.9.151)$$

or, on shifting the indices once again,

$$J_{n-1}(x) + J_{n+1}(x) = \left(\frac{2n}{x}\right) J_n(x). \qquad (7.9.152)$$

7.9.15 An Alternative Representation for $Y_n(x)$

The Bessel function of the second kind of order $\nu \neq$ integer, can be expressed in terms of $J_\nu(x)$ and $J_{-\nu}(x)$ since both of these functions are linearly independent solutions. This form for $Y_\nu(x)$ is

$$Y_\nu(x) = \frac{\cos(\pi\nu)J_\nu(x) - J_{-\nu}(x)}{\sin(\pi\nu)}. \qquad (7.9.153)$$

For $n = 0$ or an integer, $Y_n(x)$ is taken to be

$$Y_n(x) = \lim_{\nu \to n} \left[\frac{\cos(\pi\nu)J_\nu(x) - J_{-\nu}(x)}{\sin(\pi\nu)}\right]. \qquad (7.9.154)$$

Applying L'Hospital's rule gives

$$Y_n(x) = \frac{\partial}{\partial\nu}[\cos(\pi\nu)J_\nu(x) - J_{-\nu}(x)] \Big/ \frac{\partial}{\partial\nu}\sin(\pi\nu)$$

$$= \left(\frac{1}{\pi}\right)\left[\frac{\partial J_\nu(x)}{\partial\nu} - (-1)^n \frac{\partial J_{-\nu}(x)}{\partial\nu}\right]_{\nu=n}. \qquad (7.9.155)$$

7.9.16   *Equations Reducible to Bessel's Equation*

The second-order differential equation

$$x^2 y'' + (1 - 2\alpha)xy' + \left[(\beta\gamma x^\gamma)^2 + (\alpha^2 - n^2\gamma^2)\right] y = 0, \qquad (7.9.156)$$

can be transformed into the equation

$$t^2 \frac{d^2 w}{dt^2} + t \frac{dw}{dt} + (\beta^2 t^2 - n^2)w = 0, \qquad (7.9.157)$$

by means of the change of variables

$$y = x^\alpha w, \qquad t = x^\gamma. \qquad (7.9.158)$$

Equation (7.9.157) is a form of the Bessel equation and has the two linearly independent solutions $J_n(\beta t)$ and $Y_n(\beta t)$. Therefore, the general solution of Eq. (7.9.156) is

$$y(x) = x^\alpha w = x^\alpha [c_1 J_n(\beta x^\gamma) + c_2 Y_n(\beta x^\gamma)], \qquad (7.9.159)$$

where c_1 and c_2 are arbitrary constants.

To illustrate the application of the above formalism, first consider the harmonic oscillator equation

$$y'' + y = 0. \qquad (7.9.160)$$

To be of the form given by Eq. (7.9.156), we rewrite it as

$$x^2 y'' + x^2 y = 0. \qquad (7.9.161)$$

Comparison with Eq. (7.9.156) gives the relations

$$\alpha = \frac{1}{2}, \quad \beta\gamma = 1, \quad n^2\gamma^2 = \alpha^2, \quad 2\gamma = 2, \qquad (7.9.162)$$

and

$$\alpha = \frac{1}{2}, \quad \beta = 1, \quad \gamma = 1, \quad n = \pm\left(\frac{1}{2}\right). \qquad (7.9.163)$$

Therefore, the general solution is

$$y(x) = x^{1/2}[c_1 J_{1/2}(x) + c_2 J_{-1/2}(x)] \qquad (7.9.164)$$

Since, we know already that

$$y(x) = c_1 \sin x + c_2 \cos x, \qquad (7.9.165)$$

we conclude that, up to constants,

$$J_{1/2}(x) \propto \frac{\sin x}{\sqrt{x}}, \qquad J_{-1/2}(x) \propto \frac{\cos x}{\sqrt{x}}. \qquad (7.9.166)$$

Actually, the just stated argument is not entirely correct, but it is true that

$$J_{1/2}(x) = \sqrt{\frac{2}{\pi x}} \sin x, \qquad J_{-1/2}(x) = \sqrt{\frac{2}{\pi x}} \cos x. \tag{7.9.167}$$

For the second example, consider the Airy differential equation

$$y'' + xy = 0 \quad \text{or} \quad x^2 y'' + x^3 y = 0. \tag{7.9.168}$$

For this case

$$\alpha = \frac{1}{2}, \quad n^2 \gamma^2 = \alpha^2 = \frac{1}{4}, \quad \gamma = \frac{3}{2}, \quad \beta \gamma = 1, \tag{7.9.169}$$

or

$$\alpha = \frac{1}{2}, \quad \beta = \frac{2}{3}, \quad \gamma = \frac{3}{2}, \quad n = \pm \left(\frac{1}{3}\right). \tag{7.9.170}$$

Thus, the general solution for the Airy equation is

$$y(x) = x^{1/2} \left[c_1 J_{1/3}(2x^{3/2}/3) + c_2 J_{-1/3}(2x^{3/2}/3) \right]. \tag{7.9.171}$$

Problems

Section 7.1

7.1.1 Derive the differential equations for $R(r)$, $\Theta(\theta)$, $\Psi(\psi)$ and $T(t)$ as given by the expressions in Eqs. (7.1.4) to (7.1.7).

7.2.2 Construct separation of variables solution to the wave equation having axial symmetry, i.e., ϕ does not depend on ψ. Show that the differential equations obtained correspond to Eqs. (7.1.4), (7.1.5), with $m = 0$, and (7.1.7).

Section 7.2

7.2.1 Show that any finite interval, $[a, b]$, where $a < b$, can by means of a linear change of variables, be transformed to the interval $[-1, 1]$.

7.2.2 From the definition of the generating function given by Eq. (7.2.5) show that $\phi_n(x)$ is given by Eq. (7.2.6).

7.2.3 Using Eq. (7.2.8), express $\phi_n''(x)$ in terms of other COP's and not their derivatives.

Section 7.3

7.3.1 Use the generating function, Eq. (7.3.4), to show that $P_n(-x) = (-1)^n P_n(x)$. Also, calculate $P_n(0)$, $P_n(1)$ and $P_n(-1)$.

7.3.2 Use Rodrique's formula to show that $P_n(-x) = (-1)^n P_n(x)$. By direct substitution into the left side of the Legendre differential equation, prove that the representation for $P_n(x)$ given by Rodrique's formula is a solution.

Section 7.4

7.4.1 Derive the first recurrence relation in Eq. (7.4.7) by using the generating function given in Eq. (7.4.4).

7.4.2 Prove the properties of the Hermite polynomials stated in Eq. (7.4.9).

Section 7.5

7.5.1 Verify that $T_n(x)$, as given by the Rodrique's formula of Eq. (7.5.6) is a solution to the differential equation given in Eq. (7.5.1).

7.5.2 Show that $\sqrt{1-x^2}U_n(x)$ is a second linearly independent solution of the Chebyshev differential equation (7.5.1).

7.5.3 Derive the properties of $T_n(x)$ and $U_n(x)$ stated in Eq. (7.5.12).

Section 7.6

7.6.1 State why the general solution to the Laguerre differential equation, as given by Eq. (7.6.1) cannot be defined on $-\infty < x < \infty$.

7.6.2 Use the generating function for the Laguerre polynomials to derive the first of the recurrence relations in Eq. (7.6.7).

7.6.3 Calculate the general solution to the Laguerre differential equation for $n = 0$. Discuss the various interesting features of this solution. Can it be expressed in terms of elementary functions?

Section 7.7

7.7.1 Plot $Q_0(x)$ vs x for $|x| < 1$. Is $Q_0(x)$ defined for $|x| > 1$?

7.7.2 Use Eqs. (7.7.5) and (7.7.6) to calculate $P_2^m(x)$ and $Q_2^m(x)$ for all allowable values of m.

Section 7.8

7.8.1 Prove for $n =$ integer that $J_n(x)$ and $J_{-n}(x)$ are not linearly independent. Explain why the result in Eq. (7.8.5) should hold true.

7.8.2 Give arguments to show that the asymptotic representations given by Eqs. (7.8.16) and (7.8.17) hold.

7.8.3 Show directly, by substitution into the differential equation, that the two latter expressions in Eq. (7.8.23) are solutions to the Bessel equation.

Section 7.9

7.9.1 Evaluate the integral

$$\int_0^1 P_\ell(x)dx$$

by use of the generating function for the Legendre functions. Show that

$$\int_0^1 P_\ell(x)dx = \begin{cases} 1, & \text{if } \ell = 0, \\ \frac{1}{2}, & \text{if } \ell = 1, \\ 0, & \text{if } \ell \text{ is even and} \\ & \quad \text{greater than zero}, \\ \dfrac{\left(\frac{1}{2}\right)\left(\frac{1}{2}-1\right)\cdots\left[\frac{1}{2}-\left(\frac{\ell-1}{2}\right)\right]}{\left(\frac{\ell+1}{2}\right)!}, & \text{if } \ell = \text{odd.} \end{cases}$$

7.9.2 Express $(1, x, x^2, x^3, x^4)$ in terms of $H_0(x)$, $H_1(x)$, $H_2(x)$, $H_3(x)$, and $H_4(x)$. Can you determine a formula to represent an nth degree polynomial function in terms of a linear combination of the Hermite polynomials?

7.9.3 Calculate $J_{\pm 1/2}(x)$ and $J_{\pm 3/2}(x)$.

7.9.4 Evaluate the following integrals:
$\int x^n J_{n-1}(x)dx$;
$\int x^{-n} J_{n+1}(x)dx$;
$\int x^4 J_1(x)dx$.

7.9.5 Prove that

$$J_n'(x)J_{-n}(x) - J_{-n}'(x)J_n(x) = \frac{2\sin(\pi n)}{\pi x}.$$

7.9.6 Can the general solution of the differential equation

$$y'' + (\lambda - x^2)y = 0$$

be written in terms of Bessel functions?

Comments and References

[1] J. Irving and N. Mullineux, *Mathematics in Physics and Engineering* (Academic Press, New York, 1959).

[2] See the book by McBridge listed in the bibliography.

[3] R. E. Mickens, *Nonlinear Oscillations* (Cambridge University Press, New York, 1981).

[4] Eugen Merzbacher, *Quantum Mechanics* (Wiley, New York, 1961).

Bibliography

Applications and General Topics

M. Abramowitz and I. A. Stegun, editors, *Handbook of Mathematical Functions* (Dover, New York, 1964).

P. Dennery and A. Krzywicki, *Mathematics for Physicists* (Harper and Row, New York, 1967).

A. W. Erdelyi, W. Magnus, F. Oberhettinger, and F. G. Tricomi, *Higher Transcendental Functions, Vol. I, II, III* (McGraw-Hill, New York, 1981).

J. D. Jackson, *Classical Electrodynamic*, 2nd ed. (Wiley, New York, 1975).

N. N. Lebedev, *Special Functions and Their Applications* (Dover, New York, 1972).

Y. L. Luke, *Mathematical Functions and Their Approximations* (Academic Press, New York, 1975).

W. Magnus, F. Oberhettinger, and R. P. Soni, *Formulas and Theorems for the Special Functions of Mathematical Physics* (Springer-Verlag, New York, 1966).

M. E. Rose, *Elementary Theory of Angular Momentum* (Wiley, New York, 1957).

G. Sansone, *Orthogonal Functions* (Wiley-Interscience, New York, 1977).

I. N. Sneddon, *Special Functions of Mathematical Physics and Chemistry*, 3rd ed. (Longman, New York, 1980).

G. Szegö, *Orthogonal Polynomials* (AMS Colloquium, Vol. 23, 1939).

E. T. Whittaker and G. N. Watson, *A Course of Modern Analysis* (Cambridge University Press, Cambridge, 1973).

Bessel Functions

H. Jeffreys and B. S. Jeffreys, *Methods of Mathematical Physics* (Cambridge University Press, Cambridge, 1972).

G. N. Watson, *A Treatise on the Theory of Bessel Functions*, 2nd ed. (Cambridge University Press, Cambridge, 1952).

Generating Functions

E. B. McBridge, *Obtaining Generating Functions* (Springer-Verlag, New York, 1971).

Legendre Polynomials and Functions

A. R. Edmonds, *Angular Momentum in Quantum Mechanics* (Princeton University Press; Princeton, NJ; 1957).

L. Fox and I. B. Parker, *Chebyshev Polynomials in Numerical Analysis* (Oxford University Press, Oxford, 1968).

E. W. Hobson, *The Theory of Spherical and Ellipsoidal Harmonics* (Chelsea, New York, 1955).

T. M. Magnus, *Spherical Harmonics* (Methuen, London, 1947).

E. D. Rainville, *Special Functions* (Macmillan, New York, 1960).

L. J. Slater, *Confluent Hypergeometric Functions* (Cambridge University Press, Cambridge, 1960).

C. Snow, *The Hypergeometric and Legendre Functions with Applications to Integral Equations and Potential Theory*, 2nd ed. (National Bureau of Standards; Washington, DC; 1952).

Chapter 8

Perturbation Methods for Oscillatory Systems

8.1 Introduction

Few differential equations, whether linear or nonlinear, can be solved exactly in terms of a finite number of the elementary functions. Even for the case where exact solutions are obtained, they are either in implicit form or are so complex in mathematical structure that little useful information can be directly derived from them. Fortunately, for many equations modeling oscillatory behavior, a large body of calculational methods exist for determining analytical approximations to the required solutions.

The main purpose of this chapter is to examine some of these methods and apply them to differential equations modeling one-dimensional oscillatory systems. We begin with a brief discussion of the general philosophy of perturbation procedures, what such methods are required to do, and how they work in practice. This presentation is followed by sections giving the details for several important perturbation methods: first-order averaging, the Lindstedt-Poincaré method, and harmonic balance. Following each of these individual sections, we illustrate the use of each method by applying it to a number of second-order, nonlinear differential equations.

Section 8.10 is devoted to the issues related to constructing an averaging procedure for a class of second-order, nonlinear difference equations corresponding to discrete models of nonlinear oscillators. The following section 8.12 applies this method to a number of equations.

A bibliography, at the end of the chapter, gives a selected listing of books on both the general theory of perturbation procedures and their application to the nonlinear differential equations arising in the analysis of particular systems in the natural and engineering sciences.

The second-order differential equations of main interest for the perturbation methods introduced in this chapter take the form

$$\frac{d^2x}{dt^2} + x = \epsilon f\left(x, \frac{dx}{dt}\right), \tag{8.1.1}$$

where ϵ is a small positive parameter, i.e.,

$$0 \le \epsilon \ll 1, \tag{8.1.2}$$

and the function f is a polynomial function of its arguments. This type of equation is sufficiently general to cover a large number of the important nonlinear differential equations modeling one-dimensional oscillatory systems. Of particular interest will be the ability to obtain information on the frequency of the oscillations and the values of the associated amplitudes for limit-cycles.

Finally, the materials to be presented in this chapter will be better understood if the reader first reviews the elements of asymptotics given in Appendix B.

8.2 The General Perturbation Procedure

Consider a system having its dynamics determined by the equation

$$Lu = v + \epsilon N(u), \tag{8.2.1}$$

where ϵ is a small parameter, $0 < \epsilon \ll 1$, that can always be selected to be positive; L is a linear operator; N is a nonlinear operator; and v is a specified function. (It should be indicated that at this stage of the discussion the exact nature or type of these operators or functions are not essential.) The central issue is to determine u given (L, N, v, ϵ). We assume that the "problem"

$$Lu_0 = v, \tag{8.2.2}$$

can be solved. Note that Eq. (8.2.2) is that which follows from setting $\epsilon = 0$ in Eq. (8.2.1). The general concept behind the method of perturbations is to assume that u can be represented by the expression

$$u = u_0 + \epsilon u_1 + \epsilon^2 u_2 + \cdots, \tag{8.2.3}$$

determine the required equations satisfied by the u_k, and solve them.

To gain insight into what kinds of equations can be handled by perturbation procedures, we list below five examples. For all these cases, it is assumed that ϵ is small, i.e.,

$$0 < \epsilon \ll 1. \tag{8.2.4}$$

For each particular case, the corresponding operators and functions are given.

A. A nonlinear algebraic equation:

$$\epsilon u^2 + u = a, \tag{8.2.5}$$

$$L \equiv \hat{1}, \qquad v \equiv a, \qquad N(u) \equiv -u^2, \tag{8.2.6}$$

where $\hat{1}$ is the identity operator.

B. First-order differential equation:

$$\frac{du}{dt} = u - \epsilon u^3, \qquad u(0) = u_0 = \text{ given}, \tag{8.2.7}$$

$$L \equiv \frac{d}{dt} - a\hat{1}, \qquad v = 0, \qquad N(u) = -u^3. \tag{8.2.8}$$

C. Linear integral equation:

$$u(t) = g(t) + \epsilon \int_0^1 k_1(t,s)u(s)ds, \tag{8.2.9}$$

where $g(t)$ and $k_1(t,s)$ are given, known functions, and

$$L \equiv \hat{1}, \qquad v \equiv g(t), \qquad N(u) \equiv \int_0^1 k_1(t,s)u(s)ds. \tag{8.2.10}$$

Observe for this case that N is a linear operator.

D. Nonlinear integral equation:

$$u(t) = h(t) + \epsilon \int_0^1 k_2(t,s)[u(s)]^2 ds, \tag{8.2.11}$$

where $h(t)$ and $k_2(t,s)$ are given, and

$$L \equiv \hat{1}, \qquad v \equiv h(t), \qquad N(u) \equiv \int_0^1 k_2(t,s)[u(s)]^2 ds. \tag{8.2.12}$$

E. van der Pol equation:

$$\frac{d^2u}{dt^2} + u = \epsilon(1 - u^2)\frac{du}{dt}, \qquad u(0) = u_0, \qquad \frac{du(0)}{dt} = \dot{u}_0, \tag{8.2.13}$$

where the initial conditions $(u_0, \dot{u}_0)$ are given and

$$L \equiv \frac{d^2}{dt^2} + \hat{1}, \qquad v = 0, \qquad N(u) = (1 - u^2)\frac{du}{dt}. \tag{8.2.14}$$

The above examples illustrate that the possibility for use of perturbation methods is not limited to any particular type of equation. These techniques

apply in general to any nonlinear equation having the structure given by
Eq. (8.2.1).

To proceed, we need to evaluate Lu and $N(u)$ as series in ϵ using the
expression for u in Eq. (8.2.3). Since L is a linear operator, it follows that

$$Lu = Lu_0 + \epsilon Lu_1 + \epsilon^2 Lu_2 + \cdots. \tag{8.2.15}$$

Now we have to evaluate

$$N(u) = N(u_0 + \epsilon u_1 + \epsilon^2 u_2 + \cdots). \tag{8.2.16}$$

For the cases where differentiation of $N(u)$ has mathematical meaning, the
expansion in ϵ can be directly calculated and is given by the expression

$$N[u(\epsilon)] = N(u)|_{\epsilon=0} + \epsilon \frac{dN}{du} \frac{du}{d\epsilon}\Big|_{\epsilon=0} + \cdots$$
$$= N(u_0) + \epsilon u_1 N'(u_0) + \cdots, \tag{8.2.17}$$

where $N' = dN/du$. Carrying out the evaluation of more terms in the ϵ
expansion allows the following conclusion to be reached

$$N[u(\epsilon)] = N_0(u_0) + \epsilon N_1(u_0, u_1) + \epsilon^2 N_2(u_0, u_1, u_2) + \cdots$$
$$+ \epsilon^k N(u_0, u_1, \ldots, u_k) + \cdots, \tag{8.2.18}$$

i.e., N_k depends on $(u_0, u_1, \ldots, u_k)$. Note that given $N(u)$, the various
terms in Eq. (8.2.18) can be explicitly calculated. As an example, let
$N(u) = u^2$, then

$$N(u_0 + \epsilon u_1 + \epsilon^2 u_2 + \cdots) = (u_0 + \epsilon u_1 + \epsilon^2 u_2 + \cdots)^2$$
$$= u_0^2 + \epsilon(2u_0 u_1) + \epsilon^2(u_1^2 + 2u_0 u_2) + \cdots, \tag{8.2.19}$$

and, consequently,

$$N_0(u_0) = u_0^2, \qquad N_1(u_0, u_1) = 2u_0 u_1,$$
$$N_2(u_0, u_1, u_2) = u_1^2 + 2u_0 u_2, \quad \text{etc.} \tag{8.2.20}$$

Substituting Eqs. (8.2.15) and (8.2.18) into Eq. (8.2.1) gives

$$Lu_0 + \epsilon Lu_1 + \epsilon^2 Lu_2 + \cdots + \epsilon^k Lu_k + \cdots$$
$$= v + \epsilon N_0(u_0) + \epsilon^2 N_1(u_0, u_1) + \cdots$$
$$+ \epsilon^k N_k(u_0, u_1, \ldots, u_{k-1}) + \cdots. \tag{8.2.21}$$

Equating the various coefficients of the powers of ϵ leads to an infinite system of linear, inhomogeneous equations for the u_k's.

$$\begin{cases} Lu_0 = v \\ Lu_1 = N_0(u_0) \\ Lu_2 = N_1(u_0, u_1) \\ \quad\vdots \quad\vdots \\ Lu_k = N_{k-1}(u_0, u_1, \ldots, u_{k-1}) \\ \quad\vdots \quad\vdots \end{cases} \qquad (8.2.22)$$

Thus, perturbation procedures begin with a (in this case) single nonlinear equation and convert it into an infinite system of equations. The importance of this fact is that these linear equations are to be solved recursively. This means that the first equation can be solved for u_0. This u_0 is then substituted into the right-side of the second of Eqs. (8.2.22) to give a linear, inhomogeneous equation for u_1, which can be solved for u_1. Continuing, u_0 and u_1 can then be used to obtain u_2 by use of the third equation. This process can be applied to calculate the u_k to any given value of k.

From the view of practicality, perturbation methods are only of value if the first few terms give the essential features of the phenomena under investigations and this approximation to the solution is also (in some sense usually specified by the particular problem) accurate. However, the existence of symbol manipulating software for use on digital computers can permit the explicit determination of perturbation expansions to large orders [1].

With all of the above preliminary background information in hand for the general perturbation procedure, several questions immediately come to mind. In particular, the following five are of importance:

(i) Taken as an expansion in the small parameter ϵ, does the perturbation derived solution converge?

(ii) If it converges, does it converge to an actual solution of the original equation?

(iii) If it doesn't converge, can the calculated expansion still be of practical value and used as an approximation to the actual solution under the restriction $0 < \epsilon \ll 1$?

(iv) Do there exist solutions of the original equation that cannot be represented in the form given by Eq. (8.2.3)? If so, how can they be calculated for $0 < \epsilon \ll 1$?

(v) For differential equations, how should initial and/or boundary value conditions be incorporated into the terms arising in the perturbation representation?

We will not directly discuss most of the issues raised by these questions. The worked examples of the next section illustrate possible answers to them. However, several books [2; 3; 4] do give details as to what is currently known on the status of answers to these questions and many other related issues. Our general view on this subject is close to the comments stated by Bellman [2]:

> "Although (these) problems are obviously of paramount theoretical and practical import, and although a great deal of attention has been paid to their study, there exists no startlingly large collection of rigorously established results in this area. Taking this fact into account ... we ... focus our attention almost exclusively upon methods and soft pedal the proofs. In return for this freedom, we shall be able to describe a variety of powerful techniques that have worked quite well in applications. ..."

8.3 Worked Examples Using the General Perturbation Procedure

8.3.1 *Example A*

Consider the following algebraic quadratic equation

$$u = 1 + \epsilon u^2. \tag{8.3.1}$$

This equation corresponds to having

$$L = \hat{1}, \qquad v = 1, \qquad N(u) = u^2. \tag{8.3.2}$$

For this case, we have u as a function of ϵ, with

$$u(\epsilon) = u_0 + \epsilon u_1 + \epsilon^2 u_2 + \cdots \tag{8.3.3}$$

$$[u(\epsilon)]^2 = u_0^2 + (2u_0 u_1)\epsilon + (u_1^2 + 2u_0 u_2)\epsilon^2 + \cdots. \tag{8.3.4}$$

The relations satisfied by u_0, u_1 and u_2 are

$$\begin{cases} u_0 = 1, \\ u_1 = u_0^2 \ \Rightarrow \ u_1 = 1, \\ u_2 = 2u_0 u_1 \ \Rightarrow \ u_2 = 2, \\ u_3 = u_1^2 + 2u_0 u_2 \ \Rightarrow \ u_3 = 5. \end{cases} \tag{8.3.5}$$

Thus, the four-term expansion of $u(\epsilon)$ is

$$u(\epsilon) = 1 + \epsilon + 2\epsilon^2 + 5\epsilon^3 + \cdots . \qquad (8.3.6)$$

Since Eq. (8.3.1) is a quadratic equation, its *two* solutions can be explicitly calculated and are given by the following expressions

$$x^{(+)} = \frac{1 + \sqrt{1 - 4\epsilon}}{2\epsilon}, \qquad x^{(-)} = \frac{1 - \sqrt{1 - 4\epsilon}}{2\epsilon}. \qquad (8.3.7)$$

For $0 \leq \epsilon < 0.25$, the square root can be expanded in a convergent Taylor series given by

$$\sqrt{1 - 4\epsilon} = 1 - 2\epsilon - 2\epsilon^2 - 4\epsilon^3 - 10\epsilon^4 + \cdots . \qquad (8.3.8)$$

Consequently, $x^{(+)}(\epsilon)$ and $x^{(-)}(\epsilon)$ have the following expansions

$$x^{(+)}(\epsilon) = \frac{1}{\epsilon} - 1 - \epsilon - 2\epsilon^2 - 5\epsilon^3 + \cdots \qquad (8.3.9)$$

$$x^{(-)}(\epsilon) = 1 + \epsilon + 2\epsilon^2 + 5\epsilon^3 + \cdots . \qquad (8.3.10)$$

A comparison of Eqs. (8.3.6) and (8.3.10) shows that the perturbation expansion corresponds to the actual solution, $x^{(-)}(\epsilon)$, of the original quadratic equation. It is also important to observe that the second solution, $x^{(+)}(\epsilon)$, when expanded in terms of ϵ, does not have the form of our standard or regular expansion as presented in Eq. (8.2.3). It has the property that when ϵ is small, its value can become large; in fact,

$$\operatorname*{Lim}_{\epsilon \to 0} x^{(+)}(\epsilon) = \infty. \qquad (8.3.11)$$

The expression given by Eq. (8.3.9) is an example of a so-called singular perturbation expansion. Note that for $\epsilon \neq 0$, Eq. (8.3.1) is a quadratic equation, while for $\epsilon = 0$, it reduces to a linear equation. This property is a generic feature of equations for which singular perturbation expansions occur. When $\epsilon = 0$, the nature of the equation changes. Particular examples include Eq. (8.3.1) which changes from a second- to first-order algebraic equation and the singular, second-order differential equation

$$\epsilon \frac{d^2 x}{dt^2} + \frac{dx}{dt} + x = 0, \qquad (8.3.12)$$

which goes from being a second-order equation when $\epsilon \neq 0$ to a first-order differential equation for $\epsilon = 0$.

We will not treat equations that require the use of singular perturbation techniques. However, they are very important for many applications and many of the books listed in the bibliography to this chapter deal with this and related topics [5].

The example of this section does illustrate one critical point: The standard or regular perturbation expansion may only provide an appropriate approximation for some of the solutions to a particular equation. A knowledge of the other solutions may require a more complex set of procedures.

8.3.2   Example B

The second example is a first-order, nonlinear differential equation. It is often called the *logistic equation* and takes the form

$$\frac{dx}{dt} = -x + \epsilon x^2, \qquad \epsilon > 0. \tag{8.3.13}$$

Using

$$x(t, \epsilon) = x_0(t) + \epsilon x_1(t) + \epsilon^2 x_2(t) + \cdots, \tag{8.3.14}$$

$$\frac{dx(t, \epsilon)}{dt} = \frac{dx_0(t)}{dt} + \epsilon \frac{dx_1(t)}{dt} + \epsilon^2 \frac{dx_2(t)}{dt} + \cdots, \tag{8.3.15}$$

$$x^2 = x_0^2 + \epsilon(2x_0 x_1) + \cdots, \tag{8.3.16}$$

the following equations are obtained for $x_0(t)$, $x_1(t)$, and $x_2(t)$,

$$\frac{dx_0}{dt} + x_0 = 0, \tag{8.3.17}$$

$$\frac{dx_1}{dt} + x_1 = x_0^2, \tag{8.3.18}$$

$$\frac{dx_2}{dt} + x_2 = 2x_0 x_1. \tag{8.3.19}$$

To proceed, we need to specify an initial condition, which we take to be

$$x(0, \epsilon) = a, \qquad 0 < a < \frac{1}{\epsilon}. \tag{8.3.20}$$

It follows that

$$x(0, \epsilon) = a = x_0(0) + \epsilon x_1(0) + \epsilon^2 x_2(0) + \cdots \tag{8.3.21}$$

and this implies

$$x_0(0) = a, \qquad x_1(0) = 0, \qquad x_2(0) = 0, \text{ etc.} \tag{8.3.22}$$

The solution to Eq. (8.3.17), with the initial condition $x_0(0) = a$, is

$$x_0(t) = ae^{-t}. \tag{8.3.23}$$

Substituting this into the right-side of Eq. (8.3.18) gives

$$\frac{dx_1}{dt} + x_1 = a^2 e^{-2t}, \qquad x_1(0) = 0. \tag{8.3.24}$$

The solution to this equation is

$$x_1(t) = a^2 e^{-t}(1 - e^{-t}). \tag{8.3.25}$$

Likewise, $x_2(t)$ is the solution to

$$\frac{dx_2}{dt} + x_2 = 2a^3(e^{-2t} - e^{-3t}), \qquad x_2(0) = 0, \tag{8.3.26}$$

and it is given by

$$x_2(t) = a^3 e^{-t}(1 - 2e^{-t} + e^{-2t})$$
$$= a^3 e^{-t}(1 - e^{-t})^2. \tag{8.3.27}$$

Thus, to three terms, the perturbation solution for Eq. (8.3.13) is

$$x(t, \epsilon) = ae^{-t} + \epsilon a^2 e^{-t}(1 - e^{-t}) + \epsilon^2 a^2 e^{-t}(1 - e^{-t})^2 + \cdots \tag{8.3.28}$$

or

$$x(t, \epsilon) = ae^{-t}\left[1 + \epsilon a(1 - e^{-t}) + \epsilon^2 a^2(1 - e^{-t})^2 + \cdots\right]. \tag{8.3.29}$$

Inspection of Eq. (8.3.29) shows that the actual expansion parameter is not ϵ but the combination ϵa. This is a general feature of many perturbation procedures and its appearance will be illustrated again in section 8.7 on worked examples using the Lindstedt-Poincaré method.

The logistic equation can be solved exactly. To do this, rewrite Eq. (8.3.13) as

$$\frac{dx}{x(\epsilon x - 1)} = dt. \tag{8.3.30}$$

This expression can be integrated easily by using partial fractions on the left-side. Doing this gives

$$\frac{\epsilon x - 1}{x} = Ce^t, \tag{8.3.31}$$

where C is related to the constant of integration. Imposing the initial condition, $x(0) = a$, means that C is

$$C = \frac{\epsilon a - 1}{a}, \tag{8.3.32}$$

and $x(t, \epsilon)$ is

$$x(t, \epsilon) = \frac{ae^{-t}}{1 - (\epsilon a)(1 - e^{-t})}. \tag{8.3.33}$$

Now $0 < a\epsilon < 1$, by the assumption in Eq. (8.3.20). Also, it is clear that

$$0 \le 1 - e^{-t} < 1, \qquad t > 0. \tag{8.3.34}$$

Consequently, we conclude that

$$0 \le (\epsilon a)(1 - e^{-t}) < 1, \qquad t > 0. \tag{8.3.35}$$

This last result means that the right-side of Eq. (8.3.33) has the expansion

$$x(t, \epsilon) = ae^{-t} \sum_{n=0}^{\infty} \left[(\epsilon a)(1 - e^{-t}) \right]^n. \tag{8.3.36}$$

Comparing Eq. (8.3.29) with Eq. (8.3.26), we see that the first three terms of the perturbation expansion are equal to the first three terms of the expansion of the exact solution.

Finally, inspection of the convergent series expansion of Eq. (8.3.36) shows the validity of the following two points:

(i) Considered as an expansion in ϵ, the actual expansion parameter is ϵa and it satisfies the inequalities $0 \le a\epsilon < 1$.

(ii) The first term provides a good approximation to $x(t, \epsilon)$ for all values of the initial condition satisfying the condition $0 \le a\epsilon < 1$.

8.4   First-Order Method of Averaging

The method of first-order averaging was developed by Krylov and Bogoliubov [6]. A major advantage of the technique is the ease with which the amplitude and frequency of the oscillations can be calculated. In addition, for systems having dissipation, the method allows the determination of the transitory behavior toward either a fixed-point (equilibrium solution) or a periodic solution (limit-cycle). Further, if limit-cycles exist, the method is able to calculate its (linear) stability properties.

8.4.1   *The Method*

Consider the second-order, nonlinear differential equation

$$\frac{d^2x}{dt^2} + x = \epsilon f \left(x, \frac{dx}{dt} \right), \qquad 0 \le \epsilon \ll 1. \tag{8.4.1}$$

For $\epsilon = 0$, this equation reduces to the linear equation

$$\frac{d^2x}{dt^2} + x = 0, \tag{8.4.2}$$

for which the general solution and its derivative are

$$x = a\cos(t + \phi), \qquad \frac{dx}{dt} = -a\sin(t + \phi), \qquad (8.4.3)$$

where a and ϕ are arbitrary integration constants. We now assume, for $0 < \epsilon \ll 1$, that Eq. (8.4.1) has a solution of the form

$$x(t, \epsilon) = a(t, \epsilon)\cos[t + \phi(t, \epsilon)], \qquad (8.4.4)$$

with the derivative given by

$$\frac{dx}{dt} = -a(t, \epsilon)\sin[t + \phi(t, \epsilon)]. \qquad (8.4.5)$$

Note that $a(t, \epsilon)$ and $\phi(t, \epsilon)$ are functions of both t and ϵ.

Differentiating Eq. (8.4.4) gives, for $\psi = t + \phi$,

$$\begin{aligned}\frac{dx}{dt} &= \frac{da}{dt}\cos\psi - a\sin\psi - a\frac{d\phi}{dt}\sin\psi \\ &= \left[\frac{da}{dt}\cos\psi - a\frac{d\phi}{dt}\sin\psi\right] - a\sin\psi. \qquad (8.4.6)\end{aligned}$$

Comparison of Eqs. (8.4.5) and (8.4.6) gives the relation

$$\frac{da}{dt}\cos\psi - a\frac{d\phi}{dt}\sin\psi = 0. \qquad (8.4.7)$$

Now differentiating Eq. (8.4.5) gives

$$\frac{d^2x}{dt^2} = -\frac{da}{dt}\sin\psi - a\cos\psi - a\frac{d\phi}{dt}\cos\psi. \qquad (8.4.8)$$

Substituting Eqs. (8.4.4), (8.4.5) and (8.4.8) into Eq. (8.4.1) leads to

$$\frac{da}{dt}\sin\psi + a\frac{d\phi}{dt}\cos\psi = -\epsilon f(a\cos\psi, -a\sin\psi). \qquad (8.4.9)$$

Observe that Eqs. (8.4.7) and (8.4.9) are linear in da/dt and $a\,d\phi/dt$. Consequently, solving for them gives the expressions

$$\frac{da}{dt} = -\epsilon f(a\cos\psi, -a\sin\psi)\sin\psi, \qquad (8.4.10)$$

$$\frac{d\phi}{dt} = -\left(\frac{\epsilon}{a}\right) f(a\cos\psi, -a\sin\psi)\cos\psi, \qquad (8.4.11)$$

$$\psi(t, \epsilon) = t + \phi(t, \epsilon). \qquad (8.4.12)$$

These are exact equations for a and ϕ, when the assumed solution for $x(t, \epsilon)$ and its derivative take the forms given by Eqs. (8.4.4) and (8.4.5). They are two coupled, nonlinear, first-order differential equations and can have quite complex functional forms on their right-sides.

Now inspection of Eqs. (8.4.10) and (8.4.11) shows that the expressions on the right-sides are periodic functions of ψ with period 2π. Therefore, the following Fourier series representations exist

$$f \sin \psi = K_0(a) + \sum_{m=1}^{\infty} [K_m(a) \cos(m\psi) + L_m(a) \sin(m\psi)], \qquad (8.4.13)$$

$$f \cos \psi = P_0(a) + \sum_{m=1}^{\infty} [P_m(a) \cos(m\psi) + Q_m(a) \sin(m\psi)], \qquad (8.4.14)$$

where

$$K_0(a) = \left(\frac{1}{2\pi}\right) \int_0^{2\pi} f \sin \psi \, d\psi, \qquad (8.4.15)$$

$$K_m(a) = \left(\frac{1}{\pi}\right) \int_0^{2\pi} f \sin \psi \cos(m\psi) d\psi, \qquad (8.4.16)$$

$$L_m(a) = \left(\frac{1}{\pi}\right) \int_0^{2\pi} f \sin \psi \sin(m\psi) d\psi, \qquad (8.4.17)$$

$$P_0(a) = \left(\frac{1}{2\pi}\right) \int_0^{2\pi} f \cos \psi \, d\psi, \qquad (8.4.18)$$

$$P_m(a) = \left(\frac{1}{\pi}\right) \int_0^{2\pi} f \cos \psi \cos(m\psi) d\psi, \qquad (8.4.19)$$

$$Q_m(a) = \left(\frac{1}{\pi}\right) \int_0^{2\pi} f \cos \psi \sin(m\psi) d\psi. \qquad (8.4.20)$$

For a particular differential equation, the function $f(x, dx/dt)$ is known and it follows

$$f(x, dx/dt) \to f(a \cos \psi, -a \sin \psi) \qquad (8.4.21)$$

is also known. In practice, for f's that are polynomial functions, the Fourier representations are easily calculated by direct expansion of the trigonometric polynomials rather than the use of the Fourier coefficient representations given above. The examples of the next section will illustrate this procedure.

Except for the first terms in Fourier series expansions, the remaining terms for $f \sin \psi$ and $f \cos \psi$ oscillate with average values of zero. The first approximation of Krylov and Bogoliubov consists of neglecting all the terms on the right-sides of Eqs. (8.4.13) and (8.4.14) except for the first; this gives

$$\frac{da}{dt} = -\epsilon K_0(a), \qquad \frac{d\phi}{dt} = -\left(\frac{\epsilon}{a}\right) P_0(a), \qquad (8.4.22)$$

or using Eqs. (8.4.15) and (8.4.18),

$$\frac{da}{dt} = - \left(\frac{\epsilon}{2\pi} \right) \int_0^{2\pi} f(a \cos \psi, -a \sin \psi) \sin \psi \, d\psi, \qquad (8.4.23)$$

$$\frac{d\phi}{dt} = - \left(\frac{\epsilon}{2\pi a} \right) \int_0^{2\pi} f(a \cos \psi, -a \sin \psi) \cos \psi \, d\psi. \qquad (8.4.24)$$

In summary, the first approximation of Krylov and Bogoliubov, also known as the first-order method of averaging, provides an approximation to the solution of the differential equation

$$\frac{d^2x}{dt^2} + x = \epsilon f \left(x, \frac{dx}{dt} \right), \qquad 0 < \epsilon \ll 1, \qquad (8.4.25)$$

in the form

$$x(t, \epsilon) \simeq a(t, \epsilon) \cos \left[t + \phi(t, \epsilon) \right], \qquad (8.4.26)$$

where the functions $a(t, \epsilon)$ and $\phi(t, \epsilon)$ are the solutions to Eqs. (8.4.23) and (8.4.24). These equations can be written as

$$\frac{da}{dt} = \epsilon A(a), \qquad \frac{d\phi}{dt} = \epsilon B(a). \qquad (8.4.27)$$

Thus, the general procedure consists of solving the first equation for $a(t, \epsilon)$ and then substituting this result into the second equation and solving for $\phi(t, \epsilon)$.

8.4.2 Two Special Cases for $f(x, dx/dt)$

For the first case, let f be a function only of x, i.e.,

$$f(x, dx/dt) = f_1(x). \qquad (8.4.28)$$

Therefore, the equation for a is

$$\begin{aligned}
\frac{da}{dt} &= - \left(\frac{\epsilon}{2\pi} \right) \int_0^{2\pi} f_1(a \cos \psi) \sin \psi \, d\psi \\
&= - \left(\frac{\epsilon}{2\pi} \right) \int_{-\pi}^{\pi} f_1(a \cos \psi) \sin \psi \, d\psi \\
&= 0. \qquad (8.4.29)
\end{aligned}$$

Consequently,

$$a(t, \epsilon) = a_0 = \text{constant}, \qquad (8.4.30)$$

and

$$\phi(t, \epsilon) = \epsilon \Omega(a_0)t + \phi_0,$$

where ϕ_0 is a constant and

$$\Omega(a_0) = -\left(\frac{1}{2\pi a_0}\right) \int_0^{2\pi} f_1(a_0 \cos\psi) \cos\psi \, d\psi. \qquad (8.4.31)$$

Thus, for the situation where f depends only on x, the first order averaging solution is

$$x(t,\epsilon) \simeq a_0 \cos\{[1 + \epsilon\Omega(a_0)]\, t + \phi_0\}. \qquad (8.4.32)$$

This case corresponds to a classical conservative oscillator [8]. The major effect of the nonlinearity is to modify the frequency of the oscillation, i.e.,

$$w(\epsilon, a_0) = 1 + \epsilon\Omega(a_0). \qquad (8.4.33)$$

The frequency depends on the amplitude.

For the second case, let f depend only on dx/dt, i.e.,

$$f(x, dx/dt) = f_2\left(\frac{dx}{dt}\right). \qquad (8.4.34)$$

We have

$$\frac{d\phi}{dt} = -\left(\frac{\epsilon}{2\pi a}\right) \int_0^{2\pi} f_2(-a \sin\psi) \cos\psi \, d\psi. \qquad (8.4.35)$$

If $f_2(v)$ is an odd function of v, then

$$\frac{d\phi}{dt} = 0 \Rightarrow \phi(t,\epsilon) = \phi_0 = \text{constant}, \qquad (8.4.36)$$

$$\frac{da}{dt} = -\left(\frac{\epsilon}{2\pi}\right) \int_0^{2\pi} f_2(-a \sin\psi) \sin\psi \, d\psi = \epsilon A(a). \qquad (8.4.37)$$

If $A(a) = 0$ has real, positive zeros, then they correspond to the amplitudes of limit cycles. If we denote these values by $(\bar{a}_i : i = 1, 2, \ldots, I)$, then the first approximation to the ith limit-cycle is

$$x^{(i)}(t,\epsilon) = \bar{a}_i \cos(t + \phi_0), \qquad i = 1, 2, \ldots, I, \qquad (8.4.38)$$

where I is the total number of positive solutions to $A(a) = 0$. Finally, note that for this case, the angular frequency for all the limit-cycle can be written as

$$\omega_i(\epsilon) = 1 + O(\epsilon^2). \qquad (8.4.39)$$

8.4.3 Stability of Limit-Cycles

The variation of the amplitude with time (in the first-approximation) is given by

$$\frac{da}{dt} = \epsilon A(a); \tag{8.4.40}$$

see Eqs. (8.4.23) and (8.4.27). Let $\bar{a}$ be a non-negative zero of $A(a)$, i.e.,

$$A(\bar{a}) = 0. \tag{8.4.41}$$

As noted above, $a = \bar{a}$ corresponds to a limit-cycle. Let $a(t)$ be a solution close to $\bar{a}$, i.e.,

$$a(t) = \bar{a} + \beta(t), \qquad |\beta(0)| \ll \bar{a}. \tag{8.4.42}$$

Substituting this relationship into Eq. (8.4.40) and retaining only linear terms gives

$$\frac{d\beta}{dt} = \epsilon R \beta, \tag{8.4.43}$$

where

$$R \equiv \left. \frac{dA}{da} \right|_{a=\bar{a}}. \tag{8.4.44}$$

The solution to Eq. (8.4.43) is

$$\beta(t) = \beta(0)e^{\epsilon R t}. \tag{8.4.45}$$

Note that a small initial perturbation $\beta(0)$ increases with time if $R > 0$, while it decreases if $R < 0$. Consequently, the stationary amplitude $a = \bar{a}$ has the following (linear) stability properties:

$$\begin{cases} a(t, \epsilon) = \bar{a}, & \text{stable if } R < 0; \\ a(t, \epsilon) = \bar{a}, & \text{unstable if } R > 0. \end{cases} \tag{8.4.46}$$

These results also hold for the case that $\bar{a} = 0$.

8.5 Worked Examples for First-Order Averaging

8.5.1 Example A

Consider a conservative system modeled by the differential equation

$$\frac{d^2x}{dt^2} + x + \epsilon x^3 = 0. \tag{8.5.1}$$

For this case, $f(x, dx/dt) = -x^3$ and it follows from the results of section 8.4.2 that

$$a(t, \epsilon) = \text{constant} = a_0, \tag{8.5.2}$$

and $\phi(t)$ is determined by the equation

$$\frac{d\phi}{dt} = \left(\frac{\epsilon}{2\pi a_0}\right) \int_0^{2\pi} (a_0 \cos \psi)^3 \cos \psi \, d\psi. \tag{8.5.3}$$

Using the formula

$$(\cos \psi)^4 = \left(\frac{3}{8}\right) + \left(\frac{1}{2}\right) \cos 2\psi + \left(\frac{1}{8}\right) \cos 4\psi, \tag{8.5.4}$$

it follows that

$$\frac{d\phi}{dt} = \frac{3\epsilon a_0^2}{8} \Rightarrow \phi(t, \epsilon) = \left(\frac{3\epsilon a_0^2}{8}\right) + \phi_0. \tag{8.5.5}$$

Therefore, the first-order average method gives the following approximation to the solution of Eq. (8.5.1)

$$x(t, \epsilon) = a_0 \cos \left[\left(1 + \frac{3\epsilon a_0^2}{8}\right) t + \phi_0\right]. \tag{8.5.6}$$

The angular frequency is, to this order of approximation,

$$\omega(\epsilon, a_0) = 1 + \frac{3\epsilon a_0^2}{8}, \tag{8.5.7}$$

and depends both on the parameter ϵ and the amplitude a_0.

8.5.2   *Example B*

An important example of a linear oscillator is the differential equation

$$\frac{d^2 x}{dt^2} + 2\epsilon \frac{dx}{dt} + x = 0, \qquad 0 < \epsilon \ll 1, \tag{8.5.8}$$

which models a weakly damped simple harmonic oscillator. For this case,

$$f\left(x, \frac{dx}{dt}\right) = -2\frac{dx}{dt} \to +2a \sin \psi. \tag{8.5.9}$$

With this result, the equations determining the phase and amplitude are

$$\frac{d\phi}{dt} = 0, \tag{8.5.10}$$

$$\frac{da}{dt} = -\left(\frac{\epsilon}{2\pi}\right) \int_0^{2\pi} (2a \sin \psi) \sin \psi \, d\psi = -\epsilon a. \tag{8.5.11}$$

The corresponding solutions are

$$\phi(t, \epsilon) = \phi_0, \qquad a(t, \epsilon) = a_0 e^{-\epsilon t}, \tag{8.5.12}$$

and the approximation to the solution of Eq. (8.5.8) is

$$x(t, \epsilon) = a_0 e^{-\epsilon t} \cos(t + \phi_0). \tag{8.5.13}$$

This result is to be compared to the exact solution for Eq. (8.5.8),

$$x(t, \epsilon) = a_0 e^{-\epsilon t} \cos\left[\left(1 - \frac{\epsilon^2}{4}\right)^{1/2} t + \phi_0\right]. \tag{8.5.14}$$

Note that the amplitude functions are the same, but the exact angular frequency has a dependence on ϵ, i.e.,

$$\omega_{\text{approximate}} = 1, \qquad \omega_{\text{exact}} = \left(1 - \frac{\epsilon^2}{4}\right)^{1/2} = 1 + O(\epsilon^2). \tag{8.5.15}$$

8.5.3 *Example C*

The van der Pol equation is

$$\frac{d^2 x}{dt^2} + x = \epsilon(1 - x^2)\frac{dx}{dt}. \tag{8.5.16}$$

For this equation

$$f\left(x, \frac{dx}{dt}\right) = (1 - x^2)\frac{dt}{dt}, \tag{8.5.17}$$

and

$$f(a \cos \psi, -a \sin \psi) = (1 - a^2 \cos^2 \psi)(-a \sin \psi). \tag{8.5.18}$$

The amplitude and phase equations are

$$\frac{da}{dt} = \left(\frac{\epsilon}{2\pi}\right) \int_0^{2\pi} a(1 - a^2 \cos^2 \psi) \sin^2 \psi \, d\psi. \tag{8.5.19}$$

$$\frac{d\phi}{dt} = \left(\frac{\epsilon}{2\pi}\right) \int_0^{2\pi} (1 - a^2 \cos^2 \psi) \sin \psi \cos \psi \, d\psi. \tag{8.5.20}$$

The integrals can be easily evaluated to give

$$\frac{da}{dt} = \left(\frac{\epsilon a}{2}\right)\left(1 - \frac{a^2}{4}\right), \qquad \frac{d\phi}{dt} = 0. \tag{8.5.21}$$

The solution to the ϕ equation is

$$\phi(t, \epsilon) = \phi_0. \tag{8.5.22}$$

Define the variable z to be

$$z = a^2. \tag{8.5.23}$$

Then multiply the first of Eqs. (8.5.21) by $2a$, to obtain

$$\frac{dz}{dt} = \epsilon z \left(1 - \frac{z^2}{4} \right). \tag{8.5.24}$$

This is a separable equation and using the method of partial fractions the following solution is found

$$z = a^2 = \frac{z_0 e^{\epsilon t}}{1 + \left(\frac{z_0}{4} \right) (e^{\epsilon t} - 1)}, \tag{8.5.25}$$

where $z_0 = a_0^2$ is the integration constant. Therefore, the first approximation to the solution of the van der Pol equation based on the Krylov-Bogoliubov procedure or the method of first-order averaging is

$$x(t, \epsilon) = \frac{a_0 e^{\epsilon t/2} \cos(t + \phi_0)}{\left[1 + \left(\frac{a_0^2}{4} \right) (e^{\epsilon t} - 1) \right]^{1/2}}. \tag{8.5.26}$$

Examination of this expression shows that for any initial condition, the solution approaches the function $2\cos(t + \phi_0)$, i.e.,

$$x(t, \epsilon) \underset{\text{large } t}{\longrightarrow} 2\cos(t + \phi_0). \tag{8.5.27}$$

Consequently, the amplitude and angular frequency are (to this level of approximation)

$$\bar{a} = 2, \qquad \omega = 1. \tag{8.5.28}$$

The stability of the limit-cycle can be determined from Eq. (8.5.21). Note that

$$\begin{cases} \frac{da}{dt} > 0 & \text{for } a < 2, \\ \frac{da}{dt} < 0 & \text{for } a > 2, \end{cases} \tag{8.5.29}$$

consequently, the limit-cycle at $\bar{a} = 0$ is stable.

8.6   The Lindstedt-Poincaré Method

8.6.1   *Secular Terms*

This section presents one of the first perturbation methods applied to differential equations modeling oscillatory systems. The differentiation equation is taken to be

$$\frac{d^2 x}{dt^2} + x = \epsilon f \left(x, \frac{dx}{dt} \right), \qquad 0 < \epsilon \ll 1. \tag{8.6.1}$$

The starting point for the perturbation procedure is the assumption that periodic solutions of Eq. (8.6.1) can be represented as

$$x(t, \epsilon) = \sum_{m=0}^{n} \epsilon^m x_m(t) + O(\epsilon^{n+1}). \tag{8.6.2}$$

In general, this expression is an asymptotic expansion and the original justification for this form was given by Poincaré [9].

To illustrate the method, consider the nonlinear equation

$$\frac{d^2 x}{dt^2} + x + \epsilon x^3 = 0, \qquad 0 < \epsilon \ll 1. \tag{8.6.3}$$

If y is defined as $y = dx/dt$, then the equation has a first integral given by

$$\frac{y^2}{2} + \frac{x^2}{2} + \epsilon \frac{x^4}{4} = E \geq 0, \tag{8.6.4}$$

which is the total energy. In the (x, y) phase-space, Eq. (8.6.4) corresponds to a one parameter set of closed curves. This implies that all the solutions of Eq. (8.6.3) are periodic. The substitution of Eq. (8.6.2) into Eq. (8.6.3) gives

$$\left(\frac{d^2 x_0}{dt^2} + \epsilon \frac{d^2 x_1}{dt^2} + \epsilon^2 \frac{d^2 x_2}{dt^2} + \cdots \right)$$
$$+ (x_0 + \epsilon x_1 + \epsilon^2 x_2 + \cdots) + \epsilon(x_0 + \epsilon x_1 + \epsilon^2 x_2 + \cdots)^3 = 0, \quad (8.6.5)$$

and

$$\left(\frac{d^2 x_0}{dt^2} + x_0 \right) + \epsilon \left(\frac{d^2 x_1}{dt^2} + x_1 + x_0^3 \right)$$
$$+ \epsilon^2 \left(\frac{d^2 x_2}{dt^2} + x_2 + 3x_0^2 x_1 \right) + \cdots = 0. \tag{8.6.6}$$

Setting the coefficients of the powers of ϵ to zero leads to the following system of linear, inhomogeneous equations,

$$\begin{cases} \dfrac{d^2 x_0}{dt^2} + x_0 = 0, \\[2mm] \dfrac{d^2 x_1}{dt^2} + x_1 = -x_0^3, \\[2mm] \dfrac{d^2 x_2}{dt^2} + x_2 = -3x_0^2 x_1, \\[2mm] \vdots \qquad \vdots \end{cases} \tag{8.6.7}$$

To solve these equations, initial conditions must be selected. We take them to be

$$x(0) = A, \qquad \frac{dx(0)}{dt} = 0, \tag{8.6.8}$$

and this choice gives

$$\begin{cases} x_0(0) = A; x_i(0) = 0, & \text{for } i \geq 1, \\ \dfrac{dx_k(0)}{dt} = 0, & \text{for } k \geq 0. \end{cases} \qquad (8.6.9)$$

The solution to the first of Eqs. (8.6.7)

$$x_0(t) = A \cos t. \qquad (8.6.10)$$

Substituting this into the right-side of the differential equation for x_1 gives

$$\frac{d^2 x_1}{dt^2} + x_1 = -x_0^3 = -\left(\frac{3A^2}{4}\right) \cos t - \left(\frac{A^3}{4}\right) \cos 3t. \qquad (8.6.11)$$

The complete solution to Eq. (8.6.11), which incorporates the initial conditions $x_1(0) = 0$ and $dx_1(0)/dt = 0$, is

$$x_1(t) = \left(\frac{A^3}{32}\right)(\cos 3t - \cos t) - \left(\frac{3A^2}{8}\right) t \sin t. \qquad (8.6.12)$$

Thus, to terms of order ϵ, the "solution" to Eq. (8.6.3) is

$$x(t, \epsilon) = x_0(t) + \epsilon x_1(t) + O(\epsilon^2)$$

$$= A \cos t + \epsilon \left(\frac{A^3}{32}\right)[(\cos 3t - \cos t) - 12t \sin t] + O(\epsilon^2). \qquad (8.6.13)$$

Inspection of Eq. (8.6.4) shows that $x_1(t)$ is not only nonperiodic, but, is also unbounded as $t \to \infty$! Thus, a direct, naive application of the expansion of Eq. (8.6.2) leads to serious difficulties if the goal is to calculate analytic periodic approximations to the solutions of Eq. (8.6.3). Since Eq. (8.6.3) is a special case of Eq. (8.6.1), we expect that this is a generic problem with use of this type of expansion.

Terms such as $t^m \cos t$ or $t^m \sin t$ are called *secular terms*. The technique needed to avoid the presence of secular terms was developed by Lindstedt [10]. Later Poincaré [9] proved that the expansions obtained by Lindstedt's method are both asymptotic and uniformly valid. However, the genesis of secular terms can be fully understood from the following physical arguments:

(i) The expansion in Eq. (8.6.2) implicitly takes the angular frequency of the periodic solution to Eq. (8.6.7) to be $\omega = 1$.

(ii) However, we are already aware, from the calculation using the method of first-order averaging, that to order ϵ, the angular frequency depends on both ϵ and the initial condition $x(0) = A$. From Eq. (8.5.7), we have

$$\omega(\epsilon, A) = 1 + \frac{3\epsilon A^2}{8}. \qquad (8.6.14)$$

(iii) Generalizing the result in Eq. (8.6.14), we see that $\omega(\epsilon, A)$ should have an expansion of the form

$$\omega(\epsilon, A) = 1 + \epsilon\omega_1 + \epsilon^2\omega_2 + \cdots, \tag{8.6.15}$$

where the ω_i depend on A.

(iv) All of these results suggest that the independent variable is not the time, t, but another variable, θ, related to the time.

The next section presents the details as to how this modified expansion procedure is to be done.

8.6.2 The Formal Procedure

The Lindstedt-Poincaré method to be presented is based on a central observation made in the previous section: The nonlinear terms in Eq. (8.6.3) changes the angular frequency of the equation away from the value given by the linear system, i.e., when $\epsilon = 0$. To take into account this change, we introduce a new variable

$$\theta = \omega t$$
$$= (1 + \epsilon\omega_1 + \epsilon^2\omega_2 + \cdots + \epsilon^n\omega_n + \cdots), \tag{8.6.16}$$

where, for the moment, the ω_i are unknown parameters. Likewise, in place of $x(t, \epsilon)$, we introduce

$$x(\theta, \epsilon) = x_0(\theta) + \epsilon x_1(\theta) + \epsilon^2 x_2(\theta) + \cdots + \epsilon^n x_n(\theta) + \cdots, \tag{8.6.17}$$

and impose the following periodicity condition

$$x(\theta + 2\pi) = x(\theta), \tag{8.6.18}$$

which implies

$$x_n(\theta + 2\pi) = x_n(\theta), \qquad n \geq 0. \tag{8.6.19}$$

Let us now define the following notation:

$$\dot{x} \equiv \frac{dx}{d\theta}, \qquad \ddot{x} \equiv \frac{d^2x}{d\theta^2}$$

$$F_x(x, \dot{x}) \equiv \frac{\partial F(x, \dot{x})}{\partial x}, \qquad F_{\dot{x}} \equiv \frac{\partial F(x, \dot{x})}{\partial \dot{x}}. \tag{8.6.20}$$

If Eqs. (8.6.16) and (8.6.17) are substituted into the differential equation

$$\frac{d^2x}{dt^2} + x = \epsilon f\left(x, \frac{dx}{dt}\right), \tag{8.6.21}$$

and the coefficients of the various powers of ϵ are set equal to zero, then the following equations are obtained for the $x_n(\theta)$:

$$\ddot{x}_0 + x_0 = 0, \tag{8.6.22}$$

$$\ddot{x}_1 + x_1 = -2\omega_1\ddot{x}_0 + f(x_0, \dot{x}_0), \tag{8.6.23}$$

$$\ddot{x}_2 + x_2 = -2\omega_1\ddot{x}_1 - (\omega_1^2 + 2\omega_2)\ddot{x}_0 + f_x(x_0, \dot{x}_0)x_1$$
$$+ f_{\dot{x}}(x_0, \dot{x}_0)(\omega\dot{x}_0 + \dot{x}_1) \tag{8.6.24}$$

$$\vdots \quad \vdots \qquad\qquad \vdots$$

$$\ddot{x}_n + x_n = g_n(x_0, x_1, \ldots, x_{n-1}; \dot{x}_0, \dot{x}_1, \ldots, x_{n-1}) \tag{8.6.25}$$

$$\vdots \quad \vdots \qquad\qquad \vdots$$

Note that if f is a polynomial function of x and dx/dt, then the g_n are also polynomial functions of their arguments.

Now if Eq. (8.6.17) is to be an approximation to the periodic solutions of Eq. (8.6.21), then the right-sides of Eqs. (8.6.23), (8.6.24) and (8.6.25) must not contain constant multiples of either $\sin\theta$ or $\cos\theta$, otherwise, secular terms will exist. Therefore, if $x_n(\theta)$ is to be periodic, two conditions must be satisfied at each step of the calculation, i.e., the coefficients of the $\sin\theta$ and $\cos\theta$ on the right-sides of these equations must be such that they can be set equal to zero. The only consistent way to do this is to have available two free parameters. A detailed study of Eqs. (8.6.23), etc., shows that one of the parameters is ω_n. The only other place where a second constant can be introduced is from the initial condition on x_{n-1}, if we are calculating the solution for x_n. This implies that the initial conditions should take the form

$$\begin{cases} x(0, \epsilon) = A_0 + \epsilon A_1 + \epsilon^2 A_2 + \cdots \\ \dot{x}(0, \epsilon) = 0, \end{cases} \tag{8.6.26}$$

where the A_n are, a priori, unspecified constants. Thus, the periodicity requirement for $x_1(\theta)$ forces the right-side of Eq. (8.6.23) to have a term linear in ω_1 and another nonlinear in A_0, which occur in the coefficients involving $\sin\theta$ and $\cos\theta$. This means that setting these coefficients equal to zero allows all secular terms to be eliminated and also permits both ω_1 and A_0 to be calculated. In a similar manner, for $n \geq 2$, the periodicity condition on $x_n(\theta)$ gives a pair of equations for ω_n and A_{n-1}. Again, the elimination of secular terms allows both quantities to be determined. Consequently, at the nth level of calculation $(\omega_n, A_{n-1}, x_n(\theta))$ are simultaneously found.

There does exist a situation for which the preceding analysis can be simplified. This occurs when the function f depends only on x and/or is an even function of dx/dt. For this case, $x(t)$ can always be chosen to be an even function of t by use of the initial conditions

$$x(0) = A, \qquad \frac{dx(0)}{dt} = 0. \tag{8.6.27}$$

This means that both $x(\theta)$ and $x_n(\theta)$ are even functions of θ, and, as a consequence, the right-sides of Eqs. (8.6.23) to (8.6.25) contain no terms in $\sin \theta$. For this situation, only one free parameter is needed to ensure that the $\cos \theta$ term is eliminated. This result implies that the initial conditions become

$$x(\theta = 0) = A = A_0, \qquad \frac{dx(0)}{d\theta} = 0, \tag{8.6.28}$$

or

$$\begin{cases} x_0(0) = A, & x_i(0) = 0 \quad i \geq 1, \\ \dot{x}_j(0) = 0, & j \geq 0. \end{cases} \tag{8.6.29}$$

In summary, the $(n+1)$th approximation to the solution of Eq. (8.6.21), according to the Lindstedt-Poincaré perturbation method is given by

$$x(\theta, \epsilon) = \sum_{m=0}^{n} \epsilon^m x_m(\theta) + O(\epsilon^{n+1}), \tag{8.6.30}$$

where

$$\theta(\epsilon, t) = \left[1 + \epsilon \omega_1 + \cdots + \epsilon^n \omega_n + O(\epsilon^{n+1}) \right] t, \tag{8.6.31}$$

and the ω_i's will in general be functions of the $(A_0, A_1, \ldots, A_n)$.

It should be clear from the comments made earlier in this section that the Lindstedt-Poincaré perturbation method will only give approximations to periodic solutions. In particular, this means that if an equation has a limit-cycle solution then the parameters defining the limit-cycle may be determined, but not the transitory behavior near the limit-cycle.

8.7 Worked Examples Using the Lindstedt-Poincaré Method

8.7.1 *Example A*

Consider the differential equation

$$\frac{d^2 x}{dt^2} + x + \epsilon x^3 = 0. \tag{8.7.1}$$

Since $f = -x^3$, the equations for $x_0(\theta)$, $x_1(\theta)$ and $x_2(\theta)$ follow directly from Eqs. (8.6.22), (8.6.23) and (8.6.24):

$$\ddot{x}_0 + x_0 = 0, \tag{8.7.2}$$

$$\ddot{x}_1 + x_1 = -2\omega_1\ddot{x}_0 - x_0^3, \tag{8.7.3}$$

$$\ddot{x}_2 + x_2 = -2\omega_1\ddot{x}_1 - (\omega_1^2 + 2\omega_2)\ddot{x}_0 - 3x_0^2 x_1. \tag{8.7.4}$$

The initial conditions are

$$\begin{cases} x_0(0) = A, \quad x_1(0) = x_2(0) = 0, \\ \dot{x}_0(0) = \dot{x}_1(0) = \dot{x}_2(0) = 0. \end{cases} \tag{8.7.5}$$

The solution to Eq. (8.7.2) is $x_0 = A\cos\theta$. If this is substituted into the right-side of Eq. (8.7.3) and the resulting expression is simplified, then the following result is obtained

$$\ddot{x}_1 + x_1 = \left(2\omega_1 A - \frac{3A^3}{4}\right)\cos\theta - \left(\frac{A^3}{4}\right)\cos 3\theta. \tag{8.7.6}$$

Secular terms can be eliminated if

$$\omega_1 = \frac{3A^2}{8}. \tag{8.7.7}$$

The solution of

$$\ddot{x}_1 + x_1 = -\left(\frac{A^3}{4}\right)\cos 3\theta, \qquad x_1(0) = 0, \qquad \dot{x}_1(0) = 0, \tag{8.7.8}$$

is

$$x_1(\theta) = \left(\frac{A^3}{32}\right)(-\cos\theta + \cos 3\theta). \tag{8.7.9}$$

If $x_0(\theta)$ and $x_1(\theta)$ are now substituted into the right-side of Eq. (8.7.4), this equation becomes

$$\ddot{x}_2 + x_2 = \left(\frac{21A^4}{128} + 2\omega_2\right)A\cos\theta + \left(\frac{3A^5}{16}\right)\cos 3\theta - \left(\frac{3A^5}{128}\right)\cos 5\theta. \tag{8.7.10}$$

Again, we see that secular terms are eliminated if

$$\omega_2 = -\frac{21A^4}{256}. \tag{8.7.11}$$

The complete solution to the resulting differential equation for $x_2(\theta)$, subject to $x_2(0) = 0$ and $\dot{x}_2(0) = 0$, is

$$x_2(\theta) = \left(\frac{A^5}{1024}\right)(23\cos\theta - 24\cos 3\theta + \cos 5\theta). \tag{8.7.12}$$

Putting all this together, the third approximation to the solution of Eq. (8.7.1) is

$$x(\theta, \epsilon) = A\cos\theta + \epsilon\left(\frac{A^3}{32}\right)(-\cos\theta + \cos 3\theta)$$

$$+ \epsilon^2\left(\frac{A^5}{1024}\right)(23\cos\theta - 24\cos 3\theta + \cos 5\theta) + O(\epsilon^2), \quad (8.7.13)$$

where $\theta = \omega t$ and

$$\omega(\epsilon) = 1 + \epsilon\left(\frac{3A^3}{8}\right) - \epsilon^2\left(\frac{21A^4}{256}\right) + O(\epsilon^3). \quad (8.7.14)$$

Note that to terms of $O(\epsilon)$, the method of first order averaging, see Eq. (8.5.7) and the Lindstedt-Poincaré method give the same result for the angular frequency $\omega(\epsilon)$.

8.7.2 Example B

The van der Pol equation provides an interesting application of the Lindstedt-Poincaré method. For this case, the differential equation is

$$\frac{d^2x}{dt^2} + x = \epsilon(1 - x^2)\frac{dx}{dt}, \quad (8.7.15)$$

and

$$f\left(x, \frac{dx}{dt}\right) = (1 - x^2)\frac{dx}{dt}. \quad (8.7.16)$$

The equations for x_0, x_1, and x_2 can be easily found by making use of Eqs. (8.6.22), (8.6.23), and (8.6.24). They are, along with their associated initial conditions, given by the expressions:

$$\ddot{x}_0 + x_0 = 0, \quad x_0(0) = A_0, \quad \dot{x}_0(0) = 0, \quad (8.7.17)$$

$$\ddot{x}_1 + x_1 = -2\omega_1\ddot{x}_0 + (1 - x_0^2)\dot{x}_0, \quad x_1(0) = A_1, \quad \dot{x}_1(0) = 0, \quad (8.7.18)$$

$$\ddot{x}_2 + x_2 = -2\omega_1\ddot{x}_1 - (\omega_1^2 + 2\omega_2)\ddot{x}_0 - 2x_0x_1\dot{x}_0$$

$$+ (1 - x_0^2)(\dot{x}_1 + \omega_1\dot{x}_0), \quad x_2(0) = A_2, \quad \dot{x}_2(0) = 0. \quad (8.7.19)$$

Substituting $x_0(\theta) = A_0\cos\theta$ into Eq. (8.7.18) and simplifying the resulting expression gives

$$\ddot{x}_1 + x_1 = 2\omega_1A_0\cos\theta - A_0\left(1 - \frac{A_0^2}{4}\right)\sin\theta + \left(\frac{A_0^3}{4}\right)\sin 3\theta. \quad (8.7.20)$$

Elimination of secular terms requires

$$A_0 = 2, \quad \omega_1 = 0, \quad (8.7.21)$$

and gives

$$\ddot{x}_1 + x_1 = 2\sin 3\theta. \tag{8.7.22}$$

Subject to the initial conditions, $x_1(0) = A_1$ and $\dot{x}_1(0) = 0$, the solution to Eq. (8.7.22) is

$$x_1(\theta) = A_1 \cos\theta + \left(\frac{1}{4}\right)(3\sin\theta - \sin 3\theta). \tag{8.7.23}$$

Now substituting Eqs. (8.7.21) and (8.7.23), along with $x_0(\theta)$, into the right-side of Eq. (8.7.19) gives

$$\ddot{x}_2 + x_2 = \left(4\omega_2 + \frac{1}{4}\right)\cos\theta + 2A_1\sin\theta - \left(\frac{3}{2}\right)\cos 3\theta$$

$$+ 3A_1\sin 3\theta + \left(\frac{5}{4}\right)\cos 5\theta. \tag{8.7.24}$$

Again, the requirement of no secular terms gives

$$A_1 = 0, \qquad \omega_2 = -\frac{1}{16}, \tag{8.7.25}$$

and

$$\ddot{x}_2 + x_2 = -\left(\frac{3}{2}\right)\cos 3\theta + \left(\frac{5}{4}\right)\cos 5\theta, \quad x_2(0) = A_2, \quad \ddot{x}_2(0) = 0, \tag{8.7.26}$$

for which the solution is

$$x_2(\theta) = \left(A_2 - \frac{13}{96}\right)\cos\theta + \left(\frac{1}{96}\right)(18\cos 3\theta - 5\cos 5\theta). \tag{8.7.27}$$

Continuing this procedure, it can be shown $A_2 = 0$ and

$$x(\theta, \epsilon) = 2\cos\theta + \left(\frac{\epsilon}{4}\right)(3\sin\theta - \sin 3\theta)$$

$$+ \left(\frac{\epsilon^2}{96}\right)(-13\cos\theta + 18\cos 3\theta - 5\cos 5\theta)$$

$$+ O(\epsilon^3), \tag{8.7.28}$$

where $\theta = \omega t$ and

$$\omega(\epsilon) = 1 - \frac{\epsilon^2}{16} + O(\epsilon^3). \tag{8.7.29}$$

8.8   Harmonic Balance

One of the most useful general techniques for calculating analytical approximations to the periodic solutions of nonlinear differential equations is the method of harmonic balance. This procedure generates solutions by making use of a truncated Fourier series representation for the required periodic function approximating the solution. An important advantage of the method is that it can be applied to nonlinear oscillatory problems for which the nonlinear terms need not be small, i.e., a perturbation parameter may not exist. Like the Lindstedt-Poincaré perturbation method, it only gives the periodic solutions; consequently, for systems having limit-cycles, the transitory behavior of the approach to (or away) from a limit-cycle cannot be determined.

There is a rather large literature on the method of harmonic balance. This includes its generalization to expansion functions other than the trigonometric sine's and cosine's, and its incorporation into other types of perturbation techniques. The references [12]–[17] provide a good, general introduction to these formulations and their application to specific equations.

The mathematical basis of harmonic balancing procedures has been investigated by several persons. The papers by Borges et al. [18], Leipholz [19], and Miletta [20] give excellent introductions to various issues relating to convergence of the approximations and error bounds. Two other useful publications are [21; 22].

8.8.1   *Direct Harmonic Balance*

Consider the general second-order differential equation

$$F\left(x, \frac{dx}{dt}, \frac{d^2x}{dt^2}\right) = 0, \tag{8.8.1}$$

and assume that it is of odd parity, i.e.,

$$F(-u, -v, -w) = -F(u, v, w). \tag{8.8.2}$$

This particular restriction causes no actual loss of generality since a large class of nonlinear oscillator systems can be modeled by this type of differential equation. An important feature of these equations is that the Fourier expansions of their solutions only contain the odd numbered harmonics [23], i.e.,

$$x(t) = \sum_{m=1}^{\infty} \left[A_m \cos(2m-1)\omega t + B_m \sin(2m-1)\omega t\right]. \tag{8.8.3}$$

The Nth order method of direct harmonic balance approximates the exact solution $x(t)$ to Eq. (8.8.1) by the expression

$$x_N(t) = \sum_{m=1}^{N} \left[\bar{A}_m^N \cos(2m-1)\bar{\omega}_N t + \bar{B}_m^N \sin(2m-1)\bar{\omega}_N t \right], \qquad (8.8.4)$$

where $\bar{A}_m^N$, $\bar{B}_m^N$, and $\bar{\omega}_N$ are approximations to the exact A_m, B_m, and ω. To proceed, we need to examine two cases.

First, let the differential equation represent a conservative oscillator. For this case, Eq. (8.8.1) generally takes the form

$$\frac{d^2 x}{dt^2} + f(x, \lambda) = 0 \qquad (8.8.5)$$

where $f(-x, \lambda) = -f(x, \lambda)$ and λ represents the parameters defining the force function $f(x, \lambda)$. The initial conditions

$$x(0) = A, \qquad \frac{dx(0)}{dt} = 0, \qquad (8.8.6)$$

translations to $x(t)$ containing only cosine terms, i.e.,

$$x_N(t) = \sum_{m=1}^{N} \bar{A}_m^N \cos(2m-1)\bar{\omega}_N t, \qquad (8.8.7)$$

where it is expected that the $\bar{A}_m^N$ and $\bar{\omega}_N$ will depend on both A and the parameters λ. This expression for $x_N(t)$ contain $(N+1)$ unknowns: The N coefficients and the angular frequency. They can be determined as follows:

(1) Substitute Eq. (8.8.7) into Eq. (8.8.5) and write this result as

$$\sum_{m=1}^{N} H_m \cos(2m-1)\bar{\omega}_n t + \text{HOH} \simeq 0, \qquad (8.8.8)$$

where HOH stands for higher-order harmonics and the H_m are functions of $\bar{A}_m^N$, $\bar{\omega}_N$, and λ.

Note that we assume the first and second derivatives of $x(t)$ are also approximated by taking the appropriate derivatives of $x_N(t)$ as given by Eq. (8.8.7).

(2) Set the functions H_m equal to zero, i.e.,

$$H_m(\bar{A}_1^n, \bar{A}_2^n, \dots, \bar{A}_n^n, \bar{\omega}_n, \lambda) = 0, \qquad m = 1, 2, \dots, N, \qquad (8.8.9)$$

and solve for $\bar{A}_2^n, \bar{A}_3^n, \dots, \bar{A}_n^n$, and $\bar{\omega}_n$ in terms of the amplitude $\bar{A}_1^n$.

(3) $\bar{A}_1^n$ can now be written as a function of the initial condition, by use of the relation

$$x_n(0) = A = \bar{A}_1^n + \sum_{m=2}^{N} \bar{A}_m^N(\bar{A}_1^n, \lambda), \qquad (8.8.10)$$

where we have explicitly indicated that $\bar{A}_m^N$, for $2 \leq m \leq N$, depend on $\bar{A}_1^N$ and λ.

Note that for each value of $N = 1, 2, 3, \ldots$, we require that the initial conditions of Eqs. (8.8.6) be satisfied, i.e.,

$$x_N(0) = A, \qquad dx_N(0)/dt = 0. \qquad (8.8.11)$$

(4) From (3), it follows that A_1^N will be given in terms of A and λ. Thus, an approximation to the periodic solutions of Eq. (8.8.5) can now be written down in the form expressed in Eq. (8.8.7).

The second case is for non-conservative oscillators where dx/dt will appear to some "odd power" in Eq. (8.8.1). For many systems, the relevant differential equation takes the form

$$\frac{d^2x}{dt^2} + f(x, \lambda_1) = g(x, \lambda_2)\frac{dx}{dt}\,, \qquad (8.8.12)$$

where λ_1 and λ_2 are parameters defining the respective functions in which they appear; and f and g have the properties

$$f(-x, \lambda_1) = -f(x, \lambda_1), \qquad g(-x, \lambda_2) = +g(x, \lambda_2). \qquad (8.8.13)$$

For this case, limit-cycles may exist and the initial conditions, in general, cannot be *a priori* specified. The Nth order method of harmonic balance approximates the periodic solutions of Eq. (8.8.12) by the expression

$$x_N(t) = \bar{A}_1^N \cos \bar{\omega}_N t$$

$$+ \sum_{m=2}^{N} \bar{A}_m^N \cos(2m-1)\bar{\omega}_N t + \bar{B}_m^N \sin(2m-1)\bar{\omega}_N t. \qquad (8.8.14)$$

The $2N$ unknowns

$$\begin{cases} \bar{A}_1^N, \bar{A}_2^N, \ldots, \bar{A}_n^N, \\ \bar{\omega}_N, \bar{B}_2^N, \ldots, \bar{B}_n^N, \end{cases} \qquad (8.8.15)$$

are determined as follows:

(1) Substitute Eq. (8.8.14) into Eq. (8.8.12) and write the resulting expression as

$$\sum_{m=1}^{N} [H_m \cos(2m-1)\bar{\omega}_N t + L_m \sin(2m-1)\bar{\omega}_N t] + \text{HOH} \simeq 0, \qquad (8.8.16)$$

where the coefficients H_m and L_m will be functions of the $(2N-1)$ amplitudes and the angular frequency $\bar{\omega}_n$.

(2) Set the $2N$ functions H_m and L_m equal to zero, i.e.,

$$H_m = 0, \qquad L_m = 0; \quad m = 1, 2, \ldots, N; \qquad (8.8.17)$$

and solve for all the quantities listed in Eq. (8.8.15). In general, they will be expressed in terms of the system parameters λ_1 and λ_2.

In the next section, we apply the harmonic balance method to obtain approximations to the periodic solutions for several differential equations modeling nonlinear oscillators.

8.9 Worked Examples for Harmonic Balance

8.9.1 *Example A*

The following differential equation

$$\frac{d^2x}{dt^2} + x^3 = 0, \tag{8.9.1}$$

has all periodic solutions. This can easily be seen by noting that it has the first-integral ($y = dx/dt$),

$$\frac{y^2}{2} + \frac{x^4}{4} = E \geq 0. \tag{8.9.2}$$

In the (x, y) phase-plane, Eq. (8.9.2) represents a one-parameter family of closed curves. Consequently, all solutions are periodic.

Note that this equation has no harmonic oscillator limit. Thus, none of the two previously discussed methods, averaging and Lindstedt-Poincaré, can be applied to obtain approximations to the periodic solutions.

Let us first calculate a first approximation to the solution of Eq. (8.9.1). For the initial conditions

$$x(0) = A, \qquad \frac{dx(0)}{dt} = 0, \tag{8.9.3}$$

and the harmonic balance approximations

$$x(t) = A_1 \cos \omega t, \quad \frac{dx}{dt} = -\omega A_1 \sin \omega t, \quad \frac{d^2x}{dt^2} = -\omega^2 A_1 \cos \omega t, \tag{8.9.4}$$

we obtain on substitution into Eq. (8.9.1) the expression

$$-\omega^2 A_1 \cos \omega t + A_1^3 (\cos \omega t)^3 \simeq 0 \tag{8.9.5}$$

or

$$\left(-\omega^2 + \frac{3A_1^2}{4} \right) A_1 \cos \omega t + \text{HOH} \simeq 0, \tag{8.9.6}$$

where we have used

$$(\cos \theta)^3 = \left(\frac{3}{4} \right) \cos \theta + \left(\frac{1}{4} \right) \cos 3\theta. \tag{8.9.7}$$

Setting the coefficient of $\cos \omega t$ to zero, we obtain

$$\omega = \sqrt{\frac{3}{4}} A_1. \tag{8.9.8}$$

Comparison of the form for $x(t)$ with the initial conditions, gives

$$A_1 = A. \tag{8.9.9}$$

Therefore, the first approximation to the periodic solution of Eq. (8.9.1), based on the harmonic balance method, is

$$x(t) = A \cos\left(\sqrt{\frac{3}{4}} At\right). \tag{8.9.10}$$

An improved, second approximation can be found by using the following form for $x(t)$,

$$x(t) = A_1 \cos\theta + A_2 \cos 3\theta, \qquad \theta = \omega t, \tag{8.9.11}$$

with

$$\frac{dx(t)}{dt} = -\omega A_1 \sin\theta - 3\omega A_2 \sin 3\theta, \tag{8.9.12}$$

$$\frac{d^2 x(t)}{dt^2} = -\omega A_1 \cos\theta - 9\omega^2 A_2 \cos 3\theta. \tag{8.9.13}$$

Substitution of these three expressions into Eq. (8.9.1) and simplifying the resulting expression, gives

$$H_1(A_1, A_2, \omega) \cos\theta + H_2(A_1, A_2, \omega) \cos 3\theta + \text{HOH} \simeq 0, \tag{8.9.14}$$

where

$$H_1(A_1, A_2, \omega) = A_1 \left[\omega^2 - \left(\frac{3}{4}\right) A_1^2 - \left(\frac{3}{4}\right) A_1 A_2 - \left(\frac{3}{2}\right) A_2^2\right], \tag{8.9.15}$$

$$H_2(A_1, A_2, \omega) = -9A_2\omega^2 + \left(\frac{1}{4}\right) A_1^3 + \left(\frac{3}{2}\right) A_1^2 A_2 + \left(\frac{3}{4}\right) A_2^3. \tag{8.9.16}$$

The harmonic balancing condition leads to the two equations

$$H_1(A_1, A_2, \omega) = 0, \qquad H_2(A_1, A_2, \omega) = 0. \tag{8.9.17}$$

Solving the first equation for $\omega(A_1, A_2)$ gives

$$\omega = \sqrt{\frac{3}{4}} A_1 (1 + x + 2x^2)^{1/2}, \tag{8.9.18}$$

where

$$x = \frac{A_2}{A_1}. \qquad (8.9.19)$$

If this value for ω is now substituted into the second of the equations, in Eq. (8.9.17), then, after some algebraic manipulation, the following equation is obtained for x.

$$51x^3 + 27x^2 + 21x - 1 = 0. \qquad (8.9.20)$$

Note that this equation is cubic; however, the desired root for our purposes is the real one for which

$$|x| \ll 1. \qquad (8.9.21)$$

An excellent approximation for this root is

$$\bar{x} \simeq \frac{1}{21} = 0.0476. \qquad (8.9.22)$$

Hence,

$$A_2 \simeq \left(\frac{1}{21}\right) A_1, \qquad (8.9.23)$$

and it follows that the amplitude A_2 has a value about 5% of the value of the fundamental amplitude.

At this stage of the calculation, we can write for the second approximation

$$x(t) = A_1[\cos\theta + \bar{x}\cos 3\theta]. \qquad (8.9.24)$$

Since $x(0) = A$, it follows

$$A_1 = \frac{A}{1+\bar{x}} \simeq \left(\frac{21}{22}\right) A. \qquad (8.9.25)$$

Note that ω is, from Eq. (8.9.18), given by

$$\omega \simeq \sqrt{\frac{3}{4}} A \left(\frac{21}{22}\right) \left[1 + \frac{1}{21} + 2\left(\frac{1}{21}\right)^2\right]^{1/2} \simeq \left(\sqrt{\frac{3}{4}}A\right)\left(\frac{43}{44}\right), \qquad (8.9.26)$$

which corresponds to a change in ω of about 2% as compared to the first method of approximation.

These values for A_1 and ω can be substituted into Eq. (8.9.24) to obtain the second approximation to the periodic solution to Eq. (8.9.1).

The exact solution to Eq. (8.9.1) is given in terms of the Jacobi elliptic function [7],

$$x(t) = A \cdot \text{cn}\left(At; 1/\sqrt{2}\right). \qquad (8.9.27)$$

The Fourier expansion of the exact solution gives a ratio of the amplitudes of the $\cos 3\omega t$ and $\cos \omega t$ terms to be 0.0450778, as compared to the value given by Eq. (8.9.21). (Actually, we can greatly improve on the value of $\bar{x}$ by retaining the quadratic term in Eq. (8.9.20) and obtain $\bar{x} \simeq 0.0450$.) These results show that the method of harmonic balance provides an excellent approximation to the solution of Eq. (8.9.1).

8.9.2 Example B

Consider now the oscillator modeled by

$$\frac{d^2x}{dt^2} + x + \epsilon x^3 = 0, \qquad \epsilon > 0. \tag{8.9.28}$$

Again, we can calculate a first-integral and obtain the result $(y = dx/dt)$,

$$\frac{y^2}{2} + \frac{x^2}{2} + \epsilon \frac{x^4}{4} = E \geq 0, \tag{8.9.29}$$

and from this conclude that all solutions to the differential equation are periodic.

For a first approximation to these periodic solutions we substitute Eqs. (8.9.4) into Eq. (8.9.28) to obtain

$$-\omega^2 A_1 \cos \omega t + A_1 \cos \omega t + \epsilon A_1^3 (\cos \omega t)^3 \simeq 0, \tag{8.9.30}$$

or

$$\left[-\omega^2 + 1 + \frac{3\epsilon A_1^2}{4} \right] A_1 \cos \omega t + \text{HOH} \simeq 0. \tag{8.9.31}$$

It is clear that the initial conditions

$$x(0) = A, \qquad \frac{dx(0)}{dt} = 0 \Rightarrow A_1 = A, \tag{8.9.32}$$

and setting the coefficient of the $\cos \omega t$ term to zero in Eq. (8.9.31) gives

$$\omega \simeq \sqrt{1 + \frac{3\epsilon A^2}{4}}. \tag{8.9.33}$$

Hence, the first approximation to the periodic solution of Eq. (8.9.28) is

$$x(t) = A \cos \left(\sqrt{1 + \frac{3\epsilon A^2}{4}} \, t \right). \tag{8.9.34}$$

It should be indicated that the derivation of this result does not depend on the assumption that ϵ is small. However, for $0 \leq \epsilon \ll 1$, we have

$$\omega = \sqrt{1 + \frac{3\epsilon A^2}{4}} \simeq 1 + \frac{3\epsilon A^2}{8}, \tag{8.9.35}$$

which is just the result gotten from use of either the first-order averaging method or the Lindstedt-Poincaré technique.

8.9.3   *Example C*

Our last example is the van der Pol equation

$$\frac{d^2x}{dt^2} + x = \epsilon(1 - x^2)\frac{dx}{dt}, \qquad \epsilon > 0. \tag{8.9.36}$$

Substituting the expressions in Eq. (8.9.4) into, respectively, the left- and right-sides of Eq. (8.9.36) gives

$$\frac{d^2x}{dt^2} + x \simeq (-\omega^2 + 1)A_1 \cos \omega t \tag{8.9.37}$$

$$\epsilon(1 - x^2)\frac{dx}{dt} \simeq \epsilon(1 - A_1^2 \cos^2 \omega t)(-A_1\omega \sin \omega t)$$

$$= -\epsilon A_1\omega \left[\left(1 - \frac{A_1^2}{4}\right) \sin \omega t + \text{HOH} \right]. \tag{8.9.38}$$

Placing these expressions into Eq. (8.9.36) gives

$$(-\omega^2 + 1)A_1 \cos \omega t + \epsilon A_1\omega \left(1 - \frac{A_1^2}{4}\right) \sin \omega t + \text{HOH} \simeq 0. \tag{8.9.39}$$

Setting the coefficients of $\cos \omega t$ and $\sin \omega t$ to zero and solving for ω and A_1 gives two solutions,

$$A_1 = 0, \qquad \omega = \text{arbitrary}, \tag{8.9.40}$$

$$A_1 = 2, \qquad \omega = 1. \tag{8.9.41}$$

The first solution corresponds to the fixed-point at $(x, y = dx/dt) = (0, 0)$, while the second solution gives approximations to the parameters for the limit-cycle solution, i.e.,

$$x(t) = 2 \cos t. \tag{8.9.42}$$

Note that this is exactly the first term in the Lindstedt-Poincaré derived perturbation solution and is also the same result obtained using the first-order averaging method for large times.

8.10   Averaging for Difference Equations

Difference equations [24] provide a source of discrete models for differential equations [25]. Just as for the class of nonlinear differential equations

$$\frac{d^2x}{dt^2} + x = \epsilon f\left(x, \frac{dx}{dt}\right), \tag{8.10.1}$$

corresponding methods of perturbation can be derived for second-order difference equations taking the form [26; 27]

$$\Gamma x_k = \epsilon F(x_{k+1}, x_k, x_{k-1}),\tag{8.10.2}$$

where ϵ is a small parameter, i.e.,

$$0 \leq \epsilon \ll 1;\tag{8.10.3}$$

k takes on integer values; the operator Γ is defined by the relation

$$\Gamma x_k \equiv \frac{x_{k+1} - 2x_k + x_{k-1}}{4\sin^2\left(\frac{h}{2}\right)} + x_k,\tag{8.10.4}$$

and h is a parameter that can be identified with the step-size, $h = \Delta t$.

This section presents an averaging type perturbation method for calculating $O(\epsilon)$ solutions to Eq. (8.10.2). A Lindstedt-Poincaré type perturbation method can also be constructed. See Mickens [27] for details of the procedure and an application.

We begin by introducing two discrete variables: k and $s = \epsilon k$, and assume that the solution to Eq. (8.10.2) takes the form [26]

$$x_k \equiv x(k, s, \epsilon) = x_0(k, s) + \epsilon x_1(k, s) + O(\epsilon^2),\tag{8.10.5}$$

where x_k is assumed to have at least a first partial derivative with respect to s. Therefore, we have

$$x_{k+1} = x(k+1, s+\epsilon, \epsilon) = x_0(k+1, s+\epsilon) + \epsilon x_1(k+1, s+\epsilon)$$
$$+ O(\epsilon^2),\tag{8.10.6}$$

$$x_0 = x(k+1, s+\epsilon) = x_0(k+1, s) + \epsilon\frac{\partial x_0(k+1, s)}{\partial s} + O(\epsilon^2),\tag{8.10.7}$$

$$x_1 = x(k+1, s+\epsilon) = x_1(k+1, s) + O(\epsilon).\tag{8.10.8}$$

Therefore, it follows that

$$x_{k+1} = x_0(k+1, s) + \epsilon\left[x_1(k+1, s) + \frac{\partial x_0(k+1, s)}{\partial s}\right] + O(\epsilon^2),\tag{8.10.9}$$

$$x_{k-1} = x_0(k-1, s) + \epsilon\left[x_1(k-1, s) - \frac{\partial x_0(k-1, s)}{\partial s}\right] + O(\epsilon^2).\tag{8.10.10}$$

Substituting these relations into Eq. (8.10.2) and setting the coefficients of the ϵ^0 and ϵ^1 terms equal to zero, gives the following two equations,

$$\Gamma x_0(k, s) = 0,\tag{8.10.11}$$

$$\Gamma x_1(k, s) = \left[\frac{1}{4\sin^2\left(\frac{h}{2}\right)}\right]\left[\frac{\partial x_0(k-1, s)}{\partial s} - \frac{\partial x_0(k+1, s)}{\partial s}\right]$$

$$+ F[x_0(k+1,s), x_0(k,s), x_0(k-1,s)]. \qquad (8.10.12)$$

The first of these equations has the general solution

$$x_0(k,s) = A(s)\cos(hk) + B(s)\sin(hk), \qquad (8.10.13)$$

where $A(s)$ and $B(s)$ are, at the present, unknown functions of s.

The substitution of Eq. (8.10.13) into the right-side of Eq. (8.10.12) gives

$$\Gamma x_1(k,s) = \left[\lambda \frac{dA}{ds} + M_1(A,B,h)\right]\sin(hk)$$

$$+ \left[-\lambda \frac{dB}{ds} + N_1(A,B,h)\right]\cos(hk)$$

$$+ \text{(higher-order harmonics)}, \qquad (8.10.14)$$

where

$$\lambda = \frac{\sin(h)}{2\sin^2\left(\frac{h}{2}\right)}, \qquad (8.10.15)$$

and M_1 and N_1 are obtained from the Fourier series expansion of the function

$$F[x_0(k+1,s), x_0(k,s), x_0(k-1,s)]$$

$$= \sum_{\ell=1}^{\infty}[M_\ell \sin(\ell nk) + N_\ell \cos(\ell hk)]. \qquad (8.10.16)$$

Note that the higher-order harmonics term on the right-side of Eq. (8.10.14) is the sum in Eq. (8.10.16) for $\ell \geq 2$. If $x_1(k,s)$ is to be bounded, i.e., contain no secular terms, then the coefficients of $\sin(hk)$ and $\cos(hk)$ must be zero. If otherwise, then the complete solution for $x_0(k,s)$ will contain terms such as $k\sin(hk)$ and $k\cos(hk)$. Therefore, we obtain two coupled, first-order, differential equations to be solved for the functions $A(s)$ and $B(s)$:

$$\lambda \frac{dA}{ds} + M_1(A,B,h) = 0, \qquad (8.10.17)$$

$$\lambda \frac{dB}{ds} - N_1(A,B,h) = 0. \qquad (8.10.18)$$

Substitution of these functions into Eq. (8.10.13) provides the required first approximation to the solution to Eq. (8.10.2).

Note that this procedure is similar to the averaging method for differential equations given in section 8.4.

8.11 Worked Examples for Difference Equations

8.11.1 *Example A*

Consider the following equation

$$\Gamma x_k + \epsilon x_k^3 = 0. \tag{8.11.1}$$

This is a discrete model for the differential equation [25]

$$\frac{d^2 x}{dt^2} + x + \epsilon x^3 = 0. \tag{8.11.2}$$

Thus, for this problem

$$F(x_{k+1}, x_k, x_{k-1}) = -x_k^3, \tag{8.11.3}$$

and

$$
\begin{aligned}
[x_0(k, s)]^3 &= [\cos(hk) + B \sin(hk)]^3 \\
&= \left(\frac{3}{4}\right) A(A^2 + B^2) \cos(hk) + \left(\frac{3}{4}\right) B(A^2 + B^2) \sin(hk) \\
&\quad + \left(\frac{1}{4}\right) A(A^2 - 3B^2) \cos(3hk) \\
&\quad + \left(\frac{3}{4}\right) B(3A^2 - B^2) \sin(3hk).
\end{aligned} \tag{8.11.4}
$$

By inspection, it follows

$$M_1 = -\left(\frac{3}{4}\right) B(A^2 + B^2), \qquad N_1 = -\left(\frac{3}{4}\right) A(A^2 + B^2), \tag{8.11.5}$$

and

$$\lambda \frac{dA}{ds} = \left(\frac{3}{4}\right) B(A^2 + B^2), \tag{8.11.6}$$

$$\lambda \frac{dB}{ds} = -\left(\frac{3}{4}\right) A(A^2 + B^2), \tag{8.11.7}$$

where λ is given by Eq. (8.10.15). Multiplying Eqs. (8.11.6) and (8.11.7), respectively, by A and B, and adding the resulting expressions gives

$$\frac{d}{ds} A(^2 + B^2) = 0, \tag{8.11.8}$$

or

$$A^2 + B^2 = \bar{a}^2 = \text{constant}. \tag{8.11.9}$$

If ω is defined as

$$\omega \equiv \frac{3\bar{a}^2}{4\lambda}, \qquad (8.11.10)$$

then Eqs. (8.11.6) and (8.11.7) become

$$\frac{dA}{ds} = \omega B, \qquad \frac{dB}{ds} = -\omega A. \qquad (8.11.11)$$

The solutions for these equations are

$$A(s) = \bar{a}\sin(\omega s + \phi), \qquad B(s) = \bar{a}\cos(\omega s + \phi), \qquad (8.11.12)$$

where ϕ is an arbitrary constant. If these results are now substituted into Eq. (8.10.13), the following is obtained

$$x_0(k, s) = \bar{a}\sin(\omega s + hk + \phi). \qquad (8.11.13)$$

Using $s = \epsilon k$, we finally get

$$x_k = x_0(k, s) + O(\epsilon)$$

$$= \bar{a}\sin\left\{\left[1 + \left(\frac{3\bar{a}^2\epsilon}{2}\right)\frac{\sin^2(h/2)}{h\sin(h)}\right]t_k + \phi\right\} + O(\epsilon), \qquad (8.11.14)$$

where $t_k = hk$. Observe that as

$$h \to 0, \quad k \to \infty, \quad hk = t = \text{fixed}, \qquad (8.11.15)$$

the result of Eq. (8.11.14) goes to the function

$$x(t) = \bar{a}\sin\left[\left(1 + \frac{3\bar{a}^2\epsilon}{8}\right)t + \phi\right], \qquad (8.11.16)$$

which is just the first-order perturbation solution for Eq. (8.11.2).

8.11.2 Example B

A discrete model for the van der Pol equation

$$\frac{d^2x}{dt^2} + x = \epsilon(1 - x^2)\frac{dx}{dt}, \qquad (8.11.17)$$

is

$$\Gamma x_k = \epsilon(1 - x_k^2)\left(\frac{x_{k+1} - x_{k-1}}{2h}\right). \qquad (8.11.18)$$

A direct, but long, calculation gives

$$\frac{dA}{ds} = -\left(\frac{\lambda_2}{2}\right)A(A^2 + B^2 - 4), \qquad (8.11.19)$$

$$\frac{dB}{ds} = -\left(\frac{\lambda_2}{2}\right) B(A^2 + B^2 - 4), \tag{8.11.20}$$

where

$$\lambda_2 = \frac{\sin(h)}{2\lambda h}. \tag{8.11.21}$$

If we define z as

$$z = A^2 + B^2,$$

multiply Eq. (8.11.19) and (8.11.20), respectively, by A and B, and add the resulting expressions, then the following differential equation is obtained for z,

$$\frac{dz}{ds} = -\lambda_2 z(z - 4). \tag{8.11.22}$$

For $z(0) = z_0$, the solution is

$$z(s) = \frac{4z_0}{[z_0 - (z_0 - 4)\exp(-4\lambda_2 s)]}. \tag{8.11.23}$$

Now,

$$x_0(k, s) = A(s)\cos(hk) + B(s)\sin(hk) = a(s)\cos[hk + \phi(s)], \tag{8.11.24}$$

where

$$a^2 = A^2 + B^2 = z, \qquad \tan\phi = \frac{B}{A}. \tag{8.11.25}$$

Since,

$$\frac{dA}{dB} = \frac{A}{B}, \tag{8.11.26}$$

we have

$$A(s) = cB(s), \tag{8.11.27}$$

where c is an arbitrary constant. Consequently,

$$\tan\phi = -\frac{1}{c}, \tag{8.11.28}$$

and the first-order solution to the discrete van der Pol equation is

$$x_k = \frac{2a_0 \cos(hk + \phi_0)}{\sqrt{a_0^2 - (a_0^2 - 4)\exp(-4\lambda_2 s)}}, \tag{8.11.29}$$

where ϕ_0 is an arbitrary constant. Note that

$$x_k \xrightarrow[\text{large } k]{} 2\cos(t_k + \phi_0). \tag{8.11.30}$$

Problems

Section 8.2

8.2.1 Consider the following differential-integral equation

$$\frac{dx}{dt} + 3x = 10\cos\left(\sqrt{2}t\right) + \epsilon \int_0^1 [\sin(s-t)]^2 [x(s)]^3 \, ds.$$

Identify L, v, and N for this equation.

8.2.2 For the van der Pol differential equation $N(u)$ is

$$N(u) = (1 - u^2)\frac{du}{dt}.$$

Using the expansion

$$u = u_0 + \epsilon u_1 + \epsilon^2 u_2 + \cdots,$$

calculate $N_0(u_0)$, $N_1(u_0, u_1)$ and $N_2(u_0, u_1, u_2)$ in

$$N(u_0 + \epsilon u_1 + \epsilon^2 u_2 + \cdots) = N_0 + \epsilon N_1 + \epsilon^2 N_2 + \cdots.$$

Section 8.3

8.3.1 For the Example A, carry out the calculation such that terms to u_5 are obtained.

8.3.2 Plot $x^{(+)}(\epsilon)$ and $x^{(-)}(\epsilon)$ versus ϵ for positive values of ϵ. Their explicit forms are given in Eq. (8.3.7).

8.3.3 Discuss the qualitative properties of the solutions to the logistic Eq. (8.3.13) for all non-negative initial conditions. For the perturbation calculations, why is the condition on a given in Eq. (8.3.20) needed? What happens for the case $a\epsilon > 1$?

8.3.4 Work out the details of the calculation to go from Eq. (8.3.30) to (8.3.31).

Section 8.4

8.4.1 Use a physical argument to justify the derivation of the first approximation of Krylov and Bogoliubov. Hint: See ref. [6] and [7].

8.4.2 Prove the result given by Eq. (8.4.29).

8.4.3 For the case where $R = 0$, how should Eq. (8.4.43) be modified? What can one conclude in this case?

Section 8.5

Construct solutions to the following three equations using the first-order averaging method. For each case plot $x(t)$ vs t, $dx(t)/dt$ vs t, and $y(t) = dx/dt$ vs $x(t)$. The latter plot is of the trajectories in phase space. Use physical reasoning to explain what you find.

8.5.1 $\dfrac{d^2x}{dt^2} + x + \epsilon x^3 = -2\epsilon \dfrac{dx}{dt}$.

8.5.2 $\dfrac{d^2x}{dt^2} + x = -\epsilon \left| \dfrac{dx}{dt} \right| \dfrac{dx}{dt}$.

8.5.3 $\dfrac{d^2x}{dt^2} + x = \epsilon(1 - |x|)\dfrac{dx}{dt}$. This is the Lewis equation [8] and has a limit-cycle. Is the limit-cycle stable or unstable?

8.5.4 Evaluate the integrals on the right-sides of Eqs. (8.5.19) and (8.5.20).

8.5.5 Use the arguments presented in section 8.4.3 to show that the limit-cycles for both the Lewis and van der Pol equations are stable.

Section 8.6

8.6.1 Show that Eq. (8.6.12) is the complete solution to Eq. (8.6.11)

8.6.2 Work out the details to obtain the results given in Eqs. (8.6.22) to (8.6.24).

Section 8.7

8.7.1 Apply the Lindstedt-Poincaré method to the linear damped oscillator equation

$$\frac{d^2x}{dt^2} + 2\epsilon \frac{dx}{dt} + x = 0.$$

Explain your results both mathematically and in terms of the physics of a system modeled by this differential equation.

8.7.2 Examine closely Eq. (8.7.13) and (8.7.14), and show that the "true" expansion parameter is ϵA^2 rather than ϵ.

8.7.3 Do the calculations necessary to determine A_2 appearing in Eq. (8.7.27).

Section 8.8

8.8.1 Explain why the truncated Fourier series for Eq. (8.8.5), under the conditions given by Eq. (8.8.16) only contain cosine terms.

8.8.2 Select a set of initial conditions for Eq. (8.8.5) such that only sine terms will appear in the Fourier series solutions.

8.8.3 What is the reason for $\bar{B}_1^n = 0$ in Eq. (8.8.14)?

Section 8.9

8.9.1 Derive the expressions given in Eqs. (8.9.15) and (8.9.16) and then to calculate the results for ω and x of, respectively, Eqs. (8.9.18) and (8.9.20).

8.9.2 Calculate the exact values for the three roots of Eq. (8.9.20). Compare the magnitude of the smallest root with its approximate value given in Eq. (8.9.22)

8.9.3 Using a mathematics handbook, locate the Fourier representation for $x(t) = A \cdot \mathrm{cn}(At; 1\sqrt{2})$ and compare its first two terms with the result of Eq. (8.9.24). Carry out a comparison of the exact and approximate angular frequencies.

8.9.4 Construct a first-order harmonic balance solution to the equation

$$\frac{d^2x}{dt^2} + |x|x = 0.$$

Hint: See Mickens and Mixon [16].

8.9.5 Consider the differential equation of the previous problem and construct a second-order harmonic balance approximation to the solution using the rational form

$$x(t) \simeq \frac{A_1 \cos \omega t}{1 + B_1 \cos 2\omega t},$$

with the initial conditions $x(0) = A$ and $dx(0)/dt = 0$. Hint: See Mickens [7], section 4.5.2.

8.9.6 The following differential equation is called the Duffing-harmonic oscillator

$$\frac{d^2x}{dt^2} + \frac{x^3}{1 + x^2} = 0, \qquad x(0) = A, \qquad \frac{dx(0)}{dt} = 0.$$

Determine the first integral and use this to show that all solutions are periodic. Apply first-order harmonic balance to obtain an approximation for the periodic solution. Plot $\omega(A)$ vs A and examine its behavior for both small and large A.

Section 8.10

8.10.1 If x_k is taken to be an approximation for $x(t_k)$ where $t_k = hk$, then show that

$$\underset{\substack{h \to 0 \\ k \to 0 \\ hk=t=\text{fixed}}}{\text{Lim}} \frac{x_{k+1} - 2x_k + x_{k-1}}{4 \sin^2 \left(\frac{h}{2}\right)} = \frac{d^2 x}{dt^2}.$$

8.10.2 Derive the results given by Eqs. (8.10.9) and (8.10.10).

8.10.3 Show that the equation for determining $x_1(k, s)$ has the structure given by Eq. (8.10.14).

8.10.4 Prove that if the conditions given by Eqs. (8.10.17) and (8.10.18) does not hold, then the complete solution for $x_1(k, s)$ contains terms of the form $k \sin(hk)$ and $k \cos(hk)$.

Section 8.11

8.11.1 Take the limits: $h \to 0$, $k \to \infty$, $hk = t = $ fixed, for Eq. (8.11.1) and show that it reduces to Eq. (8.11.2).

8.11.2 Solve Eqs. (8.11.11) for $A(s)$ and $B(s)$.

8.11.3 Derive the result given in Eq. (8.11.13).

8.11.4 Another valid discrete model for Eq. (8.11.2) is

$$\Gamma x_k + \epsilon \left(\frac{x_{k+1} + x_{k-1}}{2}\right) x_k^2 = 0.$$

Construct the solution $x_0(k, s)$ for this model. How do these results compare with that given by Eq. (8.11.14)?

8.11.5 Show that

$$\underset{\substack{h \to 0 \\ k \to \infty \\ hk=t=\text{fixed}}}{\text{Lim}} \frac{x_{k+1} - x_{k-1}}{2h} = \frac{dx}{dt}.$$

8.11.6 Derive the differential equation for z, as given by Eq. (8.11.22) from Eqs. (8.11.19) and (8.11.20). Solve this equation to obtain (8.11.23).

8.11.7 Show that the result of Eq. (8.11.27) is a consequence of Eq. (8.11.26).

8.11.8 A possible discrete model for the linear damped oscillator

$$\frac{d^2 x}{dt^2} + 2\epsilon \frac{dx}{dt} + x = 0,$$

is

$$\Gamma x_k = -2\epsilon \left(\frac{x_{k+1} - x_{k-1}}{2h}\right).$$

Calculate $x_0(k, s)$ and compare with the exact solution for the differential equation.

Comments and References

[1] The following books by Rand and his collaborators provide an indication of what can be done for perturbation analysis with the use of computer algebra:

(i) R. H. Rand, *Computer Algebra in Applied Mathematics: An Introduction to MACSYMA* (Pitman, Boston, 1984).

(ii) R. H. Rand and D. Armbruster, *Perturbation Methods, Bifurcation Theory and Computer Algebra* (Springer, Berlin, 1987).

(iii) R. H. Rand, *Topics in Nonlinear Dynamics with Computer Algebra* (Gordon and Breach; Langhorne, PA; 1994).

[2] R. Bellman, *Perturbation Techniques in Mathematics, Physics, and Engineering* (Holt, Reinhart and Winston; New York, 1964). See pp. 5.

[3] A. W. Bush, *Perturbation Methods for Engineers and Scientists* (CRC Press; Boca Raton, FL; 1992).

[4] J. A. Murdock, *Perturbations: Theory and Methods* (Wiley-Interscience, New York, 1991).

[5] P. A. Lagerstrom, *Matched Asymptotic Expansions* (Springer-Verlag, New York, 1988).

[6] N. Krylov and N. Bogoliubov, *Introduction to Nonlinear Mechanics* (Princeton University Press; Princeton, NJ; 1943).

[7] R. E. Mickens, *Oscillations in Planar Dynamic Systems* (World Scientific, Singapore, 1996).

[8] J. B. Lewis, *Trans. Am. Institute of Electrical Engineers, Part II* **72**, 449 (1953).

[9] H. Poincaré, *New Methods in Celestial Mechanics, Vols. I, II, and III* (English translation, NASA Reports TTF-450, -451, -452; 1967).

[10] A Lindstedt, *Astron. Nach.* **103**, 211 (1882).

[11] K. Huseyin and R. Lin, *Int. J. Nonlinear Mechanics* **26**, 727 (1991).

[12] F. F. Seelig, *Z. Naturforsch* **35a**, 1054 (1980); **38a**, 636 (1983); **38a**, 729 (1983).

[13] J. Garcia-Margallo et al., *J. Sound and Vibration* **116**, 591 (1987); **125**, 13 (1988); **136**, 453 (1990).

[14] S. Bravo Yuste, *J. Sound and Vibration* **130**, 33 (1989); **145**, 381 (1991).

[15] J. L. Summers and M. D. Savage, *Phil. Trans. Royal Soc. London A* **340**, 473 (1992).

[16] R. E. Mickens and M. Mixon, *J. Sound and Vibration* **159**, 546 (1992).

[17] See ref. [7], Chapter 4: Harmonic Balance.

[18] C. A. Borges, L. Cesari, and D. A. Sanchez, *Q. Appl. Math.* **32**, 457 (1975).

[19] H. Leipholz, *Direct Variational Methods and Eigenvalue Problems in Engineering* (Noordhoff International Publishing, Leyden, 1975). See section 4.2.

[20] P. Miletta, in R. Chuagui (editor), *Analysis, Geometry and Probability* (Marcel Dekker, New York, 1985). See pps. 1–12.

[21] M. Urable, *Arch. Rat. Mech. Analy.* **20**, 120 (1965).

[22] A. Stokes, *J. Diff. Eq.* **12**, 535 (1972).

[23] R. E. Mickens, *J. Sound and Vibration* **258**, 398 (2000).

[24] R. E. Mickens, *Difference Equations: Theory and Applications* (Chapman and Hall, New York, 1990).

[25] R. E. Mickens, *Nonstandard Finite Difference Models of Differential Equations* (World Scientific, Singapore, 1994).

[26] R. E. Mickens, *J. Franklin Institute* **324**, 263 (1987).

[27] R. E. Mickens, *J. Difference Eqs. and Applications* **6**, 337 (2000).

Bibliography

R. Bellman, *Perturbation Techniques in Mathematics, Physics, and Engineering* (Holt, Reinhart and Winston; New York, 1964).

C. M. Bender and S. A. Orszag, *Advanced Mathematical Methods for Scientists and Engineering* (McGraw-Hill, New York, 1978).

A. W. Bush, *Perturbation Methods for Engineering and Scientists* (CRC Press; Boca Raton, FL; 1992).

J. D. Cole, *Perturbation Methods in Applied Mathematics* (Blaisdell; Waltham, MA; 1968).

M. Farkas, *Periodic Motions* (Springer-Verlag, New York, 1994).

J. Kevorkian and J. D. Cole, *Perturbation Methods in Applied Mathematics* (Springer-Verlag, New York, 1981).

R. E. Mickens, *Nonlinear Oscillations* (Cambridge University Press, New York, 1981).

A. H. Nayfeh, *Perturbation Methods* (Wiley, New York, 1973).

A. H. Nayfeh, *Introduction to Perturbation Techniques* (Wiley, New York, 1981).

R. E. O'Malley, *Introduction to Singular Perturbations* (Academic Press, New York, 1974).

D. R. Smith, *Singular Perturbation Theory* (Cambridge University Press, Cambridge, 1985).

J. J. Stokes, *Nonlinear Vibrations* (Wiley-Interscience, New York, 1950).

M. Urabe, *Nonlinear Autonomous Oscillations* (Academic, New York, 1967).

M. Van Dyke, *Perturbation Methods in Fluid Mechanics* (Parabolic Press; Palo Alto, CA; 1975).

Chapter 9

Approximations of Integrals and Sums

Functions defined by integrals often appear in the investigation of dynamical systems. For example, in chemistry, reaction rate coefficients take the form [1]

$$K(\beta) = A(\beta) \int_0^\infty \sigma(x) e^{-\beta x} dx,$$

where, $\beta > 0$, is inversely related to the temperature and the functions $A(\beta)$ and $\sigma(x)$ are specified for a given type of reaction. This particular expression has been used to study the behaviors of $K(\beta)$ in the limits $\beta \to 0$ and $\beta \to \infty$ [2].

Other integrals having this same structure also appear in a variety of situations arising in the natural and physical sciences, and applied mathematics. Part of the task of this chapter is to present results on functions defined by these kind of integrals and give formula for determining the leading term in their expansions when the relevant variable becomes large. We begin by briefly presenting the major results on the inverse-power set of expansion functions, i.e., $\{x^{-n} : n = 0, 1, 2, \dots\}$, and how they can be used to construct large-x representations for many of the functions defined by integrals. Next, we show that asymptotic series can often be obtained by the repeated use of integration by parts. Several examples are used to illustrate his procedure. In section 9.3, the general Laplace method is discussed along with Watson's lemma. Finally, we present the Euler-Maclaurin sum formula and show how it can be used to evaluate various sums. In all of these discussions, no proofs are provided. However, we do indicate good, introductory references which do go into the details of the required proofs and the restrictions that must be applied for their application.

9.1   Resume of Asymptotics

Consider the infinite sequence of continuous functions, $\{\phi_n(x) : n = 1, 2, \ldots\}$, and let x_0 be a point lying on the interval of definition of this sequence of functions. This sequence is called an asymptotic sequence if for each fixed n

$$\phi_{n+1}(x) = o(\phi_n(x)), \qquad x \to x_0. \tag{9.1.1}$$

Let $f(x)$ be a given function, then the following is called an asymptotic series for $f(x)$ at x_0, if

$$f(x) = a_0\phi_0(x) + a_1\phi_1(x) + \cdots + a_n\phi_n(x) + +O(\phi_{n+1}(x)). \tag{9.1.2}$$

The notation

$$f(x) \sim \sum_{n=0}^{\infty} a_n\phi_n(x), \tag{9.1.3}$$

indicates that the formal series on the right-side is the asymptotic series for $f(x)$ at x_0.

Note that the expression on the right-side of Eq. (9.1.3) may not converge and this property is generally not required for the construction and evaluation of asymptotic series. If we write Eq. (9.1.4) as

$$f(x) = f_n(x) + O(\phi_{n+1}(x)), \tag{9.1.4}$$

then it follows that

$$\operatorname*{Lim}_{x \to x_0} \left[\frac{f(x) - f_n(x)}{\phi_n(x)} \right] = 0, \tag{9.1.5}$$

and the coefficients, $\{a_n : n = 0, 1, 2, \ldots\}$, are formally calculated by use of the relation

$$a_n = \operatorname*{Lim}_{x \to x_0} \left[\frac{f(x) - f_{n-1}(x)}{\phi_n(x)} \right], \qquad f_{-1}(x) = 0. \tag{9.1.6}$$

Asymptotic sequences can take many forms and generally depend on the value of x_0. For example, if $x_0 = 0$, then an asymptotic sequence is

$$x_0 = 0 : \phi_n(x) = x^n, \qquad (n = 0, 1, 2, \ldots); \tag{9.1.7}$$

however, for $x_0 = \infty$, a useful asymptotic sequence is

$$x_0 = \infty : \phi_n(x) = \frac{1}{x^n}, \qquad (n = 0, 1, 2, \ldots). \tag{9.1.8}$$

The following properties of asymptotic series are of importance to the issues considered in this chapter:

(i) For a given asymptotic sequence, the representation of $f(x)$ is unique. However, the exact representation does depend on the particular asymptotic sequence selected. To show this, consider

$$\frac{1}{x-1} \sim \sum_{n=1}^{\infty} \frac{1}{x^n}, \qquad x \to \infty, \tag{9.1.9}$$

and

$$\frac{1}{x-1} = \frac{x+1}{x^2-1} \sim \sum_{n=1}^{\infty} \frac{x+1}{x^{2n}}, \qquad x \to \infty. \tag{9.1.10}$$

(ii) Two different functions $f_1(x)$ and $f_2(x)$ may have the same asymptotic representation. This result is a consequence of the fact that the asymptotic representation depends on the asymptotic sequence selected to represent the functions. For example,

$$f_1(x) = f(x), \qquad f_2(x) = f(x) + e^{-x} \tag{9.1.11}$$

have exactly the same asymptotic representation in terms of the asymptotic sequence $\{\phi_n(x) = x^{-n} : n = 0, 1, 2, \dots\}$, as $x \to \infty$.

(iii) The linear combining of two asymptotic representations of two functions is the asymptotic representation of the combined functions. This means that if

$$f_1(x) \sim \sum_{n=0}^{\infty} a_n \phi_n(x), \qquad f_2(x) \sim \sum_{n=0}^{\infty} b_n \phi_n(x), \tag{9.1.12}$$

then

$$c_1 f_1(x) + c_2 f_2(x) \sim \sum_{n=0}^{\infty} (c_1 a_n + c_2 b_n) \phi_n(x), \tag{9.1.13}$$

where c_1 and c_2 are arbitrary constants.

(iv) Let $f(x)$ be continuous and represented as

$$f(x) \sim \sum_{n=0}^{\infty} \frac{a_n}{x^n}, \qquad x \to \infty. \tag{9.1.14}$$

Then

$$F(x) = \int_x^{\infty} \left[f(t) - a_0 - \frac{a_1}{t} \right] dt \sim \sum_{n=1}^{\infty} \frac{a_{n+1}}{n x^n}, \qquad x \to \infty. \tag{9.1.15}$$

The proof of this last result is straightforward and we give it. If

$$f(x) \sim \sum_{n=0}^{\infty} \frac{a_n}{x^n}, \tag{9.1.16}$$

this means that

$$f(x) = \sum_{n=0}^{k} \frac{a_n}{x^n} + O\left(\frac{1}{x^{k+1}}\right). \tag{9.1.17}$$

Therefore,

$$F(x) = \int_0^\infty \left[\sum_{n=2}^{k} \frac{a_n}{t^n} + O\left(\frac{1}{t^{k+1}}\right)\right] dt$$

$$= \sum_{n=2}^{k} \frac{a_n}{(n-1)x^{n-1}} + \int_x^\infty O\left(\frac{1}{t^{k+1}}\right) dt, \tag{9.1.18}$$

and therefore a constant $A > 0$ exists such that

$$\left|\int_x^\infty O\left(\frac{1}{t^{k+1}}\right) dt\right| \leq \int_x^\infty \frac{A}{t^{k+1}} = \frac{A}{kx^k} = O\left(\frac{1}{x^k}\right), \tag{9.1.19}$$

and, consequently, we have

$$F(x) = \sum_{n=1}^{k-1} \frac{a_{n+1}}{nx^n} + O\left(\frac{1}{x^k}\right). \tag{9.1.20}$$

Based on the definition of an asymptotic expansion, see Eqs. (9.1.2) and (9.1.3), we conclude that the integral of $f(x)$ is $F(x)$.

Finally, it should be indicated that the main value of an asymptotic expansion of a function is that for computational purposes only a small number of terms need to be used to obtain numerically accurate values. This follows from the fact that if $f_n(x)$ is asymptotic to $f(x)$, then for the asymptotic sequence, $\{\phi_n(x) = x^{-n} : n = 0, 1, 2, \ldots\}$, we have

$$\lim_{x \to \infty} \{x^n[f(x) - f_n(x)]\} = 0, \qquad n = \text{fixed.} \tag{9.1.21}$$

Thus, the full asymptotic representation may or may not converge. Since we only use a finite value of n, questions related to convergence do not arise. In practice, for many problems one or two terms are all that is required to either have an accurate numerical solution or to illustrate the required physical phenomena.

9.2   Integration by Parts

For functions defined by integrals, often the most direct way of obtaining an asymptotic representation is by way of applying the method of integration by parts to the original integral. We illustrate the use of this method by considering three examples.

The exponential integral is

$$E(x) = \int_x^\infty \frac{e^{-t}}{t} \, dt. \tag{9.2.1}$$

If we integrate once by parts, then the following result is obtained

$$E(x) = \frac{e^{-x}}{x} - \int_x^\infty \frac{e^{-t}}{t^2} \, dt. \tag{9.2.2}$$

Repeating the integration by parts n-times gives

$$E(x) = e^{-x} \sum_{k=1}^n (-1)^{k+1} \left[\frac{(k-1)!}{x^k} \right] + (-1)^n n! \int_x^\infty \frac{e^{-t}}{t^{n+1}} \, dt. \tag{9.2.3}$$

Define $R_n(x)$ to be

$$R_n(x) \equiv (-1)^n n! \int_x^\infty \frac{e^{-t}}{t^{n+1}} \, dt; \tag{9.2.4}$$

then $R(x)$ is bounded for large values of x by the expression

$$|R_n(x)| \leq \left(\frac{n!}{x^{n+1}} \right) \int_x^\infty e^{-t} dt = \frac{n! e^{-x}}{x^{n+1}}, \tag{9.2.5}$$

and for n-fixed, $x \to \infty$, we have

$$|R_n(x)| = O\left(\frac{1}{x^{n+1}} \right). \tag{9.2.6}$$

Thus, the asymptotic representation for $E(x)$ is

$$E(x) \sim e^{-x} \sum_{n=1}^\infty (-1)^{n+1} \frac{(n-1)!}{x^n} \tag{9.2.7}$$

and

$$E(x) = e^{-x} \sum_{k=0}^n (-1)^{k+1} \frac{(k-1)!}{x^k} + O\left(\frac{1}{x^{n+1}} \right). \tag{9.2.8}$$

Now consider the incomplete gamma function defined as follows

$$\gamma(a, x) \equiv \int_0^x e^{-t} t^{a-1} dt, \tag{9.2.9}$$

where $a > 0$. Note that the "complete" gamma function is

$$\Gamma(a) = \int_0^\infty e^{-t} t^{a-1} dt, \qquad a > 0. \tag{9.2.10}$$

For large x, we choose to write $\gamma(a, x)$ as

$$\gamma(a, x) = \int_0^\infty e^{-t} t^{a-1} dt - \int_x^\infty e^{-t} t^{a-1} dt$$

$$= \Gamma(a) - E(a, x) \tag{9.2.11}$$

where $E(a, x)$ is the second integral in the first line of Eq. (9.2.11). Now using integration by parts, we obtain

$$E(a, x) = \int_x^\infty e^{-t} t^{a-1} dt$$

$$= e^{-x} x^{a-1} + (a - 1) \int_a^\infty e^{-t} t^{a-2} dt$$

$$= e^{-x} \{ x^{a-1} + (a - 1) x^{a-2} + \cdots$$
$$+ (a - 1)(a - 2) \cdots (a - n + 1) x^{a-n} \}$$
$$+ (a - 1)(a - 2) \cdots (a - n) \int_x^\infty e^{-t} t^{a-n-1} dt, \tag{9.2.12}$$

where the last line shows the result of n repeated integrations by parts. Let $R_n(x)$ be the last term in Eq. (9.2.12). Then for fixed $n > a - 1$, we have

$$|R_n(x)| \le (a - 1)(a - 2) \cdots (a - n) \int_x^\infty e^{-t} t^{a-n-1} dt$$

$$< (a - 1)(a - 2) \cdots (a - n) x^{a-n-1} \int_x^\infty e^{-t} dt$$

$$= (a - 1)(a - 2) \cdots (a - n) x^{a-n-1} e^{-x}$$

$$= O(x^{a-n-1} e^{-x}). \tag{9.2.13}$$

This calculation shows that the asymptotic expansion for the incomplete gamma function is

$$\gamma(a, x) \sim \Gamma(a) - e^{-x} x^a \left[\frac{1}{x} + \frac{(a - 1)}{x^2} + \frac{(a - 1)(a - 2)}{x^3} + \cdots \right]. \tag{9.2.14}$$

For a third example consider

$$I(x) = \int_0^\infty \frac{e^{-t} dt}{x + t}. \tag{9.2.15}$$

Our interest is when x is positive and large. One integration by parts gives

$$I(x) = \frac{1}{x} - \int_0^\infty \frac{e^{-t} dt}{(x + t)^2}. \tag{9.2.16}$$

Repeating n-times the integration by parts leads to the expression

$$I(x) = \frac{1}{x} - \frac{1}{x^2} + \frac{2!}{x^3} - \frac{3!}{x^4} + \cdots + (-1)^n \left[\frac{n!}{x^{n+1}} \right]$$

$$+ (1-)^{n+1} (n + 1)! \int_0^\infty \frac{e^{-t} dt}{(x + t)^{n+2}}. \tag{9.2.17}$$

Calling $R_n(x)$ the last term, we obtain

$$|R_n(x)| = (n+1)! \int_0^\infty \frac{e^{-t} dt}{(x+t)^{n+2}}.$$ (9.2.18)

The change of variable $t = xy$ gives

$$|R_n(x)| = \left[\frac{(n+1)!}{x^{n+1}} \right] \int_0^\infty \frac{e^{-xy} dy}{(1+y)^{n+2}}$$

$$\leq \left[\frac{(n+1)!}{x^{n+1}} \right] \int_0^\infty e^{-xy} dy = \frac{(n+1)!}{x^{n+2}}.$$ (9.2.19)

From the definition of an asymptotic representation, it follows from the above calculation that the asymptotic series for $I(x)$ is

$$I(x) \sim \frac{1}{x} - \frac{1}{x^2} + \frac{2!}{x^3} - \frac{3!}{x^4} + \cdots + (-1)^n \left(\frac{n!}{x^{n+1}} \right) + \cdots$$ (9.2.20)

and

$$I(x) = \left(\frac{1}{x} \right) \sum_{k=0}^n (-1)^k \left(\frac{k!}{x^k} \right) + O\left(\frac{1}{x^{n+1}} \right).$$ (9.2.21)

To show that things can go "wrong" using the integration by parts method, we now examine a problem for which this technique does not work. Consider the integral

$$B(x) = \int_0^\infty e^{-xt^2} dt > 0,$$ (9.2.22)

where

$$\frac{dB(x)}{dx} = - \int_0^\infty t^2 e^{-xt^2} dt > 0$$ (9.2.23)

and

$$B(0) = +\infty, \qquad B(\infty) = 0.$$ (9.2.24)

It follows that $B(x)$ is a monotonic, decreasing, positive function that is unbounded at $x = 0$. The integration by parts procedure can be started by writing $B(x)$ as

$$B(x) = \int_0^\infty \left(\frac{1}{-2xt} \right) (-2xte^{-xt^2}) dt$$

$$= \frac{e^{-xt^2}}{(-2xt)} \Bigg|_{t=0}^{t=\infty} - \int_0^\infty \left(\frac{e^{-xt^2}}{2xt^2} \right) dt.$$ (9.2.25)

Note that both the boundary term and the integral on the right-side do not exist. Consequently, no expansion in inverse integer powers of x exist for $B(x)$. This is in contrast to the previous three examples studied above. To highlight what the difficulty is for this case, we examine $B(x)$ again. In fact, it can be evaluated exactly by using the change of variable

$$y = xt^2 \qquad (9.2.26)$$

to obtain

$$B(x) = \left(\frac{1}{2\sqrt{x}}\right) \int_0^\infty y^{\frac{1}{2}-1} e^{-y} dy = \frac{\Gamma\left(\frac{1}{2}\right)}{2\sqrt{\pi}} = \left(\frac{1}{2}\right)\sqrt{\frac{\pi}{x}}. \qquad (9.2.27)$$

Inspection of Eq. (9.2.28) clearly shows that $B(x)$ cannot have an asymptotic expansion of the form

$$B(x) \sim \sum_{n=0}^\infty \frac{a_n}{x^n}. \qquad (9.2.28)$$

A general question can now be raised: Given a particular integral, how can we know whether or not the integration by parts methods will work? From the just studied example, it follows that whenever this procedure leads to either boundary terms or integrals that do not exist, then the integration by parts methods cannot be applied.

9.3   Laplace Methods

This section presents formulas for determining the asymptotic expansions of integrals having the form

$$I(x) = \int_a^b f(t) e^{x\phi(t)} dt, \qquad (9.3.1)$$

where the limits of integration may be unbounded. Proofs for Watson's lemma and the general Laplace integrals are given in the books by Bender and Orszag, Keener, and Murray.

9.3.1   *Watson's Lemma*

The integral

$$I(x) = \int_0^T f(t) e^{-xt} dt, \qquad T > 0, \qquad (9.3.2)$$

where T can be unbounded, is called a Laplace integral. The full asymptotic expansion for $I(x)$ can be determined provided that $f(t)$ has certain properties. The general expression for the expansion is called Watson's lemma.

Watson's Lemma Let $f(t)$ have the following properties:
(i) $f(t)$ is continuous on $0 \le t \le T$;
(ii) $f(t)$ has the asymptotic series expansion

$$f(t) \sim t^\gamma \sum_{n=0}^\infty a_n t^{\beta_n}, \qquad t \to 0^+, \qquad (9.3.3)$$

where

$$\gamma > -1; \qquad \beta_n > 0, \quad (n = 0, 1, 2, \dots). \qquad (9.3.4)$$

(iii) if $T = \infty$, then

$$|f(t)| < M e^{bt} \qquad (9.3.5)$$

for some positive constants M and c.

Under these conditions, the full asymptotic series expansion for $I(x)$, as given by Eq. (9.3.2) is

$$I(x) \sim \sum_{n=0}^\infty \frac{a_n \Gamma(\gamma + \beta_n + 1)}{x^{\gamma + \beta_n + 1}}, \qquad x \to +\infty. \qquad (9.3.6)$$

Note that if T is finite, its value does not appear in the asymptotic series representation.

9.3.2 *Laplace's Method for Integrals*

This method can be used to easily calculate the leading term in the asymptotic series for integrals having the form

$$I(x) = \int_{-\infty}^\infty f(t) e^{x\phi(t)} dt. \qquad (9.3.7)$$

Laplace's Theorem for Integrals Let $f(t)$ have the properties:
(i) $f(t)$ is continuous in a neighborhood of $t = 0$;
(ii) $f(0) \ne 0$.
Let $\phi(t)$ have the properties:
(i) $\phi(t)$ is a real continuous function;
(ii) $\phi(0) = 0$ and $\phi(t) < 0$ for $t \ne 0$;
(iii) there exist positive numbers a and b, such that $a > b$, and $\phi(t) \le -a$ when $|t| \ge b$:
(iv) there exists a neighborhood of $t = 0$ for which $\phi(t)$ has a second derivative and $\phi''(0) < 0$;
(v) the integral $I(x)$, in Eq. (9.3.7) is finite for each fixed $x > 0$.

Under these conditions, the leading term in the asymptotic series expansion for $I(x)$ is

$$I(x) \sim f(0) \left[\frac{2\pi}{-x\phi''(0)} \right]^{1/2}, \qquad x \to \infty. \qquad (9.3.8')$$

This result can be generalized as follows: Let $\phi(t)$ have a local maximum at $t = t_0$. This means that

$$\phi(t) = \phi(t_0) + \left[\frac{\phi^{(n)}(t_0)}{n!} \right] (t - t_0)^n + \cdots, \qquad (9.3.9)$$

where n is an even integer and $\phi^{(n)}(t_0) < 0$. Then,

$$I(x) \sim 2f(t_0)\Gamma\left(\frac{n+1}{n} \right) \left[\frac{n!}{-x\phi^{(n)}(t_0)} \right]^{1/n} e^{\phi(t_0)x}. \qquad (9.3.10)$$

9.4   Worked Examples

9.4.1   *Stirling's Formula*

The gamma function for $\Gamma(x + 1)$ is

$$\Gamma(x + 1) = \int_0^\infty e^{-t} t^x dt = \int_0^\infty e^{-t} e^{x \ln t} dt = \int_0^\infty e^{(x \ln t - t)} dt. \qquad (9.4.1)$$

The maximum of the integrand is at $t = x$. Making the substitution $t = (y + 1)x$ gives

$$\Gamma(x + 1) = x^{x+1} e^{-x} \int_{-1}^\infty e^{x[\ln(1+y) - y]} dy, \qquad (9.4.2)$$

and the integral is of the form given by Eq. (9.3.7) where

$$f(t) = 1, \qquad \phi(t) = \ln(1 + t) - t, \qquad (9.4.3)$$

and all the conditions, as specified for the Laplace theorem for integrals, are satisfied. Note that the integration limits are $(-1, \infty)$ rather than $(-\infty, \infty)$, but this is not relevant for our asymptotics calculation since it is only the behavior in the neighborhood of $y = 0$ that is significant. Now

$$\phi''(t) = \frac{(-1)}{(1+t)^2} \Rightarrow \phi''(0) = -1 < 0, \qquad (9.4.4)$$

and we obtain using Eq. (9.3.10)

$$\Gamma(x + 1) \sim 2x^{x+1} e^{-x} \Gamma\left(\frac{3}{2} \right) \left(\frac{2}{x} \right)^{1/2} = \sqrt{2\pi x} \left(\frac{x}{e} \right)^x, \qquad x \to \infty. \qquad (9.4.5)$$

For integer x, we have

$$n! = \Gamma(n + 1) \sim \sqrt{2\pi n} \left(\frac{n}{e} \right)^n, \qquad n \to \infty, \qquad (9.4.6)$$

and this is the leading term in the asymptotic series expansion of $n!$.

9.4.2 Integral Containing a Logarithmic Function

Watson's lemma can be used to determine the asymptotics of the integral

$$I(x) = \int_0^\infty [\ln(1 + t^2)] e^{-xt} dt, \qquad x \to \infty. \tag{9.4.7}$$

For this case, we have

$$f(t) = \ln(1 + t^2) = t^2 - \frac{t^4}{2} + \frac{t^6}{3} + \cdots$$

$$= \sum_{n=0}^\infty (-)^n \left[\frac{t^{2(n+1)}}{n+1} \right] = t^2 \sum_{n=0}^\infty (-1)^n \left(\frac{t^{2n}}{n+1} \right), \tag{9.4.8}$$

where on comparison with Eq. (9.3.3)

$$\gamma = 2, \qquad a_n = \frac{(-1)^n}{n+1}, \qquad \beta_n = 2n. \tag{9.4.9}$$

Therefore, placing these results in Eq. (9.3.6) gives the following asymptotic series representation

$$I(x) = \int_0^\infty [\ln(1 + t^2)] e^{-xt} dt$$

$$\sim \sum_{n=0}^\infty \left[\frac{(-1)^n}{n+1} \right] \left[\frac{\Gamma(2n+3)}{x^{2n+3}} \right], \qquad x \to \infty. \tag{9.4.10}$$

9.4.3 Integral Containing a Complex Exponential Structure

We now consider the rather complex looking integral

$$I(x) = \int_{-\infty}^\infty e^{-x \cosh(\theta)} d\theta, \tag{9.4.11}$$

where

$$\cosh(\theta) = \frac{e^\theta + e^{-\theta}}{2}. \tag{9.4.12}$$

With the change of variables

$$\cosh(\theta) = 1 + t^2, \tag{9.4.13}$$

the integral becomes

$$I(x) = \sqrt{2}\, e^{-x} \int_{-\infty}^\infty \frac{e^{-xt^2} dt}{\sqrt{1 + \frac{t^2}{2}}}. \tag{9.4.14}$$

Since only the small t behavior is needed for

$$f(t) = \left(1 + \frac{t^2}{2} \right)^{-1/2}, \tag{9.4.15}$$

we can apply the following Taylor series expansion to $f(t)$ and obtain

$$f(t) = \sum_{n=0}^{\infty} \frac{\Gamma\left(\frac{1}{2}\right)}{\Gamma\left(\frac{1-2n}{2}\right) n!} \left(\frac{t^2}{2}\right)^n, \qquad |t| < \sqrt{2}. \tag{9.4.16}$$

Substitution of this expression for $f(t)$ into Eq. (9.4.14) and carrying out the integration gives

$$I(x) \sim \left(\frac{2\pi}{x}\right)^{1/2} e^{-x} \sum_{n=0}^{\infty} \left[\frac{\Gamma\left(\frac{2n+1}{2}\right)}{\Gamma\left(\frac{1-2n}{2}\right)}\right] \left(\frac{1}{2x}\right)^n, \qquad x \to \infty. \tag{9.4.17}$$

This is the required asymptotic series expansion for $I(x)$ as defined in Eq. (9.4.11).

9.4.4 Cosine and Sine Integrals

The cosine and sine integrals are defined as follows

$$Ci(x) \equiv -\int_x^{\infty} \frac{\cos t}{t} \, dt, \qquad Si(t) \equiv -\int_x^{\infty} \frac{\sin t}{t} \, dt. \tag{9.4.18}$$

From these, two functions $f(x)$ and $g(x)$ can be formed and defined as

$$f(x) \equiv Ci(x) \sin x - Si(x) \cos x = \int_0^{\infty} \frac{\sin y}{x+y} \cdot dy, \tag{9.4.19}$$

$$g(x) \equiv -Ci(x) \cos x - Si(x) \sin x = \int_0^{\infty} \frac{\cos y}{x+y} \cdot dy, \tag{9.4.20}$$

where the new variable y was introduced by $y = t - x$. Using

$$e^{iy} = \cos y + i \sin y, \tag{9.4.21}$$

we have

$$g(x) + if(x) = \int_0^{\infty} \frac{e^{iy}}{x+y} \cdot dy. \tag{9.4.22}$$

If the variable u, defined as

$$u = \frac{y}{ix}, \tag{9.4.23}$$

is now used, Eq. (9.4.22) becomes

$$g(x) + if(x) = \int_0^{\infty} \frac{ie^{-xy}}{1+iu} \cdot du. \tag{9.4.24}$$

Equating $g(x)$ and $f(x)$, respectively, to the real and imaginary parts of the integral, on the right-side of Eq. (9.4.24), gives

$$f(x) = \int_0^{\infty} \frac{e^{-xu}}{1+u^2} \cdot du, \tag{9.4.25}$$

$$g(x) = \int_0^\infty \frac{ue^{-xu}}{1+u^2} \cdot du. \tag{9.4.26}$$

For $x \geq 0$, both of these integrals exist. Now make a further change of variable

$$v = xu, \tag{9.4.27}$$

to obtain

$$f(x) = \left(\frac{1}{x}\right) \int_0^\infty \frac{e^{-v}}{1 + \left(\frac{v}{x}\right)^2} \, dv. \tag{9.4.28}$$

If $(v/x)^2 < 1$, then

$$f(x) = \left(\frac{1}{x}\right) \int_0^\infty e^{-v} \sum_{n=0}^\infty (-1)^n \left(\frac{v}{x}\right)^{2n} dv$$

$$= \left(\frac{1}{x}\right) \sum_{n=0}^\infty (-1)^n \left[\frac{(2n)!}{x^{2n}}\right]. \tag{9.4.29}$$

However, we will also use this evaluation of the integral for $(v/x)^2 > 1$. This is allowable since our main goal is to calculate an asymptotic series and the contributions from $(v/c)^2 > 1$ will be negligible for large x because of the negative exponential. A similar calculation for $g(x)$ gives

$$g(x) \sim \left(\frac{1}{x^2}\right) \sum_{n=0}^\infty (-1)^n \left[\frac{(2n+1)!}{x^{2n}}\right]. \tag{9.4.30}$$

Equations (9.4.19) and (9.4.20) provide two linear equations for $Ci(x)$ and $Si(x)$. Solving for these functions and using the above expressions for $f(x)$ and $g(x)$ gives the following asymptotic series representations

$$Ci(x) \sim \left(\frac{\sin x}{x}\right) \sum_{n=0}^\infty (-1)^n \left[\frac{(2n)!}{x^{2n}}\right]$$

$$- \left(\frac{\cos x}{x}\right) \sum_{n=0}^\infty (-1)^n \left[\frac{(2n+1)!}{x^{2n}}\right], \tag{9.4.31}$$

$$Si(x) \sim - \left(\frac{\cos x}{x}\right) \sum_{n=0}^\infty (-1)^n \left[\frac{(2n)!}{x^{2n}}\right]$$

$$- \left(\frac{\sin x}{x}\right) \sum_{n=0}^\infty (-1)^n \left[\frac{(2n+1)!}{x^{2n}}\right]. \tag{9.4.32}$$

9.5   Euler-Maclaurin Sum Formula

Often it is of interest to obtain an evaluation of a sum that takes the form

$$S(n) = \sum_{i=1}^{n} f(i). \tag{9.5.1}$$

For example, in the investigation of the thermodynamic properties of a rotating diatomic molecule the following expression occurs

$$S(I) = \sum_{i=0}^{I} (2i + 1)e^{-ai(i+1)}, \tag{9.5.2}$$

where $\{i = 0, 1, 2, \ldots, I\}$ and a is a known, positive parameter. A method for determining accurate values for such sums is based on the Euler-Maclaurin sum formula. However, before we introduce this relation, a brief discussion is presented on a topic needed in the application of the method. Therefore, in the next section, we give an introduction to Bernoulli numbers and the functions from which they are derived.

9.5.1   *Bernoulli Functions and Numbers* [3]

The following function

$$G(x,t) = \frac{te^{tx}}{e^t - 1}, \tag{9.5.3}$$

is the generating functions for the Bernoulli functions. This means that if we expand $G(x,t)$ as a power series in t, then the coefficients will depend on x, and, except for a constant, are the Bernoulli polynomials. In particular, we have

$$G(x,t) = \frac{te^{tx}}{e^t - 1} = \sum_{n=0}^{\infty} B_n(x) \left(\frac{t^n}{n!}\right)$$

$$= B_0(x) + B_1(x) + B_2(x) \left(\frac{t^2}{2!}\right) + \cdots. \tag{9.5.4}$$

The functions can be determined by writing Eq. (9.5.4) as

$$e^{tx} = \left(\frac{e^t - 1}{t}\right) \sum_{n=0}^{\infty} B_n(x) \left(\frac{t^n}{n!}\right), \tag{9.5.5}$$

and expanding each term in a Taylor series in x:

$$1 + tx + \frac{t^2 x^2}{2!} + \frac{t^3 x^3}{3!} + \cdots = \left(1 + \frac{t}{2!} + \frac{t^2}{3!} + \cdots\right)$$

$$\cdot \left[B_0(x) + \frac{B_1(x)}{1!} t + \frac{B_2(x)}{2!} t^2 + \cdots \right]$$

$$= B_0(x) + \left[\frac{B_1(x)}{1!} + \frac{B_0(x)}{2!} \right] t$$

$$+ \left[\frac{B_2(x)}{2!} + \frac{B_1(x)}{(2!)(1!)} + \frac{B_0(x)}{3!} \right] t^2 + \cdots . \tag{9.5.6}$$

If the coefficients of the various powers of t are set equal, then the following expressions are obtained

$$\begin{cases} B_0(x) = 1, \quad B_1(x) + \dfrac{B_0(x)}{2} = x, \\ \dfrac{B_0(x)}{2} + \dfrac{B_1(x)}{2} + \dfrac{B_0(x)}{6} = \dfrac{x^2}{2}, \text{ etc.} \end{cases} \tag{9.5.7}$$

Therefore, solving for the $B_n(x)$ gives

$$B_0(x) = 1,$$

$$B_1(x) = x - \frac{1}{2},$$

$$B_2(x) = x^2 - x + \frac{1}{6},$$

$$B_3(x) = x^3 - \left(\frac{3}{2} \right) x^2 + \left(\frac{1}{2} \right) x, \tag{9.5.8}$$

$$B_4(x) = x^4 - 2x^3 + x^2 - \frac{1}{30},$$

$$B_5(x) = x^5 - \left(\frac{5}{2} \right) x^4 + \left(\frac{5}{3} \right) x^3 - \left(\frac{1}{6} \right) x,$$

$$B_6(x) = x^6 - 3x^5 + \left(\frac{5}{2} \right) x^4 - \left(\frac{1}{2} \right) x^2 + \frac{1}{42}.$$

The **Bernoulli numbers**, B_n, are defined as

$$B_n \equiv B_n(0), \tag{9.5.9}$$

i.e., they are the values of the Bernoulli polynomials evaluated at $x = 0$

$$B_0 = 1, \; B_1 = - \left(\frac{1}{2} \right), \; B_2 = \frac{1}{6}, \; B_4 = - \left(\frac{1}{30} \right), \; B_6 = \frac{1}{42}, \tag{9.5.10}$$

$$B_{2n+1} = 0, \qquad n \geq 1. \tag{9.5.11}$$

An interesting property of the Bernoulli polynomials is

$$B_n \equiv B_n(0) = B_n(1), \qquad n \neq 1. \tag{9.5.12}$$

Further, the Bernoulli polynomials satisfy the differential equation

$$\frac{dB_n(x)}{dx} = nB_{n-1}(x), \qquad n \geq 1. \tag{9.5.13}$$

This last result can be easily proved. We start by taking the derivative of both sides of Eq. (9.5.4) with respect to x

$$\frac{t^2 e^{tx}}{e^t - 1} = B_0'(x) + \sum_{n=1}^{\infty} B_n'(x) \left(\frac{t^n}{n!}\right) = \sum_{n=0}^{\infty} B_{n+1}'(x) \left[\frac{t^{n+1}}{(n+1)!}\right]. \tag{9.5.14}$$

But,

$$\frac{t^2 e^{tx}}{e^t - 1} = t \left(\frac{te^{tx}}{e^t - 1}\right) = t \sum_{n=0}^{\infty} B_n(x) \left(\frac{t^n}{n!}\right) = \sum_{n=0}^{\infty} B_n(x) \left(\frac{t^{n+1}}{n!}\right). \tag{9.5.15}$$

Equating the coefficients of t^{n+1} in the sums given by Eqs. (9.5.8) and (9.5.9) leads to the expression

$$\frac{B_{n+1}'(x)}{(n+1)!} = \frac{B_n(x)}{n!}, \tag{9.5.16}$$

or

$$B_{n+1}'(x) = (n+1)B_n(x), \tag{9.5.17}$$

which is just Eq. (9.5.13) with a shift in index.

9.5.2   Euler-Maclaurin Sum Formula

This formula expresses a sum in terms of a related integral plus correction terms. In certain cases, it allows an exact evaluation of the sum. However, in general, it gives an asymptotic series representation for the sum. Note that the relationship, to be given below, can also be used to obtain approximations for the integral in terms of a sum and correction terms.

The Euler-Maclaurin sum formula is [4]

$$\sum_{k=1}^{n} f(x) \sim \int_1^n f(x)dx + \left(\frac{1}{2}\right)[f(1) + f(n)]$$

$$+ \sum_{k=1}^{\infty} \left[\frac{B_{2k}}{(2k)!}\right] \left[f^{(2k-1)}(n) - f^{(2k-1)}(1)\right]. \tag{9.5.18}$$

First, note that the continuous version of $f(k)$, namely, $f(x)$, is integrable. Second, only even Bernoulli numbers appear. Third, only odd derivatives are needed for the evaluations. For most practical applications, three terms retained in the sum, on the right-side of Eq. (9.5.18), provides an excellent approximation to the sum, i.e.,

$$\sum_{k=1}^{n} f(k) \simeq \int_{1}^{n} f(x)dx + \left(\frac{1}{2}\right) [f(1) + f(n)]$$

$$+ \left(\frac{1}{12}\right) [f^{(1)}(n) - f^{(1)}(1)] - \left(\frac{1}{730}\right) [f^{(3)}(n) - f^{(3)}(1)]$$

$$+ \left(\frac{1}{30240}\right) [f^{(5)})(n) - f^{(5)}(1)]. \qquad (9.5.19)$$

It should be clear that if $f(x)$ is a polynomial function, then the sum can be calculated exactly. This follows from the fact that if $f(x) = P_n(x)$, where $P_n(x)$ is an nth degree polynomial, then $f^{(n+1)}(x)$ and higher derivatives are all zero.

The Bernoulli numbers can also be written as

$$B_{2n} = \frac{(-1)^{n-1}2(2n)!}{(2\pi)^{2n}} \xi(2n), \qquad (9.5.20)$$

where $\xi(x)$ is the Riemann zeta function. This representation shows that B_{2n} increases very fast and consequently, as expected, the Euler-Maclaurin sum formula gives in general an asymptotic series for the sum of the right-side. However, for computational purposes, the error arising from using a finite number of terms can often be shown to be less in absolute value than the first term neglected. If the derivative difference terms, $f^{(2k-1)}(n) - f^{(2k-1)}(1)$, maintain the same sign, then this is so. This is a consequence of the alternation in sign of the Bernoulli numbers which then cause the series to also have this property.

9.6   Worked Examples for the Euler-Maclaurin Sum Formula

9.6.1   *Sums of Powers*

As easy task for the Euler-Maclaurin sum formula is to sum powers, i.e.,

$$S(n) = \sum_{k=0}^{n} k^m, \qquad (m = 0, 1, 2, \dots). \qquad (9.6.1)$$

For example, if $m = 1$, then $f(k) = k$,

$$\left(\frac{1}{2}\right) [f(n) + f(0)] = \frac{n}{2}, \qquad (9.6.2)$$

$$\int_0^n x\,dx = \frac{x^2}{2}\Big|_0^n = \frac{n^2}{2}\,, \tag{9.6.3}$$

$$f^{(1)}(x) = 1; \qquad f^{(r)}(x) = 0, \quad (r = 2, 3, \ldots). \tag{9.6.4}$$

Therefore, from Eq. (9.5.18), we have

$$\sum_{k=0}^n k = \frac{n^2}{2} + \frac{n}{2} = \frac{n(n+1)}{2}\,. \tag{9.6.5}$$

However, there is a closed form for the sum in Eq. (9.6.1) that involves only the Bernoulli polynomials. We now derive this relation. To begin, we note that the Bernoulli polynomials, $B_m(k)$, are solutions to the difference equation [3]

$$B_m(k+1) - B_m(k) = mk^{m-1}. \tag{9.6.6}$$

With $\Delta B_m(k) \equiv B_m(k+1) - B_m(k)$, we have

$$\Delta B_{m+1}(k) = (m+1)k^m. \tag{9.6.7}$$

From section 5.3.4, it follows that

$$\sum_{k=0}^n [\Delta B_{m+1}(k)] = (m+1)\sum_{k=0}^n k^m. \tag{9.6.8}$$

However, if

$$\Delta F_k = f_k, \tag{9.6.9}$$

then

$$\sum_{k=a}^b f_k = F(b+1) - F(a). \tag{9.6.10}$$

(See reference [3], Theorem 1.4, on page 36.) For our case

$$F_k = B_{m+1}(k), \quad f_k = (m+1)k^m, \quad a = 0, \quad b = n, \tag{9.6.11}$$

and, therefore, Eq. (9.6.8) can be written as

$$\sum_{k=0}^n k^m = \frac{B_{m+1}(n+1) - B_{m+1}(0)}{(m+1)}\,. \tag{9.6.12}$$

This means that sums of non-negative integer powers can always be represented in closed form. Two examples will now be given to illustrate the application of Eq. (9.6.12).

For $m = 0$, we have

$$\sum_{k=0}^{n} 1 = \frac{B_1(n+1) - B_1(0)}{0+1} = \left[(n+1) - \frac{1}{2}\right] - \left(-\frac{1}{2}\right) = n+1. \quad (9.6.13)$$

For $m = 1$, we have

$$\sum_{k=0}^{n} k = \frac{B_2(n+1) - B_2(0)}{2} = \left(\frac{1}{2}\right)\left\{\left[(n+1)^2 - (n+1) + \frac{1}{6}\right] - \frac{1}{6}\right\}$$

$$= \left(\frac{1}{2}\right)\left[n^2 + 2n + 1 - n - 1 + \frac{1}{6} - \frac{1}{6}\right] = \frac{n(n+1)}{2}. \quad (9.6.14)$$

9.6.2 Evaluation of $\ln(n!)$

The function n-factorial is defined to be

$$n! = n(n-1)(n-2)\cdots 2 \cdot 1. \quad (9.6.15)$$

Therefore,

$$\ln(n!) = \sum_{k=1}^{n} \ln(k). \quad (9.6.16)$$

Applying the Euler-Maclaurin sum formula, we have with $f(x) = \ln(x)$

$$\int_1^n [\ln(x)]dx = [x\ln(x) - x]\Big|_1^n = \ln(n) - n + 1, \quad (9.6.17)$$

$$\left(\frac{1}{2}\right)[f(n) + f(1)] = \left(\frac{1}{2}\right)\ln(n), \quad (9.6.18)$$

$$\left(\frac{1}{12}\right)\left[f^{(1)}(n) - f^{(1)}(1)\right] = \left(\frac{1}{12}\right)\left(\frac{1}{n} - 1\right), \quad (9.6.19)$$

$$-\left(\frac{1}{720}\right)\left[f^{(3)}(n) - f^{(3)}(1)\right] = -\left(\frac{1}{360}\right)\left(\frac{1}{n^3} - 1\right), \quad (9.6.20)$$

and, finally,

$$\ln(n!) \simeq \sum_{k=1}^{n} \ln(k)$$

$$= n\ln(n) - n + 1 + \left(\frac{1}{2}\right)\ln(n) + \left(\frac{1}{12}\right)\left(\frac{1}{n} - 1\right)$$

$$- \left(\frac{1}{360}\right)\left(\frac{1}{n^3} - 1\right)$$

$$= n\ln(n) - n + \left(\frac{1}{2}\right)\ln(n) + \left(\frac{1}{12}\right)\frac{1}{n} - \frac{1}{360n^3} + A, \quad (9.6.21)$$

where

$$A = 1 - \frac{1}{12} + \frac{1}{360} = \frac{349}{360}. \quad (9.6.22)$$

9.6.3 $f(k) = x^{-1/2}$

We now evaluate the sum

$$S(n) = \sum_{k=1}^{n} k^{-1/2}, \qquad f(k) = k^{-1/2}. \tag{9.6.23}$$

Therefore,

$$\int_1^n \frac{dx}{\sqrt{x}} = 2\left(\sqrt{n} - 1\right), \tag{9.6.24}$$

$$\left(\frac{1}{2}\right)[f(n) + f(1)] = \left(\frac{1}{2}\right)\left[\frac{1}{n^{1/2}} + 1\right], \tag{9.6.25}$$

$$\left(\frac{1}{12}\right)[f^{(1)}(n) - f^{(1)}(1)] = -\left(\frac{1}{24}\right)\left[\frac{1}{n^{3/2}} - 1\right], \tag{9.6.26}$$

$$-\left(\frac{1}{720}\right)[f^{(3)}(n) - f^{(3)}(1)] = \left(\frac{1}{384}\right)\left[\frac{1}{n^{7/2}} - 1\right], \tag{9.6.27}$$

and

$$S(n) = \sum_{k=1}^{n} k^{-1/2}$$

$$\simeq 2\left(\sqrt{n} - 1\right) + \left(\frac{1}{2}\right)\left[\frac{1}{n^{1/2}} + 1\right]$$

$$-\left(\frac{1}{24}\right)\left(\frac{1}{n^{3/2}} - 1\right) + \left(\frac{1}{384}\right)\left(\frac{1}{n^{7/2}} - 1\right)$$

$$= 2\sqrt{n} + A + \left(\frac{1}{2}\right)\frac{1}{n^{1/2}} - \left(\frac{1}{24}\right)\frac{1}{n^{3/2}} + \left(\frac{1}{384}\right)\frac{1}{n^{7/2}}, \tag{9.6.28}$$

where

$$A = -2 + \frac{1}{2} + \frac{1}{24} - \frac{1}{384} = -\frac{561}{384}. \tag{9.6.29}$$

Problems

Section 9.2

9.2.1 The Stieltjes integral is

$$I(x) = \int_0^\infty \frac{e^{-t}}{1 + xt}\, dt.$$

Show that integration by parts can be used to find the behavior of $I(x)$ for x small, but not for $x \to \infty$.

9.2.2 Calculate the asymptotic behavior as $x \to \infty$ of the integral

$$\int_0^x e^{t^2}\, dx.$$

Hint: See Bender and Orszag, section 6.3.

Section 9.3

9.3.1 Prove Watson's lemma.

9.3.2 In the statement of Laplace's theorem for integrals, are there any advantages to writing the exponential function, in the integrand, as $\exp[-x\phi(t)]$ rather than $\exp[x\phi(t)]$?

Section 9.4

9.4.1 Complete the details of the calculations to obtain the result of Eq. (9.4.5).

9.4.2 Using the change of variables given in Eq. (9.4.13), show that $I(x)$ of Eq. (9.4.11) becomes the expression written in Eq. (9.4.14)

9.4.3 Derive the results for $f(x)$ and $g(x)$, in Eqs. (9.4.25) and (9.4.26) from Eq. (9.4.24).

9.4.4 Fully explain the reason why we can use $(v/x)^2 > 1$ in our construction of the asymptotic series for $f(x)$ and $g(x)$ in spite of the fact that

$$\frac{1}{1 + \left(\frac{v}{x}\right)^2} = 1 - \left(\frac{v}{x}\right)^2 + \left(\frac{v}{x}\right)^4 - \left(\frac{v}{x}\right)^6 + \cdots,$$

converges only for $(v/x) < 1$.

9.4.5 Is it possible to construct an asymptotic series for

$$L(x) = \int_x^\infty \frac{\ln t}{1 + t^2}\, dt?$$

9.4.6 Show that the indicated asymptotic series formulas for the following integrals are correct:

$$\int_x^\infty \sin(x^2)dx \sim \frac{\cos(x^2)}{2x} + \frac{\sin(x^2)}{2^2 x^3} - \frac{3\cos(x^2)}{2^3 x^5} - \frac{3 \cdot 5 \sin(x^2)}{2^4 x^7} + \cdots$$

$$\int_x^\infty \cos(x^2)dx \sim -\frac{\sin(x^2)}{2x} + \frac{\cos(x^2)}{2^2 x^3} + \frac{3\sin(x^2)}{2^3 x^5} - \frac{3 \cdot 5 \cos(x^2)}{2^4 x^7} + \cdots.$$

Hint: Using integration by parts and use, respectively, $x\sin(x^2)$ and $x\cos(x^2)$ as one of the factors to be integrated.

Section 9.5

9.5.1 Work out the details of the calculations which lead to Eqs. (9.5.6) and (9.5.7).

9.5.2 Use the generating function $G(x, t)$, given by Eq. (9.5.4), to calculate the Bernoulli numbers. Hint: Calculate $G(0, t)$.

9.5.3 Prove the result presented in Eq. (9.5.12).

9.5.4 Derive the Euler-Maclaurin sum formula.

Section 9.6

9.6.1 Show that the following relations hold for the Bernoulli polynomials:

(i) $B_n(x+1) = nx^{n-1} + B_n(x)$,

(ii) $B_n(1-x) = (-1)^n B_n(x)$,

(iii) $\int_a^x B_n(t)dt = \frac{B_{n+1}(x) - B_{n-1}(a)}{n+1}$.

9.6.2 Evaluate the following sums:

(i) $\sum_{k=1}^n k^2$,

(ii) $\sum_{k=1}^n \left[3 + k + \left(\frac{1}{2}\right)k^2\right]$.

9.6.3 Calculate the next two terms in the asymptotic expansion for $\ln(n!)$ and show that

$$\ln(n!) \simeq n\ln(n) - n + 1 + \left(\frac{1}{2}\right)\ln(n) + \left(\frac{1}{12}\right)\left(\frac{1}{n} - 1\right)$$

$$- \left(\frac{1}{360}\right)\left(\frac{1}{n^3} - 1\right) + \left(\frac{1}{1260}\right)\left(\frac{1}{n^5} - 1\right)$$

$$- \left(\frac{1}{1680}\right)\left(\frac{1}{n^7} - 1\right).$$

Further, show that for $n = 10$, the above expression gives 15.0992 while the exact value for $\ln(10!)$ is 15.10441.

9.6.4 Calculate three terms in the asymptotic series, $n \to \infty$, of the following sum

$$\sum_{k=1}^n \frac{\sin(k)}{k}.$$

9.6.5 Determine the first three terms in the asymptotic expansion of the function

$$f(x) = \exp\left[\left(\frac{2}{3}\right)\sum_{k=1}^n k^{1/2}\right].$$

Comments and References

[1] J. Nicholas, *Chemical Kinetics* (Halsted Press, New York, 1976).

[2] R. E. Mickens, *Chemical Physics Letters* **121**, 334 (1985).

[3] Good summaries of the major properties of the Bernoulli polynomials and numbers are given in the following text:

* R. E. Mickens, *Difference Equations: Theory and Applications* (Chapman and Hall, New York, 1990); see section 2.3

[4] Proofs of the Euler-Maclaurin sum formula are presented in the books:

* W. G. Kelley and A. C. Peterson, *Difference Equations*, 2nd ed. (Academic Press, New York, 2001).

* L. Brand, *Differential and Difference Equations* (Wiley, New York, 1966); see section 175.

See also the paper:

* A. Sidi, *Numerical Mathematics* **98**, 1371 (2004).

Bibliography

C. M. Bender and S. A. Orszag, *Advanced Mathematical Methods for Scientists and Engineers* (McGraw-Hill, New York, 1978).

N. Bleistein and R. A. Handelsman, *Asymptotic Expansions of Integrals* (Holt, Rinehart and Winston; New York, 1975).

E. T. Copson, *Asymptotic Expansions* (Cambridge University Press, Cambridge, 1967).

R. B. Dingle, *Asymptotic Expansions: Their Derivation and Interpretation* (Academic Press, New York, 1973).

A. Erdelyi, *Asymptotic Expansions* (Dover, New York, 1956).

H. Jeffreys, *Asymptotic Approximations* (Clarendon Press, Oxford, 1962).

H. Jeffreys and B. S. Jeffreys, *Methods of Mathematical Physics*, 3rd ed. (Cambridge University Press, Cambridge, 1956).

J. P. Keener, *Principles of Applied Mathematics* (Addison-Wesley; Reading, MA; 1995).

J. D. Murray, *Asymptotic Analysis* (Springer-Verlag, New York, 1984).

F. W. J. Olver, *Asymptotics and Special Functions* (Academic Press, New York, 1974).

G. N. Watson, *Theory of Bessel Functions*, 2nd ed. (Cambridge University Press, Cambridge, 1952).

Chapter 10

Some Important Nonlinear Partial Differential Equations

Many of the equations modeling complex dynamical systems in the natural and engineering sciences are partial differential equations. For these equations the independent variables are usually the time and one or more space variables. It is important to note that a vast number of scientific and technologically interesting phenomena can be modeled in terms of a relatively small set of linear and nonlinear partial differential equations. Of further significance is the fact that several techniques exist to calculate special solutions to these equations and these particular solutions are often the ones needed to analyze the phenomena of interest.

The major goals of this chapter are to introduce some special nonlinear partial differential equations and the techniques needed for determining particular solutions to them. The bibliography, at the end of the chapter, provides a list of books on both the theory and application of linear and nonlinear partial differential equations. Two excellent introductions to this subject are the books by Logan and Myint-U. Several books on specific application topics are also given:
* combustion (Bebernes and Eberly; Fickett),
* fluid flow (Courant and Friedrichs; Kreiss and Cole),
* mathematical biosciences (Britton; Murray; Okubo);
* perturbation methods (Kevorkian and Cole);
* reaction-diffusion processes (Britton; Fife; Smoller).
The book of Lin and Segel is well-written and contains many applications of both linear and nonlinear differential equations to interesting problems from several of the natural sciences. Another book giving the model based genesis of certain partial differential equations and various methods for determining solutions and/or their mathematical properties is the one by Whitham.

For the remainder of this chapter the abbreviations ODE and PDE will be used to denote ordinary and partial differential equations. This chapter begins with a brief discussion of linear wave equations. The purpose for this is to provide a review of certain topics related to linear PDE's and some methods for calculating exact solutions. Section 10.2 gives definitions of two special classes of solutions to PDE's, namely, traveling waves and solitons. The next eight sections are devoted to presentations on the elementary properties of eight nonlinear PDE's for which important applications exist. Our goal is to give a broad overview of these particular PDE's without going into the depth needed for a formal study of these equations, which would include the stating and proof of various theorems.

The following notation is used throughout this chapter to denote derivatives:

$$\frac{\partial u(x,t)}{\partial x} \equiv u_x, \qquad \frac{\partial u(x,t)}{\partial t} \equiv u_t, \quad \text{etc.}$$

10.1   Linear Wave Equations

The most elementary linear wave equation is one modeling propagation in one space dimension and in a definite direction. For example, the first-order PDE

$$u_t + u_x = 0, \tag{10.1.1}$$

represents a wave propagating in the direction of increasing x. A general solution to this equation is

$$u(x,t) = f(x-t), \tag{10.1.2}$$

as can be directly verified by substitution into the left-side of Eq. (10.1.1). For the initial value problem, i.e., where

$$u(x,0) = F(x) = \text{given}, \tag{10.1.3}$$

the solution is

$$u(x,t) = F(x-t). \tag{10.1.4}$$

Thus, if the initial form is

$$u(x,0) = F(x) = e^{-x^2} \tag{10.1.5}$$

which corresponds to a single "hump," whose maximum is at the origin, then

$$u(x,t) = F(x-t) = e^{-(x-t)^2} \tag{10.1.6}$$

is the rigid propagation of this "hump" to the right with a speed of one unit. Note that the shape of the wave form does not change, it merely gets translated.

In many physical systems, dissipative forces exist. The inclusion of a linear damping term in Eq. (10.1.11) gives

$$u_t + u_x = -\lambda u. \tag{10.1.7}$$

The solution to this equation is

$$u(x,t) = e^{-\lambda t} f(x-t), \tag{10.1.8}$$

where $f(z)$ is any function having a first derivative. With the initial condition, of Eq. (10.1.5), the solution is

$$u(x,t) = e^{-\lambda t} F(x-t). \tag{10.1.9}$$

This represents a rigid translation of the form $F(x)$ to the right along the x-axis, but with an exponentially damped amplitude. Thus, for the initial shape given by Eq. (10.1.5), we have for $t > 0$,

$$u(x,t) = e^{-\lambda t} e^{-(x-t)^2}. \tag{10.1.10}$$

Finally, the equation

$$u_t - u_x = 0, \tag{10.1.11}$$

describes wave forms traveling to the left. A general solution to this equation is

$$u(x,t) = g(x+t). \tag{10.1.12}$$

For the initial form, given by Eq. (10.1.5), the solution is

$$u(x,t) = e^{-(x+t)^2}. \tag{10.1.13}$$

The linear, bi-direction wave equation is

$$u_{tt} - u_{xx} = 0. \tag{10.1.14}$$

Using the "factorization"

$$\left(\frac{\partial^2}{\partial t^2} - \frac{\partial^2}{\partial x^2} \right) u = \left(\frac{\partial}{\partial t} + \frac{\partial}{\partial x} \right) \left(\frac{\partial}{\partial t} - \frac{\partial}{\partial x} \right) u$$
$$= \left(\frac{\partial}{\partial t} - \frac{\partial}{\partial x} \right) \left(\frac{\partial}{\partial t} + \frac{\partial}{\partial x} \right) u, \tag{10.1.15}$$

it follows that both of the following expressions are solutions to Eq. (10.1.14),

$$u_1(x,t) = f(x-t), \qquad u_2(x,t) = g(x+t), \qquad (10.1.16)$$

where both $f(z)$ and $g(z)$ have second-derivatives. Since Eq. (10.1.14), also known as the "wave equation," is a linear PDE, the sum of these solutions is a solution, i.e., a general solution to the wave equation is

$$u(x,t) = f(x-t) + g(x+t). \qquad (10.1.17)$$

Let us now study the initial-value problem for the wave equation, i.e., function $F(x)$ and $G(x)$ are given such that

$$u(x,0) = F(x), \qquad u_t(x,0) = G(x). \qquad (10.1.18)$$

Using the result in Eq. (10.1.17), we have

$$f(x) + g(x) = F(x), \qquad -f'(x) + g'(x) = G(x), \qquad (10.1.19)$$

where

$$f'(x) = \frac{df}{dx}, \qquad g'(x) = \frac{dg}{dx}. \qquad (10.1.20)$$

Integration of the second equation in Eq. (10.1.19) gives

$$-f(x) + g(x) = -f(0) + g(0) + \int_0^x G(y)dy. \qquad (10.1.21)$$

Note that Eqs. (10.1.19) and (10.1.21) provide two expressions linear in $f(x)$ and $g(x)$, and when solved gives

$$f(x) = \left(\frac{1}{2}\right) F(x) - \left(\frac{1}{2}\right)[-f(0) + g(0)] - \left(\frac{1}{2}\right)\int_0^x G(y)dy, \quad (10.1.22)$$

$$g(x) = \left(\frac{1}{2}\right) F(x) + \left(\frac{1}{2}\right)[-f(0) + g(0)] + \left(\frac{1}{2}\right)\int_0^x G(y)dy. \quad (10.1.23)$$

Substituting these relations into Eq. (10.1.17) gives the solution to the initial-value problem, as specified by the conditions in Eq. (10.1.18),

$$u(x,t) = \left(\frac{1}{2}\right)[F(x-t) + F(x+t)] + \left(\frac{1}{2}\right)\int_{x-t}^{x+t} G(y)dy. \quad (10.1.24)$$

This form for the solution is called the d'Alembert solution.

To illustrate this method, consider the following set of initial conditions

$$u(x,0) = 10e^{-x^2}, \qquad u_t(x,0) = 0. \qquad (10.1.25)$$

According to the above results, the solution is

$$u(x,t) = 5e^{-(x-t)^2} + 5e^{-(x+t)^2}. \qquad (10.1.26)$$

This function represents two separate humps, both of maximum magnitude five, traveling with speeds of value one, in opposite directions along the x-axis. Note that the maximum magnitude for each hump is one-half that of the initial hump at $t = 0$.

10.2 Traveling Wave and Soliton Solutions

Two types of special solutions to nonlinear PDE's play important roles for the analysis of the properties of these equations and their application to particular problems. We discuss the nature of these solutions and in the remaining sections of this chapter show how they arise in solving nonlinear PDE's. In the following, it is assumed that the PDE's depend on one-space and time variables. This means that the solutions can be written as $u(x, t)$. It is also important to note that all of the nonlinear equations are first-order in the time derivative, i.e., they take the form

$$u_t = H(u, u_x, u_{xx}, u_{xxx}, \ldots).$$ (10.2.1)

Such PDE's are called evolutionary equations [1].

Definition A *traveling wave solution* of a PDE is a solution that can be expressed as

$$u(x, t) = f(x - ct),$$ (10.2.2)

where c is a constant.

If the initial form, $u(x, 0) = F(x)$, is given, then the solution at $t > 0$ is $u(x, t) = F(x - ct)$. Therefore, the initial form is translated along the x-axis at constant speed c. For $c > 0$, the motion is to the right, for $c < 0$, the form moves in the leftward direction. In general, for a given nonlinear PDE, the value of the speed is not known *a priori*, but is later calculated from either the initial and/or boundary conditions, or by the application of some other restriction such as the positivity of the solution [2].

Definition A *traveling wavefront solution* is a traveling wave solution, if for any fixed t, $u(x, t)$ has the properties

$$\lim_{x \to -\infty} u(x, t) = u_1, \qquad \lim_{x \to +\infty} u(x, t) = u_2,$$ (10.2.3)

where u_1 and u_2 are constants, with $u_1 \neq u_2$. In general, the solution changes monotonically between the two values u_1 and u_2.

Definition If a traveling wavefront has $u_1 = u_2$, the solution is a *pulse*.

Definition The definition of a *soliton solution* to a nonlinear PDE is ambiguous. There are several general features of these types of solutions that most researchers in the field agree on. First, a soliton represents a wave having a definite permanent form. Second, a soliton is localized in space for t fixed in value. Third, for circumstances where multi-solitons solutions exist for a particular nonlinear PDE, the solitons interact strongly with each other when they overlap, but after moving apart their individual identities reappear.

We now consider seven nonlinear PDE's and demonstrate how to construct the special solutions discussed above.

10.3 A Linear Advective, Nonlinear Reaction Equation

The nonlinear differential equation

$$u_t + u_x = u(1 - u), \tag{10.3.1}$$

with the initial value

$$u(x, 0) = f(x), \tag{10.3.2}$$

where $f(x)$ is known, can be solved exactly. This equation consists of three terms:

(i) u_t is the time evolution term;

(ii) u_x is an *advection* term which determines the behavior or motion of the solution along the x-axis;

(iii) $u(1 - u)$ is the nonlinear *reaction* term.

To construct the exact solution, we make the change of dependent variable

$$u(x, t) = \frac{1}{w(x, t)}, \tag{10.3.3}$$

and obtain

$$u_t = -w^{-2}w_t, \qquad u_x = -w^{-2}w_x. \tag{10.3.4}$$

On substitution of these expressions into Eq. (10.3.1), the following linear first-order, inhomogeneous PDE is gotten

$$w_t + w_x + w = 1. \tag{10.3.5}$$

A direct calculation, using standard methods [3], gives the general solution of Eq. (10.3.5) to be

$$w(x, t) = g(x - t)e^{-t} + 1, \tag{10.3.6}$$

where $g(z)$ is an arbitrary function of z having a first derivative. Imposing the initial condition of Eq. (10.3.2) gives

$$g(x) + 1 = \frac{1}{f(x)} \quad \text{or} \quad g(x) = \frac{1 - f(x)}{f(x)}. \tag{10.3.7}$$

Therefore, we have

$$w(x,t) = \left[\frac{1 - f(x - t)}{f(x - t)} \right] e^{-t} + 1 \tag{10.3.8}$$

or

$$u(x,t) = \frac{f(x - t)}{e^{-t} + (1 - e^{-t}) f(x - t)}. \tag{10.3.9}$$

Note that the solution to the initial value problem does not fit nicely into any of the four categories of solution types discussed in the previous section. It has some aspects of each. For example, while the traveling wave form, $f(x - t)$, appears in the functional form for $u(x,t)$, the $u(x,t)$ is clearly not a traveling wave because of the appearance of the exponential in time factor. We also have the following feature

$$\underset{\substack{t \to \infty \\ x - \text{fixed}}}{\text{Lim}} \, u(x,t) = 1. \tag{10.3.10}$$

Thus, for any given value of x, as the solution evolves in time, it reaches the value one.

10.4 Burgers' Equation

The Burgers' equation in general form is given by the following nonlinear PDE,

$$w_t + c w_y + a w w_y = D w_{yy}, \tag{10.4.1}$$

where c, a, and D are constants with $D \geq 0$. In addition to the evolution term w_t and the linear advection term $c w_y$, there is also a nonlinear advection expression, $a w w_y$, and a linear diffusion term $D w_{yy}$. The linear advection term can be eliminated by defining a new dependent variable u and a new independent space variable x, as follows

$$w(y,t) \equiv u(y - ct, t) = u(x,t). \tag{10.4.2}$$

With these changes,

$$w_t = -c u_x + u_t, \quad w_y = u_x, \quad w_{yy} = u_{xx}, \tag{10.4.3}$$

and on substitution into Eq. (10.4.1), we obtain

$$u_t + auu_x = Du_{xx}. \tag{10.4.4}$$

A rescaling of the dependent and independent variables transform this equation to the form

$$u_t + uu_x = u_{xx}. \tag{10.4.5}$$

However, for our work, we will keep $D \neq 0$ and get the standard form of the Burgers' equation,

$$u_t + uu_x = Du_{xx}. \tag{10.4.6}$$

We now demonstrate how the traveling wave solution to Eq. (10.4.6) can be calculated. This solution takes the form

$$u(x,t) = f(x - ct), \tag{10.4.7}$$

which when substituted into Eq. (10.4.6) gives

$$-cf' + ff' = Df'', \tag{10.4.8}$$

where

$$z = x - ct, \qquad f'(z) \equiv \frac{df}{dz}. \tag{10.4.9}$$

This second-order ODE can be integrated once and the resulting expression is

$$\frac{df}{dz} = \left(\frac{1}{2D}\right)(f^2 - 2cf - 2A), \tag{10.4.10}$$

where A is an arbitrary integration constant and the currently unknown speed of propagation must be determined. In fact, both A and c may be calculated by making use of the following boundary conditions

$$\begin{cases} \operatorname*{Lim}_{z \to -\infty} f(z) = u_1, & \operatorname*{Lim}_{z \to +\infty} f(z) = u_2, \\ u_1 > u_2. \end{cases} \tag{10.4.11}$$

Thus, the solution we seek is a traveling front type solution. The conditions given by Eq. (10.4.11) imply that the differential equation for $f(z)$ has the structure

$$\begin{aligned} \frac{df}{dz} &= \left(\frac{1}{2D}\right)(f - u_1)(f - u_2) \\ &= \left(\frac{1}{2D}\right)\left[f^2 - (u_1 + u_2)f + u_1 u_2\right]. \end{aligned} \tag{10.4.12}$$

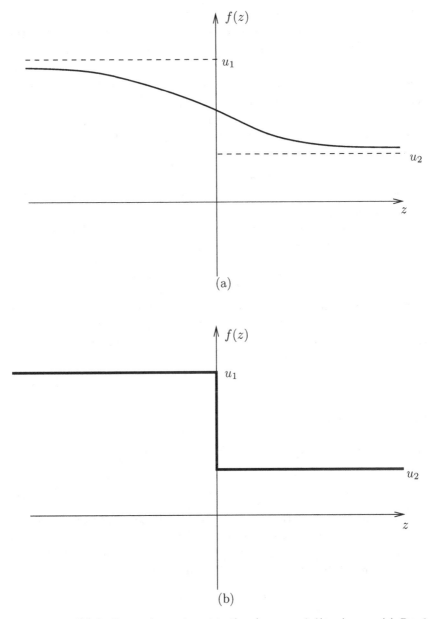

Fig. 10.4.1 $f(z)$ for Burgers' equation with $f(-\infty) = u_1$ and $f(+\infty) = u_2$. (a) $D > 0$. (b) $D = 0$.

Why must the right-side of Eq. (10.4.10) have this form? First, $f(z)$ satisfies the first-order, nonlinear ODE given by Eq. (10.4.10). Second if the solutions have asymptotic limits as $z \to \infty$, these limits must correspond to fixed-points or constant solutions of Eq. (10.4.10). By requirement, the fixed-points, using the restrictions given by Eq. (10.4.11) are u_1 and u_2. With these results, it follows that the right-side of Eq. (10.4.10) must equal the expression given on the right-side of Eq. (10.4.12).

Comparison of Eqs. (10.4.10) and (10.4.12) shows that

$$A = - \left(\frac{u_1 u_2}{2} \right), \qquad c = \frac{u_1 + u_2}{2}, \qquad (10.4.13)$$

and both the integration constant, A, and speed, c, are determined by the known asymptotic values u_1 and u_2.

Equation (10.4.12) can be solved using the method of variable separation and the solution is

$$f(z) = u_2 + \frac{(u_1 - u_2)}{1 + \exp\left[\left(\frac{u_1 - u_2}{2D}\right) z\right]}. \qquad (10.4.14)$$

Figure 10.4.1 gives $f(z)$ for two cases: $D > 0$ and $D = 0$. Observe that for $D > 0$, the solution is smooth and monotonically decreasing from its value $f(-\infty) = u_1$ to $f(\infty) = u_2$. For $D = 0$, the solution becomes a shock wave. In both cases, the traveling front solution is

$$u(x,t) = f(x - ct). \qquad (10.4.15)$$

10.5   The Fisher Equation

In terms of dimensionless variables, the Fisher PDE is given by the expression

$$u_t - u_{xx} = u(1 - u). \qquad (10.5.1)$$

Again, there is an evolution term u_t present, a linear diffusion term u_{xx}, and a nonlinear reaction term $u(1 - u)$. Hence, the Fisher equation is an example of a reaction-diffusion PDE. For one space dimension, the general reaction-(linear) diffusion class of PDE's takes the form

$$u_t = u_{xx} + f(u), \qquad (10.5.2)$$

where $f(u)$ is a nonlinear function of u [4].

We now show that the Fisher equation has traveling wave solutions. While we can demonstrate the existence of such solutions, no general functional function can be derived for them.

The traveling wave solutions take the form

$$u(x,t) = f(z), \qquad z = x - ct, \qquad (10.5.3)$$

where c or restrictions on it must be determined as the calculation proceeds. We assume that $f(z)$ has the following limits

$$\operatorname*{Lim}_{z \to -\infty} f(z) = 1, \qquad \operatorname*{Lim}_{z \to +\infty} f(z) = 0. \qquad (10.5.4)$$

Note that these limits correspond to the "fixed-points" or constant solutions to the Fisher equation (10.5.1) where $u = 0$ and $u = 1$ represent unstable and stable solutions. Also, it should be indicated that the Fisher equation was derived for systems for which the variable $u(x,t)$ satisfies the condition

$$0 \le u(x,0) \le 1, \qquad (10.5.5)$$

from which it can be proved that [5]

$$0 \le u(x,t) \le 1, \qquad t > 0. \qquad (10.5.6)$$

The substitution of $f(z)$ into Eq. (10.5.1) gives a second-order ODE,

$$-cf' - f'' = f(1-f), \qquad f' \equiv \frac{df}{dz}. \qquad (10.5.7)$$

Note that unlike Burgers' equation, we cannot integrate this equation once to obtain a first-order ODE. What this means is that the ODE must be studied using phase-space methods.

Equation (10.5.7) can be reduced to two coupled first-order ODE's by introducing a new variable $g(z)$ defined by $g = f'$. Therefore, we have

$$\frac{df}{dz} = g, \qquad \frac{dg}{dz} = -cg - f(1-f). \qquad (10.5.8)$$

This system of equations has two fixed-points. They are

$$P_1 : (\bar{f}^{(1)}, \bar{g}^{(1)}) = (0,0), \qquad P_2 : (\bar{f}^{(2)}, \bar{g}^{(2)}) = (1,0). \qquad (10.5.9)$$

The Jacobian matrix for Eq. (10.5.8) is

$$J(f,g) = \begin{pmatrix} 0 & 1 \\ 2f - 1 & -c \end{pmatrix}. \qquad (10.5.10)$$

The eigenvalues at the fixed-point P_2 are determined by the solution to

$$\det[J(1,0) - \lambda I] = 0, \qquad I = \text{2-dim unit matrix}, \qquad (10.5.11)$$

which are

$$\lambda_{1,2} = \left(\frac{1}{2}\right)\left[-c \pm \sqrt{c^2 + 4}\right]. \qquad (10.5.12)$$

Since $\lambda_2 < 0 < \lambda_1$, it follows that $P_2 : (1,0)$ is unstable and corresponds to a saddle point. In a similar manner, the eigenvalues of $J(0,0)$ are

$$\lambda_{1,2} = \left(\frac{1}{2}\right)\left[-c \pm \sqrt{c^2 - 4}\right], \tag{10.5.13}$$

and the fixed-point $P_1 : (0,0)$ is a stable node for $c^2 \geq 4$, since both eigenvalues are negative. However, for $c^2 < 4$, the eigenvalues are complex valued, with real parts negative and, as a consequence, the $P_1 : (0,0)$ fixed-point is a stable spiral and this allows the solution, $f(z)$, to become negative. But, this case violates the condition given in Eq. (10.5.6). Our conclusion is that $c^2 \geq 2$, if the correct "physical" solution is to be obtained. Note that we have found a minimum value for c, i.e., non-negative, traveling wave solutions exist only for $|c| \geq 2$. (Since the Fisher equation is invariant under $x \to -x$, it is sufficient to only consider solutions propagating in the direction of increasing (positive) x.)

Figure 10.5.1 gives the phase-plane trajectories for the two cases [6], namely, for $c > 2$ and $0 < c < 2$. For $c > 2$, there is a unique separatrix connecting the saddle point, $P_2(1,0)$, to the stable node, $P_1(0,0)$. Along this path,

$$0 \leq f(z) \leq 1, \qquad g(z) < 0, \tag{10.5.14}$$

with

$$f(-\infty) = 1, \quad f(+\infty) = 0; \quad g(-\infty) = 0, \quad g(+\infty) = 0. \tag{10.5.15}$$

The corresponding plot of $u(x,t) = f(z)$ vs z is illustrated by Figure 10.5.2(a).

For $0 < c < 2$, the phase-space trajectories are indicated in Figure 10.5.1(b), with the plot of $u(x,t) = f(z)$ given in Figure 10.5.2(b). Note that the phase-space trajectories oscillate in the neighborhood of the origin and this gives the plot in Figure 10.5.2(b).

In summary, we have the following result for the Fisher PDE:

For each value of the speed of propagation, $c \geq 2$, there is a unique traveling wave solution

$$u(x,t), \qquad f(x - ct), \tag{10.5.16}$$

to the PDE

$$u_t = u_{xx} + u(1 - u), \tag{10.5.17}$$

such that

$$\begin{cases} 0 \leq f(z) \leq 1, & -\infty < z < \infty, \\ f(-\infty) = 1, \quad f(+\infty) = 0, \quad f'(\pm\infty) = 0, \end{cases} \tag{10.5.18}$$

with $f(z)$ monotonically decreasing.

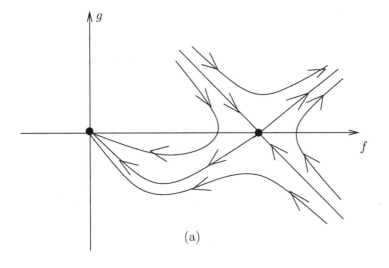

(a)

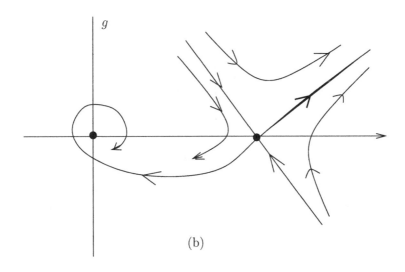

(b)

Fig. 10.5.1 Phase-plane trajectories. (a) $c > 2$. (b) $0 < c < 2$.

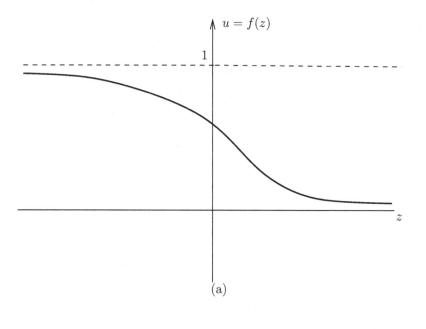

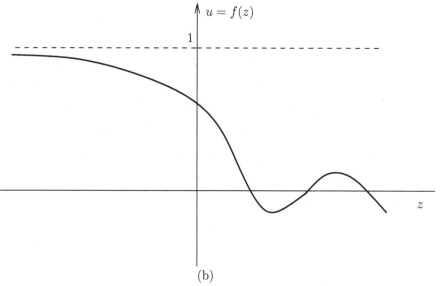

Fig. 10.5.2   Traveling wave solution to the Fisher equation. (a) $c > 2$. (b) $0 < c < 2$.

While we have given arguments in support of the existence of traveling wave solutions for the Fisher equation, no explicit solution has been derived. As briefly discussed earlier in this section, such an exact solution does not exist. However, we can calculate an approximate solution using a perturbation method for the case $c \geq 2$. We follow closely the presentation of Logan [7].

Our task is to construct a perturbation solution to Eq. (10.5.7), i.e.,

$$\begin{cases} f'' + cf' + f(1 - f) = 0 \\ f(-\infty) = 1, \qquad f(\infty) = 0, \end{cases} \qquad (10.5.19)$$

for $c \geq 2$. First, we note that $f(0)$ can be selected to be

$$f(0) = \frac{1}{2} \,. \qquad (10.5.20)$$

This follows from the fact that Eq. (10.5.19) does not depend explicitly on the independent variable z and, as a consequence, if $f(z)$ is a solution, then $f(z + z_0)$ is also a solution for any given value of the constant z_0. Since $f(z)$ is non-negative and bounded by one, $f(0)$ can be any value in this interval. Our choice is that given in Eq. (10.5.20).

We have to pick a "small" perturbation parameter ϵ. However, since the only parameter in Eq. (10.5.19) is c and since $c \geq 2$, the only possibility for ϵ is an inverse power of c, i.e.,

$$\epsilon = \frac{1}{c^a}, \qquad a > 0, \qquad (10.5.21)$$

where a is currently unknown. To proceed, make the following change of variables,

$$s = \epsilon^b z, \qquad h(s) = f\left(\frac{s}{\epsilon^b}\right). \qquad (10.5.22)$$

Therefore,

$$\frac{d}{dz} = \epsilon^b \frac{d}{ds}, \qquad \frac{d^2}{dz^2} = \epsilon^{2b} \frac{d^2}{ds^2}, \qquad (10.5.23)$$

and Eq. (10.5.19) becomes

$$\epsilon^{2b} h'' + c\epsilon^b h' + h(1 - h), \qquad (10.5.24)$$

where $h' = dh/ds$, etc. The selection

$$c\epsilon^b = 1, \qquad b = \frac{1}{2} \,, \qquad (10.5.25)$$

gives

$$\epsilon h'' + h' + h(1 - h) = 0, \qquad (10.5.26)$$

and determines a to be equal to two, i.e., the small expansion parameter for the perturbation procedure is

$$\epsilon = \frac{1}{c^2}. \tag{10.5.27}$$

The conditions given in Eq. (10.5.19) and (10.5.20) are now

$$h(-\infty) = 1, \quad h(0) = \frac{1}{2}, \quad h(+\infty) = 0. \tag{10.5.28}$$

We take the perturbation expansion of the solution $h(s, \epsilon)$, for Eq. (10.5.26), to be

$$h(s, \epsilon) = h_0(s) + \epsilon h_1(s) + O(\epsilon^2). \tag{10.5.29}$$

Substituting this expression into Eq. (10.5.26) and the conditions of Eq. (10.5.28), and setting the coefficients of ϵ^0 and ϵ equal to zero, gives

$$\epsilon^0 : \begin{cases} h_0' = -h_0(1 - h_0), \\ h_0(-\infty) = 1, \quad h_0(0) = \frac{1}{2}, \quad h_0(+\infty) = 0, \end{cases} \tag{10.5.30}$$

$$\epsilon : \begin{cases} h_1' = -h_1(1 - h_0) + h_0'', \\ h_1(-\infty) = h_1(0) = h_1(+\infty) = 0. \end{cases} \tag{10.5.31}$$

The differential equation in Eq. (10.5.30) can be solved by the use of separation of variables. Imposition of the initial condition, $h_0(0) = 1/2$, gives

$$h_0(s) = \frac{1}{1 + e^s}. \tag{10.5.32}$$

Note that the two boundary conditions of Eq. (10.5.30) are automatically satisfied! Likewise, the substitution of $h_0(s)$ into the right-side of the differential equation in Eq. (10.5.31), with the initial condition, $h_1(0) = 0$, gives an equation that can be solved for $h_1(s)$; this solution is

$$h_1(s) = \left[\frac{e^s}{(1 + e^s)^2} \right] \ln \left[\frac{4e^s}{(1 + e^s)^2} \right]. \tag{10.5.33}$$

Again, note that the boundary conditions, $h_1(-\infty) = 0$ and $h_1(+\infty) = 0$ are satisfied.

Putting all of these results together, the perturbation solution for the traveling wave of the Fisher equation is, for $c \geq 2$, given by the expression,

$$u(x, t) = f(z) = f(x - ct)$$

$$= \frac{1}{1 + \exp\left(\frac{z}{c}\right)} + \left(\frac{1}{c^2}\right) \frac{\exp\left(\frac{z}{c}\right)}{\left[1 + \exp\left(\frac{z}{c}\right)\right]^2} + O\left(\frac{1}{c^4}\right). \tag{10.5.34}$$

This approximation provides a rather accurate representation to the actual solution to Eq. (10.5.18) as can be verified by numerical integration.

10.6   The Korteweg-de Vries Equation

The first nonlinear PDE known to explicitly show solitary wave behavior is the Korteweg-de Vries equation [8]. The importance of this equation is derived in large measure from its applications to a broad range of phenomena in almost every branch of science and engineering. In the following material, we give a brief outline of how the single soliton solution can be obtained for a special set of boundary conditions. Reference [9] lists books providing additional information on this equation along with how to determine more general solutions and applications for which it arises.

The Korteweg-de Vries (KdV) equation is

$$u_t + uu_x + u_{xxx} = 0. \tag{10.6.1}$$

Note that this equation is similar looking to the Burgers' equation (see section 10.4) with the diffusion term replaced by the negative of the third-order space derivative. However, whereas the Burgers' equation had only traveling wave solutions, the KdV equation will be shown to have soliton solutions. To find this form of the solution, assume that $u(x,t)$ can be written as

$$u(x,t) = f(x - ct), \tag{10.6.2}$$

and assume that $f(z)$ and its first three derivatives vanish at $z = \pm\infty$, i.e.,

$$\underset{|z|\to\infty}{\text{Lim}} f^{(n)}(z) = 0; \qquad n = 0, 1, 2, 3. \tag{10.6.3}$$

If $u = f$ is substituted into Eq. (10.6.1), we obtain

$$-cf' + ff' + f''' = 0. \tag{10.6.4}$$

Integrating this equation once gives

$$-cf + \left(\frac{1}{2}\right) f^2 + f'' = 0, \tag{10.6.5}$$

where the constant of integration is taken to be zero so that the conditions of Eq. (10.6.3) are satisfied. Now multiply by f' and integrate once again the resulting expression; doing this gives

$$-\left(\frac{c}{2}\right) f^2 + \left(\frac{1}{6}\right) f^3 + \left(\frac{1}{2}\right) (f')^2 = 0, \tag{10.6.6}$$

which can be solved for f'. If the positive square-root is selected, then we obtain

$$\frac{\sqrt{3}}{f\sqrt{3c - f}} \cdot f' = 1. \tag{10.6.7}$$

Inspection of this expression shows that to have a real valued function, $f(z)$, then $f(z)$ must satisfy the following condition

$$0 \le f(z) \le 3c. \tag{10.6.8}$$

Now making the almost obvious, nonlinear change of variables,

$$g^2 = 3c - f, \tag{10.6.9}$$

and substituting in Eq. (10.6.7), we obtain

$$\left(\frac{2\sqrt{3}}{3c - g^2} \right) g' = -1. \tag{10.6.10}$$

This first-order ODE can be easily integrated; its solution is

$$g(z) = - \left(\sqrt{3c} \right) \tanh \left[\left(\frac{1}{2} \right) \sqrt{c}z \right]. \tag{10.6.11}$$

(The integration constant has been set to zero.) In terms of the function $f(z)$, we have

$$f(z) = (3c)\text{sech} \left[\left(\frac{1}{2} \right) \sqrt{c}z \right], \tag{10.6.12}$$

and, therefore, $u(x,t) = f(x - ct)$, is

$$u(x,t) = (3c)\text{sech}^2 \left[\left(\frac{\sqrt{c}}{2} \right) (x - ct) \right]. \tag{10.6.13}$$

Since the sech(y) function is defined to be

$$\text{sech}(y) = \frac{2}{e^y + e^{-y}}, \tag{10.6.14}$$

it is clear that $u(x,t)$, as given by Eq. (10.6.13) corresponds to a "pulse" traveling in the positive x-direction, if $c > 0$, with a speed of c and having a maximum amplitude of $3c$. A major conclusion is that large amplitude solitons (for the KdV equation) travel faster than small ones.

Finally, it should be mentioned that the KdV equation has other types of solutions including cnoidal trains of waves, multi-soliton solutions, rational solutions, etc. An excellent introduction to these and related topics are given in the book: L. Debnath, *Nonlinear Partial Differential Equations for Scientists and Engineering* (Birkhäuser, Boston, 1997); see chapter 9.

10.7   The Nonlinear Schrödinger Equation

The nonlinear Schrödinger equation is often called the cubic Schrödinger equation and takes the form

$$i\phi_t + \phi_{xx} + \gamma|\phi|^2\phi = 0, \qquad -\infty < x < \infty, \qquad t > 0, \qquad (10.7.1)$$

where $i = \sqrt{-1}$ and, for our purposes, γ is a real constants. The appearance of i leads to the conclusion that $\phi(x,t)$ is, in general, a complex valued function.

This nonlinear PDE arises in a wide variety of applications [9]. In particular, it provides a mathematical model for phenomena in the evolution of water waves, nonlinear optics, hydromagnetic and plasma waves, propagation of heat pulses in solids and of solitary waves in piezoelectric semiconductors.

The assumed form of the special solution for the nonlinear Schrödinger equation (NLSE) is taken to be

$$\phi(x,t) = f(z)e^{-i(mz-nt)}, \qquad (10.7.2)$$

where

$$z = x - ct, \qquad (10.7.3)$$

and (m,n) are constants. Note that this form is a generalization of what was assumed for the KdV equation studied in the previous section. Strictly speaking, it is not a traveling wave solution.

Substituting Eq. (10.7.2) into Eq. (10.7.1) gives

$$f'' + i(2m - c)f' + (n - m^2)f + \gamma|f|^2 f = 0, \qquad (10.7.4)$$

where $f' \equiv df/dz$. Since, (m,n,c) are not known at this stage of the calculation, we will require

$$2m - c = 0. \qquad (10.7.5)$$

The importance of doing this is that the resulting differential equation

$$f'' + (n - m^2)f + f^3 = 0 \qquad (10.7.6)$$

is now for a real valued $f(z)$. If we now define the new constant, α, by

$$\alpha = m^2 - n = \frac{c^2}{4} - n, \qquad (10.7.7)$$

then the ODE becomes

$$f'' - \alpha f + \gamma f^3 = 0. \qquad (10.7.8)$$

This ODE can be solved exactly in terms of a Jacobian elliptic function [10]. However, if we now require

$$\begin{cases} \operatorname*{Lim}_{|z|\to\infty} f^{(n)}(z) = 0; & n = 0, 1, 2, \\ \alpha > 0, & \gamma > 0, \end{cases} \qquad (10.7.9)$$

then the following simpler solution appears

$$f(z) = \left(\frac{2\alpha}{\gamma}\right)^{1/2} \operatorname{sech}\left(\sqrt{\alpha}z\right). \qquad (10.7.10)$$

Therefore, $\phi(x, t)$ is

$$\phi(x, t) = \left(\frac{2\alpha}{\gamma}\right)^{1/2} e^{-i[m(x-ct)-nt]} \cdot \operatorname{sech}\left[\sqrt{\alpha}(x - ct)\right], \qquad (10.7.11)$$

and this represents the one solitary wave solution for the NLSE.

In physical applications, the absolute square of ϕ plays an important role; this is given by the expression

$$|\phi(x, t)|^2 = \left(\frac{2\alpha}{\gamma}\right) \operatorname{sech}^2\left[\sqrt{\alpha}(x - ct)\right]. \qquad (10.7.12)$$

Note that $|\phi|^2$ is proportional to the same function as appears for the single soliton solution of the KdV equation of the last section. However, for the NLSE, the amplitude and velocity of the waveform are independent parameters.

10.8 Similarity Methods and Solutions

We now present a powerful method for determining special solutions to partial differential equations. It can be applied to both linear and nonlinear PDE's. The basic technique uses the algebraic symmetry of the PDE to reduce it to a related ordinary differential equation. The solutions to the ODE are called similarity solutions and often provide the required original solution to the PDE with its associated initial and/or boundary conditions.

A major advantage of the similarity method is that the ODE may be easier to solve than the PDE. In case no usable analytical solution is available for the ODE, the effort to numerically integrate the ODE is generally much smaller than attempting this for the PDE.

In this introduction to similarity methods and how they can be used to determine special solutions, we first give several definitions and follow these by several theorems, presented without proof. References to literature which explain these concepts, ideas, and their application with all the required detail can be found in [11].

10.8.1 Similarity Methods

This section is based on the discussion of similarity methods as presented by Logan.

Consider a first-order PDE expressed in the form

$$F(x, t, u, p, q) = 0, \tag{10.8.1}$$

where

$$u = u(x, t), \qquad p = u_x, \qquad q = u_t. \tag{10.8.2}$$

Definition A one-parameter set of stretching transformations in the (x, t, u)-space is a transformation of the form

$$x \to \bar{x} = \epsilon^a x, \quad t \to \bar{t} = \epsilon^b t, \quad u \to \bar{u} = \epsilon^c u, \tag{10.8.3}$$

where (a, b, c) are fixed, real constants and ϵ is a real parameter which belongs to an open interval I containing $\epsilon = 1$. This set of transformations will be denoted by T_ϵ.

Note that the transformation, T_ϵ, transforms the derivatives in the following way,

$$p \to \bar{p} = \epsilon^{c-a} p, \qquad q \to \bar{q} = \epsilon^{c-b} q. \tag{10.8.4}$$

Definition The PDE, of Eq. (10.8.1), is invariant under T_ϵ if and only if there exist a smooth function $h(\epsilon)$ such that

$$F(\bar{x}, \bar{t}, \bar{u}, \bar{p}, \bar{q}) = h(\epsilon) F(x, t, u, p, q), \tag{10.8.5}$$

for all ϵ in the open interval I, with

$$h(1) = 1. \tag{10.8.6}$$

Theorem 10.8.1 *If the PDE, given by Eq. (10.8.1), is invariant under T_ϵ, then the transformation*

$$u(x, t) = t^{c/b} y(z), \qquad z = \frac{x}{t^{a/b}}, \tag{10.8.7}$$

reduces the PDE to a first-order ODE in $y(z)$ of the form

$$f(z, y, y') = 0. \tag{10.8.8}$$

The new independent variable z is called a similarity variable, while the transformation in Eq. (10.8.7) is called a similarity transformation. The form for $u(x, t)$, in Eq. (10.8.7), where it is expressed in terms of t and $y(z)$, is called the self-similar form of the solution for the original PDE. To obtain it, first Eq. (10.8.8) is solved for $y(z)$ and this result is placed back into Eq. (10.8.7). Then z can be replaced by $x/t^{a/b}$ to give the self-similar solution for $u(x, t)$.

Theorem 10.8.2 *If the second-order PDE*

$$F(x, t, u, u_x, u_t, u_{xx}, u_{xt}, u_{tt}) = 0 \qquad (10.8.9)$$

is invariant under T_ϵ, *then the transformation*

$$u(x, t) = t^{c/b} y(z), \qquad z = \frac{x}{t^{a/b}}, \qquad (10.8.10)$$

reduces the PDE to a second-order ODE in $y(z)$ *of the form*

$$f(z, y, y', y'') = 0. \qquad (10.8.11)$$

Note that T_ϵ transforms the second-derivatives as follows

$$r \to \bar{r} = \epsilon^{c-2a} r, \quad s \to \bar{s} = \epsilon^{c-a-b} s, \quad v \to \bar{v} = \epsilon^{c-2b} v, \qquad (10.8.12)$$

where

$$r = u_{xx}, \quad s = u_{xt}, \quad v = u_{tt}. \qquad (10.8.13)$$

10.8.2   Examples

Example A

The diffusionless Bergers' equation is

$$u_t + u u_x = 0. \qquad (10.8.14)$$

In terms of (p, q), as defined in Eq. (10.8.2), we have

$$q + up = 0. \qquad (10.8.15)$$

From Eqs. (10.8.3) and (10.8.4), it follows that

$$\bar{q} + \bar{u}\bar{p} = \epsilon^{c-b} q + \epsilon^{2c-a} up. \qquad (10.8.16)$$

If we require

$$-b = c - a, \qquad (10.8.17)$$

then

$$\bar{q} + \bar{u}\bar{p} = \epsilon^{2c-a}(q + up), \qquad (10.8.18)$$

and we conclude that the transformation

$$x \to \bar{x} = \epsilon^a x, \quad t \to \bar{t} = \epsilon^{a-c}, \quad u \to \bar{u} = \epsilon^c u, \qquad (10.8.19)$$

for arbitrary constants a and c, leaves Eq. (10.8.16) invariant.

Note that $h(\epsilon)$, as defined in Eq. (10.8.5) is $h(\epsilon) = \epsilon^{2c-a}$ and $h(1) = 1$.

The similarity transformations can now be determined from Eq. (10.8.7); they are

$$u = t^{\left(\frac{c}{a-c}\right)} y(z), \qquad z = \frac{x}{t^{\left(\frac{a}{a-c}\right)}} . \qquad (10.8.20)$$

If these relations are now substituted into Eq. (10.8.14) and the chain rule is used to calculate u_x and u_t, then the following first-order ODE is obtained for $y(z)$,

$$yy' + \left(\frac{a}{a-c}\right) zy' + \left(\frac{c}{c-a}\right) y = 0. \qquad (10.8.21)$$

Note that this is both a nonlinear and nonautonomous equation. The constants a and b are usually selected such that certain initial and/or boundary values are satisfied. This means, in general, their values will depend on the particular problem studied.

Example B

Consider now a problem having linear diffusion. The PDE for this case is

$$u_t = Du_{xx}, \qquad (10.8.22)$$

where D is the constant, positive diffusion coefficient and we will study it for the space-time domain

$$0 \le x < \infty, \qquad t > 0, \qquad (10.8.23)$$

having the following boundary and initial conditions,

$$u(0,t) = 1, \qquad (10.8.24)$$

$$u(0,t) \to 0, \quad \text{as } x \to \infty, \quad t > 0, \qquad (10.8.25)$$

$$u(x,0) = 0, \quad 0 < x < \infty. \qquad (10.8.26)$$

The physical interpretation for this problem is that it models a system for which diffusion occurs into the region $x > 0$; the concentration in $x > 0$ is initially zero; and a constant concentration, $u = 1$, is maintained at $x = 0$.

The above linear diffusion equation can rewritten to the form (see the notation in Eqs. (10.8.2) and (10.8.13)),

$$q - Dr = 0. \qquad (10.8.27)$$

Under the stretching transformation T_ϵ, it follows that

$$\bar{q} - D\bar{r} = \epsilon^{c-b} - D\epsilon^{c-2a} r = \epsilon^{c-b}(q - Dr), \qquad (10.8.28)$$

if

$$b = 2a. \tag{10.8.29}$$

Therefore, for arbitrary a and c, Eq. (10.8.27) is invariant under T_ϵ defined as

$$x \to \bar{x} = \epsilon^a x, \quad t \to \bar{t} = \epsilon^{2a} t, \quad u \to \bar{u} = \epsilon^c u. \tag{10.8.30}$$

From this result, we obtain the similarity transformations

$$u = t^{c/2a} y(z), \quad z = \frac{x}{\sqrt{t}}, \tag{10.8.31}$$

and on substitution into Eq. (10.8.22) gives a second-order, nonautonomous ODE for $y(z)$, namely,

$$Dy'' + \left(\frac{z}{2}\right) y' - \left(\frac{c}{2a}\right) y = 0. \tag{10.8.32}$$

The conditions in Eqs. (10.8.25) and (10.8.26) force $y(z)$ to take the value

$$y(\infty) = 0. \tag{10.8.33}$$

Now imposing the requirement of Eq. (10.8.24) gives

$$u(0, t) = t^{c/2a} y(0) = 1, \quad t > 0, \tag{10.8.34}$$

and this implies that $c = 0$ and $y(0) = 1$. Therefore, the second-order ODE for $y(z)$ becomes

$$\begin{cases} y'' + \left(\frac{z}{2D}\right) y' = 0, & z > 0 \\ y(0) = 1, & y(\infty) = 0, \end{cases} \tag{10.8.35}$$

and we have a boundary value problem to solve.

This differential equation is almost trivial to solve and its solution is

$$y(z) = c_1 + c_2 \int_0^z \exp\left[-\left(\frac{s^2}{4D}\right)\right] ds, \tag{10.8.36}$$

where c_1 and c_2 are constants that can be found by the following arguments. Since $y(0) = 1$, then

$$y(0) = c_1 = 1. \tag{10.8.37}$$

The other boundary condition gives

$$y(\infty) = 1 + c_2 \int_0^\infty \exp\left[-\left(\frac{s^2}{4D}\right)\right] ds = 0. \tag{10.8.38}$$

Since the error function, $\text{erf}(z)$, is defined to be

$$\text{erf}(z) \equiv \frac{2}{\sqrt{\pi}} \int_0^z \exp(-v^2)dv, \qquad (10.8.39)$$

$y(z)$ can be written as

$$y(z) = 1 + c_2\sqrt{\pi D} \cdot \text{erf}\left(\frac{z}{\sqrt{4D}}\right). \qquad (10.8.40)$$

Hence, Eq. (10.8.38) becomes

$$1 + c_2\sqrt{\pi D} \cdot \text{erf}(\infty) = 0. \qquad (10.8.41)$$

Using $\text{erf}(\infty) = 1$, we obtain

$$c_2 = -\left(\frac{1}{\sqrt{\pi D}}\right), \qquad (10.8.42)$$

and $y(z)$ is

$$y(z) = 1 - \text{erf}\left(\frac{z}{\sqrt{4D}}\right). \qquad (10.8.43)$$

Finally, we can now write down the similarity solution for the linear diffusion equation having constant diffusion coefficient, and satisfying the conditions of Eqs. (10.8.24)–(10.8.26); it is

$$u(x,t) = 1 - \text{erf}\left(\frac{x}{\sqrt{4Dt}}\right). \qquad (10.8.44)$$

10.9   The Boltzmann Problem

Physically, the Boltzmann problem corresponds to a "detonation" event, i.e., some material of "total amount one" is introduced at $x = 0$, at time $t = 0$, and then allowed to diffuse according to the following nonlinear PDE,

$$u_t = (uu_x)_x. \qquad (10.9.1)$$

The initial, boundary, and conservation conditions are taken to be

$$u(x,0) = \delta(x), \qquad (10.9.2)$$

$$u(\pm\infty, t) = 0, \quad t > 0, \qquad (10.9.3)$$

$$\int_{-\infty}^{\infty} u(x,t)dx = 1, \quad t > 0. \qquad (10.9.4)$$

The T_ϵ set of transformations that leave Eq. (10.9.1) invariant are

$$x \to \bar{x} = \epsilon^a x, \quad t \to \bar{t} = \epsilon^b t, \quad u \to \bar{u} = \epsilon^{2b-a}u, \qquad (10.9.5)$$

and the corresponding similarity transformation is

$$u(x,t) = t^{\frac{2a-b}{b}} y(z), \qquad z = \frac{x}{t^{a/b}} \, . \tag{10.9.6}$$

To proceed, substitute $u(x,t)$ from this last equation into the condition of Eq. (10.9.4) and obtain

$$t^{\left(\frac{2a-b}{b}\right)} \int_{-\infty}^{\infty} y\left(\frac{x}{t^{a/b}}\right) dx = t^{\frac{3a-b}{b}} \int_{-\infty}^{\infty} y(z)dz = 1. \tag{10.9.7}$$

Now the integral over z is assumed to exist and have some definite, finite value. Also, the extreme right-side of this equality is independent of t and, in fact, is equal to one. The only way we can have an expression that is independent of t is to require

$$\frac{a}{b} = \frac{1}{3} \, . \tag{10.9.8}$$

This gives

$$u(x,t) = t^{-1/3} y(z) = \frac{z}{t^{1/3}} \, . \tag{10.9.9}$$

The second-order ODE for $y(z)$ can now be determined by substitution of Eq. (10.9.9) into the Boltzmann Eq. (10.9.1); the result of doing this is the equation

$$3(yy') + zy' + y = 0. \tag{10.9.10}$$

Using the fact that

$$(zy)' = zy' + y, \tag{10.9.11}$$

allows one integration of Eq. (10.9.10) to be done, giving

$$3yy' + zy = C, \tag{10.9.12}$$

where C is the integration constant. Now the nature of the problem implies that $y(z)$ must be an even function of z. This further implies that $y'(0) = 0$ and from (10.9.12) the result $C = 0$ follows, i.e.,

$$3y' + z = 0 \quad \text{or} \quad y = 0. \tag{10.9.13}$$

Clearly, the solution $y(z) = 0$ is not the one we need. The other solution is

$$y(z) = \frac{A^2 - z^2}{6} \, , \tag{10.9.14}$$

where A is an arbitrary constant. Three conditions must now be imposed on this $y(z)$:

(i) $y(z) \geq 0$, for $0 < z < +\infty$

(ii) $y(+\infty) = 0,$

(iii) $y(z)$ must be continuous for $0 \le z < \infty.$

The only way to satisfy the three conditions is to take $y(z)$ to be

$$y(z) = \begin{cases} \dfrac{A^2 - z^2}{6}, & |z| < A, \\ 0, & |z| > A. \end{cases} \tag{10.9.15}$$

The constant A can be calculated from the condition derived from Eqs. (10.9.7) and (10.9.8), namely,

$$1 = \int_{-\infty}^{\infty} y(z)dz = \int_{-A}^{A} y(z)dz = 2\int_{0}^{A} y(z)dz = \left(\frac{2}{9}\right)A^3, \tag{10.9.16}$$

or

$$A = \left(\frac{9}{2}\right)^{1/3}. \tag{10.9.17}$$

Therefore,

$$u(x,t) = t^{-1/3}y\left(\frac{x}{t^{1/3}}\right), \tag{10.9.18}$$

and, finally,

$$u(x,t) = \begin{cases} \left(\dfrac{1}{6t}\right)\left[\left(\dfrac{9}{2}\right)^{2/3}t^{2/3} - x^2\right], & |x| < \left(\dfrac{9}{2}\right)^{1/3}t^{1/3}, \\ 0, & |x| > \left(\dfrac{9}{2}\right)^{1/3}t^{1/3}. \end{cases} \tag{10.9.19}$$

Figure 10.9.1 gives plots of $u(x,t)$ for three different times, $t_1 < t_2 < t_3$. The areas of all three curves are the same. This follows directly from the condition of Eq. (10.9.4). Note that for any given value of t, there is a sharp wavefront beyond which the the density of the material is zero. At this wavefront, the function $u(x,t)$ is continuous, but its x-derivative is discontinuous. The location of the wavefront, $x_f(t)$, and the speed of the wavefront, dx_f/dt, are given by the following expressions

$$x_f(t) = \left(\frac{9}{2}\right)^{1/3}t^{1/3}, \tag{10.9.20}$$

$$\frac{dx_f(t)}{dt} = \left(\frac{1}{3}\right)\left(\frac{9}{2}\right)^{1/3}\left(\frac{1}{t^{2/3}}\right). \tag{10.9.21}$$

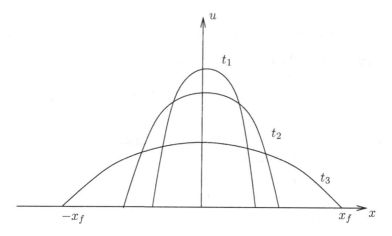

Fig. 10.9.1   The solution at three times: $t_1 < t_2 < t_3$.

10.10   The Nonlinear Diffusion Equation: $uu_t = u_{xx}$

Consider the following nonlinear PDE

$$uu_t = u_{xx}, \qquad (10.10.1)$$

which corresponds to a nonlinear heat conduction process. The initial and boundary conditions are taken to be,

$$u(x, 0) = 0, \qquad x > 0, \qquad (10.10.2)$$

$$u(\infty, t) = 0, \qquad t > 0, \qquad (10.10.3)$$

$$u_x(0, t) = -1, \qquad t > 0. \qquad (10.10.4)$$

This particular problem has been studied in great detail by Dresner [12] and arises in the study of the thermal expulsion of fluid from a long, slender heated tube. For this situation u represents the flow velocity induced in the fluid by the heating of the tube wall.

Now Eq. (10.10.1) is invariant under the T_ϵ transformation

$$x \to \bar{x} = \epsilon^a x, \quad t \to \bar{t} = \epsilon^b t, \quad u \to \bar{u} = \epsilon^{b-2a} u, \qquad (10.10.5)$$

and these give the following similarity transformation

$$u(x, t) = t^{\frac{b-2a}{b}} y(z), \qquad z = \frac{x}{t^{a/b}}. \qquad (10.10.6)$$

To determine what possibilities exist for the constants a and b, we first examine the condition given in Eq. (10.10.4). Now

$$u_x(x, t) = t^{\frac{b-3a}{b}} y\left(\frac{x}{t^{a/b}}\right) \qquad (10.10.7)$$

and

$$u_x(0,t) = t^{\frac{b-3a}{b}} y'(0) = -1. \tag{10.10.8}$$

Since the right-side does not depend on t, it follows that

$$\frac{a}{b} = \frac{1}{3}. \tag{10.10.9}$$

This means that the similarity transformation can be written as

$$u(x,t) = t^{1/3} y(z), \qquad z = \frac{x}{\sqrt{3}t^{1/3}}, \tag{10.10.10}$$

where the z-variable has been modified to include an additional factor of $\sqrt{3}$ in the denominator. (This change is made by most authors studying Eq. (10.10.1) and has the consequences of simplifying the equation for $y(z)$.)

The condition given by Eq. (10.10.3) allows the following conclusion to be reached

$$y(\infty) = 0. \tag{10.10.11}$$

Likewise, for the initial condition requirement of Eq. (10.10.2), we have

$$\operatorname*{Lim}_{t \to 0} u(x,t) = \operatorname*{Lim}_{t \to 0} t^{1/3} y\left(\frac{x}{\sqrt{3}t^{1/3}}\right) = \operatorname*{Lim}_{z \to \infty}\left(\frac{x}{\sqrt{3}z}\right) y(z) = 0. \tag{10.10.12}$$

Note that the requirement in Eq. (10.10.4), on using Eq. (10.10.10), gives the other boundary condition for $y(z)$, i.e.,

$$y'(0) = -\sqrt{3}. \tag{10.10.13}$$

If Eq. (10.10.10) is now substituted into Eq. (10.10.1), the following ODE is obtained for $y(z)$,

$$y'' - y(y - zy') = 0, \qquad z > 0. \tag{10.10.14}$$

This equation is to be solved subject to the two conditions for Eqs. (10.10.11) and (10.10.13).

The book of Dresner discusses a variety of techniques that can be applied to obtain information on the required solution to Eq. (10.10.14). In particular, the value of $y(0)$ is *a priori* unknown. Further, Eq. (10.10.14) does not have any known exact solution that can be used to help us make this determination. Dresner applies a phase-space method combined with numerical integration of some related ODE's to both estimate $y(0)$ and also obtain an approximate solution. He finds $y(0) = 1.5111$ and

$$y(z) \simeq \frac{1.5111}{1 + (1.1462)z + (0.7556)z^2}. \tag{10.10.15}$$

Problems

Section 10.1

10.1.1 Let $u(x,0)$ be defined as follows

$$u(x,0) = \begin{cases} 1, & \text{for } |x| < 1, \\ 0, & \text{for } |x| > 1. \end{cases}$$

For both the linear undamped and damped wave equations determine $u(x,t)$. Plot $u(x,0)$ vs x, and $u(x,s)$ and $u(x,10)$ vs t on the same graph.

10.1.2 Calculate a general solution to

$$u_t - u_x = -\lambda u.$$

10.1.3 Explain why in Eq. (10.1.17), no arbitrary constants multiplying f and g are required.

10.1.4 Calculate the result given in Eq. (10.1.24).

10.1.5 Explain why $u(x,0) = e^{-|x|}$ cannot be used as an initial condition for either of the linear wave equations.

Section 10.3

10.3.1 Solve Eq. (10.3.5) to obtain its solution as given by Eq. (10.3.6).

10.3.2 Consider the evolution of the solution, as given by Eq. (10.3.9), for the initial function

$$u(x,0) = f(x) = \begin{cases} 1, & |x| < 1, \\ 0, & |x| > 1. \end{cases}$$

Sketch the behavior of $u(x,t)$ for values of $t > 0$.

10.3.3 Carry out the same analysis, as that requested in the previous problem, using

$$u(x,0) = f(x) = e^{-x^2}.$$

Section 10.4

10.4.1 Calculate $\mathrm{Lim}_{D \to 0} f(z)$ for Eq. (10.4.14).

10.4.2 Show that for $D = 0$, the Burgers' equation becomes

$$u_t + u u_x = 0,$$

and that with the initial value

$$u(x,0) = g(x) \text{ given,}$$

an exact solution is the implicit expression

$$u(x,t) = g[x - u(x,t)t].$$

Section 10.5

10.5.1 Show that $u = 0$ and $u = 1$ represent, respectively, linearly unstable solutions for the Fisher PDE. Hint: Linearize the equation at these solutions and calculate the affect of a small perturbation.

10.5.2 Plot the phase-space trajectories g vs f, and $f(z)$ vs z for the case $c = 2$.

10.5.3 Solve the ODE's in Eqs. (10.5.30) and (10.5.31), respectively for $h_0(s)$ and $h_1(s)$, and show that the solutions are given by the expressions in Eqs. (10.5.32) and (10.5.33).

10.5.4 Since the small parameter ϵ multiplies the highest derivative in Eq. (10.5.26), why is this not a problem in singular perturbations? Hint: See the discussion in Logan [7].

10.5.5 With the change of notation, $f \to x$ and $z \to t$, the differential equation in Eq. (10.5.18) can be considered a 1-dimensional dynamical system moving in the potential energy

$$V(x) = \frac{x^2}{2} - \frac{x^2}{4},$$

and acted upon by a linear damping force. Analysis the behavior of the phase-space for $c \geq 0$ and compare with the corresponding analysis for the Fisher equation.

Section 10.6

10.6.1 Explain why the positive square root sign can be used to derive Eq. (10.6.7). What occurs for the case where the negative sign is selected?

10.6.2 Derive Eq. (10.6.10) from Eq. (10.6.7) using the transformation of dependent variable given in Eq. (10.6.9).

10.6.3 Discuss the soliton solution if $c < 0$.

10.6.4 Does the KdV have a soliton solution if the integration constants are not taken to be zero?

Section 10.7

10.7.1 Why is the expression given in Eq. (10.7.2) not a traveling wave solution?

10.7.2 Derive Eq. (10.7.4).

10.7.3 Study the properties of $|\phi(x,t)|^2$ for various sign possibilities for the constants α and γ.

Section 10.8

10.8.1 Show that the set of all transformations $\{T_\epsilon\}$ form a group, i.e.,
   (a) composition law: $T_a T_b = T_{ab}$,
   (b) commutative law: $T_{ab} = T_{ba}$,
   (c) associative law: $T_a(T_b T_c) = (T_a T_b)T_c$,
   (d) existence of an identity: T_1,
   (e) existence of an inverse: T_a is the inverse of $T_{a^{-1}}$,
   where (a, a^{-1}, b) are in I.

10.8.2 Prove Theorems 10.8.1 and 10.8.2.

10.8.3 Carry out the necessary calculations to derive the result given by Eq. (10.8.21).

10.8.4 Solve the differential equation in Eq. (10.8.35) and obtain $Y(z)$ given by Eq. (10.8.36).

10.8.5 Determine the set of transformations, T_ϵ, under which the following equations are invariant:
   (a) $u_t + u_x = 0$,
   (b) $u_{tt} - u_{xx} = 0$,
   (c) $u u_t + (u_x)^2 = 0$.
   Calculate the similarity transformation and the corresponding ODE associated with the PDE for each equation. If possible, solve the ODE's.

Section 10.9

10.9.1 Show that the T_ϵ set of transformations for the Boltzmann problem is given by Eq. (10.9.5). Further show that the similarity transformation is contained in Eq. (10.9.6).

10.9.2 Give the full details of the argument as to why $C = 0$ in Eq. (10.9.12).

10.9.3 Show that using Eq. (10.9.19) we have

$$\int_{-\infty}^{\infty} u(x, t)dx = 1, \qquad t > 0.$$

Section 10.10

10.10.1 Show that Eq. (10.10.1) is invariant under the transformation T_ϵ given in Eq. (10.10.5).

10.10.2 Work out the full details of the calculation given in Eq. (10.10.12).

10.10.3 Physically, Eq. (10.10.1) models a nonlinear heat conduction process. For such a system, it is expected that the following two conditions hold: $y(z) > 0$ and $y'(z) < 0$, for $0 \leq z < \infty$. Using Eq. (10.10.14) prove that

$$(-1)^n \frac{d^n y}{dz^n} > 0, \qquad 0 \leq z < \infty.$$

10.10.4 Analyze the following nonlinear diffusion problem using similarity methods [12],

$$u_t = (u_x^{1/3})_x, \quad x > 0, \quad t > 0,$$
$$u_x(0, t) = -1, \qquad t > 0,$$
$$u(x, 0) = 0, \qquad x > 0,$$
$$u(\infty, t) = 0, \qquad t > 0.$$

Comments and References

[1] For a more detailed discussion of evolutionary equations see the books of Britton, Fife, Murray, Smoller and Whitham.

[2] In the sections to follow, we will see how this plays out for particular nonlinear PDE's. A good introduction to this topic is provided by the book of Logan.

[3] For a variety of methods that can be used to solve first-order, linear, inhomogeneous PDE's see the book of Zauderer.

[4] The Fisher PDE provides models for the dynamics of many important systems that can be represented in terms of a single space dimension. Particular applications include:
* gene-culture waves: K. Aoki, *Math. Biology* **25**, 453 (1987).
* chemical waves: R. Arnold, K. Showalter, and J. J. Tyson, *J. Chem. Educ.* **64**, 740 (1987).
* early farming in Europe: A. J. Ammerman and L. L. Cavalli-Sforza, *Man* **6**, 674 (1971).
* advance of genes: R. A. Fisher, *Ann. Eugenics* **7**, 335 (1936).

[5] A. Kolmogorov, I. Petrovsky, and N. Piscunov, *Bull. Univ. Moscow, Ser. Internat. Sec. A.* **1**, 1–25 (1973).

[6] All of the results regarding the trajectories in the phase-plane can be proved rigorously. A good starting point is the following reference: P. Hartman, *Ordinary Differential Equations* (Wiley, New York, 1964).

[7] See the book of Logan, section 4.4.

[8] A. T. Filipov, *The Versatile Soliton* (Birkhäuser, Boston, 2000).

[9] The Korteweg-de Vries and nonlinear Schrödinger equations play important roles in providing models of dynamical systems. The following books are useful for the more advanced study of these interesting nonlinear PDE's:
* R. Knoebel, *An Introduction to the Mathematical Theory of Waves* (American Mathematical Society - Institute for Advanced Studies; Providence, RI; 2000).
* L. Debnath, *Nonlinear Partial Differential Equations* (Birkhäuser, Boston, 1997).
* B. Kursunoglu, et al. editors, *The Significance of Nonlinearity in the Natural Sciences* (Plenum, New York, 1977).

[10] See Debnath in [9] and P. F. Byrd and M. S. Friedman, *Handbook of Elliptic Integrals for Engineers and Physicists* (Springer-Verlag, Berlin, 1954).

[11] The following two books give excellent quick introductions to the essentials of similarity methods and their application to diffusion PDE's:

 * L. Debnath, *Nonlinear Partial Differential Equations* (Birkhäuser, Boston, 1977); section 8.11.

 * J. D. Logan, *Applied Mathematics: A Contemporary Approach* (Wiley-Interscience, New York, 1987).

 Two standard texts on this subject are the books, listed in the bibliography to this chapter, by Bluman and Kumei, and Dresner.

[12] L. Dresner, "Thermal Expulsion of Helium from a Quenching Cable-in-Conduit Conductor," in *Proc. of the* 9th *Symposium on the Engineering Problems of Fuson Research* (IEEE Publication No. 81CH1715-2NPS; Chicago, IL; Oct. 26–29, 1981); pp. 618–621.

Bibliography

J. Bebernes and D. Eberly, *Mathematical Problems from Combustion Theory* (Springer-Verlag, New York, 1989).

G. W. Bluman and S. Kumei, *Symmetries and Differential Equations* (Springer-Verlag, New York, 1989).

N. F. Britton, *Reaction-Diffusion Equations and Their Applications in Biology* (Academic Press, London, 1986).

D. Colton, *Partial Differential Equations* (Random House, New York, 1988).

R. Courant and K. O. Friedrichs, *Supersonic Flow and Shock Waves* (Springer-Verlag, New York, 1976).

L. Dresner, *Similarity Solutions of Nonlinear Partial Differential Equations* (Pitman Books, London, 1983).

W. Fickett, *Introduction to Detonation Theory* (University of California Press, Berkeley, 1985).

P. C. Fife, *Mathematical Aspects of Reacting and Diffusion Systems* (Springer-Verlag, New York, 1979).

F. John, *Partial Differential Equations*, 4th ed. (Springer-Verlag, New York, 1982).

J. Kevorkian and J. D. Cole, *Perturbation Methods in Applied Mathematics* (Springer-Verlag, New York, 1981).

H. O. Kreiss and J. Lorentz, *Initial-Boundary Value Problems and the Navier-Stokes Equations* (Academic Press, New York, 1989).

C. C. Lin and L. A. Segel, *Mathematics Applied to Deterministic Problems in the Natural Sciences* (Macmillan, New York, 1974. Reprinted by the Society for Industrial and Applied Mathematics, Philadelphia).

J. D. Logan, *An Introduction to Nonlinear Partial Differential Equations* (Wiley-Interscience, New York, 1994).

J. D. Murray, *Lectures on Nonlinear-Differential-Equation Models in Biology* (Oxford University Press, Oxford, 1977).

T. Myint-U, *Partial Differential Equations for Scientists and Engineers*, 3rd. ed. (North-Holland, New York, 1987).

A. Okubo, *Diffusion and Ecological Problems: Mathematical Models* (Springer-Verlag, New York, 1980).

J. Smoller, *Shock Waves and Reaction-Diffusion Equations* (Springer-Verlag, New York, 1983).

G. B. Whitham, *Linear and Nonlinear Waves* (Wiley-Interscience, New York, 1974).

E. Zauderer, *Partial Differential Equations of Applied Mathematics*, 2nd ed. (Wiley-Interscience, New York, 1989).

Chapter 11

Generalized Periodic Functions

11.1 Scope of this Chapter

The purpose of this chapter is to introduce various generalizations of the standard trigonometric functions. We will discover that new classes of periodic functions exist under a variety of circumstances involving, for example, the modeling of physical systems and determining solutions to so-called functional equations (see Chapter 12). As a reminder, we note that if $f(t)$ is a periodic function, with period T, then

$$f(t + T) = f(t), \quad T > 0. \tag{11.1.1}$$

In general, if $f(t)$ is defined on $0 \le t \le T$, then it may have discontinuities in this interval and may not be bounded [1].

We begin this chapter with the analysis of a dynamic system consisting of a particle undergoing elastic collisions in a one-space dimension box. The solutions are easy to calculate and they have the important feature of being periodic. Next, we consider a second dynamic system corresponding to a mechanical system for which the energy function is

$$H(x, y) = |x| + |y| = \text{constant}, \tag{11.1.2}$$

where $y = dx/dt$. Section 11.4 is devoted to the derivation of particular periodic solutions to the functional equation

$$f(t)^2 + g(t)^2 = 1, \tag{11.1.3}$$

based on the Euler relations

$$e^{i\theta(t)} = \cos\theta(t) + i\sin\theta(t). \tag{11.1.4}$$

Section 11.5 presents results on the Leah-cosine and -sine functions; these functions are related to solutions of a nonlinear oscillating system modeled by the differential equation

$$\frac{d^2x}{dt^2} + x^{1/3} = 0. \tag{11.1.5}$$

429

Finally, we close the chapter with a general discussion on periodic functions which may arise in applied mathematics and the natural sciences.

11.2   Particle-in-a-Box

One of the most elementary and famous problems in quantum mechanics is the particle-in-a-box (PIAB) [2]. We will treat the corresponding classical system.

The PIAB classical problem can be stated as follows: consider a particle, having mass m, confined to a one-dimensional space of length L. It is assumed that for $0 < x < L$, no force acts on the particle, and that at both $x = 0$ and $x = L$, the particle undergoes elastic collisions with the walls. Under these conditions determine $x(t)$.

It should be clear that $x(t)$ is periodic, with the motion completely determined by the initial conditions which are taken to be

$$x(0) = 0, \quad \frac{dx(0)}{dt} = v_0 > 0, \qquad (11.2.1)$$

where v_0 is the initial velocity.

Applying Newton's force law

$$m\frac{d^2x}{dt^2} = 0, \quad 0 < x < L, \qquad (11.2.2)$$

it follows that $x(t)$ has the form

$$x(t) = a + bt \qquad (11.2.3)$$

where the constants a and b must be determined from the conditions of the particle at the walls.

Note that the ball leaves the left wall at $t = 0$, with $x(0) = 0$ and $dx(0)/dt = v_0 > 0$, and arrives at the right wall at $t = T^*$, where T^* is

$$T^* = \frac{L}{v_0}. \qquad (11.2.4)$$

At the right wall there is a reversal of velocity and the particle takes a time T^* to return to the left wall. Hence, the total time of travel from the left wall to its return is $T = 2T^*$. Since this is a repeating motion, the period is $T = 2T^*$ or

$$T = \frac{2L}{v_0}. \qquad (11.2.5)$$

To calculate $x(t)$, we observe that from $t = 0$ to $t = T^* = T/2$, $x(t)$ must satisfy the initial conditions $x(0) = 0$ and $dx(0)/dt = v_0 > 0$. Using this in Eq. (11.2.3) gives

$$\begin{cases} x(t) = v_0 t, \\ v(t) = \frac{dx(t)}{dt} = v_0, \end{cases} \quad 0 \le t < T^* = \frac{T}{2}. \tag{11.2.6}$$

At $t = T^*$, the initial conditions are

$$x(T^*) = L, \quad \frac{dx(T^*)}{dt} = -v_0. \tag{11.2.7}$$

(Note the minus sign on the right-side of the second expression.) Therefore, we find

$$\begin{cases} x(t) = 2L - v_0 t, \\ v(t) = -v_0, \end{cases} \quad \frac{T}{2} < t \le T. \tag{11.2.8}$$

Combining the results of Eqs. (11.2.6) and (11.2.8), gives

$$x(t) = \begin{cases} v_0 t, & 0 \le t \le \frac{T}{2}, \\ 2L - v_0 t, & \frac{T}{2} \le t \le T, \end{cases} \tag{11.2.9}$$

and

$$v(t) = \begin{cases} v_0, & 0 < t < \frac{T}{2}, \\ -v_0, & \frac{T}{2} < t < T. \end{cases} \tag{11.2.10}$$

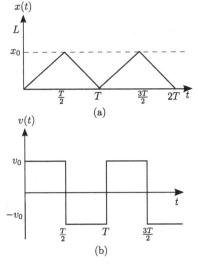

Fig. 11.2.1   Plots of $x(t)$ and $v(t)$ for the PIAB problem.

These results demonstrate that (i) both $x(t)$ and $v(t)$ are periodic with period $T = 2L/v_0$, (ii) $x(t)$ is continuous, and (iii) $v(t)$ is piecewise continuous, having finite discontinuities at times $t_k = kT/2$, where $k =$ integer.

Figure 11.2.1 give plots of $x(t)$ and $v(t)$.

This type of periodic function is often called a "triangle wave" for its triangular shape [3].

11.3 Square Periodic Functions [4]

11.3.1 *Hamiltonian Formulation*

The square periodic functions (SPF) are the periodic solutions to a classical dynamic system for which the energy function is given by the expression

$$H(x, y) = |x| + |y| = 1, \tag{11.3.1}$$

and where the equation of motion is determined by the relations

$$\frac{dx}{dt} = \frac{\partial H}{\partial y}, \quad \frac{dy}{dt} = -\frac{\partial H}{\partial x}. \tag{11.3.2}$$

$H(x, y)$ is called the Hamiltonian function [5] and the value one, on the right-side of Eq. (11.3.1) is selected so that the plot of $H(x, y)$ is a square with vertices at the points: $A(1, 0)$, $B(0, 1)$, $C(-1, 0)$, and $D(0, -1)$; see Fig. 11.3.1.

From Eqs. (11.3.1) and (11.3.2), it follows that the equations of motion are

$$\frac{dx}{dt} = \text{sgn}(y), \quad \frac{dy}{dt} = -\text{sgn}(x), \tag{11.3.3}$$

where $\text{sgn}(z)$ is the sign-function, i.e.,

$$\text{sgn}(z) = \begin{cases} 1, & z > 0, \\ 0, & z = 0, \\ -1, & z < 0. \end{cases} \tag{11.3.4}$$

In the $x - y$ phase-space, the path $[x(t), y(t)]$ is clockwise and proceeds in the directions

$$(1, 0) \to (0, -1) \to (-1, 0) \to (0, 1) \to (1, 0). \tag{11.3.5}$$

Thus, to explicitly calculate $x(t)$ and $y(t)$, we must determine them along each of the four straight line segments. For example, along $(1, 0) \to (0, -1)$, Eq. (11.3.3) become

$$\begin{cases} \frac{dx}{dt} = -1 & \frac{dy}{dt} = -1, \\ x(0) = 1, & y(0) = 0, \end{cases} \tag{11.3.6}$$

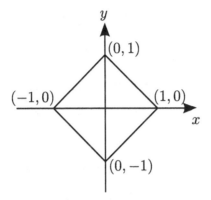

Fig. 11.3.1 Plot of $H(x, y) = |x| + |y| = 1$.

with the solutions

$$\begin{cases} x(t) = 1 - t, \\ y(t) = -t, \end{cases} \quad \text{for } 0 \le t \le 1. \tag{11.3.7}$$

Repeating this calculation for the other segments gives

$$(0, -1) \to (-1, 0) : \begin{cases} x(t) = 1 - t, \\ y(t) = -2 + t, \end{cases} \quad \text{for } 1 \le t \le 2, \tag{11.3.8}$$

$$(-1, 0) \to (0, 1) : \begin{cases} x(t) = -3 + t, \\ y(t) = -2 + t, \end{cases} \quad \text{for } 2 \le t \le 3, \tag{11.3.9}$$

$$(0, 1) \to (1, 0) : \begin{cases} x(t) = -3 + t, \\ y(t) = 4 - t, \end{cases} \quad \text{for } 3 \le t \le 4. \tag{11.3.10}$$

Note that because of the particular dynamics of this system and the fact that $H(x, y) = 1$, the motion is periodic with period $T = 4$.

All of these results can be summarized with the following equations

$$x(t + 4) = x(t) : x(t) = \begin{cases} 1 - t, & 0 \le t \le 2, \\ -3 + t, & 2 \le t \le 4, \end{cases} \tag{11.3.11}$$

$$y(t + 4) = y(t) : y(t) = \begin{cases} -t, & 0 \le t \le 1, \\ -2 + t, & 1 \le t \le 3, \\ 4 - t, & 3 \le t \le 4. \end{cases} \tag{11.3.12}$$

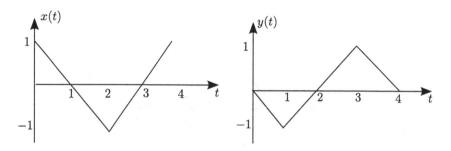

Fig. 11.3.2   Plots of $x(t)$ vs t and $y(t)$ vs t for one period.

The following is a listing of several of the important properties of the SPF's as defined by Eqs. (11.3.1) and (11.3.2):

(i) $x(t)$ and $y(t)$ are special cases of periodic traveling wave functions [3], of period -4.

(ii) $x(t)$ and $y(t)$ are bounded, i.e.,

$$|x(t)| \leq 1, \quad |y(t)| \leq 1, \tag{11.3.13}$$

and continuous.

(iii) $x(t)$ and $y(t)$ are, respectively, even and odd periodic functions, i.e.,

$$x(-t) = x(t), \quad y(-t) = -y(t). \tag{11.3.14}$$

(iv) $x(t)$ and $y(t)$ have the integral properties

$$\int_0^4 x(t)dt = 0, \quad \int_0^4 y(t)dt = 0, \tag{11.3.15a}$$

$$\int_0^4 x(t)y(t)dt = 0, \tag{11.3.15b}$$

$$\int_0^4 (x(t))^2 dt = \int_0^4 (y(t))^2 dt = \frac{4}{3}. \tag{11.3.15c}$$

11.3.2   *SPF as Generalized Trigonometric Functions*

Another form of square periodic functions (SPF), not identical to those presented in the previous Section 11.3.1, can be created [5; 6].

To begin, redraw the first quadrant of Fig. 11.3.1 as indicated in Fig. 11.3.3. Note that because the full $x-y$ phase-space curve is $|x|+|y| = 1$, then the symmetric transformations

$$\begin{cases} S_1 : x \to -x, \ y \to y, \\ S_2 : x \to x, \ y \to -y, \\ S_3 = S_1 S_2 = S_2 S_1 : x \to -x, \ y \to -y, \end{cases} \tag{11.3.16}$$

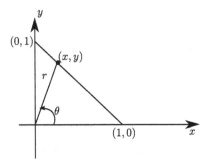

Fig. 11.3.3 Geometrical notation.

allows the following arguments to apply to all the four quadrants.

Note that for a given point (x, y) on the square function curve, the following restrictions hold

$$r^2 = x^2 + y^2, \quad |x| + |y| = 1. \tag{11.3.17}$$

We now define three geometrical based periodic functions based on the use of the angle θ as the independent variable

$$\text{Square cosine}(\theta) = Sqc(\theta) \equiv x = r(\theta)\cos\theta, \tag{11.3.18}$$

$$\text{Square sine}(\theta) = Sqs(\theta) \equiv r(\theta)\sin\theta, \tag{11.3.19}$$

$$\text{Square dine}(\theta) = Sqd(\theta) \equiv r(\theta). \tag{11.3.20}$$

From the second of the relations in Eq. (11.3.17), we find

$$r(\theta) = \frac{1}{|\cos\theta| + |\sin\theta|}, \tag{11.3.21}$$

and, as a consequence, it follows that

$$Sqc(\theta) = \frac{\cos\theta}{|\cos\theta| + |\sin\theta|}, \tag{11.3.22}$$

$$Sqs(\theta) = \frac{\sin\theta}{|\cos\theta| + |\sin\theta|}, \tag{11.3.23}$$

$$Sqd(\theta) = \frac{1}{|\cos\theta| + |\sin\theta|}. \tag{11.3.24}$$

These definitions of the three geometrical based square periodic functions follow the general methodology used by Schwalm [7] in his construction of the Jacobi elliptic functions.

It is straightforward to show that these functions have the following properties:

(i) $Sqc(\theta)$, $Sqs(\theta)$ and $Sqd(\theta)$ are periodic, with respective periods of $(2\pi, 2\pi, \pi/2)$.

(ii) $Sqc(\theta)$ and $Sqd(\theta)$ are even functions, while $Sqs(\theta)$ is odd, i.e.,

$$Sqc(-\theta) = Sqc(\theta), \quad Sqs(-\theta) = -Sqs(\theta), \quad Sqd(-\theta) = Sqd(\theta).$$
$$(11.3.25)$$

(iii) These functions satisfy two nonlinear relations

$$|Sqc(\theta)| + |Sqs(\theta)| = 1, \tag{11.3.26}$$

$$[Sqc(\theta)]^2 + [Sqs(\theta)]^2 = [Sqd(\theta)]^2. \tag{11.3.27}$$

(iv) The following bounds are satisfied

$$-1 \le Sqc(\theta) \le +1, \quad -1 \le Sqs(\theta) \le \pm 1, \tag{11.3.28a}$$

$$\frac{1}{\sqrt{2}} \le Sqd(\theta) \le 1. \tag{11.3.28b}$$

(v) The zeros and min/max values for $Sqc(\theta)$ and $Sqs(\theta)$ are the same as for the standard trigonometric functions $\cos\theta$ and $\sin\theta$. The minimum values for $Sqd(\theta)$ occur for odd multiples of $\pi/4$; the maximum values for even multiples of $\pi/4$.

(vi) The first-derivatives are given by the expressions:

$$\frac{d\,Sqc(\theta)}{d\theta} = \begin{pmatrix} -1, & 0 < \theta < \frac{\pi}{2} \\ -1, & \frac{\pi}{2} < \theta < \pi \\ +1, & \pi < \theta < \frac{3\pi}{2} \\ +1, & \frac{3\pi}{2} < \theta < 2\pi \end{pmatrix} [Sqd(\theta)]^2, \tag{11.3.29a}$$

$$\frac{d\,Sqs(\theta)}{d\theta} = \begin{pmatrix} +1, & 0 < \theta < \frac{\pi}{2} \\ -1, & \frac{\pi}{2} < \theta < \pi \\ -1, & \pi < \theta < \frac{3\pi}{2} \\ +1, & \frac{3\pi}{2} < \theta < 2\pi \end{pmatrix} [Sqd(\theta)]^2, \tag{11.3.29b}$$

$$\frac{d\,Sqd(\theta)}{d\theta} = \begin{pmatrix} Sqs(\theta) - Sqc(\theta), & 0 < \theta < \frac{\pi}{2} \\ -Sqs(\theta - Sqc(\theta), & \frac{\pi}{2} < \theta < \pi \\ -Sqs(\theta) + Sqc(\theta), & \pi < \theta < \frac{3\pi}{2} \\ Sqs(\theta) + Sqc(\theta), & \frac{3\pi}{2} < \theta < 2\pi \end{pmatrix} Sqd(\theta). \tag{11.3.29c}$$

Note the derivatives are not defined at $\theta = (0, \pi/2, \pi, 3\pi/2)$.

(vii) Figure 11.3.4 are plots of the functions $Sqc(\theta)$, $Sqs(\theta)$ and $Sqd(\theta)$.

Following the work and analysis of Schwalm [7], it turns out that θ is not the natural independent variable for Sqc, Sqs, and Sqd. The fundamental independent variable is $u(\theta)$

$$u(\theta) \equiv \int_0^\theta r(\psi)d\psi, \tag{11.3.30}$$

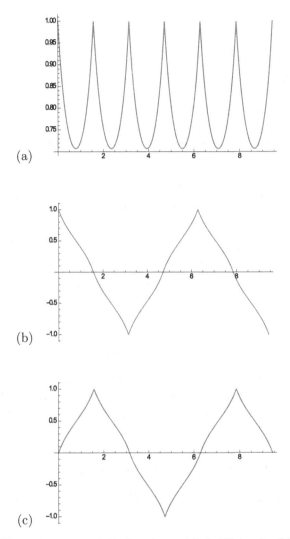

Fig. 11.3.4 Plots of square periodic functions. (a) $Sqd(\theta)$ vs θ. (b) $Sqc(\theta)$ vs θ.
(c) $Sqd(\theta)$ vs θ.

where $r(\theta)$ is given in Eq. (11.3.21). Given that $r(\theta)$ is an even function of θ, it follows that

$$u(-\theta) = -u(\theta). \tag{11.3.31}$$

Further, substituting the bounds from Eq. (11.3.28b) into the right-side of

Eq. (11.3.30) gives

$$\left(\frac{1}{\sqrt{2}}\right)\theta < u(\theta) < \theta. \tag{11.3.32}$$

Using the $r(\theta)$ from Eq. (11.3.21) also gives

$$u(\theta) = \int_0^\theta \frac{d\psi}{|\cos\theta| + |\sin\theta|}, \tag{11.3.33}$$

which can be explicitly integrated over the interval, $0 \le \theta \le \frac{\pi}{2}$, to produce the results [8]

$$u(\theta) = \left(\frac{1}{\sqrt{2}}\right) \mathrm{Ln}\left[\frac{\tan\left(\frac{\theta}{2} + \frac{\pi}{8}\right)}{\tan\left(\frac{\pi}{8}\right)}\right], \quad 0 \le \theta \le \frac{\pi}{2}. \tag{11.3.34}$$

The values of $u(\theta)$ for the intervals $\left(\frac{\pi}{2}, \pi\right)$, $\left(\pi, \frac{3\pi}{2}\right)$, and $\left(\frac{3\pi}{2}, 2\pi\right)$ can be obtained by using the shift and reflection properties of the trigonometric functions.

The period of the square functions, Sqc, Sqs, and Sqd, in terms of the variable u is

$$T = 4u\left(\frac{\pi}{2}\right) \tag{11.3.35}$$

$$= 4.985\ 801\ 922.$$

It is interesting to observe that there are at least two different types of square periodic functions. The first were constructed based on treating the square function, $|x| + |y| = 1$, as a Hamiltonian system, while the second set followed directly from the geometrical properties of $|x| + |y| = 1$.

11.4 Analytical Periodic Solutions to $f(t)^2 + g(t)^2 = 1$

Consider functions $f(t)$ and $g(t)$, for which both are assumed to have continuous second derivatives. If there exists a T such that

$$f(t + T) = f(t), \quad g(t + T) = g(t), \tag{11.4.1}$$

and if

$$f(t)^2 + g(t)^2 = 1, \tag{11.4.2}$$

then we say that $f(t)$ and $g(t)$ are generalized periodic functions.

Note that the standard trigonometric functions, $\sin(at)$ and $\cos(at)$, and the Jacobian functions, $cn(bt)$ and $sn(bt)$, are examples of this class of periodic functions.

There also exist functions which satisfy Eqs. (11.4.1) and (11.4.2) which do not have a continuous second derivative. An almost trivial example is

$$f_1(t) = \begin{cases} +1, & -1 < t \leq 0, \\ 0, & 0 < t < 1, \end{cases} \tag{11.4.3a}$$

$$g_1(t) = \begin{cases} 0, & -1 < t \leq 0, \\ +1, & 0 < t < 1. \end{cases} \tag{11.4.3b}$$

The periodic extension of these functions is

$$f_1(t+2) = f_1(t), \quad g_1(t+2) = g_1(t). \tag{11.4.4}$$

In this section, we explore the construction of periodic solutions to Eq. (11.4.2) under the above stated assumptions. The obtained results are based on an application of the Euler formula. The following is a brief summary of the major results of this analysis [8]:

(i) There exist periodic solutions to Eq. (11.4.2) for any $T > 0$.

(ii) These periodic solutions are determined by a function, $\theta(t)$, which is strongly constrained by the periodicity condition.

(iii) In addition to cosine- and sine-like functions, there exists a third periodic function and its value is in general non-negative.

(iv) The functions $f(t)$ and $g(t)$ satisfy the same linear, second-order differential equations having the mathematical structure of a Mathieu equation [9].

(v) The methodology for constructing $f(t)$ and $g(t)$ forces them to have many of the essential features of the standard trigonometric functions and, in this sense, $f(t)$ and $g(t)$, may be taken as direct generalizations of the cosine and sine functions.

11.4.1 *Generalized Cosine and Sine Functions*

Define the functions $c(t)$ and $s(t)$ as follows

$$e^{i\theta(t)} = c(t) + is(t), \tag{11.4.5}$$

where $\theta(t)$ is a real function of t and assumed to have a continuous second derivative. One consequence of this definition is

$$c(t)^2 + s(t)^2 = 1, \tag{11.4.6}$$

with further constraint

$$-1 \leq c(t) \leq +1, \quad -1 \leq s(t) \leq +1. \tag{11.4.7}$$

The $f(t)$ and $g(t)$ functions may be identified, respectively, with $c(t)$ and $s(t)$.

Note that since

$$e^{i\theta(t)} = \cos\theta(t) + i\sin\theta(t), \tag{11.4.8}$$

we have

$$c(t) = \cos\theta(t), \quad s(t) = \sin\theta(t). \tag{11.4.9}$$

If we use for $\theta(t)$ the function

$$\theta(t) = \left(\frac{2\pi}{T}\right)t, \quad T > 0, \tag{11.4.10}$$

then

$$c(t) = \cos\left(\frac{2\pi t}{T}\right), \quad s(t) = \sin\left(\frac{2\pi t}{T}\right), \tag{11.4.11}$$

and these are the usual cosine and sine functions. Inspection of Eq. (11.4.10) indicates that $\theta(t)$ is an odd function of t. In all the work to follow, we assume this condition to hold, i.e.,

$$\theta(-t) = -\theta(t). \tag{11.4.12}$$

An immediate consequence of Eqs. (11.4.9) and (11.4.12) is

$$c(-t) = c(t), \quad s(-t) = -s(t), \tag{11.4.13}$$

i.e., $c(t)$ and $s(t)$ are, respectively, even and odd functions of t.

If the time derivative of the expression in Eq. (11.4.6) is taken, then we obtain

$$c(t)\frac{dc(t)}{dt} + s(t)\frac{ds(t)}{dt} = 0, \tag{11.4.14}$$

for which it follows that

$$\frac{dc(t)}{dt} = -d(t)s(t), \quad \frac{ds(t)}{dt} = d(t)c(t), \tag{11.4.15}$$

and, where for the time being, $d(t)$ is not known.

For the special case, $c(t) = \cos t$ and $s(t) = \sin t$, $d(t) = 1 > 0$. This particular $d(t)$ has the properties

$$d(t) > 0, \quad d(-t) = d(t), \tag{11.4.16}$$

i.e., $d(t)$ is positive and an even function of t. A similar result can be determined for the general case. To demonstrate this, square the two expressions in Eq. (11.4.15), add them together, and use the result in Eq. (11.4.6), to obtain

$$d(t)^2 = \left(\frac{dc}{dt}\right)^2 + \left(\frac{ds}{dt}\right)^2. \tag{11.4.17}$$

If the function $d(t)$ is taken as the positive square root, then

$$d(t) = \sqrt{\left(\frac{dc}{dt}\right)^2 + \left(\frac{ds}{dt}\right)^2}, \tag{11.4.18}$$

and both restrictions given in Eq. (11.4.16) are satisfied.

Observe that given $\theta(t)$, all three functions, $(c(t), s(t), d(t))$, are uniquely determined.

11.4.2 Calculation of Phase Function – $\theta(t)$

In the natural and engineering sciences, $\theta(t)$ is called a phase-function. We will now calculate the general mathematical structure of $\theta(t)$ to insure that $c(t)$ and $s(t)$ are periodic, with period T.

The periodicity requirement for $c(t)$ and $s(t)$ is

$$c(t + T) = c(t), \quad s(t + T) = s(t). \tag{11.4.19}$$

Therefore,

$$\begin{aligned} e^{i\theta(t+T)} &= c(t + T) + is(t + T) \\ &= c(t) + is(t) \\ &= e^{i\theta(t)}. \end{aligned} \tag{11.4.20}$$

Since

$$1 = e^{2\pi k i}, \quad k = \text{integer}, \tag{11.4.21}$$

we have

$$\theta(t + T) = \theta(t) + 2\pi, \tag{11.4.22}$$

where $k = 1$ has been selected because it corresponds to the fundamental period. The general solution of Eq. (11.4.21) is [10]

$$\theta(t) = A(t) + \left(\frac{2\pi}{T}\right)t, \tag{11.4.23}$$

where $A(t)$ has the properties

$$A(-t) = -A(t), \quad A(t + T) = A(t). \tag{11.4.24}$$

To prove this, note that $A(t)$ has to be an odd function of t since $\theta(t)$ is an odd function. Therefore, using the ansatz of Eq. (11.4.23), we find

$$\theta(t+T) = A(t+T) + \left(\frac{2\pi}{T}\right)(t+T)$$

$$= A(t+T) + \left(\frac{2\pi}{T}\right)t + 2\pi, \qquad (11.4.25)$$

and if it is to be correct, then $A(t+T) = A(t)$. Therefore, it follows that

$$\theta(t+T) = A(t) + \left(\frac{2\pi}{T}\right)t + 2\pi$$

$$= \theta(t) + 2\pi, \qquad (11.4.26)$$

which shows that the expression in Eq. (11.4.23) is a solution to Eq. (11.4.22).

The phase function, $\theta(t)$, is assumed to have a continuous second-derivative and this implies that the same holds for $A(t)$. Since $A(t)$ is periodic, with period T, and since it is also odd, it follows that $A(t)$ has a sine Fourier representation [11] which takes the form

$$A(t) = \sum_{k=1}^{\infty} a_k \sin\left[k\left(\frac{2\pi}{T}\right)t\right]. \qquad (11.4.27)$$

Therefore, $\theta(t)$ has the form given in Eq. (11.4.23) with $A(t)$ from Eq. (11.4.27).

11.4.3 *Differential Equations for $f(t)$ and $g(t)$*

Taking the derivative of Eq. (11.4.5) and setting the real and imaginary parts each side equal to each other, gives

$$\frac{dc(t)}{dt} = -\left(\frac{d\theta(t)}{dt}\right)s(t), \quad \frac{ds(t)}{dt} = \left(\frac{d\theta(t)}{dt}\right)c(t). \qquad (11.4.28)$$

Comparison of these expressions with the result in Eq. (11.4.15) allows the following result to be obtained

$$d(t) = \frac{d\theta(t)}{dt}$$

$$= \frac{dA(t)}{dt} + \frac{2\pi}{T}. \qquad (11.4.29)$$

Since $d(t) > 0$, the latter equation places a restriction on the derivatve of $A(t)$, namely

$$\frac{dA(t)}{dt} > -\left(\frac{2\pi}{T}\right). \qquad (11.4.30)$$

Differential equations can be obtained for $c(t)$ and $s(t)$ by eliminating one or the other variables in Eq. (11.4.15) and using the first result in Eq. (11.4.29); doing this gives

$$w'' - \left(\frac{\theta''}{\theta'}\right) w' + (\theta')^2 w = 0, \qquad (11.4.31)$$

where the prime denotes taking the appropriate time derivative, i.e., $w' = dw/dt$ and $w'' = d^2w/dt^2$, with the same for θ, and $w(t)$ stands for either $c(t)$ or $s(t)$.

Note that since $\theta(t)$ is a priori given, then Eq. (11.4.31) is a second-order, linear, nonautonomous differential equation with periodic coefficients. It corresponds to a generalized Mathieu equation.

11.4.4 An Explicit Example

The simplest example to study is one for which $A(t)$, see Eq. (11.4.27) contains only one term, i.e.,

$$A(t) = a_1 \sin\left(\frac{2\pi}{T}\right) t \qquad (11.4.32)$$

and the corresponding $\theta(t)$ is

$$\theta(t) = a_1 \sin\left(\frac{2\pi}{T}\right) t + \left(\frac{2\pi}{T}\right) t. \qquad (11.4.33)$$

Therefore,

$$c(t) = \cos\left[a_1 \sin\left(\frac{2\pi}{T}\right) t + \left(\frac{2\pi}{T}\right) t\right], \qquad (11.4.34a)$$

$$s(t) = \sin\left[a_1 \sin\left(\frac{2\pi}{T}\right) t + \left(\frac{2\pi}{T}\right) t\right], \qquad (11.4.34b)$$

$$d(t) = \left(\frac{2\pi}{T}\right) \left[1 + a_1 \cos\left(\frac{2\pi}{T}\right) t\right], \qquad (11.4.34c)$$

with

$$|a_1| < 1. \qquad (11.4.34d)$$

Letting $T = 2\pi$, to simplify the subsequent equations, the Fourier series for $c(t)$ and $s(t)$ are

$$c(t) = \cos(t + a_1 \sin t)$$

$$= -J_1(a_1) + \sum_{k=1}^{\infty} [J_{2k-1}(a_1) - J_{2k+1}(a_1)] \cos(2kt)$$

$$+ \sum_{k=0}^{\infty} [J_{2k}(a_1) + J_{2k+2}(a_1)] \cos(2k+1)t, \qquad (11.4.35a)$$

$$s(t) = \sin(t + a_1 \sin t)$$

$$= \sum_{k=1}^{\infty} [J_{2k-1}(a_1) - J_{2k+1}(a_1)] \sin(2kt)$$

$$+ \sum_{k=0}^{\infty} [J_{2k}(a_1) + J_{2k+2}(a_1)] \sin(2k+1)t. \qquad (11.4.35b)$$

Both of these results were derived using the following relations[12]:

$$\cos(z \sin t) = J_0(z) + 2 \sum_{k=1}^{\infty} J_{2k}(z) \cos(2kt), \qquad (11.4.36a)$$

$$\sin(z \sin t) = 2 \sum_{k=0}^{\infty} J_{2k+1}(z) \sin(2k+1)t. \qquad (11.4.36b)$$

The graphs of $c(t)$, $s(t)$ and $d(t)$ versus t are similar to those of the Jacobi elliptic functions: $cn(t)$, $sn(t)$, and $dn(t)$ [13].

11.4.5   *Jacobi Elliptic Functions*

If $\theta(t) = t$, then Eq. (11.4.31) becomes

$$w'' + w = 0, \qquad (11.4.37)$$

with solutions

$$w_1(t) = \cos t, \quad w_2(t) = \sin t, \qquad (11.4.38)$$

and $d(t) = 1$. These are just the standard trigonometric functions.

Consider now the Jacobi elliptic functions: $cn(t)$, $sn(t)$ and $dn(t)$. Assume $dn(t)$ is given and let

$$\theta'(t) = dn(t). \qquad (11.4.39)$$

Therefore, it follows that

$$\theta(t) = \int_0^t dn(z)dz \qquad (11.4.40)$$

and

$$e^{i\theta(t)} = C(t) + iS(t)$$

$$= \cos\left[\int_0^t dn(z)dz\right] + i\sin\left[\int_0^t dn(z)dz\right], \qquad (11.4.41)$$

which gives

$$C(t) = \cos\left[\int_0^t dn(z)dz\right], \qquad (11.4.42\text{a})$$

$$S(t) = \sin\left[\int_0^t dn(z)dz\right]. \qquad (11.4.42\text{b})$$

Now taking the derivatives on both sides of Eq. (11.4.42), we obtain

$$\frac{dC(t)}{dt} = -dn(t)S(t), \quad \frac{dS(t)}{dt} = dn(t)C(t), \qquad (11.4.43)$$

which are the defining equations for the Jacobi elliptic functions and we can conclude that

$$C(t) = cn(t), \quad S(t) = sn(t). \qquad (11.4.44)$$

11.4.6 *Comments*

Starting with the functional equation (see Chapter 12)

$$f(t)^2 + g(t)^2 = 1, \qquad (11.4.45)$$

and assuming that $f(t)$ and $g(t)$ are related by means of the formula

$$e^{i\theta(t)} = f(t) + i\,g(t), \qquad (11.4.46)$$

it follows that a number of general conclusions can be reached if $\theta(t)$ is odd, with a continuous second derivative:

(i) $f(t)$ and $g(t)$ are, respectively, even and odd.

(ii) $f(t)$ and $g(t)$ are bounded in magnitude by the value one.

(iii) $f(t)$ and $g(t)$ are periodic, with period T, if $\theta(t)$ has the mathematical structure given in Eqs. (11.4.23) and (11.4.24).

(iv) There exists a third periodic function $d(t)$, which is related to the derivatives of $f(t)$ and $g(t)$; see Eq. (11.4.18), where $f(t) = c(t)$ and $g(t) = s(t)$.

(v) Both $f(t)$ and $g(t)$ satisfy the same second-order, linear, nonautonomous differential equation; see Eq. (11.4.31). This differential equation belongs to the class of damped Mathieu type equations.

In summary, given a proper $\theta(t)$, the functions $f(t) = c(t)$, $g(t) = s(t)$, and $d(t)$ can be calculated.

11.5   Leah-Cosine and -Sine Functions

Consider a particle of unit mass acted on by a force

$$F(x) = -x^{1/3}. \tag{11.5.1}$$

The equation of motion is

$$\frac{d^2x}{dt^2} + x^{1/3} = 0. \tag{11.5.2}$$

We have named this second-order, nonlinear, autonomous equation the *Leah differential equation* (LDE).

This equation can be written as a system of two first-order equations

$$\frac{dx}{dt} = y, \quad \frac{dy}{dt} = -x^{1/3}. \tag{11.5.3}$$

Therefore, in the 2-dim, (x, y), phase plane the trajectories, $y = y(x)$, are solutions to the first-order equation

$$\frac{dy}{dx} = -\frac{x^{1/3}}{y}. \tag{11.5.4}$$

Using the initial conditions

$$x(0) = 1, \quad y(0) = 0, \tag{11.5.5}$$

the solution to Eq. (11.5.4) is

$$\frac{y^2}{2} + \left(\frac{3}{4}\right) x^{4/3} = \frac{3}{4}, \tag{11.5.6}$$

and this is easily seen to correspond to a non-intersecting closed curve in the (x, y) phase-plane. However, this result implies that the motion, i.e., $(x(t), y(t))$, in the phase-plane are periodic [15].

Inspection of the result in Eq. (11.5.6), which is called a first-integral of Eq. (11.5.4), shows that it is invariant with respect to the following three transformations

$$\begin{cases} T_1 : x \to -x, \ y \to y, \\ T_2 : x \to x, \ y \to -y, \\ T_3 = T_1 T_2 = T_2 T_1 : x \to -x, \ y \to -y, \end{cases} \tag{11.5.7}$$

where

(i) T_1 corresponds to reflection in the y-axis;
(ii) T_2 is reflection in the x-axis;
(iii) T_3 is reflection through the origin.

Our definition of the Leah-cosine and -sine functions will not be based on the consideration of x and y as being dynamic variables depending on t. These and related functions will be introduced and defined in terms of the geometry of the phase-plane curve given by Eq. (11.5.6); see Fig. 11.5.1. In particular, we wish expressions of them in terms of the angle θ.

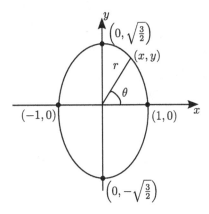

Fig. 11.5.1   Plot of Eq. (11.5.6) with the definitions of θ and r.

The variables (r, θ) and their relationships to an arbitrary point, (x, y), on the phase-plane first-integral curve are

$$x(\theta) = r(\theta)\cos\theta, \quad y(\theta) = r(\theta)\sin\theta, \tag{11.5.8}$$

$$r^2 = x^2 + y^2, \quad \left(\frac{2}{3}\right) y^2 + x^{4/3} = 1. \tag{11.5.9}$$

The task is now to make appropriate definitions of the Leah (periodic) functions, and express all of them in terms of θ.

11.5.1   *Three Leah Periodic Functions*

There are three Leah periodic functions and we define them in the following manner

$$\begin{cases} \text{Leah-cosine} = Lcn(\theta) \equiv x = r(\theta)\cos\theta, \\ \text{Leah-sine} = Lsn(\theta) \equiv \sqrt{\tfrac{2}{3}}y = \sqrt{\tfrac{2}{3}}r(\theta)\sin\theta, \\ \text{Leah-dine} = Ldn(\theta) \equiv \sqrt{\tfrac{2}{3}}r(\theta). \end{cases} \tag{11.5.10}$$

Using the implications following from the results of Eq. (11.5.7), the relationships among (r, θ) and (x, y), and the definitions of the Leah periodic functions, the following results are obtained:

(a) $r(\theta)$ has the bounds

$$1 \leq r(\theta) \leq \sqrt{\frac{3}{2}}. \tag{11.5.11}$$

(b) For x and y

$$-1 \leq x \leq 1, \quad -\sqrt{\frac{3}{2}} \leq y \leq \sqrt{\frac{3}{2}}. \tag{11.5.12}$$

(c) The Leah functions have the properties

$$-1 \leq Lcn(\theta) \leq 1, \quad -1 \leq Lsn(\theta) \leq 1, \tag{11.5.13a}$$

$$\sqrt{\frac{2}{3}} \leq Ldn(\theta) \leq 1; \tag{11.5.13b}$$

$$Lcn(-\theta) = Lcn(\theta), \quad Lsn(-\theta) = -Lsn(\theta), \tag{11.5.14a}$$

$$Ldn(-\theta) = Ldn(\theta); \tag{11.5.14b}$$

$$Lcn(\theta + 2\pi) = Lcn(\theta), \quad Lsn(\theta + 2\pi) = Lsn(\theta), \tag{11.5.15a}$$

$$Ldn(\theta + \pi) = Ldn(\theta). \tag{11.5.15b}$$

(d) Max/min values and zeros

$$Lcn(0) = 1, \quad Lcn\left(\frac{\pi}{2}\right) = 0, \quad Lcn(\pi) = -1, \quad Lcn\left(\frac{3\pi}{2}\right) = 0;$$
$$\tag{11.5.16a}$$

$$Lsn(0) = 0, \quad Lsn\left(\frac{\pi}{2}\right) = 1, \quad Lsn(\pi) = 0, \quad Lsn\left(\frac{3\pi}{2}\right) = -1;$$
$$\tag{11.5.16b}$$

$$Ldn(0) = \sqrt{\frac{2}{3}}, \quad Ldn\left(\frac{\pi}{2}\right) = 1, \quad Ldn(\pi) = \sqrt{\frac{2}{3}}, \quad Ldn\left(\frac{3\pi}{2}\right) = 1.$$
$$\tag{11.5.16c}$$

In summary, $Lcn(\theta)$ and $Ldn(\theta)$ are even functions, while $Lsn(\theta)$ is an odd function. The Leah-cosine and -sine functions have their max/min values and zeros at the same locations of θ as the standard trigonometric functions. Also, both $Lcn(\theta)$ and $Lsn(\theta)$ are periodic with period 2π; however, $Ldn(\theta)$ has period π. Further, $Ldn(\theta)$ is always non-negative.

11.5.2 *Determination of* $r(\theta)$

From the definitions of $Lcn(\theta)$, $Lsn(\theta)$, and $Ldn(\theta)$, it follows that if $r(\theta)$ is known, then these three functions can be determined completely in terms of θ. Using Eq. (11.5.8) in the second formula of Eq. (11.5.9), we find that $r(\theta)$ satisfies the relation

$$F(r,\theta) = (\sin\theta)^2 r^2 + \left[\left(\frac{3}{2} \right) (\cos\theta)^{4/3} \right] r^{4/3} - \frac{3}{2} = 0. \qquad (11.5.17)$$

Defining

$$R(\theta) = [r(\theta)]^{2/3}, \qquad (11.5.18)$$

allows the rewriting of Eq. (11.5.17) to the form

$$(\sin\theta)^2 R^3 + \left[\left(\frac{3}{2} \right) (\cos\theta)^{4/3} \right] R^2 - \frac{3}{2} = 0. \qquad (11.5.19)$$

If we now apply the standard procedures for determining the three roots of a cubic equation [16], then the following conclusions are reached:

(1) Eq. (11.5.19) has one real, non-negative root; denote it as $R^{(+)}(\theta)$.

(2) The other two roots are either both real and negative, or complex conjugate with negative real parts.

(3) Since $r(\theta) > 0$, then

$$r(\theta) = \left[R^{(+)}(\theta) \right]^{3/2}. \qquad (11.5.20)$$

Note that the functional dependence of $r(\theta)$ and θ is expected to be complicated.

11.5.3 *Comments*

It should be pointed out that the graphs of the three Leah periodic functions have structures very similar to those of the three Jacobi functions.

An alternative construction of the Leah-cosine and -sine functions has been given by Rucker [14]. This analysis starts by defining the Leah-cosine function as the solution to the initial-value problem

$$\frac{d^2 x}{dt^2} + x^{1/3} = 0, \quad x(0) = 1, \quad \frac{dx(0)}{dt} = 0. \qquad (11.5.21)$$

The corresponding Leah-sine function is then defined by means of the functional relation

$$f(t_1 - t_2) = f(t_1)f(t_2) + g(t_1)g(t_2), \qquad (11.5.22)$$

where $f(t)$ is taken to be the Leah-cosine function, obtained as the solution to Eq. (11.5.21), and $g(t)$ is identified with the Leah-sine function. Note that for this way of characterizing the Leah-functions, the independent variable is the dynamic variable t, time, rather than the geometric variable θ.

Problems

Section 11.2

11.2.1 Consider a box where the walls are located at $x = (\pm)L$, and the initial conditions are

$$x(0) = -1, \quad v(0) = 1.$$

Determine $x(t)$ and $v(t)$.

11.2.2 Calculate the Fourier series for $x(t)$ and $v(t)$ given in Eqs. (11.2.9) and (11.2.10).

Section 11.3

11.3.1 Show that the results of Eq. (11.3.3) are correct.

11.3.2 Derive the results of Eqs. (11.3.11) and (11.3.12).

11.3.3 Evaluate the integrals given in Eqs. (11.3.15).

11.3.4 Derive the bounds for $Sqd(\theta)$ given by Eq. (11.3.28b).

11.3.5 Show that the period of $Sqd(\theta)$ is $\frac{\pi}{2}$.

11.3.6 Derive the results for the derivatives of the geometrical square periodic functions given in Eqs. (11.3.29).

11.3.7 Prove Eq. (11.3.31) from Eq. (11.3.30).

11.3.8 Use the shift and reflection properties of the trigonometric function to extend $u(\theta)$, given in Eq. (11.3.34) to the angular intervals $\left(\frac{\pi}{2}, \pi\right)$, $\left(\pi, \frac{3\pi}{2}\right)$, and $\left(\frac{3\pi}{2}, 2\pi\right)$.

11.3.9 Explain the result of Eq. (11.3.35).

11.3.10 There are six standard trigonometric functions: sine, cosine, tangent, cosecant, secant, cotangent. What are the corresponding generalizations for the two types of square periodic functions?

Section 11.4

11.4.1 Prove that $\theta(t)$ must be odd in order for $c(t)$ and $s(t)$ to have periodic solutions.

11.4.2 Derive the results of Eq. (11.4.28).

11.4.3 Using the results stated in Eq. (11.4.36) to obtain the Fourier series expressions given in Eqs. (11.4.35a,b).

Section 11.5

11.5.1 Derive Eq. (11.5.4) from Eqs. (11.5.3).

11.5.2 Prove that Eq. (11.5.6) corresponds to a simple, closed curve in the $x - y$ phase-plane.

11.5.3 Derive the results presented in Eqs. (11.5.11) to Eq. (11.5.16).

11.5.4 Why does Eq. (11.5.19) have only one real, positive root?

11.5.5 Graph the three Leah functions as functions of θ.

Comments and References

[1] An elementary example of such a function is tan(t); it is periodic with period π, and unbounded at $t = \pi/2$.

[2] An analysis of this system is presented in every elementary textbook on quantum mechanics or quantum theory. See, for example, S. Borowitz, *Fundamentals of Quantum Mechanics* (W. A. Benjamin, New York, 1967).

[3] See, for example, https://en.wikipedia.org/wiki/Triangle_wave.

[4] This section is primarily based on work reported by Torina Lewis and Ronald Mickens at the 7th International Conference on Dynamics and Applications held at Morehouse College; Atlanta, GA; May 27–30, 2015. The published papers appear in *Proceedings of Dynamic System and Applications* **7** (2016):

- Torina Lewis and Ronald Mickens, "Square functions as a dynamic system,"
- Ronald Mickens, "Some properties of square (periodic) functions,"

[5] A good discussion of the history and application of Hamiltonian dynamics as applied to classical point particle systems is presented in H. Goldstein, *Classical Mechanics*, 2nd edition (Addison-Wesley, Reading, MA, 1980).

[6] An excellent presentation and summary of the major problems of so-called "generalized trigonometric functions" are made in the following publications:

- P. Lindqvist, "Some remarkable sine and cosine functions," *Ricerche Mat.* **44**, 269 (1995).
- H. Mahdi, M. Elatrash, and S. Elmadhoun, "On generalized trigonometric functions," *Journal of Mathematical Sciences and Applications* **2**, 33 (2014).
- Li Yin and Li-Guo Huang, "Inequalities for the generalized trigonometric and hyperbolic functions with two parameters," *Journal of Nonlinear Science and Applications* **8**, 315 (2015).

Using these publications and the references contained in them, the history of generalized trigonometric and related functions can be traced to their genesis.

[7] W. Schwalm, "Elliptic functions *sn*, *cn*, *dn* as trigonometry": http://www.und.edu/instruct/schwalm/MAAPresentation10-02/handout.pdf.2005.

[8] The material in this section is taken directly from the paper: R. E. Mickens, "Periodic solutions of the functional equation $f(t)^2 + g(t)^2 = 1$," *Journal of Difference Equations and Applications* **22**, 67 (2016).

[9] N. W. McLachlan, *Theory and Application of Mathieu Functions* (Clarendon Press, Oxford, 1947).

[10] R. E. Mickens, *Difference Equations: Theory, Applications and Advanced Topics*, 3rd edition (CRC Press, New York, 2015).

[11] See Körner.

[12] See Gradshteyn and Ryzhik, p. 973, formulas (8.511.3) and (8.511.4).

[13] See Byrd and Friedman, p. 26, Fig. 10.

[14] The following two papers give good introduction to the several different definitions of the Leah-cosine and -sine functions and their genesis:

- R. E. Mickens, *Truly Nonlinear Oscillations* (World Scientific, Hackensack, NJ, 2010).

- S. A. Rucker, "Leah-cosine and -sine functions: Definitions and elementary properties," pp. 265–280, in Abba B. Gumel (editor), *Mathematics in Continuous and Discrete Dynamical Systems* (American Mathematical Society, Contemporary Mathematics Series, Providence, RI, 2014).

- S. A. Rucker and R. E. Mickens, "The Leah cosine and sine functions: Geometric definitions," to appear in *Proceedings of Dynamic System and Applications* **7** (2016).

[15] J. H. Liu, *A First Course in Qualitative Theory of Differential Equations* (Pearson, New York, 2003). See Chapter 11.

[16] N. B. Conkwright, *Introduction to the Theory of Equations* (Ginn, Boston, 1941). See sections 34–38.

Bibliography

M. Abramowitz and I. A. Stegun, *Handbook of Mathematical Functions* (Dover, New York, 1965).

F. Bowman, *Introduction to Elliptic Functions with Applications* (Dover, New York, 1961).

P. F. Byrd and M. D. Friedman, *Handbook of Elliptic Integrals for Engineers and Physicists* (Springer-Verlag, Berlin, 1954).

R. V. Churchill, *Fourier Series and Boundary Value Problems* (McGraw-Hill, New York, 1941).

M. Farkas, *Periodic Motions* (Springer-Verlag, New York, 1994).

I. S. Gradshteyn and I. M. Ryzhik, *Tables of Integrals, Series, and Products*, 5th ed. (Academic Press, Boston, 1994).

T. Gowers, editor, *The Princeton Companion to Mathematics* (Princeton University Press, Princeton and Oxford, 2008).

T. W. Körner, *Fourier Analysis* (Cambridge University Press, Cambridge, 1988).

J. D. Lang and D. E. Edmunds, *Eigenvalues, Embeddings and Generalized Trigonometric Functions* (Lecture Notes in Mathematics, Vol. 2016, Springer-Verlag, Berlin, 2011).

R. E. Mickens, *Oscillations in Planar Dynamic Systems* (World Scientific, Singapore, 1996).

H. Sagan, *Boundary and Eigenvalue Problems in Mathematical Physics* (Wiley, New York, 1961).

A. Zygmund, *Trigonometric Series* (Cambridge University Press, Cambridge, 1977).

Chapter 12

Functional Equations

A functional equation is an equation for which the unknowns are one or more functions. In general, such an equation may relate the value of these functions at one point to values at other points.

Functional equations arise in the modeling and analysis of many systems in the natural, social, and engineering sciences. Several of the books listed in the Bibliography discuss applications of functional equations to these areas; in particular see Aczél (1966, 1987), Castille and Ruiz-Cobo (1992), and Small (2007).

The main purposes of this chapter are to introduce a number of elementary functional equations and illustrate how solutions may be obtained. In general, a given functional equation can have more than one type of solution as indicated by their mathematical structures. However, the solutions of greatest value to the scientist are often the "simplest" set of such solutions and generally those which are continuous in the appropriate variables.

In Section 12.1, we introduce and determine solutions for a number of rather elementary functional equations. This will help the reader acquire an appreciation for the scope of the subject matter. Section 12.2 is devoted to an especially important set of functional equations, namely, the four Cauchy functional equations. In Section 12.3, a listing is made of a number of named functional equations with a presentation of (some) continuous solutions. We end the chapter with several applications to the mathematical and natural sciences.

12.1 Miscellaneous Elementary Functional Equations

The worked examples presented in this section, while elementary, illustrate the general lack of the existence of an overall general methodology which can be used to solve functional equations.

12.1.1 $f(-x) = f(x)$ and $g(-x) = -g(x)$

Let $f(x)$ and $g(x)$ be, respectively, members of the classes of even and odd functions; then they are characterized as follows:

$$\text{even function}: f(-x) = f(x), \tag{12.1.1a}$$

$$\text{odd function}: g(-x) = -g(x). \tag{12.1.1b}$$

Particular examples include

$$f(x): \begin{cases} x^2, \\ \frac{1}{x^4}, & x \neq 0, \\ \frac{x^2}{1+x^2}, \\ e^{-x^2}, \\ \cos x, \end{cases} \tag{12.1.2a}$$

$$g(x): \begin{cases} x, \\ \frac{x}{1+x^4}, \\ \frac{1}{x} & x \neq 0, \\ \sin x, \\ x(e^{-x} + e^x). \end{cases} \tag{12.1.2b}$$

Note that while Eqs. (12.1.1) are functional equations, they do not have unique solutions. In fact, each equation has unlimited solutions.

12.1.2 $f(x + T) = f(x)$, $T > 0$ and Fixed

Let $f(x)$ be a periodic function with (fundamental) period denoted by a fixed, positive constant T; then $f(x)$ satisfies the functional equation

$$f(x + T) = f(x). \tag{12.1.3}$$

One consequence of Eq. (12.1.3) is that if $f(x)$ is specified on any interval of length T, then $f(x)$ can be determined for any x not in the original interval.

For example, consider the function

$$f_1(x) = \sin(ax), \quad a = \text{constant}, \tag{12.1.4}$$

with period

$$T = \frac{2\pi}{a}. \tag{12.1.5}$$

If values of $f_1(x)$ are given on the natural interval $[0, \frac{2\pi}{a}]$, then $f_1(y)$, where

$$\left(\frac{2\pi}{a}\right) k \leq y \leq \left(\frac{2\pi}{a}\right)(k+1), \quad k = \text{integer}, \tag{12.1.6}$$

can be calculated, i.e., writing y as

$$y = \left(\frac{2\pi}{a}\right)\bar{k} + x, \quad 0 < x < \frac{2\pi}{a}, \tag{12.1.7}$$

for some $k = \bar{k}$, then

$$f_1(y) = f_1(x). \tag{12.1.8}$$

12.1.3 $f(x + a) = f(x)$, a *is Arbitrary*

If a is an arbitrary constant, then it is clear that the functional equation

$$f(x + a) = f(x), \tag{12.1.9}$$

implies that

$$f(x) = c = \text{constant}. \tag{12.1.10}$$

One way of achieving this result is to assume that $f(x)$ has at least a first derivative. From this fact, it follows from Eq. (12.1.9) that

$$f(x + a) = f(x) + f'(x)a + O(a^2) = f(x), \tag{12.1.11}$$

and from this we conclude

$$f'(x) = 0 \Rightarrow f(x) = \text{constant}. \tag{12.1.12}$$

These results may also be interpreted as stating that the only function which can have an arbitrary period is the constant function.

12.1.4 $f(x) - f(y) = x - y$

Rewrite this equation to the form

$$f(x) - x = f(y) - y. \tag{12.1.13}$$

Since x and y may be selected such that $x \neq y$, we may conclude that

$$f(x) - x = c, \quad c = \text{constant}, \tag{12.1.14}$$

and

$$f(x) = x + c. \tag{12.1.15}$$

12.1.5 $(x - y)f(x + y) - (x + y)f(x - y) = 4xy(x^2 - y^2)$

A somewhat more complex case which is similar to the functional equation of the previous section is

$$(x - y)f(x + y) - (x + y)f(x - y) = 4xy(x^2 - y^2). \qquad (12.1.16)$$

To proceed, make a change to the new variables, u and v, defined as

$$u = x + y, \quad v = x - y, \qquad (12.1.17)$$

or

$$x = \frac{u + v}{2}, \quad y = \frac{u - v}{2}. \qquad (12.1.18)$$

Substituting Eqs. (12.1.18) in Eq. (12.1.16) gives

$$vf(u) - uf(v) = (u^2 - v^2)uv \qquad (12.1.19)$$

or

$$\frac{f(u)}{u} - u^2 = \frac{f(v)}{v} - v^2, \qquad (12.1.20)$$

and this implies

$$\frac{f(x)}{x} - x^2 = c = \text{constant}. \qquad (12.1.21)$$

Thus, Eq. (12.1.16) has the solution

$$f(x) = x^3 + cx. \qquad (12.1.22)$$

12.1.6 $f\left(\sqrt{x - 6}\right) = x$

This functional equation allows us to conclude that for $x = 6$, $f(0) = 6$, and x must be restricted to $x \geq 6$. Let

$$u\sqrt{x - 6}, \quad x \geq 6 \Rightarrow x = u^2 + 6, \qquad (12.1.23)$$

then it follows that $f(u) = u^2 + 6$, and

$$f(x) = x^2 + 6, \quad x \geq 6. \qquad (12.1.24)$$

12.1.7 $f(x) + 2f\left(\frac{1}{x}\right) = x^2$

First, let $x = y$, then

$$f(y) + 2f\left(\frac{1}{y}\right) = y^2. \qquad (12.1.25)$$

Next, let $x = \frac{1}{y}$, and obtain

$$f\left(\frac{1}{y}\right) + 2f(y) = \frac{1}{y^2}. \tag{12.1.26}$$

These two equations are linear in $f(y)$ and $f(\frac{1}{y})$, and, consequently, can be solved for them; doing this gives

$$f(y) = \frac{2}{3y^2} - \frac{y^2}{3}, \tag{12.1.27}$$

or finally,

$$f(x) = \frac{2}{3x^2} - \frac{x^2}{3}. \tag{12.1.28}$$

Note that the solution for $f(\frac{1}{y})$ is

$$f\left(\frac{1}{y}\right) = \left(\frac{2}{3}\right)y^2 - \left(\frac{1}{3y^2}\right), \tag{12.1.29}$$

or replacing y by $\frac{1}{x}$, we obtain the result expresssed in Eq. (12.1.28).

12.1.8 $f(\mathrm{Ln}\,x) = x^2 + 2x + 1$

A solution is obtained by making a transformation of variable from x to y, as follows

$$x = e^y \quad \text{or} \quad \mathrm{Ln}\,x = y, \quad x > 0. \tag{12.1.30}$$

Therefore, the functional equation becomes

$$f(y) = e^{2y} + 2e^y + 1, \tag{12.1.31}$$

and a solution to $f(\mathrm{Ln}\,x) = x^2 + 2x + 1$, is

$$\begin{aligned} f(x) &= e^{2x} + 2e^x + 1 \\ &= (e^x + 1)^2. \end{aligned} \tag{12.1.32}$$

12.1.9 $f(ax, ay) = f(x, y)$ and $f(ax, ay) = a^\beta f(x, y)$ [1]

The first equation

$$f(ax, ay) = f(x, y) \tag{12.1.33}$$

where a is an arbitrary constant, has the obvious solution

$$f(x, y) = F\left(\frac{x}{y}\right), \tag{12.1.34}$$

where the situation $y = 0$ may need to be examined. Since the solution can also be expressed in the form

$$f(x, y) = G\left(\frac{y}{x}\right),$$ (12.1.35)

we may also need to examine the case of $x = 0$. Within the context of physics, where x and y may represent space coordinates, the constant a has the physical units of inverse distance. Consequently, a^{-1} can be taken as a length scale and for a system in which the property given in Eq. (12.1.33) holds, we may say that such a system has no intrinsic length scale or is (length) scale invariant.

Note that $F(z)$ and $G(z)$ may, in general, be taken as arbitrary functions of their respective arguments.

Based on the above results and a little insight, it should be clear that the functional equation

$$f(ax, ay) = a^\beta f(x, y),$$ (12.1.36)

has solutions

$$f(x, y) = x^\beta F_1\left(\frac{x}{y}\right)$$ (12.1.37a)

and

$$f(x, y) = y^\beta F_2\left(\frac{x}{y}\right),$$ (12.1.37b)

where $F_1(z)$ and $F_2(z)$ are arbitrary functions of z.

12.1.10 $f(ax, a^\beta y) = f(x, y)$ *and* $f(ax, a^\beta y) = a^\gamma f(x, y)$ [1]

The functional equation

$$f(ax, a^\beta y) = f(x, y),$$ (12.1.38)

states that if x is replaced by ax and y is replaced by $a^\beta y$, then $f(x, y)$ remains the same, i.e., $f(x, y)$ is invariant under the pair of transformations. Since y/x^β is invariant under this transformation, i.e.,

$$\frac{y}{x^\beta} \to \frac{a^\beta y}{a^\beta x^\beta} = \frac{y}{x^\beta},$$ (12.1.39)

it follows that a solution to Eq. (12.1.38) is

$$f(x, y) = F\left(\frac{y}{x^\beta}\right).$$ (12.1.40)

The second equation

$$f(ax, a^\beta y) = a^\gamma f(x, y) \tag{12.1.41}$$

has a solution

$$f(x, y) = x^\gamma G\left(\frac{y}{x^\beta}\right). \tag{12.1.42}$$

Again, $F(z)$ and $G(z)$ are arbitrary functions of z.

Now consider if a solution for Eq. (12.1.41) can take the form

$$f_1(x, y) = y^\alpha H\left(\frac{y}{x^\beta}\right)? \tag{12.1.43}$$

Note that

$$f_1(x, y) \xrightarrow[y \to a^\beta y]{x \to ax} = (a^\beta y)^\alpha H\left(\frac{y}{x^\beta}\right)$$

$$= a^{\alpha\beta} y^\alpha H\left(\frac{y}{x^\beta}\right). \tag{12.1.44}$$

Comparing the right-sides of Eqs. (12.1.41) and (12.1.44) gives

$$\alpha\beta = \gamma \tag{12.1.45}$$

or

$$\alpha = \frac{\gamma}{\beta}. \tag{12.1.46}$$

Therefore, if α satisfies the condition in Eq. (12.1.46), then $f_1(x, y)$ is a solution, i.e.,

$$f_1(x, y) = y^{\gamma/\beta} H\left(\frac{y}{x^\beta}\right). \tag{12.1.47}$$

12.1.11 *Discontinuous Solution of $f(xy) = f(x)f(y)$*

The "sign" function is defined as

$$\text{sgn}(x) = \begin{cases} +1, & \text{if } x > 0, \\ 0, & \text{if } x = 0, \\ -1 & \text{if } x < 0. \end{cases} \tag{12.1.48}$$

Clearly,

$$f(x) = \text{sgn}(x), \tag{12.1.49}$$

is a solution to the functional equation

$$f(xy) = f(x)f(y), \tag{12.1.50}$$

however, it is discontinuous at $x = 0$. In Section 12.2, we discuss continuous solutions.

12.1.12   *Comments*

We conclude this section of Chapter 12 with several comments based on the above calculations regarding some rather elementary functional equations:

i) No general methodology or set of rules are known to exist for the systematic calculation of solutions to (even elementary) functional equations.

ii) However, a broad background knowledge and understanding of various areas of mathematics certainly is of value for this task.

iii) With regard to the functional equations appearing in the natural sciences and engineering, a deep understanding and an awareness of the major features and properties of a given system can provide insights into the mathematical structure of possible relevant solutions to the functional equations which arise in the modeling of such systems.

12.2   Cauchy Functional Equations [4]

Among the most studied of the linear functional equations are the four Cauchy equations:

$$f(x + y) = f(x) + f(y), \tag{12.2.1}$$

$$f(x + y) = f(x)f(y), \tag{12.2.2}$$

$$f(xy) = f(x)f(y), \tag{12.2.3}$$

$$f(xy) = f(x) + f(y). \tag{12.2.4}$$

They are denoted, respectively, the additive, exponential, multiplicative, and logarithmic Cauchy functional equations. The major task of this section is to construct continuous solutions to these equations. When needed, it will be assumed that at least one derivative exists, except at possibly a finite set of points. However, it should be stated that discontinuous solutions also exist; but, for our purposes, they will not be examined. In general, for the sciences and engineering, the continous solutions are the ones of most importance for specific applications. The full treatment of the Cauchy equations are presented in a number of books and we refer any interested reader who wishes to see the complete details of the various solution constructions and other associated issues to study these volumes [4].

12.2.1 $f(x + y) = f(x) + f(y)$

The additive Cauchy functional equation is

$$f(x + y) = f(x) + f(y). \tag{12.2.5}$$

For $x = 0$ and $y = 0$, we have

$$f(0) = 2f(0) \Rightarrow f(0) = 0. \tag{12.2.6}$$

Further, letting $y = -x$ gives

$$f(0) = f(x) + f(-x) \Rightarrow f(x) \text{ is odd.} \tag{12.2.7}$$

Now, let x be fixed and use the fact that

$$\int_0^1 dy = 1, \tag{12.2.8}$$

to obtain the result

$$
\begin{aligned}
f(x) &= \int_0^1 f(x)dy \\
&= \int_0^1 [f(x+y) - f(y)]dy \\
&= \int_0^1 f(x+y)dy - \int_0^1 f(y)dy, \tag{12.2.9} \\
&= \int_x^{1+x} f(z)dz - \int_0^1 f(y)dy. \tag{12.2.10}
\end{aligned}
$$

Note that the second term on the right-side of the last expression is a constant, while the first term was obtained by using the variable transformation $z = x + y$, i.e.,

$$\int_0^1 f(x+y)dy = \int_x^{1+x} f(z)dz. \tag{12.2.11}$$

Taking the derivative on both sides of the expression given by Eq. (12.2.11) gives

$$\frac{df(x)}{dx} = f(x+1) - f(x). \tag{12.2.12}$$

If now $f(x + 1)$ is replaced by $f(x) + f(1)$, we obtain

$$\frac{df(x)}{dx} = c, \quad c = f(1), \tag{12.2.13}$$

or

$$f(x) = cx. \tag{12.2.14}$$

Consequently, the continuous solution of the additive Cauchy functional equation is the linear function given in Eq. (12.2.14). It is a trivial matter to check that this is the solution by substituting it into Eq. (12.2.5).

12.2.2 $f(x + y) = f(x)f(y)$

The exponential Cauchy functional equation is

$$f(x + y) = f(x)f(y). \tag{12.2.15}$$

For $x = 0$ and $y = 0$, we have

$$f(0) = [f(0)]^2, \tag{12.2.16}$$

and this implies that either

$$f(0) = 0 \quad \text{and} \quad f(x) = 0, \tag{12.2.17a}$$

or

$$f(0) = 1. \tag{12.2.17b}$$

Now taking the natural logarithm of Eq. (12.2.15) gives

$$\text{Ln}[f(x + y)] = \text{Ln}[f(x)] + \text{Ln}[f(y)], \tag{12.2.18}$$

and this takes the form

$$g(x + y) = g(x) + g(y), \tag{12.2.19}$$

if $g(x)$ is defined as

$$g(x) = \text{Ln}[f(x)]. \tag{12.2.20}$$

However, inspection of Eq. (12.2.19) shows that it is a Cauchy additive functional equation, with solution

$$g(x) = cx, \quad c = g(1). \tag{12.2.21}$$

Therefore, the solution to Eq. (12.2.15) is

$$f(x) = e^{cx}. \tag{12.2.22}$$

If we set

$$f(1) = a = e^c; \tag{12.2.23}$$

then

$$f(x) = a^x, \tag{12.2.24}$$

where a is a real, positive number.

12.2.3 $f(xy) = f(x) + f(y)$

The logarithmic Cauchy functional equation is

$$f(xy) = f(x) + f(y). \qquad (12.2.25)$$

Let $x = u$ and $y = u$, then

$$f(u^2) = 2f(u). \qquad (12.2.26)$$

Now let $x = -u$ and $y = -u$, and obtain

$$f(u^2) = 2f(-u). \qquad (12.2.27)$$

Comparing the last two equations, it can be concluded that $f(x)$ is an even function, i.e.,

$$f(-x) = f(x). \qquad (12.2.28)$$

Let $x = 0$ and $y = 0$, then it follows that

$$f(0) = 2f(0) \qquad (12.2.29)$$

and either $f(0) = 0$, which implies either that $f(x) = 0$, or $f(0)$ is not defined at $x = 0$. With respect to the second possibility, assume Eq. (12.2.25) holds for $x > 0$ and $y > 0$. This means that we can represent (x, y) in terms of the variables (u, v), i.e.,

$$x = e^u, \quad y = e^v, \qquad (12.2.30a)$$

and

$$u = \operatorname{Ln} x, \quad v = \operatorname{Ln} y. \qquad (12.2.30b)$$

Substituting these results into Eq. (12.2.25) gives

$$f(e^{u+v}) = f(e^u) + f(e^v). \qquad (12.2.31)$$

Now define $F(u)$ as follows

$$F(u) = f(e^u), \qquad (12.2.32)$$

and obtain from Eq. (12.2.31), the result

$$F(u + v) = F(u) + F(v), \qquad (12.2.33)$$

where

$$F(u) = cu, \qquad (12.2.34)$$

and, thus,

$$f(e^u) = cu, \qquad (12.2.35)$$

or rewriting in terms of x

$$f(x) = c \operatorname{Ln} x. \qquad (12.2.36)$$

Note that the even-ness property in Eq. (12.2.28) can be enforced by expressing $f(x)$ as

$$f(x) = c \operatorname{Ln} |x|. \qquad (12.2.37)$$

This function is the general continuous solution to the logarithmic Cauchy equation for all x, except $x = 0$.

12.2.4 $f(xy) = f(x)f(y)$

Based on the previous naming of Cauchy functional equations, it will be obvious why the following equation is named the multiplicative Cauchy equation

$$f(xy) = f(x)f(y). \tag{12.2.38}$$

Two particular (continuous) constant solutions are readily discovered by inspection, namely,

$$f(x) = 0, \quad f(x) = 1. \tag{12.2.39}$$

Two other piecewise constant solutions are

$$f(x) = \operatorname{sgn}(x), \quad f(x) = |\operatorname{sgn}(x)|, \tag{12.2.40}$$

where $\operatorname{sgn}(x)$ is the sign-function. Note that the latter functions are discontinuous at the point $x = 0$.

A direct calculation shows that two other solutions are

$$f(x) = |x|^k, \quad f(x) = |x|^k \operatorname{sgn}(x), \tag{12.2.41}$$

where k is an arbitrary, positive real constant.

12.2.5   *Generalizations*

The Cauchy functional equations can all be generalized to multi-variables and in several differing mathematical structures. For example, the additive functional equation

$$f(x + y) = f(x) + f(y) \tag{12.2.42}$$

can be generalized to

$$f(x + y + z) = f(x) + f(y) + f(z), \tag{12.2.43}$$

or to

$$f(x_1 + y_1, x_2 + y_2) = f(x_1, x_2) + f(y_1, y_2). \tag{12.2.44}$$

Likewise, a generalization of the multiplicative Cauchy functional equation

$$f(xy) = f(x)f(y), \tag{12.2.45}$$

is

$$f(x_1 y_1, x_2 y_2) = f(x_1, x_2)f(y_1, y_2). \tag{12.2.46}$$

Similarly, $f(x + y) = f(x)f(y)$ and $f(xy) = f(x) + f(y)$ generalize to, respectively, the forms

$$f(x_1 + y_1, x_2 + y_2) = f(x_1, x_2)f(y_1, y_2), \tag{12.2.47}$$

$$f(x_1 y_1, x_2 y_2) = f(x_1, x_2) + f(y_1, y_2). \tag{12.2.48}$$

Discussions of Cauchy functional equations in several variables are given in the books by Sahoo and Kannappan, Chapter 3; and Aczél (2006), see Sections 3.1.1 and 5.1.

12.3 Named Functional Equations

Following the same general theme, as presented in Section 3.8, we list below a number of "named functional equations." The books listed in the Bibliography to this chapter provide an abundance of details regarding the genesis of these equations, the construction of various types of explicit solutions, along with discussions on associated equations and related issues.

It is important to understand that a significant reason for scientists to be aware of the existence of these particular functional equations is that such awareness enhances the numbers of tools available for use in the construction and analysis of mathematical models for many systems relevant to their work. While they may lack specific knowledge as to how to solve these equations, a rather large and growing set of published materials, both books and journal articles, exist to aid with this task.

Abel

The standard Abel functional equation is

$$f(g(x)) = f(x) + a, \tag{12.3.1}$$

however, two other equations which go under this name are

$$f(x + y) = g(xy) + h(x - y), \tag{12.3.2a}$$

$$f(x + y)h(x - y) = g(xy). \tag{12.3.2b}$$

Böttcher

$$f(g(x)) = f(x)^p \tag{12.3.3}$$

Cauchy I

$$f(x + y) = f(x) + f(y) \tag{12.3.4a}$$

Cauchy II

$$f(x + y) = f(x)f(y) \tag{12.3.4b}$$

Cauchy III

$$f(xy) = f(x) + f(y) \tag{12.3.4c}$$

Cauchy IV

Cocycle

$$f(xy) = f(x)f(y) \qquad (12.3.4d)$$

$$f(x,y) + f(x+y,z) = f(x,y+z) + f(y,z) \qquad (12.3.5)$$

Conjugacy

$$f(h(x)) = H(f(x)) \qquad (12.3.6)$$

D'Alembert-I

$$f(x+y) + f(x-y) = 2f(x)f(y) \qquad (12.3.7a)$$

D'Alembert-II

$$f(x+y)f(x-y) = f(x)^2 - f(y)^2 \qquad (12.3.7b)$$

Euler

$$f(\lambda x, \lambda y) = \lambda f(x,y) \qquad (12.3.8)$$

Hosszú

$$f(x+y-xy) + f(xy) = f(x) + f(y) \qquad (12.3.9)$$

Jensen

$$f\left(\frac{x+y}{2}\right) = \frac{f(x) + f(y)}{2} \qquad (12.3.10)$$

Lobacevskii

$$f(x)^2 = f(x+y)f(x-y) \qquad (12.3.11)$$

Mean Value Type

$$f(x) - f(y) = (x - y)h\left(\frac{x+y}{2}\right) \qquad (12.3.12)$$

Pexider I

$$f(x + y) = g(x) + h(y) \qquad (12.3.13a)$$

Pexider II

$$f(x + y) = g(x)h(y) \qquad (12.3.13b)$$

Pexider III

$$f(xy) = g(x) + h(y) \qquad (12.3.13c)$$

Pexider IV

$$f(xy) = g(x)h(y) \qquad (12.3.13d)$$

Pompeiu

$$f(x + y + xy) = f(x) + f(y) + f(xy) \qquad (12.3.14)$$

Quadratic

$$f(x + y) + f(x - y) = 2f(x) + 2f(y) \qquad (12.3.15)$$

Schröder

$$f(g(x)) = \lambda f(x) \qquad (12.3.16)$$

Trigonometric

There are a number of functional equations associated with the standard trigonometric and exponential functions and six of these are listed below.

$$f(x - y) = f(x)f(y) + g(x)g(y), \qquad (12.3.17a)$$
$$g(x + y) = g(x)f(y) + f(x)g(y), \qquad (12.3.17b)$$
$$f(x + y) = f(x)f(y) - g(x)g(y), \qquad (12.3.17c)$$
$$g(x - y) = g(x)f(y) - g(y)f(x), \qquad (12.3.17d)$$
$$f(x + y)f(x - y) = f(x)^2 - f(y)^2, \qquad (12.3.17e)$$
$$f(x + y)g(x - y) = f(x)g(x) - f(y)g(y). \qquad (12.3.17f)$$

Vincze I

$$f(x + y) = g_1(x)g_2(y) + h(y) \qquad (12.3.18a)$$

Vincze II

$$f(xy) = g_1(x)g_2(y) + h(y) \qquad (12.3.18b)$$

Whitehead

$$f(x+y+z)+f(x)+f(y)+f(z) = f(x+y)+f(y+z)+f(x+z) \quad (12.3.19)$$

Wilson I

$$f(x + y) + f(x - y) = 2f(x)g(y), \qquad (12.3.20a)$$

Wilson II

$$f(x + y) + g(x - y) = 2h(x)k(y). \qquad (12.3.20b)$$

12.4   Worked Problems and Applications

In addition to illustrating how solutions may be constructed for several specific functional equations, in this section, we also present a number of applications of such equations to the modeling, analysis, and understanding of some issues in the natural sciences and mathematics. We also show how certain functional equations are related to differential equations.

12.4.1 *Area of a Rectangle* [9]

Assume that there is a rectangle of base-length, ℓ, and height, h. The task is to determine the area of the rectangle, which is denoted by A. Clearly, the area will depend on both ℓ and h, i.e.,

$$A = f(h, \ell) \geq 0. \qquad (12.4.1)$$

Let us now divide the rectangle into two other rectangles in two different ways; see Figs. 12.4.1a and 12.4.1b. Therefore,

$$A = A_1 + A_2 : f(h, \ell) = f(h_1, \ell) + f(h_2, \ell), \qquad (12.4.2a)$$

$$A = B_1 + B_2 : f(h, \ell) = f(h, \ell_1) + f(h, \ell_2), \qquad (12.4.2b)$$

and using

$$\ell = \ell_1 + \ell_2, \quad h = h_1 + h_2, \qquad (12.4.3)$$

we obtain

$$f(h_1 + h_2, \ell) = f(h_1, \ell) + f(h_2, \ell) \qquad (12.4.4a)$$

$$f(h, \ell_1 + \ell_2) = f(h, \ell_1) + f(h, \ell_2). \qquad (12.4.4b)$$

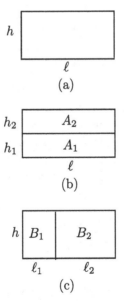

Fig. 12.4.1   (a) A rectangle. (b) The rectangle divided into two horizontal slabs. (c) The rectangle divided into two vertical slabs.

For fixed ℓ, the solution to Eq. (12.4.4a) is

$$f(h, \ell) = c(\ell)h, \qquad (12.4.5)$$

where $c(\ell)$ indicates that c is a non-negative constant whose value depends on ℓ.

If the result in Eq. (12.4.5) is used in Eq. (12.4.4b), then

$$c(\ell_1 + \ell_2)h = c(\ell_1)a + c(\ell_2)a, \qquad (12.4.6)$$

or

$$c(\ell_1 + \ell_2) = c(\ell_1) + c(\ell_2), \qquad (12.4.7)$$

and we conclude that

$$c(\ell) = a\ell, \qquad (12.4.8)$$

where a is a non-negative constant. Combining Eqs. (12.4.5) and (12.4.8) gives

$$A = f(h, \ell) = ah\ell. \qquad (12.4.9)$$

If we define the area of a square, where $h = \ell$, to be $A = h^2 = \ell^2$, then the constant a has the value one, i.e., $a = 1$, and the area of a rectangle, having base ℓ and height h, is

$$A = h\ell. \qquad (12.4.10)$$

12.4.2   Decay Phenomena [10]

There are a number of systems which exhibit decay with time of some physical substance. Particular examples include the radioactive disintegration of certain materials and the ingestion of drugs. We now derive a formula which can be used to predict the behavior of such systems.

Let $m(t)$ be the amount of some material present at time t, with $m_0 = m(0)$ the initial amount. Assume that there exists a function $f(t)$ such that

$$m(t) = m_0 f(t). \qquad (12.4.11)$$

For the moment, fix t and note that at the later time, $t + \Delta t$, we have

$$m(t + \Delta t) = m_0 f(t + \Delta t), \quad \Delta t > 0, \qquad (12.4.12)$$

and it follows that

$$m(t + \Delta t) = [m_0 f(t)]f(\Delta t) \qquad (12.4.13)$$

and

$$m_0 f(t + \Delta t) = m_0 f(t)f(\Delta t). \qquad (12.4.14)$$

Therefore, we obtain the functional equation

$$f(t + \Delta t) = f(t)f(\Delta t), \tag{12.4.15}$$

having the solution

$$f(t) = e^{at}, \tag{12.4.16}$$

where a is a constant. Since $f(t)$ is expected to be a decreasing function of t, then setting $a = -\lambda$, where $\lambda > 0$, gives

$$m(t) = m_0 e^{-\lambda t}. \tag{12.4.17}$$

This equation is known as the *decay equation*.

The corresponding differential equation, whose solution is that stated in Eq. (12.4.17) is

$$\frac{dm(t)}{dt} = -\lambda m(t), \quad m(0) = m_0. \tag{12.4.18}$$

12.4.3   *Integral Functional Equation* [11]

Consider the equation

$$\int_0^x f(t)dt = \left[\sqrt{f(0)f(x)}\right]x, \quad x \geq 0. \tag{12.4.19}$$

Observe that $f(x)$ appears in two places: first, under the square-root on the right-side, and then under the integral on the left-side. Consequently, we may think of the expression in Eq. (12.4.19) as a nonlinear *integral functional equation*. In the following, we assume that $f(x)$ has a first derivative.

Note that for $f(0) = 0$, it follows that

$$\int_0^x f(x)dt = 0, \tag{12.4.20}$$

and if this is to hold for arbitrary x, then $f(x) = 0$. Now suppose $f(0) < 0$, for x in the interval, $[0, \epsilon]$, where some $\epsilon > 0$. This assumption produces a contradiction. To see this, consider

$$\int_0^\epsilon f(t)dt = \left[\sqrt{f(0)f(x)}\right]\epsilon. \tag{12.4.21}$$

The left-side is negative, while the right-side is positive. Thus, we must conclude that $f(0) > 0$.

Now take the derivative of both sides of Eq. (12.4.19) to get

$$f(x) = \sqrt{f(0)f(x)} + x\frac{d}{dx}\left[\sqrt{f(0)f(x)}\right], \tag{12.4.22}$$

and then divide by $f(0)$ to obtain the result

$$\frac{f(x)}{f(0)} = \sqrt{\frac{f(x)}{f(0)}} + x\frac{d}{dx}\left[\sqrt{\frac{f(x)}{f(0)}}\right]. \qquad (12.4.23)$$

If the function $g(x)$ is defined as

$$g(x) = \sqrt{\frac{f(x)}{f(0)}}, \qquad (12.4.24)$$

then Eq. (12.4.23) takes the form

$$x\frac{dg(x)}{dx} + g(x) - g(x)^2 = 0, \qquad (12.4.25)$$

and this is a separable, first-order differential equation which can be integrated to the result

$$\int \frac{dg}{g(g-1)} = \int \frac{dx}{x} \qquad (12.4.26)$$

or

$$\mathrm{Ln}\,|g-1| - \mathrm{Ln}\,|g| = \mathrm{Ln}\,x + \mathrm{Ln}\,c, \qquad (12.4.27)$$

and finally

$$\left|1 - \frac{1}{g}\right| = cx, \qquad (12.4.28)$$

where the integration constant, c, is positive. From this last equation, there are two cases to consider. For

$$1 - \frac{1}{g(x)} > 0, \qquad (12.4.29)$$

we have

$$g(x) = \frac{1}{1 - cx}. \qquad (12.4.30)$$

Comment

Mathematicians will say that Eq. (12.4.30) means

$$g(x) = \frac{1}{1 - cx}, \quad x \in \left[0, \frac{1}{c}\right). \qquad (12.4.31)$$

A "good" mathematical grounded physical scientist will state that

$$g(x) = \begin{cases} \frac{1}{1-cx} > 0, & x \in [0, \frac{1}{c}), \\ \infty, & x = \frac{1}{c}, \\ \frac{(-1)}{cx-1} < 0, & x > \frac{1}{c}. \end{cases} \qquad (12.4.32)$$

The second case occurs when

$$1 - \frac{1}{g(x)} < 0. \tag{12.4.33}$$

Solving for $g(x)$ gives

$$g(x) = \frac{1}{1 + cx}, \quad x \geq 0. \tag{12.4.34}$$

Using the definition of $g(x)$, from Eq. (12.4.24), it follows that $f(x) = f(0)[g(x)]^2$, and

$$\textbf{First case: } f(x) = \begin{cases} \frac{f(0)}{(1-cx)^2}, & x \in [0, \frac{1}{c}), \\ \infty, & x = \frac{1}{c}, \\ \frac{f(0)}{(cx-1)^2}, & x > \frac{1}{c}, \end{cases} \tag{12.4.35a}$$

$$f(0) > 0, \quad c > 0. \tag{12.4.35b}$$

$$\textbf{Second case: } f(x) = \frac{f(0)}{(1 + cx)^2}, \quad x \in [0, +\infty) \tag{12.4.36a}$$

$$f(0) \geq 0, \quad c \geq 0. \tag{12.4.36b}$$

See Fig. 12.4.2 for plots of $f(x)$ vs x for these two cases.

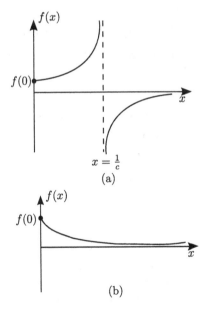

Fig. 12.4.2   (a) First case. (b) Second case.

12.4.4 $f(x+y) + f(x-y) - 2f(x) - f(y) - f(-y) = 0$ [12]

The following function equation has been investigated by Skof [12] and Szabó [12],

$$f(x+y) + f(x-y) - 2f(x) - f(y) - f(-y) = 0, \qquad (12.4.37)$$

who concluded that

$$f(x) = q(x) + h(x), \qquad (12.4.38)$$

where $q(x)$ is quadratic and $h(x)$ is additive. We produce two different methods to achieve these same results.

For the first method, begin by setting $x = 0$ and $y = 0$, giving

$$-2f(0) = 0 \Rightarrow f(0) = 0. \qquad (12.4.39)$$

Next assume that $f(x)$ is even, i.e., $f(-x) = f(x)$, and let $y = x$. This gives for Eq. (12.4.37) the expression

$$f(2x) + f(0) - 2f(x) - f(x) - f(-x) = 0, \qquad (12.4.40)$$

or

$$f(2x) = 4f(x). \qquad (12.4.41)$$

The solution to this equation is

$$f(x) = ax^2, \quad a = \text{constant}. \qquad (12.4.42)$$

Next assume that $f(x)$ is odd, i.e., $f(-x) = -f(x)$. Using this result with $y = x$ in Eq. (12.4.37) gives

$$\begin{aligned}
f(2x) + f(0) - 2f(x) - f(x) - f(-x) \\
= f(2x) + 0 - 3f(x) + f(x) \\
= f(2x) - 2f(x) \\
= 0, \qquad (12.4.43)
\end{aligned}$$

and therefore

$$f(2x) = 2f(x) \qquad (12.4.44)$$

with the solution

$$f(x) = bx, \quad b = \text{constant}. \qquad (12.4.45)$$

Since Eq. (12.4.37) is a linear equation, then adding the results from Eqs. (12.4.41) and (12.4.45) also gives a solution. Therefore

$$f(x) = ax^2 + bx, \qquad (12.4.46)$$

i.e., the solution has the form presented in Eq. (12.4.38), where

$$q(x) = ax^2, \quad h(x) = bx. \tag{12.4.47}$$

The second method begins by rewriting Eq. (12.4.37) to the form

$$f(x + y) - 2f(x) + f(x - y) = f(y) + f(-y). \tag{12.4.48}$$

If y is replaced by Δx, then

$$f(x + \Delta x) - 2f(x) + f(x - \Delta x) = f(\Delta x) + f(-\Delta x). \tag{12.4.49}$$

Assuming $f(x)$ has a Taylor series, we have

$$f(x \pm \Delta x) = f(x) \pm f'(x)(\Delta x) + \left(\frac{1}{2}\right) f''(x)(\Delta x)^2 + O[(\Delta x)^3], \tag{12.4.50}$$

and

$$f(\pm \Delta x) = f(0) \pm f'(0)(\Delta x) + \left(\frac{1}{2}\right) f''(0)(\Delta x)^2 \pm O[(\Delta x)^3]. \tag{12.4.51}$$

Substituting the two equations above into Eq. (12.4.49) and using $f(0) = 0$, the following result is obtained

$$f''(x)(\Delta x)^2 + O[(\Delta x)^4] = f''(0)(\Delta x)^2 + O[(\Delta x)^4] \tag{12.4.52}$$

or

$$f''(x) = f''(0) = c, \quad c = \text{constant}. \tag{12.4.53}$$

Therefore, $f(x)$ is given by the expression

$$f(x) = \left(\frac{1}{2}\right) cx^2 + c_1 x + c_2, \tag{12.4.54}$$

where c_1 and c_2 are arbitrary integration constants. Since $f(0) = 0$, then $c_2 = 0$ and $f(x)$ is

$$f(x) = \left(\frac{1}{2}\right) cx^2 + c_1 x. \tag{12.4.55}$$

Comparing Eqs. (12.4.46) gives $a = c/2$ and $b = c_1$.

Note that we have obtained a very interesting result: the functional equation (12.4.37) and the differential equation (12.4.53) are, in some sense, equivalent to each other.

12.4.5 $f(x + y) = f(x) + f(y) - af(x)f(y)$ [13]

This functional equation requires $a \neq 0$, otherwise, it reduces to the additive Cauchy equation. In the literature, the equation

$$f(x + y) = f(x) + f(y) - af(x)f(y), \qquad (12.4.56)$$

is also known as the equation of the theory of probability [13].

Setting $y = 0$ gives

$$0 = f(0)[1 - af(x)], \qquad (12.4.57)$$

from which the following conclusions may be reached:

i) $f(x) = 0$ is a trivial solution.

ii) $f(0) = 0$, for non-trivial solutions.

iii) A constant solution is

$$f(x) = \frac{1}{a}. \qquad (12.4.58)$$

Set $y = \Delta x$ and Eq. (12.4.56) can be rewritten to the form

$$f(x + \Delta x) - f(x) = f(\Delta x)[1 - af(x)]. \qquad (12.4.59)$$

Now assume $f(x)$ has a Taylor series, such that

$$f(\Delta x) = f'(0)(\Delta x) + O[(\Delta x)^2], \qquad (12.4.60)$$

where $f(0) = 0$ was used to eliminate the constant term. The substitution of Eq. (12.4.60) into Eq. (12.4.59) gives

$$f(x + \Delta x) - f(x) = \left\{ f'(0)(\Delta x) + O[(\Delta x)^2] \right\} [1 - af(x)], \qquad (12.4.61)$$

and

$$\frac{f(x + \Delta x) - f(x)}{\Delta x} = [f'(0) + O(\Delta x)] [1 - af(x)]. \qquad (12.4.62)$$

Taking the Limit $\Delta x \to 0$, gives the first-order, linear differential equation

$$\frac{df(x)}{dx} = f'(0) - [af'(0)]f(x), \qquad (12.4.63)$$

with the solution (remember $f(0) = 0$)

$$f(x) = \left(\frac{1}{a} \right) [1 - e^{-\beta x}], \qquad (12.4.64)$$

where

$$\beta = af'(0). \qquad (12.4.65)$$

Note that the parameter β is an arbitrary constant since $f'(0)$ is not a priori specified.

In summary, the $f(x)$ given in Eq. (12.4.64) is the nontrivial solution to the functional equation presented in Eq. (12.4.56).

12.4.6 $f(x) + g(y) = h(x+y)$

One of the Pexider functional equations is

$$f(x) + g(y) = h(x+y). \tag{12.4.66}$$

The task is to determine the unknown functions $f(x)$, $g(y)$, and $h(z)$.

To proceed [14], differentiate first with respect to x and then with respect to y; doing this gives

$$\frac{d^2 h(z)}{dz^2} = 0, \quad z = x + y, \tag{12.4.67}$$

with the solution

$$h(z) = az + b, \quad = ax + ay + b, \tag{12.4.68}$$

where (a, b) are arbitrary constants. Using this result in Eq. (12.4.66) gives

$$f(x) + g(y) = (ax + b) + ay. \tag{12.4.69}$$

Comparing left and right sides allows the determination of $f(x)$ and $g(y)$; they are

$$f(x) = ax + b + c, \quad g(y) = ay - c. \tag{12.4.70}$$

In summary, the solution to Pexider's functional equation is presented in Eqs. (12.4.68) and (12.4.70).

12.4.7 $f(x,t) = a^k f(a^m x, a^n t)$ [14]

Functional equations of this type appear as self-similar solutions to many of the nonlinear partial differential equations in the mathematical natural and engineering sciences. The (k, m, n) are *a priori* given constants and a is the positive scaling parameter. For the corresponding systems, x and t usually denote space and time variables. Note that for $a = 1$, we get $f(x,t) = f(x,t)$.

To begin the solution finding process, differentiate the equation

$$f(x,t) = a^k f(a^m x, a^n t), \tag{12.4.71}$$

with respect to the parameter a and then set $a = 1$; doing this gives the expression

$$k f(x,t) + mx \frac{\partial f(x,t)}{\partial x} + nt \frac{\partial f(x,t)}{\partial t} = 0. \tag{12.4.72}$$

Next we determine the related characteristic system of ordinary differential equations [15]. For our problem, they are given by the following relations

$$\frac{dx}{mx} = \frac{dt}{nt} = -\frac{df}{kf},$$ (12.4.73)

or

$$\frac{dx}{dt} = \left(\frac{m}{n}\right)\left(\frac{x}{t}\right), \quad \frac{df}{dt} = -\left(\frac{k}{n}\right)\left(\frac{t}{f}\right).$$ (12.4.74)

Integrating these equations gives

$$\frac{x}{t^{m/n}} = C_1, \quad t^{k/n}f(x,t) = C_2.$$ (12.4.75)

Therefore, the solution is

$$f(x,t) = t^{-(k/n)}F\left(\frac{x}{t^{m/n}}\right),$$ (12.4.76)

where $F(z)$ is an arbitrary function of z.

12.4.8 Relationship between Differential and Functional Equations

In a very general sense, differential equations may be considered to belong to the class of functional equations as we have presented and discussed them in this chapter. We now demonstrate that a direct connection does exist between these two types of equations, at least for first-order, ordinary differential equations.

The results to come generalize a remark made in the book of Aczél [16].

Consider, for simplicity, the first-order differential equation

$$\frac{df(x)}{dx} = H(f(x)),$$ (12.4.77)

where $H(f(x))$ is given. A discretization of this equation, based on the method of finite differences, takes the form [17]

$$\frac{f(x+h) - f(x)}{h} = H(f(x)) + H_1(f(x), h)h,$$ (12.4.78)

where $h = \Delta x$ and $H_1(u, v)$, in principle, can be arbitrary, but may be constrained to satisfy certain conditions such as being continuous. Letting $y = h$, this equation can be rewritten to the form

$$f(x+y) = f(x) + H(f(x))y + H_1(f(x), y)y^2.$$ (12.4.79)

Observe that the latter equation is a nonlinear functional equation. We now list several direct consequences of what this results mean:

a) Since $H_1(z, y)$ is essentially arbitrary, it follows that a given first-order differential equation can be represented by an uncounterable number of functional equations.

b) By the manner in which the functional equation was constructed, it follows that in the limit when $y \to 0$, the functional equation reduces to the original differential equations [17].

c) There may exist functional equations for which no differential equation limiting form exists.

d) The issue in a) is related to the genesis of numerical instabilities in the finite difference discretizations of differential equations [17].

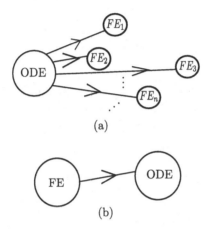

(a)

(b)

Fig. 12.4.3 Relations between differential equations (ODE) and functional equations (FE). (a) One ODE produces many FE's. (b) A given FE has only one ODE limiting equation (if it exists).

12.4.9 *Pomeranchuk Equation*

The Pomeranchuk functional equation arises in the mathematical and theoretical physics analysis of the scattering of elementary particles at very high energies [18; 19]. This equation takes the form

$$\alpha(t) = N\alpha\left(\frac{t}{N^2}\right) - (N - 1), \qquad (12.4.80)$$

where N is an integer,

$$N = 1, 2, 3, \ldots, \qquad (12.4.81)$$

and $\alpha(t)$ has the properties:

a) $\alpha(t)$ is real, for $t \le 0$;

b) $d\alpha(t)/dt < 0$, for $t < 0$.

Note, first, that Eq. (12.4.80) is a linear equation. Second, $\alpha(t)$ will not depend on N; i.e., compare the left- and right-sides of Eq. (12.4.80).

To proceed, let's check for a constant solution, i.e., $\alpha(t) = A = $ constant. Substituting this into the Pomeranchuk equation gives

$$A = NA - (N - 1) \Rightarrow A = 1. \tag{12.4.82}$$

Next, assume that

$$\alpha(t) = 1 + \beta(t). \tag{12.4.83}$$

If this is a solution, then $\beta(t)$ satisfies the functional equation

$$\beta(t) = N\beta\left(\frac{t}{N^2}\right). \tag{12.4.84}$$

Inspection of Eq. (12.4.84) suggests the ansatz

$$\beta(t) = b\sqrt{-t}, \quad t \leq 0, \tag{12.4.85}$$

where b is a real number. This can be verified by substitution into Eq. (12.4.84), i.e.,

$$N\beta\left(\frac{t}{N^2}\right) = Nb\sqrt{-\left(\frac{t}{N^2}\right)} = b\sqrt{-t} = \beta(t). \tag{12.4.86}$$

Hence, the solution to the Pomeranchuk functional equation is

$$\alpha(t) = 1 + b\sqrt{-t}. \tag{12.4.87}$$

To obtain the sign of b, we must have $b < 0$, since the derivative has the constraint, for $t < 0$,

$$\frac{d\alpha(t)}{dt} < 0 \Rightarrow b < 0. \tag{12.4.88}$$

Thus, $\alpha(t)$ is

$$\alpha(t) = 1 - a\sqrt{-t}, \quad a = -b > 0. \tag{12.4.89}$$

A generalization of the Pomeranchuk functional equation is to consider the equation

$$\alpha(t) = N\alpha\left(\frac{t}{N^m}\right) - (N - 1), \tag{12.4.90}$$

where m is real and positive. A direct calculation provides the following solution

$$\alpha(t) = 1 - \gamma\left(|t|^{1/m}\right), \quad \gamma > 0. \tag{12.4.91}$$

Inspection of both Eqs. (12.4.89) and (12.4.91) shows that the solutions do not depend on N.

12.5   Problems

Section 12.1

12.1.1 Consider the functional equation [2]
$$f(x+y) - f(x) - yf(x) = yg(x,y)$$
where $g(x,0) = 0$. Find a solution.
Hint: Rewrite this equation as
$$\frac{f(x+y) - f(x)}{y} - f(x) = g(x,y).$$
12.1.2 Determine the solution to the equation
$$f(x+y) = ry + f(x)$$
where r is a constant [2].
12.1.3 Determine all solutions to the equation [2]
$$f(xy) = y^r f(x),$$
for r real.
12.1.4 Obtain a solution to the function equation [3]
$$f(x,y) = a^k f(a^m x, a^n y).$$
12.1.5 Show that a solution to
$$f(x,y)f(y,z) = f(x,z)$$
is
$$f(x,y) = F(y)/F(x),$$
where $F(x)$ is an arbitrary function.
12.1.6 Consider the equation
$$f(x,y) + f(y,z) = f(x,z)$$
and demonstrate that the solution takes the form
$$f(x,y) = G(x) - G(y),$$
where $G(x)$ is an arbitrary function.
12.1.7 Given the functional equation
$$y(y(x)) = x,$$
shows that four particular solutions are
$$y(x) = \begin{cases} x, \\ c_1 - x, \\ \frac{c_2}{x}, \\ \frac{c_3 - x}{1 + c_4 x}, \end{cases}$$
where (c_1, c_2, c_3, c_4) are arbitrary constants.

Section 12.2

12.2.1 For the functional equation

$$f(x + y) = A^y f(x) + A^x f(y),$$

where A is a non-negative constant, show that solution is [5]

$$f(x) = cxA^x, \quad c = \text{constant}.$$

12.2.2 Determine the solutions to the equation

$$f(x + y) = A^{xy} f(x) f(y),$$

where A is a non-negative constant.
Answer [6]:

$$\begin{cases} 0, \\ b^x \, A^{x^2/2}, \end{cases}$$

where b is a positive constant.

12.2.3 Find the solutions to the functional equation [7]

$$f(x + y + a) + b = f(x) + f(y).$$

12.2.4 Show that the solution to

$$f(x) + g(y) = h(x + y)$$

is

$$f(x) = c_1 x + c_2, \quad g(y) = c_1 y + c_3, \quad h(z) = c_1 z + c_2 + c_3.$$

12.2.5 Find the solution to the functional equation

$$f(x + y) = f(x) + f(y) + a(1 - b^x)(1 - b^y),$$

where a and b are real constants, with $b > 0$.
Answer [8]: The solution is

$$f(x) = a(b^x - 1) + cx.$$

Section 12.4

12.4.1 Detail the reasoning for the results given in Eqs. (12.4.2).

12.4.2 Generalize the arguments of Section 12.4.1 to the case of a rectangular solid.

12.4.3 Complete the details required to obtain Eq. (12.4.27) from Eq. (12.4.26).

12.4.4 Prove that the solution to $f(2x) = 4f(x)$ is $f(x) = ax^2$, where a is an arbitrary constant.

12.4.5 Provide the steps in going from Eq. (12.4.49) to Eq. (12.4.53).

12.4.6 Show that Eq. (12.4.64) is the solution to Eq. (12.4.63).

12.4.7 Explain the appearance of the constant c in Eq. (12.4.70).

12.4.8 How does the result given in Eq. (12.4.76) follow from Eq. (12.4.75)?

12.4.9 Explain how the result given in Eq. (12.4.78) follows from Eq. (12.4.77).

12.4.10 The $\alpha(t)$, given in Eq. (12.4.89) is complex valued for $t > 0$. Plot $\operatorname{Re}\alpha(t)$ and $\operatorname{Im}\alpha(t)$ for $-\infty < t < +\infty$.

12.4.11 Show that Eq. (12.4.91) is a solution to Eq. (12.4.90).

Comments and References

[1] To get some fundamental understanding of the complexity of determining solutions to these particular functional equations and to such equations in general, see Section 5.2.1, of the book by Aczél (2006).

[2] See Efthimiou, section 2.2.

[3] See section 1.2 of A. D. Polyanin and A. I. Zhurov, Solution of functional equations and functional-differential equation by the differentiation method," EqWorld: http://eqworld.ipmnet.ru.

[4] A huge amount of effort has been devoted to the analysis of the four Cauchy functional equations. Good starting points are the books of Efthimiou (see Chapter 5), Aczél (1966, section 2.1), and Sahoo and Kannappan (Chapters 1–3).

[5] See Efthimiou, p. 100, Worked Problem 5.8.

[6] See Efthimiou, p. 102, Worked Problem 5.11.

[7] See Efthimiou, p. 105, Worked Problem 5.20.

[8] See Efthimiou, p. 105, Worked Problem 5.19, and R. D. Boswell, "On two functional equations," American Mathematical Monthly **66**, 716 (1959).

[9] See Aczél (2006), section 2.3.2; and Sahoo and Kannappan, section 5.2.

[10] The decay equation appears in all of the natural, medical, social, and engineering sciences. An excellent summary, along with references, is given at the website https://en.wikipedia.org/wiki/Exponential_decay.

[11] See Efthimiou, section 19.1, Problem 19.4.

[12] See F. Skof, "About a remarkable functional equation on some restricted domains," pp. 249–262, in Z. Daróczy and Z. Páles (editors), *Functional Equations – Results and Advances* (Klumer, Boston, 2002); and Gy. Szabó, "Some functional equations related to quadratic functions," *Glasnik Matematiciki* **38**, 107 (1983).

[13] See "Functional Equations" EqWorld: eqworld.ipmnet.ru/en/solutions.fe.htm.

[14] A. D. Polyanin and A. I. Zhurov, "Solution of functional equations and functional-differential equations by the differentiation method," EqWorld, 2004 (http://eqworld.ipmnet.ru/eqworld/en/methods/fe/art01.pdf).

[15] E. Zauderer, *Partial Differential Equations of Applied Mathematics* (Wiley,

New York, 1983). In particular, see Sections 3.2 and 3.3.

[16] See Aczél (2006), footnote 3. p. 2.

[17] R. E. Mickens, *Nonstandard Finite Difference Models of Differential Equations* (World Scientific, Singapore, 1994). See Chapter 3, sections 3.2 and 3.4.

[18] R. E. Mickens, "Self-consistent vacuum trajectory," *Il Nuove Cimento* **56**, 799 (1968).

[19] H. Fujisaki, "Self-reproducing conditions on the Pomeranchuk singularity," *Progress of Theoretical Physics* **44**, 219 (1970).

Bibliography

J. Aczél, *Lectures on Functional Equations and Their Applications* (Academic Press, New York-London, 1966).

J. Aczél, *A Short Course on Functional Equations Based upon Recent Applications to Social and Behavioral Sciences* (Reidel, Dortrecht-Boston-Tokyo, 1987).

J. Aczél and J. Dhombres, *Functional Equations in Several Variables* (Cambridge University Press, Cambridge, 1989).

E. Castillo and M. R. Ruiz-Cobo, *Functional Equations and Modeling in Science and Engineering* (Marcel Dekkers, New York, 1992).

S. Czerwik, *Functional Equations and Inequalities in Several Variables* (World Scientific, Singapore, 2002).

C. Efthimiou, *Introduction to Functional Equations* (Mathematical Sciences Research Institute, Berkeley, CA/American Mathematical Society, Providence, RI, 2011).

M. Kuczma, *Functional Equations in a Single Variable* (Polish Scientific Publishers, Warsaw, 1968).

M. Kuczma, *An Introduction to the Theory of Functional Equations and Inequalities* (Polish Scientific Publishers, Warsaw, 1985).

M. Kuczma, B. Choczewski, and R. Ger, *Iterative Functional Equations* (Cambridge University Press, Cambridge, 1990).

B. Ramachandran and Ka-Sing Lau, *Functional Equations in Probability Theory* (Academic Press, San Diego, 1991).

C. R. Rao and D. N. Shanbhag, *Choquet-Deny Type Functional Equations with Applications to Stochastic Models* (Wiley, New York, 1994).

I. Risteski and V. Covachev, *Complex Vector Functional Equations* (World Scientific, Singapore, 2002).

T. L. Saaty, *Modern Nonlinear Equations* (Dover, New York, 1981).

P. K. Sahoo and P. Kannappan, *Introduction to Functional Equations* (CRC Press, New York, 2011).

P. K. Sahoo and T. Riedel, *Mean Value Theorems and Functional Equations* (World Scientific, Singapore, 1998).

C. G. Small, *Functional Equations and How to Solve Them* (Springer, New York, 2007).

J. Smital, *On Functions and Functional Equations* (Hilger, Bristol-Philadelphia, 1988).

L. Szekelyhidi, *Convolution Type Functional Equations on Topological Abelian Groups* (World Scientific, Singapore, 1991).

Chapter 13

Miscellaneous Topics

This chapter presents a broad range of topics related to the natural and engineering sciences, especially as they relate to the modeling and analysis of the resulting mathematical equations. In each section, several references are given to the essential literature on the particular issue being studied. Also, it should be noted that the references are generally given in the introduction section to each topic rather than at the end of the chapter.

13.1 Calculation of $\operatorname{Im} f(z)$ from $\operatorname{Re} f(z)$

It is often the case that in the study of a particular phenomena a physical function of interest may be considered as a function of a complex (independent) variable. This is the situation occurring for the high energy scattering of elementary particles (Block and Cahn), the scattering of electromagnetic radiation (Jackson), and acoustic holographic imaging (Zhang). If we denote the function as $f(z)$, generally, only the $\operatorname{Re} f(z)$ or $\operatorname{Im} f(z)$, where $z = x + iy$, can be readily measured. Thus, to obtain, say $\operatorname{Im} f(z)$, from a knowledge of $\operatorname{Re} f(z)$, some mathematically relationship must be established between them. The purpose of this section is to provide such a connection. Our presentation relies heavily on the result presented in the article of Shaw.

The references for this section are:

(1) L. V. Ahlfors, *Complex Analysis* (McGraw-Hill, New York, 1979, 3rd edition).
(2) M. M. Block and R. N. Cahn, "High energy $\bar{p}p$ and pp forward elastic scattering and total cross sections," *Reviews of Modern Physics* **57**, 563 (1985).

(3) J. D. Jackson, *Classical Electrodynamics* (Wiley, New York, 1999, 3rd edition).

(4) W. T. Shaw, "Recovering holomorphic functions from their real or imaginary parts without the Cauchy-Riemann equation," *SIAM Review* **46**, 717 (2004).

(5) F.-C. Zhang, "Inherent relation between the real and imaginary part of a complex holographic function in acoustic holographic imaging: Digital reconstruction method without conjugate image by using only the real part (or imaginary part) of a complex holographic function," *Journal of the Acoustical Society of America*, **72**, 1492 (1982).

Before proceeding, we briefly cover several important properties of a function of a complex variable (Ahlfors).

Let

$$z = x + iy, \quad f(z) = u(x,y) + iv(x,y), \tag{13.1.1}$$

where $i \equiv \sqrt{-1}$, and $u(x,y)$ and $v(x,y)$ are, respectively called the "real" and "imaginary" parts of $f(z)$. These real functions of (x,y) satisfy the Cauchy-Riemann relations

$$\frac{\partial u(x,y)}{\partial x} = \frac{\partial v(x,y)}{\partial y}, \quad \frac{\partial u(x,y)}{\partial y} = -\frac{\partial v(x,y)}{\partial x}, \tag{13.1.2}$$

and the Laplace equation

$$\frac{\partial^2 u(x,y)}{\partial x^2} + \frac{\partial^2 u(x,y)}{\partial y^2} = 0, \quad \frac{\partial^2 v(x,y)}{\partial x^2} + \frac{\partial^2 v(x,y)}{\partial y^2} = 0. \tag{13.1.3}$$

Note that for the function

$$g(z) = if(z)$$
$$= -v(x,y) + iu(x,y)$$
$$= U(x,y) + iV(x,y), \tag{13.1.4}$$

the functions $U(x,y)$ and $V(x,y)$ also satisfy the Cauchy-Riemann relations and are solutions to the Laplace equation. A consequence of this result is that we can assume $u(x,y)$ to be known and then use this to determine $v(x,y)$, or we can assume $v(x,y)$ to be known and then calculate $u(x,y)$. For our purposes we take $u(x,y)$ as given.

13.1.1 *Method-I*

Method-I, for determining $v(x, y)$ from $u(x, y)$, see Shaw, is to carry out the following procedure:

(i) Check to confirm that $u(x, y)$ satisfies the Laplace equation. If the given $u(x, y)$ is not a solution, then there is no reason to proceed.

(ii) Next integrate the two partial differential equations $v(x, y)$ and obtain the expressions

$$v(x, y) = \int \frac{\partial u(x, y)}{\partial x} dy + \phi_1(x), \quad x = \text{fixed}, \qquad (13.1.5a)$$

$$v(x, y) = -\int \frac{\partial u(x, y)}{\partial y} dx + \phi_2(y), \quad y = \text{fixed}, \qquad (13.1.5b)$$

where $\phi_1(x)$ and $\phi_2(y)$ are arbitrary functions.

(iii) By comparing these two results for $v(x, y)$, select the functions $\phi_1(x)$ and $\phi_2(y)$ to get consistency.

(iv) Finally, using $f(z) = u(x, y) + iv(x, y)$, rearrange the terms on the right-side so that it depends only on the variable z.

We now illustrate this procedure by means of several examples.

13.1.2 *Worked Examples Using Method-I*

Consider a function

$$u(x, y) = x. \qquad (13.1.6)$$

It clearly is a solution of the Laplace equation. Also, we have

$$\frac{\partial v}{\partial y} = \frac{\partial u}{\partial x} = 1, \quad \frac{\partial v}{\partial x} = -\frac{\partial u}{\partial y} = 0, \qquad (13.1.7)$$

which when integrated gives

$$v(x, y) = y + \phi_1(x), \quad v(x, y) = \phi_2(y). \qquad (13.1.8)$$

Inspection of these two evaluations for $v(x, y)$ allows the conclusion

$$\phi_1(x) = 0, \quad v(x, y) = y. \qquad (13.1.9)$$

Therefore, $f(z)$ is

$$f(z) = x + iy = z. \qquad (13.1.10)$$

Now take $u(x, y)$ to be

$$u(x, y) = e^{-y} \cos x. \qquad (13.1.11)$$

Therefore,

$$\begin{cases} \frac{\partial u}{\partial x} = e^{-y}(-\sin x), & \frac{\partial u}{\partial y} = -e^{-y}(\cos x), \\ \frac{\partial^2 u}{\partial x^2} = e^{-y}(-\cos x), & \frac{\partial^2 u}{\partial y^2} = e^y(\cos x), \end{cases} \qquad (13.1.12)$$

and we have $\partial^2 u/\partial x^2 + \partial^2 u/\partial y^2 = 0$; thus this expression for $u(x,y)$ is a solution to the Laplace equation. Substituting $u(x,t)$ into the Cauchy-Riemann relations gives

$$\begin{cases} \frac{\partial v(x,y)}{\partial y} = -e^{-y}(\sin x) \Rightarrow v(x,y) = e^{-y}\sin x + \phi_1(x), \\ \frac{\partial v(x,y)}{\partial x} = e^{-y}(\cos x) \Rightarrow v(x,y) = e^{-y}\sin x + \phi_2(y), \end{cases} \qquad (13.1.13)$$

and, on comparison, allows us to conclude that

$$\phi_1(x) = \phi_2(y) = 0. \qquad (13.1.14)$$

Therefore,

$$\begin{aligned} f(z) &= u(x,y) + iv(x,y) \\ &= e^{-y}\cos x + ie^{-y}\sin x \\ &= e^{-y}(\cos x + i\sin x) = e^{-y}e^{ix} \\ &= e^{i(x+iy)} \\ &= e^{iz}. \end{aligned} \qquad (13.1.15)$$

Consider the following $u(x,y)$

$$u(x,y) = x^3 - 3xy^2. \qquad (13.1.16)$$

A direct calculation shows that it is a solution to the Laplace equation. Further, using the Cauchy-Riemann relations, it follows that

$$\frac{\partial v(x,y)}{\partial x} = 6xy, \quad \frac{\partial v(x,y)}{\partial y} = 3x^2 - 3y^2, \qquad (13.1.17)$$

and these may be integrated to give

$$v(x,y) = 3x^2 y + \phi_1(y), \quad v(x,y) = 3x^2 y + \phi_2(x), \qquad (13.1.18)$$

for which consistency requires that

$$\phi_1(y) = -y^3, \quad \phi_2(x) = 0. \qquad (13.1.19)$$

Therefore, we have

$$f(z) = (x^3 - 3xy^2) + i(3x^2 y - y^3) = z^3. \qquad (13.1.20)$$

A more complicated example is where $u(x,y)$ is taken to be

$$u(x,y) = \left[\exp\left(\frac{x}{x^2+y^2}\right)\right]\left[\cos\left(\frac{y}{x^2+y^2}\right)\right]. \qquad (13.1.21)$$

To determine $v(x, y)$ is possible, but not by use of Method-I. The main reason is that the various derivatives are not defined when $x = 0$ and $y = 0$. However, let's see what can be accomplished with the use of a "good memory" and some insight into the "elementary features" of complex numbers.

To begin, "remember" that for real numbers, a and b, the following holds

$$\frac{1}{a + ib} = \frac{1}{a + ib}\left(\frac{a - ib}{a - ib}\right) = \frac{a - ib}{a^2 + b^2}$$

$$= \left(\frac{a}{a^2 + b^2}\right) + i\left(\frac{-b}{a^2 + b^2}\right).$$ (13.1.22)

This result suggests that it may be useful to consider $1/z$, since

$$\frac{1}{z} = \frac{1}{x + iy} = \left(\frac{x}{x^2 + y^2}\right) + i\left(\frac{-y}{x^2 + y^2}\right).$$ (13.1.23)

$$e^{1/z} = \left[\exp\left(\frac{x}{x^2 + y^2}\right)\right]\left[\exp\left(\frac{-iy}{x^2 + y^2}\right)\right]$$

$$= \left[\exp\left(\frac{x}{x^2 + y^2}\right)\right]\left\{\left[\cos\left(\frac{y}{x^2 + y^2}\right)\right]\right.$$

$$\left. - i\left[\sin\left(\frac{y}{x^2 + y^2}\right)\right]\right\},$$ (13.1.24)

and conclude that $v(x, y)$ is

$$v(x, y) = -\left[\exp\left(\frac{x}{x^2 + y^2}\right)\right]\left[\sin\left(\frac{y}{x^2 + y^2}\right)\right].$$ (13.1.25)

Overall, this is not a satisfactory way of obtaining a $v(x, y)$ from given $u(x, y)$, since it is based on luck or prior knowledge. In a later subsection, we now present a more mathematically centered method for this type of problem.

13.1.3 *Method-II*

For a complex function, $f(z)$, which is analytic or holomorphic (Ahlfors) at $z = 0$, Shaw has shown how to construct $v(x, y)$ from $u(x, y)$. His result is summarized in the theorem (Shaw, pp. 719–720):

Theorem 13.1.1 *Let $f(z)$ be analytic in a neighborhood of $z = 0$. Denote the real and imaginary parts respectively, by $u(x, y)$ and $v(x, y)$. Then, with $z = x + iy$,*

$$f(z) = \begin{cases} 2u\left(\frac{z}{2}, \frac{z}{2i}\right) - f(0)^*, \\ 2iv\left(\frac{z}{2}, \frac{z}{2i}\right) + f(0)^*, \end{cases} \tag{13.1.26}$$

where the symbol means taking the complex conjugate, i.e., $i \to (-i)$.

13.1.4 Worked Examples Using Method-II

For the real part

$$u(x, y) = x, \tag{13.1.27}$$

it follows that

$$\begin{aligned} f(z) &= 2u\left(\frac{z}{2}, \frac{z}{2i}\right) - f(0)^* \\ &= 2\left(\frac{z}{2}\right) - f(0)^* \\ &= z - f(0)^*. \end{aligned} \tag{13.1.28}$$

Since $u(0, 0) = 0$, then $f(0)$ must be pure imaginary, i.e., $f(0) = ia$, where a is an arbitrary real number. Therefore,

$$f(z) = z + ia. \tag{13.1.29}$$

Consider now

$$u(x, y) = x^3 - 3xy^2, \tag{13.1.30}$$

then

$$\begin{aligned} f(z) &= 2u\left(\frac{z}{2}, \frac{z}{2i}\right) - f(0)^* \\ &= 2\left[\left(\frac{z}{2}\right)^3 - 3\left(\frac{z}{2}\right)\left(\frac{z}{2i}\right)^2\right] - f(0)^* \\ &= z^3 + ia, \end{aligned} \tag{13.1.31}$$

and the same reasoning was applied to obtain $f(0)$ as in the previous example.

If $u(x, y)$ is the function

$$u(x, y) = e^{-y}\cos x, \tag{13.1.32}$$

then

$$f(z) = 2e^{-z/2i}\cos\left(\frac{z}{2}\right) - 1 + ia. \tag{13.1.33}$$

Note that the constant term $f(0)$ is derived from the fact that

$$u(0,0) = 1 \Rightarrow f(0) = 1 + ia, \qquad (13.1.34)$$

where a is an arbitrary real number. But, $f(z)$ can be rewritten as follows

$$f(z) = 2e^{-z/2i} \left[\frac{e^{iz/2} + e^{-iz/2}}{2} \right] - 1 + ia$$

$$= e^{iz} + 1 - 1 + ia$$

$$= e^{iz} + ia. \qquad (13.1.35)$$

Returning to $u(x,y)$, given in Eq. (13.1.21), it is seen that $u(0,0)$ is not defined. So how should we proceed for this case. An answer is provided by Shaw. The next section summarizes this result.

13.1.5 Method-III, $f(0)$ Not Defined

Suppose that $u(0,0)$ does not exist. For this situation Shaw (p. 722) presents the theorem:

Theorem 13.1.2 *Let $f(z)$ be analytic in the neighborhood of $z = z_0 \neq 0$. Let $u(x,y)$ and $v(x,y)$ be, respectively, the real and imaginary parts of $f(z)$. Then*

$$f(z) = \begin{cases} 2u\left(\frac{z+z_0^*}{2}, \frac{z-z_0^*}{2i}\right) - f(z_0)^*, \\ 2iv\left(\frac{z+z_0^*}{2}, \frac{z-z_0^*}{2i}\right) - f(z_0)^*, \end{cases} \qquad (13.1.36)$$

Note that if $u(0,0)$ does not exist at $z = 0$, then generally $f(z)$ is not analytic at $z = 0$. Some particular functions having this behavior are

$$f(z) = \begin{cases} \frac{1}{z^2}, \\ e^{1/z}, \\ \cos\left(\frac{1}{z}\right), \\ \mathrm{Ln}(z). \end{cases} \qquad (13.1.37)$$

13.1.6 Worked Examples Using Method-III

The function

$$u(x,y)) = \frac{x}{x^2 + y^2} \qquad (13.1.38)$$

satisfies the Laplace equation and is therefore the real part of a complex function. Using the fact that

$$x^2 + y^2 \to \left(\frac{z + z_0^*}{2}\right)^2 + \left(\frac{z - z_0^*}{2i}\right)^2 = z_0^* z, \qquad (13.1.39)$$

it follows that the $f(z)$, for which Eq. (13.1.38) is the real part, is

$$f(z) = 2\left(\frac{z + z_0^*}{2}\right)\left(\frac{1}{z_0^* z}\right) - f(z_0)^*$$

$$= \frac{1}{z} + \frac{1}{z_0^*} - f(z_0)^*. \tag{13.1.40}$$

At $z = z_0$, we have

$$f(z_0) + f(z_0)^* = \frac{1}{z_0} + \frac{1}{z_0^*}. \tag{13.1.41}$$

Now write $z_0 = a + ib$, where (a, b) are real; then

$$\frac{f(z_0) + f(z_0)^*}{2} = \operatorname{Re} f(z_0) = u(a, b) = \frac{a}{a^2 + b^2}, \tag{13.1.42}$$

and

$$f(z) = \frac{1}{z} + i\beta, \tag{13.1.43}$$

where β is real.

For a second example consider the function

$$u(x, y) = \operatorname{Ln}\sqrt{x^2 + y^2} = \left(\frac{1}{2}\right)\operatorname{Ln}(x^2 + y^2). \tag{13.1.44}$$

Making the substitution presented in Eq. (13.1.36) gives

$$f(z) = 2\left[\left(\frac{1}{2}\right)\operatorname{Ln}(z_0^* z)\right] - f(z_0)^*$$

$$= \operatorname{Ln}(z) + \operatorname{Ln}(z_0^*) - f(z_0)^*, \tag{13.1.45}$$

from which it follows that

$$f(z_0) + f(z_0)^* = \operatorname{Ln}(z_0) + \operatorname{Ln}(z_0^*), \tag{13.1.46}$$

and, as a consequence

$$\operatorname{Re} f(z_0) = \operatorname{Re}\operatorname{Ln}(z_0), \tag{13.1.47}$$

which implies that $f(z)$ is

$$f(z) = \operatorname{Ln}(z) + i\beta, \tag{13.1.48}$$

where β is an arbitrary real number.

As a final example consider the function

$$u(x, y) = \frac{x^2 - y^2}{(x^2 + y^2)^2}, \tag{13.1.49}$$

which gives for $f(z)$ the result

$$f(z) = 2\left(\frac{1}{z_0^* z}\right)^2\left[\left(\frac{z + z_0^*}{2}\right)^2 - \left(\frac{z - z_0^*}{2i}\right)^2\right] - f(z_0)^*$$

$$= \left(\frac{1}{z_0^{*2} z^2}\right)(z^2 + z_0^{*2}) - f(z_0)^*$$

$$= \frac{1}{z^2} + \frac{1}{z_0^{*2}} - f(z_0)^*, \tag{13.1.50}$$

and finally,

$$f(z) = \frac{1}{z^2} + i\beta, \tag{13.1.51}$$

where β is an arbitrary real constant.

13.1.7 $f(z)$ in Modulus-Argument Form

An alternative way of representing a complex function is

$$f(z) = R(x, y)\exp[i\Theta(x, y)], \tag{13.1.52}$$

where $R(x, y)$ is the modulus of $f(z)$ and $\Theta(x, y)$ is the argument or phase of the function. Note, for z in polar form, i.e.,

$$z = x + iy = re^{i\theta}, \tag{13.1.53}$$

we have

$$r(x, y) = \sqrt{x^2 + y^2}, \quad \theta = \tan^{-1}\left(\frac{y}{x}\right). \tag{13.1.54}$$

Since

$$f(z) = u(x, y) + iv(x, y) = R(x, y)\exp i\Theta(x, y), \tag{13.1.55}$$

it follows that

$$R(x, y) = \sqrt{u(x, y)^2 + v(x, y)^2} \tag{13.1.56a}$$

$$\Theta(x, y) = \tan^{-1}\left[\frac{v(x, y)}{u(x, y)}\right]. \tag{13.1.56b}$$

Defining the function $g(z)$ as

$$g(z) = \operatorname{Ln} f(z)$$
$$= \operatorname{Ln}[R(x, y)] + i\Theta(x, y), \tag{13.1.57}$$

and applying Method-III gives the results

$$f(z) = \begin{cases} f(z_0)^* \exp\left[2i\Theta\left(\frac{z+z_0^*}{2}, \frac{z-z_0^*}{2i}\right)\right], \\ \left[\frac{1}{f(z_0)^*}\right] R^2\left(\frac{z+z_0^*}{2}, \frac{z-z_0^*}{2i}\right). \end{cases} \tag{13.1.58}$$

13.2 Exponential Functions

There may be no area of the sciences for which exponential functions do not appear. Such widespread use in applications is a clear indication of their importance to the mathematical modeling of a broad range of phenomena. The main purpose of this section is to define exponential functions (EF), present their most significant properties, and briefly discuss several applications. The initial parts of this section are directly based on material in Section 10.3 of Sahoo and Kannappan.

The following articles and books provide detailed background on both the mathematical properties and some of the applications of the exponential functions:

1) J. Aczél, *Lectures on Functional Equations and Their Applications* (Academic, New York, 1966).

2) G. Beylkin and L. Monzón, "On approximation of functions by exponential sums," *Applied and Computational Harmonic Analysis* **19**, 17 (2005).

3) R. P. Boas, Jr., "Signs of derivatives and analytic behavior," *American Mathematical Monthly* **78**, 1085 (1971).

4) S. G. Krantz, "The Exponential and Applications," Section 1.2, in *Handbook of Complex Analysis* (Birkhäuser, Boston, 1999).

5) R. E. Mickens, "Temperature dependence of three-body association results," *Chemical Physics Letters* **121**, 334 (1985).

6) J. Nicholas, *Chemical Kinetics: A Modern Survey of Gas Reactions* (Wiley, New York, 1976). See Chapter 8.

7) A. Ruhe, "Fitting empirical data by positive sums of exponentials," *SIAM Journal on Scientific and Statistical Computing* **1** 481 (1980).

8) P. K. Sahoo and P. Kannappan, *Introduction to Functional Equations* (CRC Press, Boca Raton, FL, 2011).

9) D. V. Widder, *The Laplace Transform* (Princeton University Press, Princeton, NJ, 1941).

10) E. Yeramian and P. Claverie, "Analysis of multiexponential functions without a hypothesis as to the number of components," *NATURE* **36**, 169 (1987).

Definition A function $E(x)$ is an exponential function if it is a solution to the functional equation

$$E(x + y) = E(x)E(y), \qquad (13.2.1)$$

for all real x and y. Observe that

$$E(x) = 0, \quad E(x) = 1, \qquad (13.2.2)$$

are trivial solutions to Eq. (13.2.1) and that the non-constant continuous solutions are

$$E(x) = e^{\lambda x}, \qquad (13.2.3)$$

where λ is an arbitrary complex number.

13.2.1 *Properties of Exponential Functions*

Property-1. If $E(0) = 0$, then $E(x) = 0$.
Proof. Set $y = 0$ in Eq. (13.2.1) and obtain

$$E(x) = E(x)E(0) = 0. \qquad (13.2.4)$$

∎

Property-2. If $E(x) \neq 0$, then $E(0) = 1$.
Proof. Setting $x = 0$ and $y = 0$ in Eq. (13.2.1) gives

$$E(0) = E(0)E(0) \Rightarrow E(0)[E(0) - 1] = 0. \qquad (13.2.5)$$

Since $E(0) = 0$ implies $E(x) = 0$, then it follows that for $E(x) \neq 0$, we must have $E(0) = 1$. ∎

Property-3. Assume there is an x_0 such that $E(x_0) = 0$, then $E(x) = 0$ for all x.
Proof. Write $E(x)$ as follows

$$E(x) = E[(x - x_0) + x_0], \qquad (13.2.6)$$

then, using Eq. (13.2.1), we obtain

$$\begin{aligned} E(x) &= E[(x - x_0) + x_0] \\ &= E(x - x_0)E(x_0) = E(x - x_0)(0), \qquad (13.2.7) \end{aligned}$$

and from this we conclude that $E(x) = 0$ for all x. ∎

For the remainder of this section it will be assumed that $E(x) \not\equiv 0$.
Property-4. Define $\bar{E}(x)$ as

$$\bar{E}(x) = \frac{1}{E(x)}. \qquad (13.2.8)$$

Then

$$\bar{E}(x) = E(-x). \tag{13.2.9}$$

Proof. Set $y = -x$ in Eq. (13.2.1) and obtain

$$E(0) = E(x)E(-x). \tag{13.2.10}$$

Since $E(0) = 1$, it follows that

$$E(-x) = \frac{1}{E(x)}. \tag{13.2.11}$$

Comparing Eqs. (13.2.8) and (13.2.11) gives $\bar{E}(x) = E(-x)$.   ∎
Property-5.

$$\bar{E}(x + y) = \bar{E}(x)\bar{E}(y). \tag{13.2.12}$$

Proof. We have

$$\bar{E}(x + y) = \frac{1}{E(x + y)} = \frac{1}{E(x)E(y)}$$
$$= \bar{E}(x)\bar{E}(y). \tag{13.2.13}$$

∎

Comments

(i) All of the above properties could have been directly verified by use of the explicit formula for $E(x)$ given in Eq. (13.2.3).

(ii) Define $E^{(+)}(x)$ and $E^{(-)}(x)$, respectively, by the formulas

$$E^{(+)}(x) \equiv \frac{E(x) + E(-x)}{2}, \quad E^{(-)}(x) = \frac{E(x) - E(-x)}{2}. \tag{13.2.14}$$

Note that for a real

$$\lambda = ia : E^{(+)}(x) = \cos(ax), \quad E^{(-)}(x) = i\sin(ax), \tag{13.2.15a}$$
$$\lambda = a : E^{(+)}(x) = \cosh(ax), \quad E^{(-)}(x) = i\sinh(ax), \tag{13.2.15b}$$

and while these functions are linear combinations of EF's, they are not themselves EF's.

(iii) For general complex $\lambda = \lambda_1 + i\lambda_2$, where λ_1 and λ_2 are real, we have

$$e^{\lambda x} = e^{\lambda_1 x}e^{i\lambda_2 x}$$
$$= e^{\lambda_1 x}\cos(\lambda_2 x) + ie^{\lambda_1 x}\sin(\lambda_2 x), \tag{13.2.16a}$$
$$e^{-\lambda x} = e^{-\lambda_1 x}e^{-i\lambda_2 x}$$
$$= e^{-\lambda_1 x}\cos(\lambda_2 x) - ie^{-\lambda_1 x}\sin(\lambda_2 x). \tag{13.2.16b}$$

13.2.2 Differential Equation Derivation

We now demonstrate that the solution, given in Eq. (13.2.3), of Eq. (13.2.1), can be gotten from constructing and solving a differential equation associated with the FE, $E(x + y) = E(x)E(y)$.

First, rewrite the FE as

$$f(x + y) = f(x)f(y), \tag{13.2.17}$$

and subtract $f(x)$ from both sides, i.e.,

$$f(x + y) - f(x) = f(x)f(y) - f(x)$$
$$= f(x)[f(y) - 1]. \tag{13.2.18}$$

Next, assume that $f(x)$ has a Taylor series expansion. It can be written as

$$f(x) = 1 + f'(0)x + g(x)x^2, \tag{13.2.19}$$

using the fact that $f(0) = 1$ and where for a given $f(x)$, the function $g(x)$ can be determined. Replacing x by $y = h$ in Eq. (13.2.18), and dividing by h gives

$$\frac{f(x + h) - f(x)}{h} = [f'(0) + g(h)h]f(x). \tag{13.2.20}$$

Since Eq. (13.2.17) holds for all h, the same is true for (13.2.20). Therefore, taking the limit as $h \to 0$, produces the differential equation

$$\frac{df(x)}{dx} = \lambda f(x), \tag{13.2.21}$$

where $f'(0)$ has been set equal to λ.

The solution to this linear, first-order, constant coefficient differential equation, with $f(0) = 1$, is

$$f(x) = e^{\lambda x}, \tag{13.2.22}$$

and this is exactly the same function as given in Eq. (13.2.3).

13.2.3 Exponential Representations of Non-Negative Functions

Experimental data is often summarized in the form of fitting functions. If properly selected, these functional forms can provide insight and understanding the phenomena behind the collected data; see, for example,

Beylkin and Monzón, Krantz, and Yeramian and Claverie. Often, the fitting functions are taken to be EFs, where this choice is dictated by *a priori* theoretical considerations. A general representation may take the form

$$f(x) = \sum_{k=1}^{\infty} a_k e^{-\lambda_k x}, \qquad (13.2.23)$$

where the $(\lambda_k : k = 1, 2, \dots)$ have positive real parts; the $(a_k : k = 1, 2, \dots)$ may be complex valued; and x is the independent variable, which is often the time. In practice, we may have data on $f(x)$ for a finite number of x-values, i.e.,

$$[x_1, f(x_1)], [x_2, f(x_2)], \dots, [x_m, f(x_m)], \qquad (13.2.24)$$

and the assumption is made that these data can be fitted by a finite set of EF's, i.e.,

$$f(x) \simeq \sum_{k=1}^{n} \bar{a}_k e^{-\bar{\lambda}_k x}, \qquad (13.2.25)$$

where $(\bar{a}_k, \bar{\lambda}_k; k = 1, 2, \dots, n)$ are parameters to be determined. Note that m may differ than n.

A continuous analogue for $f(x)$ is to use a Laplace transform representation, i.e.,

$$f(x) = \int_0^{\infty} \phi(t) e^{-xt} dt. \qquad (13.2.26)$$

It is of interest to note that many phenomena exist such that $\phi(t)$ is required to be non-negative, i.e.,

$$\phi(t) \geq 0, \quad 0 \leq t < \infty. \qquad (13.2.27)$$

A particular but very important example is the chemical reaction rate coefficient; see Mickens, and Nicholas.

Functions, which are defined by the mathematical structure given in Eq. (13.2.26) and for which the condition of Eq. (13.2.27) holds, are called completely monotone functions. Four of their important properties are

(i) $f(x)$ is bounded and positive.

(ii) $f(x)$ has derivative of all orders, $f^{(n)}(x)$, and they satisfy the relations

$$(-1)^n f^{(n)}(x) \geq 0, \quad x > 0; \quad n = 0, 1, 2, \dots. \qquad (13.2.28)$$

(iii) $f(0)$ is defined at $x = 0$ and is positive

$$f(0) > 0. \qquad (13.2.29)$$

(iv) the limit as $x \to \infty$ exists, i.e.,

$$\lim_{x \to \infty} f(x) = f(\infty) = 0. \tag{13.2.30}$$

All of the above features are based on the assumption that $\phi(t)$ is non-negative and that the Laplace integral, Eq. (13.2.26) converges for all $x \geq 0$.

The mathematical details, along with rigorous proofs, are given in the works by Boas, and Widder.

13.3 Some Useful Fourier Series

We presented some elementary discussions of Fourier series and their applications previously in Sections 2.5 to 2.9. The purpose of this section is to give the Fourier series of several important functions which have direct use in the analysis of nonlinear oscillating phenomena; see, for example, Chapter 8 of this text, and the books by Kovacic and Brennan, and Mickens, and the paper of Kovacic as listed below:

(1) I. S. Gradshteyn and I. M. Ryzhik, *Tables of Integrals, Series, and Products* (Academic, New York, 1980, 4th edition). See Section 8.511.

(2) R. W. Hamming, *Numerical Methods for Scientists and Engineers* (Dover, New York, 1987, 2nd edition). See pp. 612–613.

(3) I. Kovacic, "The method of multiple scales for forced oscillators with non-negative real-power nonlinearities and different damping mechanisms," *Chaos, Solitons and Fractals* **44**, 891 (2011). See Appendix: On some Fourier series expansions.

(4) I. Kovacic and M. J. Brennan, editors, *The Duffing Equation: Nonlinear Oscillators and Their Behavior* (Wiley, Chichester, UK, 2011).

(5) R. E. Mickens, *Oscillations in Planar Dynamic Systems* (World Scientific, Singapore, 1996).

(6) R. E. Mickens, *Truly Nonlinear Oscillations: Harmonic Balance, Parametric Expansions, Iteration, and Averaging Methods* (World Scientific, Singapore, 2010).

The need for these Fourier series arises in the calculations required to determine the solutions to the differential equations arising from the modeling of oscillating systems. In particular, these series play fundamental role in the methodologies based on averaging, harmonic balancing, and multi-time techniques; see Kovacic, Kovacic and Brennan, and Mickens.

13.3.1 $|\cos\theta|$ *and* $|\sin\theta|$, *for* $-\pi < \theta < \pi$

The Fourier series for these two functions are easy to calculate and are given by the expressions

$$|\cos\theta| = \frac{2}{\pi} + \left(\frac{4}{\pi}\right)\sum_{k=1}^{\infty}\left[\frac{(-1)^k}{1 - 4k^2}\right]\cos(2k\theta), \qquad (13.3.1a)$$

$$|\sin\theta| = \frac{2}{\pi} + \left(\frac{4}{\pi}\right)\sum_{k=1}^{\infty}\left(\frac{1}{1 - 4k^2}\right)\cos(2k\theta). \qquad (13.3.1b)$$

Another familiar expansion is for the function $|\theta|$. It is

$$|\theta| = \frac{\pi}{2} - \left(\frac{4}{\pi}\right)\sum_{k=1}^{\infty}\left[\frac{1}{(2k-1)^2}\right]\cos(2k-1)\theta, \qquad (13.3.2)$$

where $-\pi < \theta < \pi$.

13.3.2 $\cos(z\cos\theta)$ *and* $\sin(z\sin\theta)$

The standard references for the Fourier series of these functions is Gradshteyn and Ryzhik. They give the following representations

$$\cos(z\cos\theta) = J_0(z) + 2\sum_{k=1}^{\infty}J_{2k}(z)\cos(2k\theta) \qquad (13.3.3a)$$

$$\sin(z\sin\theta) = 2\sum_{k=0}^{\infty}J_{2k+1}(z)\sin(2k+1)\theta, \qquad (13.3.3b)$$

where z is a real parameter and $J_n(z)$ is a Bessel function of the first kind of order n.

13.3.3 $sign(\sin(m\theta))$ *and* $sign(\cos(m\theta))$

If the sign-function is taken to be

$$\mathrm{sign}(\sin\theta) = \begin{cases} +1, & 0 \le \theta < \pi, \\ -1, & \pi \le \theta < 2\pi, \end{cases} \qquad (13.3.4)$$

then

$$\mathrm{sign}(\sin\theta) = \left(\frac{4}{\pi}\right)\sum_{k=0}^{\infty}\left(\frac{1}{2k+1}\right)\sin(2k+1)\theta. \qquad (13.3.5)$$

This result is given by Hamming, who does not provide an expression for $\mathrm{sign}(\cos\theta)$. We now calculate it.

To begin, start with the fact that if $\theta \to \theta + \frac{\pi}{2}$, then

$$\sin\left(\theta + \frac{\pi}{2}\right) = \cos\theta, \tag{13.3.6}$$

and Eq. (13.3.5) becomes

$$\text{sign}\left(\sin\left(\theta + \frac{\pi}{2}\right)\right) = \left(\frac{4}{\pi}\right) \sum_{k=0}^{\infty} \left(\frac{1}{2k+1}\right) \sin\left[(2k+1)\theta + (2k+1)\left(\frac{\pi}{2}\right)\right].$$
$$\tag{13.3.7}$$

Now using

$$\sin(\theta_1 + \theta_2) = \cos\theta_1 \sin\theta_2 + \sin\theta_1 \cos\theta_2, \tag{13.3.8}$$

with $\theta_1 = (2k+1)\theta$ and $\theta_2 = (2k+1)\pi/2$, gives

$$\sin\left[(2k+1)\theta + (2k+1)\left(\frac{\pi}{2}\right)\right] = [\cos(2k+1)\theta]\sin\left[(2k+1)\left(\frac{\pi}{2}\right)\right]$$
$$+ [\sin(2k+1)\theta]\cos\left[(2k+1)\left(\frac{\pi}{2}\right)\right]$$
$$= (-1)^k \cos(2k+1)\theta, \tag{13.3.9}$$

where we have used

$$\cos\left[(2k+1)\frac{\pi}{2}\right] = 0, \quad \sin\left[(2k+1)\left(\frac{\pi}{2}\right)\right] = (-1)^k. \tag{13.3.10}$$

Therefore, upon substituting these results, namely, Eqs. (13.3.6) and (13.3.10), into Eq. (13.3.7), we obtain

$$\text{sign}(\cos\theta) = \left(\frac{4}{\pi}\right) \sum_{k=0}^{\infty} \left[\frac{(-1)^k}{2k+1}\right] \cos(2k+1)\theta. \tag{13.3.11}$$

Note that making the transformation $\theta \to m\theta$, where m is a real number, we have the general relations

$$\text{sign}(\sin(m\theta)) = \left(\frac{4}{\pi}\right) \sum_{k=0}^{\infty} \left(\frac{1}{2k+1}\right) \sin(2k+1)m\theta, \tag{13.3.12a}$$

$$\text{sign}(\cos(m\theta)) = \left(\frac{4}{\pi}\right) \sum_{k=0}^{\infty} \left(\frac{(-1)^k}{2k+1}\right) \cos(2k+1)m\theta. \tag{13.3.12b}$$

13.3.4   *[sign(z)]|z|^α, z = cos θ or sin θ*

The next two Fourier series are from the paper of Kovacic. We only quote the results, and refer the reader to her work for the details of the calculations.

$$[\text{sign}(\cos\theta)]|\cos\theta|^\alpha = \sum_{k=1}^{\infty} b_{(2k-1)}^{(\alpha)} \cos(2k-1)\theta \tag{13.3.13}$$

where $\alpha > 0$ and

$$b_1^{(\alpha)} = \left(\frac{2}{\sqrt{\pi}}\right) \frac{\Gamma\left(1 + \frac{\alpha}{2}\right)}{\Gamma\left(\frac{3+\alpha}{2}\right)}, \tag{13.3.14a}$$

and for $k \geq 2$,

$$b_{(2k-1)}^{(\alpha)} = \left\{ \frac{(\alpha - 1)(\alpha - 3)\cdots[\alpha - (2k-3)]}{(\alpha + 3)(\alpha + 5)\cdots[\alpha + (2k+1)]} \right\} b_1^{(\alpha)}. \tag{13.3.14b}$$

$$[\operatorname{sign}(\sin\theta)]|\sin\theta|^\beta = \sum_{k=1}^{\infty} b_{(2k-1)}^{(\beta)} \sin(2k-1)\theta \tag{13.3.15}$$

where $\beta > 0$ and

$$b_1^{(\beta)} = \left(\frac{2}{\sqrt{\pi}}\right) \frac{\Gamma\left(1 + \frac{\beta}{2}\right)}{\Gamma\left(\frac{3+\beta}{2}\right)}, \tag{13.3.16a}$$

where for $k \geq 2$,

$$b_{(2k-1)}^{(\beta)} = (-1)^{k-1} \left\{ \frac{(\beta - 1)(\beta - 3)\cdots[\beta - (2k-3)]}{(\beta + 3)(\beta + 5)\cdots[\beta + (2k+1)]} \right\} b_1^{(\beta)}. \tag{13.3.16b}$$

The Fourier coefficients are related and this connection can be stated as

$$b_{(2k-1)}^{(\beta)} = (-1)^k b_{(2k-1)}^{(\alpha)}, \quad \alpha = \beta. \tag{13.3.17}$$

13.3.5 $(\cos\theta)^{-1}$ *and* $(\sin\theta)^{-1}$

The following "derivation" of the Fourier series for $(\cos\theta)^{-1}$ is being reproduced by me from an unknown source for which I cannot determine its citation items. If anyone has seen this derivation and/or knows the reference to its publication, please contact me. In any case, the procedure is very clever and clearly depends on having a deep and fundamental knowledge of trigonometry.

To start, if $(\cos\theta)^{-1}$ has a Fourier series, it must be a cosine series since $(\cos\theta)^{-1}$ is an even function. Since $(\cos\theta)^{-1}$ is not defined at $\theta = \pi/2$, its Fourier series will be defined on the open interval $(-\frac{\pi}{2}, \frac{\pi}{2})$. Next, it seems reasonable to start from one of the most elementary mathematical relations, $1 = 1$, and see where this leads.

The following is "true"

$$1 = 1$$

$$= 1 + (\cos 2\theta - \cos 2\theta) + (\cos 4\theta - \cos 4\theta)$$

$$+ (\cos 6\theta - \cos 6\theta) + \cdots$$
$$+ [\cos(2k\theta) - \cos(2k\theta)] + \cdots . \tag{13.3.18}$$

Also, we have

$$\begin{cases} \cos 2\theta + \cos 4\theta = 2\cos\theta\cos 3\theta \\ \cos 4\theta + \cos 6\theta = 2\cos\theta\cos 5\theta \\ \quad\vdots \qquad\qquad \vdots \\ \cos 2(k-1)\theta + \cos 2k\theta = 2\cos\theta\cos(2k-1)\theta \\ \quad\vdots \qquad\qquad \vdots \end{cases} \tag{13.3.19}$$

Regrouping the terms on the right-side of Eq. (13.3.18) and using the results in Eq. (13.3.19), we obtain

$$\frac{1}{\cos\theta} = 2\big[\cos\theta - \cos 3\theta + \cos 5\theta - \cos 7\theta + \cdots$$
$$+ (-1)^k \cos(2k+1)\theta + \ldots\big], \tag{13.3.20}$$

and this holds for $-\left(\frac{\pi}{2}\right) < \theta < \frac{\pi}{2}$.

The Fourier series for $(\sin\theta)^{-1}$ can be obtained by rewriting Eq. (13.3.20) to the form

$$\frac{1}{\cos\theta} = 2\sum_{k=0}^{\infty} (-1)^k \cos(2k+1)\theta, \tag{13.3.21}$$

and then use the relations

$$\begin{cases} \cos\left(\theta + \frac{\pi}{2}\right) = -\sin\theta, \\ \cos\left[(2k+1)\theta + (2k+1)\left(\frac{\pi}{2}\right)\right] = (-1)^{k+1}\sin(2k+1)\theta, \end{cases} \tag{13.3.22}$$

in Eq. (13.3.21) to finally obtain

$$\frac{1}{\sin\theta} = 2[\sin\theta + \sin 3\theta + \sin 5\theta + \cdots + \sin(2k+1)\theta + \cdots], \tag{13.3.23}$$

and this holds for the open interval, $(0, \pi)$.

13.4 Dynamic Consistency

This section introduces the concept of "dynamic consistency" and its role in the construction of mathematical models of systems and some related issues. The following references are of direct relevance to these topics.

1) R. Aris, *Mathematical Modeling Techniques* (Dover, New York, 1994).

2) E. A. Bender, *An Introduction to Mathematical Modeling* (Dover, New York, 2000).

3) R. Frigg and S. Hartmann, "Models in science," *The Stanford Encyclopedia of Philosophy* (Fall 2012 Edition), Edward N. Zalta (editor), <http://plato.stanford.edu/fall2012/entries/models-science/>.

4) N. Gershenfeld, *The Nature of Mathematical Modeling* (Cambridge University Press, New York, 1998).

5) C. C. Lin and L. A. Segel, *Mathematics Applied to Deterministic Problems in the Natural Sciences* (SIAM, Philadelphia, 1988).

6) R. E. Mickens, "Dynamic consistency," *Journal of Difference Equations and Applications* **11**, 645 (2005).

13.4.1 *Modeling*

A mathematical model of a system is its description in terms of mathematical concepts and language, and the process of developing such a model is called mathematical modeling; see, http://en.wikipedia.org/wiki/Mathematical_model. The purposes of modeling is to analyze, understand, and predict the behavior of the system. While there do not exist strict rules for the modeling process, Fig. 13.4.1 provides the essential steps carried out for the general modeling methodology. In summary, the following steps are followed:

(i) There exists a system for which knowledge is needed about its future behavior and, in addition, there is also the requirement to understand its past behavior.

(ii) By probing the system, i.e., doing experiments on it, data is collected on its various properties.

(iii) From the data, a mathematical model is constructed.

(iv) The mathematical model is then studied as a mathematical structure which then helps in the analysis and interpretation of the data, and the prediction of features of the system not readily observed from the original data.

(v) The mathematical model can then be generalized to include additional phenomena not previously included in the original system.

This is a very coarse view of the modeling process and does not take into consideration the iterative nature of the general modeling methodology. For example, after the construction of a mathematical model, it may become obvious that new types of data are needed or that the original data may be

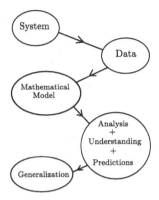

Fig. 13.4.1 The basic steps in the modeling process.

flawed. Further, it may become clear that the mathematical structures used to construct the original mathematical model may need to be changed to another type of mathematics. The references discuss these issues in detail.

13.4.2 *Dynamic Consistency Defined*

A system may be characterized by a number of features and properties. Dynamic consistency is concerned with the inclusion of these features and properties into mathematical models of the system. The following is a short list of such items (many other possibilities may exist):

- boundedness of the system
- the discreteness or not of the system
- monoticity with regard to the time evolution of some property
- the existence of special states of the system
- stability properties of these special states
- etc.

An important fact, often overlooked is that in the overall modeling process, there are several stages for which modeling is required. To see this, consider Fig. 13.4.2. The (at least initial) mathematical model is based on data coming from doing experiments on the system. This procedure gives mathematical model equations which, in general, are nonlinear. However, even for the situation where the equations are linear, in almost all instances exact solutions do not exist in any useful form. This means that one has to use numerical techniques to determine numerical values for the solutions of interest. However, these numerical solutions are given by some process of discretization of the original equations appearing in the mathematical model, and a major difficulty is that the numerical solutions, whose calculated values are obtained by implementation on a digital device, may contain numerical instabilities. Numerical instabilities are solutions to the modeling equations which do not correspond to any realistic behavior of either the original system or to the equations themselves.

It should also be indicated that the link between the numerical model and the implementation model can also be important and plays major roles in the interpretation of the results coming from the mathematical model. The actual calculation of numerical values is generally implemented on a digital device and it may have internal rules and procedures which do not correspond exactly with the discretization methodology used to create the numerical models for the equations.

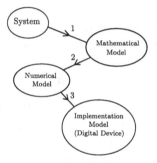

Fig. 13.4.2   An illustration of a chain of modeling

We are now ready to define the concept of "dynamic consistency." It can be applied to all three of the links illustrated in Fig. 13.4.2.

Definition Dynamic Consistency Consider a system S and a mathematical model of it, denoted by M. Let the system S have a property P. If M also contains this property, then M is said to be dynamic consistent with S, with respect to P.

Note that S can have a number of properties $(P_1, P_2, \ldots, P_N)$, only some of which are incorporated into particular mathematical models for S. For simplicity, Fig. 13.4.3 shows the possibilities when $N = 3$. This figure shows that seven mathematical models are possible; but they are not equivalent. The models, (M_1, M_2, M_3) will have the least validity, while the best model, under these conditions, is M_7.

For a general system, the essential properties may not all be known at a particular time and, consequently, only partial mathematical models will be constructed. As more properties are discovered, the quality and complexity of the models increase.

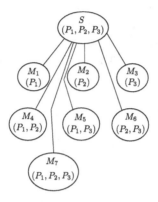

Fig. 13.4.3 A given system S, with known features, (P_1, P_2, P_3) may have multi-mathematical models.

13.4.3 *Comments*

Please keep in mind the following issues:

(i) The mathematical structures (differential equations, difference equations, integral equations, etc.) to be used may differ among the various models.

(ii) For a given set of system properties, (P_1, P_2, P_3), more than one mathematical model may exist.

For example, in considering radioactive decay, we can use a discrete-time representation

$$\begin{cases} M_{k+1} = rM_k, \quad 0 < r < 1, \ M_0 = \text{given}, \\ \quad\quad\quad k = 1, 2, 3, \ldots, \end{cases} \tag{13.4.1}$$

or a continuous-time representation

$$\frac{dM(t)}{dt} = -\lambda M(t), \quad \lambda > 0, \ M(0) = \text{given}. \tag{13.4.2}$$

In these equations, r and λ are parameters which characterize the decay process, $t_k = (\Delta t)k$, and M_k and $M(t)$ is the amount of radioactive material at, respectively, time t_k and t. Note that the solutions to these two mathematical models are

$$M_k = M_0 r^k, \quad M(t) = M(0)e^{-\lambda t}. \tag{13.4.3}$$

It should be indicated that these two models are expected to breakdown at such times for which $M(t) \lll 1$ or $M_k \lll 1$. For these situations some form of stochastic equations may be required.

(iii) Even if all the essential properties of a system are known, accurate mathematical models may be possible by not making use of all the properties, and it may be the case that some of the properties are consequences of the existence of one or more other properties.

(iv) Examination of Fig. 13.4.2 indicates that the concept of dynamic consistency can also be applied to the transitions 2 and 3.

13.5   New Law of Cooling

Experience shows that hot and cold objects cool down or warm up to the temperature of their local surroundings. Let

$$T(t) = \text{temperature of the object},$$

$$T_e = \text{temperature of the environment}$$

$$\text{surrounding the object},$$

where it is assumed that the object does not influence the temperature of the surrounding which is taken to be constant. If $T(0) = T_0$ is the initial temperature of the object and if the following condition holds

$$\left| \frac{T_0 - T_e}{T_e} \right| \ll 1, \tag{13.5.1}$$

then to a very good approximation the rate of change of $T(t)$ is

$$\frac{dT(t)}{dt} = -\lambda(T(t) - T_e), \tag{13.5.2}$$

where λ is a positive parameter having the physical units of $(\text{time})^{-1}$. This differential equation is called Newton's Law of Cooling. Note that the solution to Eq. (13.5.2) is

$$T(t) = T_e + (T_0 - T_e)e^{-\lambda t}. \tag{13.5.3}$$

Comments

(i) Denoting

$$\Delta(t) = T(t) - T_e, \tag{13.5.4}$$

it is seen that when $\Delta(0) > 0$, the object cools and for $\Delta(0) < 0$, the object heats up.

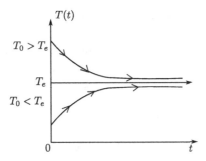

Fig. 13.5.1   General behaviors of the change in temperature for Newton's Law of Cooling.

(ii) It takes an unlimited time for the "equilibrium" temperature T_e to be achieved.

(iii) However, observations are consistent with the conclusion that the temperature T_e is reached in a finite time, i.e., there exists a time t^* such that

$$T(t^*) = T_e, \quad t \geq t^*. \qquad (13.5.5)$$

(iv) Thus, the issue now is to see if Newton's Law of Cooling can be modified to a form for which the result given in Eq. (13.5.5) holds.

The purpose of this section is to demonstrate that such a law of cooling can be formulated.

13.5.1   *Requirements for Law of Cooling*

The consideration below is an extension of the results summarized in the abstract:

• R. E. Mickens, "A new law of cooling," *Georgia Journal of Science* **67**, 55 (2009).

Note that within the context of dynamic consistency (see Section 13.4), we wish to construct a law

$$\frac{dT}{dt} = F(T, T_e, \lambda), \qquad (13.5.6)$$

such that its solution goes to T_e in a finite time, t^*. Further, this time, t^*, should be expressible in terms of T_0, the initial temperature of the object; T_e, the constant temperature of the surroundings; and (at least one) parameter, λ, which appears in the function, $F(T, T_e, \lambda)$. Note that T_0 does not appear in $F(\cdots)$.

It should be clear that $F(\cdots)$ will only depend on $T(t)$ and T_e, in the combination given by Eq. (13.5.4). Therefore, Eq. (13.5.6) has the form

$$F(T, T_e, \lambda) = f(T - T_e, \lambda) = f(\Delta, \lambda), \qquad (13.5.7)$$

and this yields for Eq. (13.5.6), the result

$$\frac{d\Delta(t)}{dt} = f(\Delta, \lambda). \qquad (13.5.8)$$

Now, what are the essential mathematical properties of $f(\Delta, \lambda)$: A moment's reflection gives

$$\begin{cases} f(\Delta, \lambda) < 0, & \Delta > 0; \\ f(0, \lambda) = 0, & \Delta = 0; \\ f(-\Delta, \lambda) = -f(\Delta, \lambda). \end{cases} \qquad (13.5.9)$$

So, the question to be answered is, What is $f(\Delta, \lambda)$?

Perhaps the simplest function is the following

$$f(\Delta, \lambda) = -\lambda[\text{sign}(\Delta)]|\Delta|^\alpha, \qquad (13.5.10a)$$

where

$$0 < \alpha < 1. \qquad (13.5.10b)$$

As we will show in the next section, this $f(\Delta, \lambda)$, when substituted into Eq. (13.5.8) has solutions for which the dynamics is finite.

The α in Eq. (13.5.10) lies in the open interval $(0, 1)$. However, since no experiment allows a measurement of infinite precision, we will take for α the expression

$$\alpha = \frac{2n + 1}{2m + 1}; \quad n = 0, 1, 2, \ldots; \ m = 1, 2, 3, \ldots; \ n < m. \qquad (13.5.11)$$

Note that Δ^p where $p = \alpha$, in Eq. (13.5.11) is an odd function of Δ, i.e.,

$$(-\Delta)^p = -(\Delta)^p, \qquad (13.5.12)$$

consequently, all the conditions stated in Eq. (13.5.9) are satisfied.

13.5.2 *Solution of $d\Delta/dt = -\lambda(\Delta)^p$*

The solution of the first-order, nonlinear differential equation

$$
\begin{cases}
\frac{d\Delta(t)}{dt} = -\lambda[\Delta(t)]^p, \\
\Delta(0) = \Delta_0 = T_0 - T_e,
\end{cases}
\tag{13.5.13}
$$

will now be determined. First, note that this is a separable equation which can be written as

$$
\Delta^{-p} d\Delta = -\lambda dt.
\tag{13.5.14}
$$

Integrating both sides and imposing the initial value condition gives

$$
\frac{\Delta^{(1-p)}}{(1-p)} = c - \lambda t,
\tag{13.5.15}
$$

$$
\Delta^{(1-p)} = c_1 - \lambda(1-p)t,
\tag{13.5.16}
$$

with

$$
\Delta(0)^{(1-p)} = c_1,
\tag{13.5.17}
$$

and

$$
\Delta^{(1-p)} = \Delta_0^{(1-p)} - (1-p)\lambda t.
\tag{13.5.18}
$$

Finally, solving for Δ gives

$$
\Delta(t) = \left[\Delta_0^{(1-p)} - (1-p)\lambda t\right]^{\left(\frac{1}{1-p}\right)}.
\tag{13.5.19}
$$

The following very important observation must be made. For $p = \alpha$, defined by Eq. (13.5.11), Δ^p is an odd function, while $\Delta^{(1-p)}$ is an even function, i.e.,

$$
(-\Delta)^p = -(\Delta)^p, \quad (-\Delta)^{(1-p)} = (\Delta)^{(1-p)}.
\tag{13.5.20}
$$

Using the definition of $\Delta(t) = T(t) - T_e$, Eq. (13.5.19) becomes

$$
T(t) = \begin{cases}
T_e + \left[(T_0 - T_e)^{(1-p)} - (1-p)\lambda t\right]^{\left(\frac{1}{1-p}\right)}, & T_0 > T_e; \\
T_e, & \text{if } T_0 = T_e; \\
T_e - \left[(T_e - T_0)^{(1-p)} - (1-p)\lambda t\right]^{\left(\frac{1}{1-p}\right)}, & T_0 < T_e;
\end{cases}
\tag{13.5.21}
$$

for times

$$
0 < t \le t^* = \frac{(T_0 - T_e)^{\frac{1}{1-p}}}{(1-p)\lambda},
\tag{13.5.22}
$$

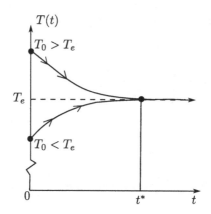

Fig. 13.5.2   Plots of the solutions to $d\Delta/dt = -\lambda\Delta^p$.

and is

$$T(t) = T_e, \quad t > t^*. \tag{13.5.23}$$

Thus, $T(t)$ is a piece-wise continuous function of time and its behavior is illustrated in Fig. 13.5.2.

Comments

(i) For $T_0 = T_e$, the object is already at its surrounding temperature and no change in temperature occurs.

(ii) If $T_0 \neq T_e$, then the body heats up for $T_0 < T_2$, and cools down for $T_0 > T_e$, to the temperature T_e. For both cases, the heating/cooling times to achieve T_e are finite and given by Eq. (13.5.22).

(iii) Within the context of dynamic consistency, it follows that the constructed law of cooling, given by Eq. (13.5.13) is dynamic consistent with the physically observed property of reaching the temperature T_e in a finite time.

(iv) The above proposed law of cooling, Eq. (13.5.13) can be generalized to the form

$$\frac{d\Delta(t)}{dt} = -\lambda[\Delta(t)]^p - k\Delta(t), \tag{13.5.24}$$

where (λ, k) are non-negative parameters, and p is an "odd fraction" (see Problem 13.5.4). A very interesting and, at the same time, important feature of this equation is that it can be solved explicitly for its exact solutions. Further, these solutions have all the same general properties as the case where only the first term occurs. Equation (13.5.24) is a *Bernoulli differential equation*.

13.6   Dissipative/Drag Forces

Forces which reduce or restrict the motion of an object are called dissipative or drag forces. These forces arise or are caused by energy in the object being transferred to its surroundings. Examples of such forces include the influence of a fluid on the motion of an object moving in the fluid and the effect of air resistance on the amplitude of the oscillations of a pendulum. Within the context of the natural sciences, a broad range of dissipative/drag forces may be encountered. The following is a short listing of references which discuss this topic and related issues in detail:

(1) I. Kovacic, "The method of multiple scales for forces oscillators with some real-power nonlinearities in the stiffness and damping forces," *Chaos Solitons and Fractals* **44**, 891 (2011).

(2) A. K. Mallik, "Forced harmonic vibration of a Duffing oscillator with different damping mechanics," Chapter 6 in, I. Kovacic and M. J. Brannan (editors), *The Duffing Equations* (Wiley, Chichester, 2011).

(3) R. E. Mickens, *Oscillations in Planar Dynamic Systems* (World Scientific, Singapore, 1996). See Section 1.2.7.

(4) R. E. Mickens, R. Bullock, W. E. Collins, and K. Oyedeji, "Nonlinear damped Duffing oscillators having finite time dynamics," *Engineering Mathematics Letters* 2014, paper 2014.6. Also, see arXiv: 1404.5596 [nlin.CD].

(5) G. Schmidt and A. Toondl, *Non-Linear Vibrations* (Cambridge University Press, New York, (1986). See pages 25 and 52.

(6) S. Zaitsev, O. Shtempluck, E. Buks, and O. Gottlieb, "Nonlinear damping in a micromechanical oscillator," *Nonlinear Dynamics* **67**, 859 (2012).

If we consider an object of unit mass acted on by several different types of forces, then Newton's equation of motion (for one space dimension) is

$$\frac{d^2 x}{dt^2} = (\text{damping/dissipative forces}) + (\text{others})$$

$$= F_{DD}\left(x, \tfrac{dx}{dt}\right) + (\text{others}), \qquad (13.6.1)$$

where the damping/dissipative (DD) force is denoted by $F_{DD}(x, dx/dt)$ and may depend on both x and $v = dx/dt$, the velocity. Note that F_{DD} always depends on v, since a force depending just on x is an elastic force and, consequently, can be placed in the second term in Eq. (13.6.1).

The following is a selected list of functional forms for F_{DD}, which have appeared in the research literature:

$$F_{DD}(x, v) = \begin{cases} -k_1 \text{sign}(v), & (13.6.2a) \\ -k_2 v, & (13.6.2b) \\ -k_3 v^\alpha, \quad \alpha > 0, & (13.6.2c) \\ -(a_0 + a_1 x^2 + a_2 v^3)v. & (13.6.2d) \end{cases}$$

The main goal of this section is to study in detail the mathematical properties of a DD force corresponding to a single power law term. We show that not all mathematical possible exponentials lead to solutions that are dynamic consistent with the data.

13.6.1   The System

Consider a small ball (bullet) fired into a gel. We take the entering velocity to be $v_0 > 0$, and assume that the only force acting on the ball is a DD force of the form

$$F_{DD} = -\lambda v^\alpha, \quad \lambda > 0, \; \alpha \geq 0, \quad (13.6.3)$$

where (α, λ) are non-negative parameters. Observations indicate that the ball continuously slows and stops in a finite time, t^*, at a distance L from the starting point. Our task is to see what behavior is actually obtained by applying Newton's force equation to this system and then solving it to determine $v(t) = dx(t)/dt$ and $x(t)$. Figure 13.6.1 gives the physical set-up for this situation.

If the mass of the ball is taken to be "one", in whatever physical units we select, then Newton's force equation is

$$\frac{dv(t)}{dt} = -\lambda v(t)^\alpha, \quad v(0) = v_0 > 0. \quad (13.6.4)$$

There are six separate cases to investigate, with respect to the values of $\alpha \geq 0$; they are

$$\alpha = 0 \quad \text{(dry or Coulomb friction)}, \quad (13.6.5a)$$
$$0 < \alpha < 1, \quad (13.6.5b)$$
$$\alpha = 1 \quad \text{(viscous damping)}, \quad (13.6.5c)$$
$$1 < \alpha < 2, \quad (13.6.5d)$$
$$\alpha = 2 \quad \text{(aerodynamic drag)}, \quad (13.6.5e)$$
$$\alpha > 2. \quad (13.6.5f)$$

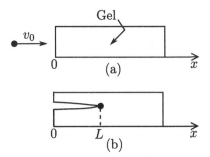

Fig. 13.6.1   A ball fired into a gel with $x(0) = 0$ and $v(0) = v_0 > 0$. (a) The situation at $t = 0$. (b) The ball at $t \geq t^*$.

Note that certain values of α are associated with standard names and we have given them at the appropriate places.

Finally, if $v(t)$ is given, then $x(t)$ can be determined by means of the relation

$$x(t) = \int_0^t v(z)\,dz. \tag{13.6.6}$$

13.6.2   *Solutions to Eqs.* (13.6.4) *and* (13.6.6)

$\alpha = 0$.

For this case, the equation of motion is

$$\frac{dv}{dt} = -\lambda, \quad v(0) = v_0 > 0, \tag{13.6.7}$$

and a solution is the expression

$$v(t) = v_0 - \lambda t. \tag{13.6.8}$$

Note that at the time, t^*, where

$$t^* = \frac{v_0}{\lambda}, \tag{13.6.9}$$

$v(t^*) = 0$ and for $t > t^*$, the ball cannot move, since, otherwise, it would violate energy conservation. This means that $v(t)$ is

$$v(t) = \begin{cases} v_0 - \lambda t, & 0 < t \leq t^* = \frac{v_0}{\lambda}, \\ 0, & t > t^*. \end{cases} \tag{13.6.10}$$

To calculate $x(t)$, we substitute this result in Eq. (13.6.6) and find

$$x(t) = \begin{cases} v_0 t - \left(\frac{\lambda}{2}\right) t^2, & 0 < t \leq t^*, \\ \frac{v_0}{2\lambda}, & t > t^*. \end{cases} \tag{13.6.11}$$

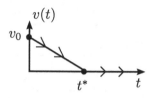

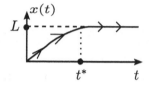

Fig. 13.6.2   Plots of $v(t)$ and $x(t)$ for $F_{DD} = -\lambda$. We have $t^* = v_0/\lambda$ and $L = v_0^2/2\lambda$.

Plots for $v(t)$ and $x(t)$ are given in Fig. 13.6.2.
Note that $x(t^*) = L$, where

$$L = \frac{v_0^2}{2\lambda}. \tag{13.6.12}$$

To summarize we have found:
 (i) $v(t)$ goes to zero in a time $t^* = v_0/\lambda$.
 (ii) $x(t)$ achieves a maximum value of $L = v_0^2/2\lambda$ at the time $t = t^*$.
 (iii) Both $v(t)$ and $x(t)$ are piece-wise-continuous functions of t.
 (iv) The $\alpha = 0$ case is consistent with the observed behavior of our physical system.

$0 < \alpha < 1.$

For this case, the solution, $v(t)$ is

$$v(t) = \begin{cases} \left[v_0^{(1-\alpha)} - \lambda(1-\alpha)t\right]^{\left(\frac{1}{1-\alpha}\right)}, & 0 < t \le t^*, \\ 0, & t > t^* = \frac{v_0^{(1-\alpha)}}{\lambda(1-\alpha)}, \end{cases} \tag{13.6.13}$$

and the corresponding functional form for $x(t)$ is given by the expression

$$x(t) = \begin{cases} \left[\frac{1}{\lambda(2-\alpha)}\right]\left\{v_0^{(2-\alpha)} - \left[v_0^{(1-\alpha)} - \lambda(1-\alpha)t\right]^{\left(\frac{2-\alpha}{1-\alpha}\right)}\right\}, & 0 < t \le t^*; \\ 0, & t > t^* = \frac{v_0^{(1-\alpha)}}{\lambda(1-\alpha)}. \end{cases} \tag{13.6.14}$$

Plots of $v(t)$ and $x(t)$ are given in Fig. 13.6.3.

In summary, we have:

(i) $v(t)$ becomes zero in a finite time, t^*, where $t^* = v_0^{(1-\alpha)}/\lambda(1-\alpha)$.

(ii) The ball travels a finite distance, L, where $L = v_0^{(2-\alpha)}/\lambda(2-\alpha)$.

(iii) Both $v(t)$ and $x(t)$ are continuous functions of t.

(iv) This case, $0 < \alpha < 1$, is consistent with the observed dynamics of this physical system.

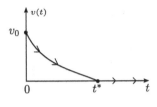

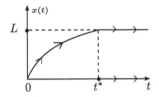

Fig. 13.6.3   Plots of $v(t)$ and $x(t)$ for $F_{DD} = -\lambda v^\alpha$, for $0 < \alpha < 1$, $t^* = v_0^{(1-\alpha)}/\lambda(1-\alpha)$ and $L = v_0^{(2-\alpha)}/\lambda(2-\alpha)$.

$\alpha = 1.$

For this case, the equation of motion and its solution are

$$\frac{dv}{dt} = -\lambda v, \quad v(t) = v_0 e^{-\lambda t} \tag{13.6.15}$$

and the corresponding $x(t)$ is

$$x(t) = \left(\frac{v_0}{\lambda}\right)(1 - e^{-\lambda t}). \tag{13.6.16}$$

Inspection of Eqs. (13.6.15) and (13.6.16) allows the following conclusions to be made:

(i) $v(t)$ only becomes zero at $t = \infty$.

(ii) $x(t)$ has a maximum value, i.e., the stopping distance, $L = v_0/\lambda$, but it takes an unlimited amount of time to achieve this value.

$1 < \alpha < 2$.

The functional forms for $v(t)$ and $x(t)$ are

$$v(t) = \frac{v_0}{\{1 + [\lambda(v_0)^\epsilon]t\}^{(1/\epsilon)}}, \qquad (13.6.17)$$

$$x(t) = \left[\frac{v_0^{(1-\epsilon)}}{\lambda}\right]\left(\frac{\epsilon}{1-\epsilon}\right)\left\{1 - \frac{1}{[1 + \lambda(v_0)^\epsilon t]^{\frac{1-\epsilon}{\epsilon}}}\right\}, \qquad (13.6.18)$$

where $\alpha = 1 + \epsilon$, and, consequently, $0 < \epsilon < 1$. Examination of the latter two equations shows the following to hold:

(i) $v(t)$ decreases to zero, but takes on infinite amount of time to become zero.

(ii) $x(t)$ reaches a maximum value, L, i.e.,

$$L = \left[\frac{v_0^{(1-\epsilon)}}{\lambda}\right]\left(\frac{1}{1-\epsilon}\right), \quad 0 < \epsilon < 1, \qquad (13.6.19)$$

but, in an infinite time.

$\alpha = 2$.

The differential equation for this case and its solution are

$$v(t) = \frac{v_0}{1 + (\lambda v_0)t}, \qquad (13.6.20a)$$

$$x(t) = \left(\frac{v_0}{\lambda}\right)\text{Ln}(1 + \lambda v_0 t). \qquad (13.6.20b)$$

Note that $v(t)$ becomes zero as $t \to \infty$, but $x(t)$ becomes unbounded at $t \to \infty$.

$\alpha > 2$.

If we write α as

$$\alpha = 2 + \beta, \qquad (13.6.21)$$

then

$$v(t) = \frac{v_0}{[1 + \lambda v_0^{(1+\beta)}t]^{\frac{1}{1+\beta}}}, \qquad (13.6.22a)$$

$$x(t) = \left(\frac{1}{\lambda v_0^\beta}\right)\left(\frac{1+\beta}{\beta}\right)\left\{[1 + \lambda v_0^{(1+\beta)}t]^{\frac{\beta}{1+\beta}} - 1\right\}, \qquad (13.6.22b)$$

and

$$\lim_{t \to \infty}\left(\frac{v(t)}{x(t)}\right) = \left(\frac{0}{\infty}\right). \qquad (13.6.23)$$

See Figs. 13.6.4 and 13.6.5 for the general behavior of $v(t)$ and $x(t)$ vs t, for $\alpha \geq 2$.

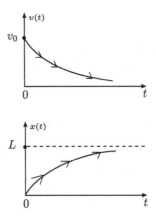

Fig. 13.6.4   Plots of $v(t)$ and $x(t)$ for $F_{DD} = -\lambda v^{\alpha}$, for $1 \leq \alpha < 2$, where $v(\infty) = 0$ and L is finite. See Eqs. (13.6.16) and (13.6.19).

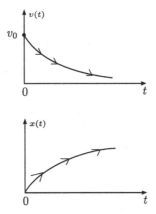

Fig. 13.6.5   Plots of $v(t)$ and $x(t)$ for $F_{DD} = -\lambda v^{\alpha}$, for $\alpha \geq 2$, where $v(\infty) = 0$ and $x(t)$ increases without bound.

13.6.3   *Discussion*

We can now answer the following question: Assuming that a ball is fired into a gel and assuming that only a power-law drag force acts on it, see Eq. (13.6.4), which values of α give rise to mathematical behaviors that are dynamic consistent with the actual motion of the ball? As a reminder, the ball travels a finite distance into the gel and stops at a finite time.

If we impose the mathematical conditions that $v(t)$ and $x(t)$ are continuous with continuous first derivatives, then a detailed examination of the results presented in the previous section shows that only α values in the interval, $0 < \alpha < 1$, are suitable.

Note that the finite-time dynamic behavior is only influenced by the behavior of $F_{DD}(v)$ for "small" velocities. The details of what takes place will be discussed in the section on the topic of "dominant balance."

13.7 Applications of Dimensional Analysis

The purpose of this section is to demonstrate the power, applicability, and usefulness of dimensional analysis to the sciences and mathematics. We do this by means of a number of examples involving concepts and issues from these areas. For simplicity and clarity, a direct approach, not dependent on the powerful tools usually associated with dimensional analysis, such as the theorem of Buckingham, is used to formulate and discuss each problem.

An excellent definition of dimensional analysis was given by Bridgemann, in his 1969 article in *Encyclopaedia Britannica* (Vol. 7, pp. 439–449):

> "The principal use of dimensional analysis is to deduce from a study of the dimensions of the variables in any physical system certain limitations on the form of any possible relationship between those variables. The method is of great generality and mathematical simplicity."

The following short list of references are of direct relevance to the topics of this section. In addition to often giving fuller details of issues we present, they also provide a multitude of other worked examples in the natural, medical, and engineering sciences. Further, many contain references to the original research literature.

(1) G. I. Barenblatt, *Dimensional Analysis* (Gordon and Breach, New York, 1987).

(2) M. P. Brenner, *Physical Mathematics* (Harvard University, Cambridge, 2010).

(3) B. Günther, "Dimensional analysis and theory of biological similarity," *Physiological Reviews,* **55**, 659 (1975).

(4) H. E. Huntley, *Dimensional Analysis* (MacDonald, London, 1952).

(5) E. and M. Isaacson de St Q, *Dimensional Methods in Engineering and Physics* (Edward Arnold, London, 1975).

(6) R. Kurt, *Dimensional Analysis and Group Theory in Astrophysics* (Pergamon Press, Oxford, 1972).

(7) J. Lehman and E. Craig, Dimensional analysis in applied psychological research, *Journal of Psychology* **55**, 223 (1963).

(8) A. and J. Menkes, "The application of dimensional analysis to learning theory," *Psychological Review* **64**, 8 (1957).

(9) R. E. Mickens, *Nonlinear Oscillations* (Cambridge University Press, New York, 1981).

(10) P. J. Nahin, *In Praise of Simple Physics* (Princeton University Press, Princeton, NJ, 2016).

(11) B. Schepartz, *Dimensional Analysis in the Biomedical Sciences* (Charles C. Thomas Publishers, Springfield, IL, 1980).

(12) T. Szirtes, *Applied Dimensional Analysis and Modeling* (McGraw-Hill, New York, 1998).

(13) E. Taylor, *Dimensional Analysis for Engineers* (Clarendon Press, Oxford, 1974).

(14) L. Weinstein and J. A. Adam, *Guestimation: Solving the World's Problems on the Back of a Cocktail Napkin* (Princeton University Press, Princeton, NJ, 2008).

(15) J. Zierep, *Similarity Laws and Modeling* (Marcel Dekker, New York, 1971).

Finally, it should be indicated that a topic closely related to dimensional analysis is the "Fermi question." Such a question is one in the sciences which seeks a quick, approximation or estimation of some quantity which may be difficult, impossible or inconvenient to measure directly [10, 14].

For the calculations to follow, all physical quantities will be expressed in units of M-mass, length-L, and time-T. The particular units, such as Kilogram for mass, meter for length, and second for T, will play no essential roles.

13.7.1 *Oscillations: Harmonic Oscillator and Pendulum*

Consider a spring attached to heavy horizontal support, with a mass, m, connected to the other end; see Fig. 13.7.1(a). Our task is to construct a formula for the period of the oscillations.

The following is a list of possible variables:

- $\bar{T}$ = period of oscillations
- m = mass attached to spring
- k = spring constant
- g = acceleration due to gravity (at the surface of the earth).

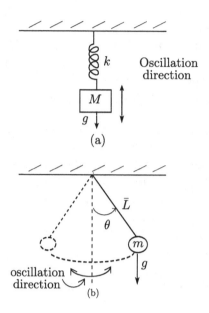

Fig. 13.7.1   (a) Simple harmonic oscillator. (b) Simple pendulum.

Therefore, the required relation for the formula of the period is

$$\bar{T} = \lambda_1 m^a k^b g^c, \tag{13.7.1}$$

where λ_1 is an (unknown) dimensionless constant, and the variables have the physical units

$$[\bar{T}] = T, \quad [m] = M, \quad [k] - \frac{M}{T^2}, \quad [g] = \frac{L}{T^2}. \tag{13.7.2}$$

Note that given some quantity, Q, then $[Q]$ denotes its physical units. Further, the (a, b, c) are a priori unknown constants to be determined by the requirement that the physical dimensions of both sides of Eq. (13.7.1) be exactly the same.

Observe that the physical dimensions of the spring constant follows from the fact that the force is $F(x) = -kx$, and, consequently, $[k] = [F(x)/x]$.

Using the results of Eq. (13.7.2) in Eq. (13.7.1) gives

$$T = M^a \left(\frac{M}{T^2} \right)^b \left(\frac{L}{T^2} \right)^c$$
$$= M^{(a+b)}(L)^c(T)^{-2(b+c)}. \tag{13.7.3}$$

Now, comparing left- and right-sides, we find

$$\begin{cases} M : 0 = a + b, \\ L : 0 = c, \\ T : 1 = -2b - 2c, \end{cases} \tag{13.7.4}$$

and consequently,

$$a = \frac{1}{2}, \quad b = -\frac{1}{2}, \quad c = 0. \tag{13.7.5}$$

Therefore, the period of oscillations for the harmonic oscillator is

$$\bar{T} = \lambda_1 \sqrt{\frac{m}{k}}. \tag{13.7.6}$$

At this stage of the investigation, the constant λ_1 cannot be calculated and therefore, must be determined by experiment. However, since the harmonic oscillator equation of motion is

$$m\frac{d^2x}{dt^2} + kx = 0, \tag{13.7.7}$$

a direct calculation gives

$$T = 2\pi\sqrt{\frac{m}{k}}, \tag{13.7.8}$$

and we can conclude that $\lambda_1 = 2\pi$.

A similar analysis can be done for the simple pendulum; see, Fig. 13.7.1(b). For this case, the variables are

- $\bar{T}$ = period of oscillation
- m = mass of attached ball
- $\bar{L}$ = length of (massless) rod
- g = acceleration due to gravity.

Note, the angle θ, will not appear since it is a dimensionless variable. Therefore, we obtain the following relation for the period

$$\bar{T} = \lambda_2 m^a \bar{L}^b g^c, \tag{13.7.9}$$

where putting in the physical units gives

$$\begin{aligned} T &= M^a L^b \left(\frac{L}{T^2}\right)^c \\ &= M^a L^{b+c} T^{-2c}, \end{aligned} \tag{13.7.10}$$

and

$$
\begin{cases}
M : 0 = a, \\
L : 0 = b + c, \\
T : 1 = -2c.
\end{cases}
\tag{13.7.11}
$$

Therefore

$$
a = 0, \quad b = \frac{1}{2}, \quad c = -\frac{1}{2}, \tag{13.7.12}
$$

and the period of a pendulum is

$$
\bar{T} = \lambda_2 \sqrt{\frac{\bar{L}}{g}}. \tag{13.7.13}
$$

As for the harmonic oscillator, the pure constant λ_2 must either be determined by a theory or from experiment.

Comment

The primary reason for using dimensional analysis to investigate a particular system is that either no fundamental theory exists to calculate the required formula or that if such a theory exists, it is very difficult to carry out the actual calculation. For this situation the application of dimensional analysis may provide the needed formula, up to an overall constant. Experiments can then be done to obtain the related parameter.

13.7.2 *Drag on a Body Moving in a Fluid*

Two important examples of fluids are bodies of water and the air. There are a number of important systems that move in these fluids, submarines submerged in water and airplanes flying in the air. One consequence of these motions is that drag forces operate on objects moving in the fluids. In the following, two different drag force formulas will be derived, one for low velocities and the other for higher velocities.

The following are variables which can be considered to be important for determining the drag force, F_d, on a body moving in a fluid:

- $v =$ velocity of the body
- $\rho =$ density of the fluid
- $A =$ cross sectional area of the body
- $\mu =$ viscosity of the fluid.

With these variables, F_d can be written in two possible ways

(1) For small velocities the inertial aspects of the fluid should not play a critical role in determining F_d. Since the inertial effects are determined by the density, then the relevant variables to use for the dimensional analysis are (v, A, μ), for the drag force $F_d^{(L)}$.

(2) For high velocities inertial effects will certainly be important and the relevant variables will now be (v, A, ρ), for the drag force $F_d^{(H)}$.

If this way of looking at the system is correct, then a clear prediction can be immediately made. It is

$$F_d^{(L)}(v) \propto v^{p_1}, \quad F_d^{(H)}(v) \propto v^{p_2}, \quad 0 < p_1 < p_2, \tag{13.7.14}$$

and this allows $F_d^{(L)}$ and $F_d^{(H)}$ to dominate, respectively, at low and high velocities.

For $F_d^{(L)}$, we have

$$F_d^{(L)} = \lambda_3 \mu^a A^b v^c, \tag{13.7.15}$$

and

$$\frac{ML}{T^2} = \left(\frac{M}{LT}\right)^a (L^2)^b \left(\frac{L}{T}\right)^c$$
$$= M^a L^{(-a+2b+c)} T^{-(a+c)}, \tag{13.7.16}$$

and, we must have

$$\begin{cases} M : 1 = a, \\ L : 1 = -a + 2b + c \\ T : -2 = -(a + c). \end{cases} \tag{13.7.17}$$

The solutions for (a, b, c) are

$$a = 1, \quad b = \frac{1}{2}, \quad c = 1, \tag{13.7.18}$$

and we obtain

$$F_d^{(L)} = \lambda_3 \mu A^{1/2} v. \tag{13.7.19}$$

Likewise, we have

$$F_d^{(H)} = \lambda_4 A^a \rho^b v^c \tag{13.7.20}$$

and

$$\frac{ML}{T^2} = L^{2a} \left(\frac{M}{L^3}\right)^b \left(\frac{L}{T}\right)^c$$
$$= M^b L^{2a-3b+c} T^{-c}, \tag{13.7.21}$$

and

$$\begin{cases} M : 1 = b, \\ L : 1 = 2a - 3b + c, \\ T : -2 = -c. \end{cases} \qquad (13.7.22)$$

Therefore,

$$a = 1, \quad b = 1, \quad c = 2$$

and

$$F_d^{(H)} = \lambda_4 A \rho v^2. \qquad (13.7.23)$$

If we assume that the above derived formulae for $F_d^{(L)}$ and $F_d^{(H)}$ are jointly valid over some range of velocity values, then the total drag force, F_d, can be written as

$$F_d(v) = F_d^{(L)}(v) + F_d^{(H)}(v)$$
$$= \lambda_3 (\mu A^{1/2}) v + \lambda_4 (A\rho) v^2. \qquad (13.7.24)$$

Note that the transition point, i.e., the velocity, v^*, where

$$F_d^{(L)}(v^*) = F_d^{(H)}(v^*), \qquad (13.7.25)$$

is

$$v^* = \left(\frac{\lambda_3}{\lambda_4} \right) \frac{\mu}{A^{1/2} \rho}. \qquad (13.7.26)$$

13.7.3   Checking Correctness of Formulas

Dimensional analysis can be used to check the correctness of formulas appearing, for example, in handbooks. These formulae may contain typos and thus lead the user to believe that a given relationship is correct when it is not. The use of dimensional analysis will often allow the detection of the most obvious errors. We will illustrate how this is done by way of several examples.

Suppose one does not know the integral of e^{ax} and turns to a table of integrals and finds the result

$$\int e^{ax} dx = a e^{ax}. \qquad (13.7.27)$$

(In the following discussions, we omit all additive constants of integration.) For this elementary integral, it is easy to show that it is incorrect. To do so assign to the variable x, the physical unit of length, L. Since the argument of exponential function must be pure numbers, then

$$[ax] = [a][x] = L^0, \qquad (13.7.28)$$

and we conclude that $[a] = L^{-1}$. Also, since

$$[dx] = L, \qquad (13.7.29)$$

it follows that the physical units of the left-side of Eq. (13.7.27) is

$$\left[\int e^{ax} dx\right] = [e^{ax}][dx] = L^0 L = L. \qquad (13.7.30)$$

However, for the right-side, we have

$$[ae^{ax}] = [a][e^{ax}] = (L^{-1})(L^c) = L^{-1}. \qquad (13.7.31)$$

Therefore, it can be concluded that the result on the right-side is not correct because the expression given in Eq. (13.7.27) is not dimensional consistent. To correct this error, the factor "a" must be replaced by a^{-1} and this gives the right answer

$$\int e^{ax} dx = \left(\frac{1}{a}\right) e^{ax}. \qquad (13.7.32)$$

Comment

Note that while dimensional considerations can provide information on the dependence of paramters, this method does not directly give any information as to the value of numerical coefficients. For example

$$\int e^{ax} dx = \left(\frac{3}{a}\right) e^{ax}, \qquad (13.7.33)$$

is certainly dimensional consistent, but the numerical coefficient is wrong.

Now consider the integral result

$$\int \frac{x\,dx}{a + bx} = \frac{x}{b} - \left(\frac{a}{b^2}\right) \mathrm{Ln}(a + bx), \qquad (13.7.34)$$

and let's check it to see if it is dimensional consistent.

The way to proceed is as follows. Start by making the denominator on the left-side dimensionless. This implies that if $[x] = L$, then

$$[a] = L^0, \quad [b] = L^{-1}, \qquad (13.7.35)$$

and

$$[x\,dx] = [x][dx] = L^2. \tag{13.7.36}$$

With this information, we see that the argument of the Ln-function, on the right-side, is of dimension L^0, as it should. (Remember, in general, the arguments of exponential, trigonometric, logarithm functions should all be of dimension L^0.) This means that the left-side has dimension L^2. Now, for the right-side

$$\left[\frac{x}{b}\right] = \frac{L}{L^{-1}} = L^2, \quad \left[\frac{a}{b^2}\right] = \frac{L^0}{L^{-2}} = L^2. \tag{13.7.37}$$

Therefore, we conclude that the result given in Eq. (13.7.34) is dimensional consistent.

A third example is (for $ab > 0$)

$$\int \frac{dx}{ae^{mx} + be^{-mx}} = \left(\frac{1}{m\sqrt{ab}}\right) \arctan\left(\sqrt{\frac{a}{b}}\,e^{mx}\right). \tag{13.7.38}$$

In order for the argument of the arctan have dimension L^0, we must have

$$[a] = [b], \quad [m] = L^{-1}. \tag{13.7.39}$$

But the dimension of the left-side is

$$\frac{[dx]}{[a]} = \frac{L}{[a]}, \tag{13.7.40}$$

and the dimension of the right-side is

$$\left[\frac{1}{m\sqrt{ab}}\right] = \frac{1}{[m][a]} = \frac{L}{[a]}. \tag{13.7.41}$$

Comparing Eqs. (13.7.40) and (13.7.41), we conclude that the formula of Eq. (13.7.38) is dimensional consistent for any dimensional units for the constants a and b.

For a final example, consider the following formula

$$\int x^2 \sin(ax)dx = \left(\frac{2x}{a^2}\right) \sin(ax) + \left(\frac{2x}{a^3}\right) \cos(ax) - \left(\frac{x^2}{a}\right) \cos(ax). \tag{13.7.42}$$

Let us check to see if it is dimensional consistent. First, note that since ax appears as the argument of trigonometric functions, then

$$[ax] = L^0 \Rightarrow [a] = L^{-1}. \tag{13.7.43}$$

Second, for the left-side

$$[x^2 \sin(ax)dx] = [x^2][\sin(ax)][dx]$$
$$= (L^2)(L^0)(L)$$
$$= L^3, \qquad (13.7.44)$$

and this must be the dimension of each term on the right-side of Eq. (13.7.42). The calculated dimensions for the three terms are

$$\left[\left(\frac{2x}{a^2}\right)\sin(ax)\right] = \left(\frac{L}{L^{-2}}\right)L^0 = L^3, \qquad (13.7.45a)$$

$$\left[\left(\frac{2x}{a^3}\right)\cos(ax)\right] = \left(\frac{L}{L^{-3}}\right)L^0 = L^4, \qquad (13.7.45b)$$

$$\left[\left(\frac{x^2}{a}\right)\cos(ax)\right] = \left(\frac{L^2}{L^{-1}}\right)L^0 = L^3. \qquad (13.7.45c)$$

Therefore, we conclude that the second-term on the right-side of Eq. (13.7.42) must have a factor of a^{-2} rather than a^{-3}, assuming the functional x-dependence is correct.

As a nice illustration of the use of elementary calculus, we now derive the correct formula for the integral appearing on the left-side of Eq. (13.7.42). To do this start with the result

$$F(a,x) \equiv \int \sin(ax)dx = -\left(\frac{1}{a}\right)\cos(ax). \qquad (13.7.46)$$

Therefore,

$$\frac{\partial F(a,x)}{\partial a} = \left(\frac{1}{a^2}\right)\cos(ax) + \left(\frac{x}{a}\right)\sin(ax)$$

$$= \int x\cos(ax)dx \qquad (13.7.47)$$

and

$$\frac{\partial^2 F(a,x)}{\partial a^2} = \left(\frac{x^2}{a}\right)\cos(ax) - \left(\frac{2x}{a^2}\right)\sin(ax) - \left(\frac{2}{a^3}\right)\cos(ax)$$

$$= -\int x^2\sin(ax)dx. \qquad (13.7.48)$$

Thus, from the last expression, we find

$$\int x^2\sin(ax)dx = \left(\frac{2}{a^3}\right)\cos(ax) + \left(\frac{2x}{a^2}\right)\sin(ax) - \left(\frac{x^2}{a}\right)\cos(ax). \qquad (13.7.49)$$

13.7.4 Surface Gravitational Acceleration of a Celestial Body

Celestial bodies of sufficient mass are to a very good approximation spherical. The current task is to estimate the surface gravitational acceleration, i.e., the acceleration one would experience if standing on the surface of such a body. For the earth, this is denoted by g and has a value close to $9.81 \ m/s^2$ (m = meter, s = second).

To proceed, consider a celestial body of mass m_b and radius R_b, and denote by g_b its surface acceleration due to its gravitational interaction. To estimate g_b, it is reasonable to assume that g_b will be related not only to (m_b, R_b), but also to G, the universal gravitational constant, whose value is $G = (6.674)10^{-11} m^3 kg^{-1} s^{-2}$. Therefore, we assume

$$g_b = \lambda (m_b)^a (R_b)^b G^c,$$ (13.7.50)

where the constant (a, b, c) are to be determined, and λ is a dimensionless constant. The expression involving the physical dimensions is

$$\frac{L}{T^2} = M^a L^b \left(\frac{L^3}{MT^2} \right)^c$$

$$= M^{(a-c)} L^{b+3c} T^{-2c}.$$ (13.7.51)

Comparing left- and right-sides gives

$$\begin{cases} M : 0 = a - c \\ L : 1 = b + 3c \\ T : -2 = -2c \end{cases}$$ (13.7.52)

and

$$a = 1, \quad b = -2, \quad c = 1.$$ (13.7.53)

Therefore, we conclude that the surface gravitational acceleration is given by

$$g_b = \lambda \left(\frac{Gm_b}{R_b^2} \right).$$ (13.7.54)

The exact value for g_b can be calculated by application of Newton's force equation. Suppose we consider a small mass, $\mu \lll m_b$, at the surface of the celestial body; then from Newton's Law of Gravitation and his force law, we have

$$-\left(\frac{Gm_b\mu}{R_b^2} \right) = \mu g_b$$ (13.7.55)

and

$$g_b = -\left(\frac{Gm_b}{R_b^2}\right). \tag{13.7.56}$$

The minus sign indicates that the acceleration is in the direction toward the center of the celestial body. Comparing Eqs. (13.7.54) and (13.7.56), we see that $\lambda = 1$.

Consider the case of celestial bodies having the same constant density

$$\rho = \frac{m_b}{R_b^3} = \rho_0 = \text{constant}. \tag{13.7.57}$$

It follows that

$$g_b = -G\rho_0 R_b. \tag{13.7.58}$$

This implies that under this condition the surface gravitational acceleration is directly proportional to the size (radius) of the body.

13.7.5   *Proof of Pythagorean Theorem*

One of the best known results in mathematics is the Pythagorean theorem. What we present below is a proof of this result using techniques from dimensional analysis.

Consider a right-triangle having the configuration shown in Fig. 13.7.2. The right-angle is located at vertex C and the angle at vertex B is denoted θ. The dashed line segment CD is the perpendicular from vertex C to side AB. Using the similarity of triangles, it follows that angle ACD is also equal to θ.

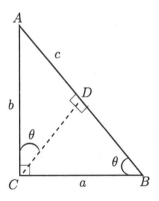

Fig. 13.7.2   Pythagorean theorem.

The following is clearly true

$$\text{area } \triangle ABC = \text{area } \triangle ACD + \text{area } \triangle BCD. \tag{13.7.59}$$

Further, since we know the length of one side for each of these three triangles and also know the values of the angles, all of which are expressible in terms of θ, and the corresponding areas take the form

$$\begin{cases} \text{area } \triangle ABC = f(\theta)c^2, \\ \text{area } \triangle ACD = f(\theta)b^2, \\ \text{area } \triangle BCD = f(\theta)a^2, \end{cases} \tag{13.7.60}$$

where $f(\theta)$ is a function only of θ. Substituting the results from Eq. (13.7.60) into Eq. (13.7.59) gives

$$f(\theta)c^2 = f(\theta)b^2 + f(\theta)a^2, \tag{13.7.61}$$

which upon cancelling the common factor of $f(\theta)$ gives

$$c^2 = a^2 + b^2, \tag{13.7.62}$$

the standard equation expressing the Pythagorean theorem.

Note that this result has been derived without explicit knowledge of the function $f(\theta)$. The fact that all of the triangles are right-triangles is critical to the proof since this forces the function $f(\theta)$ to be the same for each triangle.

13.7.6 *Hydroplaning of a Car*

During or after a major rainfall, automobiles may experience hydroplaning. This phenomena takes place when an automobile drives over a wet surface, covered with sufficient water, such that its tires lose contact with the hard surface. In other words, the car is literally surfing on the water on top of the surface.

In general, there exists a critical speed, v_c for this to take place, i.e., for speeds less than v_c, no hydroplaning occurs. We now use dimensional analysis to formulate a relationship between v_c and other assumed relevant variables.

So, let's consider the variables associated with the water. The only property which can be easily determined is its density. Other variables, such as the thickness of the water layer between the tires and the surface are, for all practical purposes, unmeasureable. Therefore, we hope that useful results can be derived from the assumption that the density, ρ, of the water is its major characteristic feature needed to analyze hydroplaning.

With regard to the automobile, the tire pressure, P, should play a role. We also assume that the tire pressure is equal to the contact pressure of the layer of the water on the surface in contact with the tire. This assumption is consistent with the assumed lack of a vertical motion for the automobile.

Are there other variables which should be included in the analysis? The natural choice is the mass, m, of the automobile. Let's put all this together and see what outcome is realized.

The critical velocity formula is taken to be

$$v_c = \lambda(P)^a(\rho)^b(m)^c, \qquad (13.7.63)$$

and in terms of physical units, we have

$$\frac{L}{T} = \left(\frac{M}{T^2L}\right)^a \left(\frac{M}{L^3}\right)^b M^c$$
$$= M^{a+b+c}L^{-a-3b}T^{-2a}. \qquad (13.7.64)$$

Therefore,

$$\begin{cases} M : 0 = a + b + c \\ L : 1 = -a - 3b, \\ T : 1 = 2a, \end{cases} \qquad (13.7.65)$$

and

$$a = \frac{1}{2}, \quad b = -\left(\frac{1}{2}\right), \quad c = 0, \qquad (13.7.66)$$

and, finally, for the critical speed we obtain

$$v_c = \lambda\sqrt{\frac{P}{\rho}}. \qquad (13.7.67)$$

Note that ρ, the density of water, is to a very good approximation constant. Thus, one major conclusion is that the critical speed is proportional to the square-root of the tire pressure. Another interesting fact is that v_c does not depend on the mass of the automobile.

The lesson for driving, in conditions where the surface is wet enough to allow hydroplaning, is to inflate the tires to the highest permitted pressure.

13.7.7 *Oscillations of a Liquid Sphere*

Suppose we have an isolated sphere of liquid in equilibrium, i.e., the various parts are in static positions relative to each other. If the sphere is now perturbed on its surface, the surface will oscillate relative to the initial spherical equilibrium position. The question arises: What is the period of these oscillations?

Comment

In general, after the perturbation is applied, a large number of oscillatory modes may exist. The calculations to follow probably only provide information on the largest period of oscillation.

To proceed, we make a list of (possible) relevant variables:

$\bar{T}$ = period of oscillations

R = radius of liquid sphere

ρ = density of liquid

G = gravitational force constant

σ = coefficient of surface tension.

A momentum's reflection indicates that there are two limiting cases:

(1) For the first case, surface tension effects dominate the self gravitational interactions of the sphere. This corresponds to the sphere being smaller than some critical value, R_c, of the radius.

(2) The second case is the reverse situation, i.e., the liquid sphere is large enough so that internal gravitational interactions dominate the influences of surface tension.

With these thoughts in hand, we see the necessity to derive two formulas for the period of the oscillations, one for each of the two limiting cases.

Case-1: Relevant variables (R, ρ, σ)

We have for this situation

$$\bar{T} = \lambda_1 R^a \rho^b \sigma^c, \tag{13.7.68}$$

where λ_1 is a pure number. Putting in the required physical units, where $[\sigma] = MT^{-2}$, gives

$$T = L^a \left(\frac{M}{L^3}\right)^b \left(\frac{M}{T^2}\right)^c$$
$$= L^{(a-3b)} M^{(b+c)} T^{-2c}, \tag{13.7.69}$$

from which it follows that

$$\begin{cases} M : 0 = b + c, \\ L : 0 = a - 3b \\ T : 1 = -2c, \end{cases} \tag{13.7.70}$$

and

$$a = \frac{3}{2}, \quad b = \frac{1}{2}, \quad c = -\left(\frac{1}{2}\right). \tag{13.7.71}$$

Therefore,

$$\bar{T}^{(s)} = \lambda_1 R^{3/2} \rho^{1/2} \sigma^{-1/2}$$

$$= \lambda_1 \sqrt{\frac{R^3 \rho}{\sigma}}, \tag{13.7.72}$$

and $\bar{T}^{(s)}$ indicates that this is a dimensional analysis derived formula for small spheres.

Case-2: Relevant variables (G, ρ, R)

For this case

$$\bar{T}^{(\ell)} = \lambda_2 R^a \rho^b G^c, \tag{13.7.73}$$

where $\bar{T}^{(\ell)}$ is the period for large values of R. Therefore,

$$T = L^a \left(\frac{M}{L^3}\right)^b \left(\frac{L^3}{MT^2}\right)^c$$

$$= L^{(a-3b+3c)} M^{(b-c)} T^{-2c}, \tag{13.7.74}$$

and

$$\begin{cases} 0 = b - c \\ 0 = a - 3b + 3c \\ 1 = -2c \end{cases} \tag{13.7.75}$$

and

$$a = 0, \quad b = -\left(\frac{1}{2}\right), \quad c = -\left(\frac{1}{2}\right). \tag{13.7.76}$$

The final result is

$$\bar{T}^{(\ell)} = \lambda_2 \left(\frac{1}{\sqrt{G\rho}}\right). \tag{13.7.77}$$

Inspection of this formula shows that $\bar{T}^{(\ell)}$ is independent of R, the radius of the liquid sphere.

Note that since density = mass/volume, then $\bar{T}^{(s)}$ can be rewritten to the form

$$\bar{T}^{(s)} = \bar{\lambda}_2 \sqrt{\frac{m}{\sigma}}, \tag{13.7.78}$$

where $m =$ total mass of sphere and $\bar{\lambda}_2$ is another pure number.

Also, $\bar{T}^{(\ell)}$ can be rearranged and expressed as

$$\bar{T}^{(\ell)}\sqrt{\rho} = \text{constant}. \qquad (13.7.79)$$

This relationship has been applied to the analysis of a class of giant stars, known as the *Cepheid variables*, where $\bar{T}^{(\ell)}$ is the period of their oscillations and the ρ refers to their mean density. An early reference to this topic is C. H. Payne, "On the relation of period to mean density for Cepheid variables," *Harvard College Observatory Bulletin*, No. 876, pp. 28–30 (1930).

We can now ask if there is a single formula which extrapolates between $\bar{T}^{(s)}$, for small values of R, to $\bar{T}^{(\ell)}$, at large R. One possibility is the function $\bar{T}$ given below

$$\bar{T}(R) = \frac{\bar{T}^{(\ell)}\bar{T}^{(s)}}{\bar{T}^{(\ell)} + \bar{T}^{(s)}}$$

$$= \frac{\left(\frac{\lambda_2}{\sqrt{G\rho}}\right)\left[\lambda_1\sqrt{\frac{R^3\rho}{\sigma}}\right]}{\left(\frac{\lambda_2}{\sqrt{G\rho}}\right) + \left[\lambda_1\sqrt{\frac{R^3\rho}{\sigma}}\right]}. \qquad (13.7.80)$$

Note that while this functional form for $\bar{T}$ has no actual theoretical basis, it may provide some general results with regard to the behavior of $\bar{T}$ as R changes; see Fig. 13.7.3.

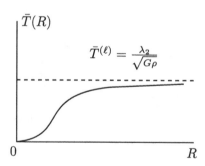

Fig. 13.7.3   Plot of $\bar{T}(R)$ vs R, as given in Eq. (13.7.80).

13.7.8   Analysis of a Nuclear Bomb Explosion

The investigation of various features of the explosioin of a nuclear bomb by means of dimensional analysis is one of its outstanding application achievements. Consequently, as an important example it fully illustrates the power of the technique and appears in some form in almost all texts on the subject of dimensional analysis. The critical reference to this problem is the article: G. I. Taylor, "The formation of a blast wave by a very intense explosion. The atomic explosion of 1945," *Proceedings of the Royal Society London* **A201**, 159 (1910). In that work Taylor's task was to estimate the energy of the July 16, 1945 Trinity nuclear bomb explosion using published photographs of the fireball taken at specific times and for which the radii of the fireballs could be determined.

Since Taylor was an authority on the subject of high energy explosions, he applied his deep and fundamental knowledge of this subject to the nuclear bomb problem. He assumed that the radius of the fireball depended only of three variables:

- E_0, the released energy of the nuclear bomb;
- ρ, the density of the air into which the fireball expanded;
- t, the time at which the radius of the fireball is $R(t)$.

Consequently, in dimensional form, he obtained the following result

$$R(t) = \lambda E_0^a \rho^b t^c, \tag{13.7.81}$$

where λ is a pure number. If we now substitute into this expression the proper dimensional units, we obtain

$$L = \left(\frac{ML^2}{T^2}\right)^a \left(\frac{M}{L^3}\right)^b T^c$$
$$= M^{(a+b)} L^{(2a-3b)} T^{(-2a+c)}, \tag{13.7.82}$$

which for dimensional consistency requires

$$\begin{cases} M : 0 = a + b, \\ L : 1 = 2a - 3b, \\ T : 0 = -2a + c, \end{cases} \tag{13.7.83}$$

and

$$a = \frac{1}{5}, \quad b = -\left(\frac{1}{5}\right), \quad c = \frac{2}{5}. \tag{13.7.84}$$

Therefore, $R(t)$ is

$$R(t) = \lambda \left(\frac{E_0}{\rho}\right)^{1/5} t^{2/5}. \tag{13.7.85}$$

At this point λ is undetermined. However, Taylor had evidence from experiments that if the variables (E_0, ρ, t) are expressed in the MKS system of physical units (meters, kilograms, seconds), then λ is very close to one. This means that Eq. (13.7.85) becomes

$$R(t) = \left(\frac{E_0}{\rho}\right)^{1/5} t^{2/5}. \tag{13.7.86}$$

While this gives the behavior of fireball radius as a function of time, t, what was really wanted was the value of E_0. If Eq. (13.7.86) is solved for E_0, the following result is obtained

$$E_0 = \frac{\rho R(t)^t}{t^2} = \text{constant} \tag{13.7.87}$$

and this relation can be used to provide an estimate for E_0, if data can be determined for the value $R(t_i)$ at the time $t = t_i$. If N such pairs of data are given, then $\bar{E}_0$, the estimated value for E_0, is

$$\bar{E}_0 = \left(\frac{1}{N}\right) \sum_{i=1}^{N} \frac{\rho R_i^5}{t_i^2}$$

$$= \left(\frac{\rho}{N}\right) \sum_{i=1}^{N} \frac{R_i^5}{t_i^2}, \tag{13.7.88}$$

where $R_i = R(t_i)$.

13.7.9   *Black Holes*

Black holes are a topic of immense interest for many individuals. In perhaps the most elementary interpretation, they are considered to be bodies so massive that light cannot escape from their surfaces. In this section, we investigate various properties of black holes (BH) using dimensional analysis and other related heuristic techniques. A number of very interesting features of these objects will be "discovered." Note that all of the derived results are suggestive of what might be correct, but rigorous calculations are needed to check the validity of any obtained results.

To begin, consider the gravitational interaction between two bodies having masses M and μ, where

$$u \lll M. \tag{13.7.89}$$

The iteraction energy is

$$V(r) = -\frac{GM\mu}{r}, \tag{13.7.90}$$

and this can be rewritten as

$$V(r) = -\frac{(GM/c^2)}{r}(\mu c^2)$$

$$= -\left(\frac{R_{BH}}{r}\right)(\mu c^2), \qquad (13.7.91)$$

where c is the speed of light and where R_{BH} is defined to be the black hole radius of the massive body, i.e.,

$$R_{BH} = \frac{GM_{BH}}{c^2}, \quad M_{BH} = M, \qquad (13.7.92)$$

and G is the usual gravitational constant.

Let's examine the result expressed by Eq. (13.7.91) in more detail:

(i) The factor μc^2 is the relativistic rest energy of the smaller mass, μ.

(ii) The factor (R_{BH}/r) is a scaling factor and is dimensionless. It represents the ratio of the BH radius to the distance of μ from the center of the more massive body $M = M_{BH}$.

(iii) Let R_0 be the physical radius of the massive body M. Then the weakness of the gravitational interaction is a direct consequence of

$$R_0 \ggg R_{BH}. \qquad (13.7.93)$$

It should be indicated that the standard (Schwarzschild) radius of a BH is

$$R_{BH}^{(S)} = \frac{2M_{BH}G}{c^2}. \qquad (13.7.94)$$

Since we are using dimensional arguments, the factor two difference between R_{BH} and $R_{BH}^{(S)}$ will have no essential consequences for the discussions presented below.

To illustrate the significance of the result given in Eq. (13.7.93), we have for the sun and earth, the following values:

$$\text{sun}: R_0 = (6.96)10^5 \text{ km}, \quad R_{BH} = 1.5 \text{ km}$$

$$\text{earth}: R_0 = (6.37)10^3 \text{ km}, \quad R_{BH} = (4.5)10^{-6} \text{ km}.$$

With the value of R_{BH}, given by Eq. (13.7.92), we will now estimate several properties of BH's.

Surface Area of BH.

Assuming a spherical shape, the area of a BH is

$$A_{BH} = 4\pi(R_{BH})^2$$

$$= \left(\frac{4\pi G^2}{c^4}\right)(M_{BH})^2. \qquad (13.7.95)$$

Volume of BH.

$$V_{BH} = \left(\frac{4}{3}\right) \pi (R_{BH})^3$$

$$= \left(\frac{4\pi G^3}{c^6}\right) (M_{BH})^3. \tag{13.7.96}$$

Density of BH.

The density is defined as mass/volume; therefore,

$$\rho_{BH} = \frac{M_{BH}}{\left(\frac{4\pi G^3}{3c^6}\right) (M_{BH})^3}$$

$$= \left(\frac{3c^6}{4\pi G^3}\right) \cdot \frac{1}{(MB_{BH})^2}. \tag{13.7.97}$$

A very interesting consequence of Eq. (13.7.97) is that the density of BH can be made arbitrarily small by a sufficient increase in its mass.

Temperature of BH.

It has been suggested by a number of scientists that BH's are not totally "black;" they can radiate as a thermodynamic black body with effective temperature T_{BH}. This process is often called Hawking radiation. Our task is to estimate T_{BH} using dimensional analysis, along with various insights from modern physics.

First, we must determine what are the relevant variables required to produce a valid dimensional analysis of the Hawking radiation phenomena. We select the following items along with a brief discussion of why they are chosen:

- C = speed of light
- G = gravitational constant
- M_{BH} = mass of BH
- K_B = Boltzmann's constant
- $\hbar = h/2\pi$, where h is Planck's constant
- T_{BH} = effective temperature of BH.

The first three items are expected to be present in any investigation of a BH. Temperature is not usually expressed in terms of the fundamental physical units (mass, length, time). However, K_B, is the natural conversion factor between temperature and energy, i.e., for a temperature T (in degree Kelvin), $K_B T$ has the physical units of an energy.

What about $\hbar$? It appears since classical BH's cannot radiate, i.e., their effective temperature is $T = O^\circ K$. Radiation effects are a consequence of quantum phenomena and, therefore, must involve Planck's constant.

To proceed, we assume that $K_B T_{BH}$ can be expressed as

$$K_B T_{BH} = \lambda C^a (\hbar)^b G^c (M_{BH})^d$$
$$= L^{(a+2b+3c)} M^{(b-c+d)} {}_j T^{(-a-b-2c)} \tag{13.7.98}$$

and from this we have three equations to determine the four unknowns (a, b, c, d); they are

$$\begin{cases} a + 2b + 3c = 2 \\ \quad\quad b + d = 1 + d \\ a + b + 2c = 2, \end{cases} \tag{13.7.99}$$

and, we obtain

$$a = 2 - c, \quad b = -c, \quad d = 1 + 2c. \tag{13.7.100}$$

Note that we have selected the undetermined variable to be c. Using these values in Eq. (13.7.98) gives

$$K_B T_{BH} = \lambda C^{(2-c)} \hbar^{(-c)} G^c M^{(1+2c)}. \tag{13.7.101}$$

What can be inferred about the value of c? First, as previously stated, $\hbar$ appears because a BH having a non-zero temperature implies that quantum effects are in play. Therefore, our formula must have the property

$$\lim_{\hbar \to 0} T_{BH} = 0, \tag{13.7.102}$$

and this implies that

$$c < 0. \tag{13.7.103}$$

Further, there is overwhelming evidence that massive BH's exist and no detectable radiation has been discovered for them. This strongly suggests that

$$1 + 2c < 0 \Rightarrow c < -\left(\frac{1}{2}\right). \tag{13.7.104}$$

For clarity, replace c by $(-f)$, where $f > 0$, and Eq. (13.7.101) becomes

$$K_B T_{BH} = \lambda C^{(2+f)} (\hbar)^f \Big/ G^f M_{BH}^{(2f-1)}. \tag{13.7.105}$$

Note that for $f > \frac{1}{2}$, this formula states that the temperature of a BH increases with a decrease in its mass!

A more rigorous calculation of T_{BH} gives the result

$$K_B T_{BH} = \frac{\hbar c^3}{8\pi G M_{BH}}. \tag{13.7.106}$$

In summary, the application of dimensional analysis, and the use of insights into the physical processes thought to be taking place in BH's, allows a characterization of many of the important properties of these objects.

13.7.10 *Résumé*

One of the best summary statements on dimensional analysis is that of Brenner (p. 34):

"Dimensional analysis is a powerful method to understand the relation between physical quantities in a given problem. The basic idea is, all scientifically interesting results are expressed in terms of dimensionless quantities, independent of the system units you are using. There are two consequences of this idea: first, one can often guess a reasonable form of an answer just by thinking of the dimensions in complex physical situations, and test the answer by experiments or more developed theories. Second, dimensional analysis is routinely used to check the plausibility of derived equations or computations."

13.8 Lambert-W Function

The elementary functions are the logarithm, exponential, power, trigonometric and hyperbolic functions. (Note that the latter two classes of functions are actually linear combinations of the exponential functions.) If an equation can have its solutions expressed in terms of the elementary functions, then the equation is (usually) said to have an exact analytical solution. However, this situation, in practice, is somewhat more complex. An example is that of the Jacobi sine and cosine functions. They clearly are (in a physical science sense, i.e., oscillations of systems) generalizations of the standard trigonometric functions; but, most individuals would not call them elementary functions.

In this section, we introduce and briefly discuss a "truly new addition" to the elementary functions. It is the Lambert-W function, $W(x)$. This function is a solution to the equation

$$x = W(x)e^{W(x)}. \tag{13.8.1}$$

A large number of articles discuss its mathematical properties, applications, and provide guidance to its implementary on various commercial softwares. Its status and recognition as an elementary function is now fixed by its listing as such in one of the important standard handbooks on special functions; see the reference list below, F. W. J. Olver et al., Section 4.13.

If $W(x)$ is required to be real, then for $x \geq 0$, there is only one real solution. However, on the x-interval $(-e^{-1}, 0)$ there are two real solutions. Fig. 13.8.1 is a plot of $W(x)$ over the interval $(-e^{-1}, \infty)$. Note that if W is required to satisfy the condition $W \geq -1$, then a single-valued function

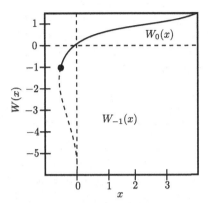

Fig. 13.8.1   The real branches, $W_0(x)$ and $W_1(x)$ of the Lambert-W function.

$W_0(x)$ is defined and for this function

$$W_0(0) = 0, \quad W_0(-1/e) = -1, \tag{13.8.2}$$

with $W_0(x)$ increasing for $x \geq (-1/e)$.

The other, lower branch has $W \leq -1$ and is designated $W_{-1}(x)$. It has the values

$$W_{-1}(-1/e) = -1, \quad W_{-1}(0) = -\infty, \tag{13.8.3}$$

and decreases in the x-interval, $(-1/e, 0)$.

The following list of references provide detailed information on the properties, applications, and possible extensions of the Lambert-W function:

(1) R. M. Corless, G. H. Gonnet, D. E. G. Hare, J. D. Jeffrey, and D. E. Knuth, "On the Lambert W function," *Advances in Computational Mathematics* **5**, 239 (1996).

(2) B. Hayes, "Why W?," *American Scientist* **93**, 104 (2005).

(3) I. Mezo and A. Baricz, "On the generalization of Lambert W function." arXiv:1408.3999v2 [math.CA] 22 Jun 2015.

(4) I. Mezo and G. Keady, "Some physical applications of generalized Lambert functions." arXiv:1505.01555v2 [math.CA] 22 Jun 2015.

(5) R. E. Mickens, *Difference Equations: Theory, Applications and Advanced Topics*, (CRC Press/Chapman and Hall, Boca Raton, 2015). See Section 8.5.

(6) F. W. J. Olver, D. W. Lozier, R. F. Boisert, and C. W. Clark, editors, *NIST Handbook of Mathematical Functions* (Cambridge University Press, Cambridge, 2010). See Section 4.13.

(7) S. M. Stewart, "A new elementary function for our curricula?," *Australian Senior Mathematics Journal* **19**, 8 (2005).

(8) S. R. Valluri, D. J. Jeffrey, and R. M. Corless, "Some applications of the Lambert W function to physics," *Canadian Journal of Physics* **78**, 823 (2000).

(9) D. Veberic, "Lambert W functions for applications in physics," *Computer Physics Communications* **183**, 2622 (2012).

The next section lists a number of important properties of the Lambert-W function. The following three sections then give two examples of the applications of these functions and a possible generalization. Again, for the reader who wishes more details of the related mathematical issues and who also wants to see additional applications, the references provide these items.

13.8.1 Important Properties

To begin, take the derivative of Eq. (13.8.1) and obtain

$$1 = [1 + W(x)]\frac{dW(x)}{dx}e^{W(x)}, \qquad (13.8.4)$$

which upon solving for $dW(x)/dx$ gives

$$\frac{dW(x)}{ds} = \frac{W(x)}{x[1 + W(x)]}, \quad x \neq -\left(\frac{1}{e}\right) \text{ and } 0. \qquad (13.8.5)$$

It follows from Eq. (13.8.4) that

$$\frac{dW(0)}{dx} = 1. \qquad (13.8.6)$$

Then Eqs. (13.8.5) and (13.8.6) may be taken as the defining differential equation for the Lambert-W function.

Now consider the first integral of $W(x)$. On making the transformation of variable

$$u = W(x), \quad x = ue^u, \qquad (13.8.7)$$

we obtain

$$\int W(x)dx = \int u\,dx = \int u[e^u + ue^u]du$$

$$= \int ue^u du + \int u^2 e^u du$$

$$= e^u(u - 1) + e^u(u^2 - 2u + 2) + c$$

$$= u(ue^u) - (ue^u) + e^u + c$$

$$= xW - x + \frac{x}{W} + c, \tag{13.8.8}$$

or, finally,

$$\int W(x)dx = x\left[W(x) - 1 + \frac{1}{W(x)}\right] + c. \tag{13.8.9}$$

The Lambert-W function has, at $x = 0$, Taylor series

$$W_0(x) = \sum_{k=1}^{\infty} \left[\frac{(-k)^{k-1}}{k!}\right] x^k. \tag{13.8.10}$$

This series has a radius of convergence $(1/e)$, a result following from the fact that $W(x)$ has a singularity at $x = (-1/e)$. In the general case where $r \in \mathbb{Z}$, then

$$[W_0(x)]^r = \sum_{k=r}^{\infty} \left[\frac{(-r)(-k)^{k-r-1}}{(k-r)!}\right] x^k, \tag{13.8.11}$$

and for $r \in \mathbb{C}$

$$[W_0(x)]^r = x^r e^{-rW_0(x)}$$

$$= \sum_{k=0}^{\infty} \left[\frac{r(n+r)^{k-1}}{k!}\right] (-x)^k. \tag{13.8.12}$$

13.8.2 $A^{ax+b} = cx + d$

For this expression, (A, a, b, c, d) are taken as constants, with the restrictions

$$A > 0, \quad a \neq 0, \quad c \neq 0. \tag{13.8.13}$$

The transformation

$$u = -ax - \left(\frac{ad}{c}\right), \tag{13.8.14}$$

gives

$$uA^u = R = -\left(\frac{a}{c}\right) A^{\left(b - \frac{ad}{c}\right)}, \tag{13.8.15}$$

and

$$u = \frac{W(R \operatorname{Ln} A)}{\operatorname{Ln} A}. \tag{13.8.16}$$

Therefore, the solution to

$$A^{ax+b} = cx + d, \tag{13.8.17}$$

is

$$x = (-)\frac{W\left[-\left(\frac{a\operatorname{Ln} A}{c}\right) A^{\left(b - \frac{ad}{c}\right)}\right]}{a \operatorname{Ln} A} - \left(\frac{d}{c}\right). \tag{13.8.18}$$

13.8.3   *Combustion Model*

An elementary model for combustion is

$$\frac{dx}{dt} = x^2(1 - x), \quad x(0) = \epsilon, \ 0 < \epsilon < 1, \tag{13.8.19}$$

where, in practice, $0 < \epsilon \ll 1$. Note that with the indicated initial condition, $x(t)$ increases monotonic with time from x_0 to $x(\infty) = 1$. This first-order differential equation is separable, i.e., it can be rewritten to the form

$$\int \frac{dx}{x^2(1 - x)} = \int dt, \tag{13.8.20}$$

and the left-side can be solved by the method of partial fractions; doing this gives

$$\mathrm{Ln} \left| \frac{x}{1 - x} \right| - \frac{1}{x} = t + c, \tag{13.8.21}$$

and this expression can be solved in terms of the Lambert-W function, i.e.,

$$x(t) = \frac{1}{W_0[e^{(-t-1-c)}] + 1}. \tag{13.8.22}$$

Using $x(0) = \epsilon$, and defining

$$u = \frac{1}{\epsilon - 1}, \tag{13.8.23}$$

$x(t)$ becomes

$$x(t) = \frac{1}{W_0(ue^{u-t}) + 1}. \tag{13.8.24}$$

See Fig. 13.8.2.

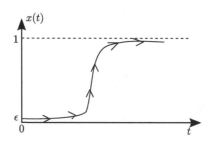

Fig. 13.8.2   Plot of the solution to the combusion equation

13.8.4   *Omega Constant*

There is a special number associated with the Lambert-W function; it is named the omega constant, Ω, and corresponds to the value of $W_0(1)$, i.e.,

$$e^{-\Omega} = \Omega. \qquad (13.8.25)$$

An accurate numerical calculation gives

$$\Omega = 0.567143290\ldots. \qquad (13.8.26)$$

The omega constant, Ω, can be associated with the Lambert-W function in the same sense that e has a connection to the exponential and logarithmic functions, and π with the trigonometric functions.

13.8.5   *Delay Differential Equation*

Consider the following linear, first-order differential-delay equation

$$\frac{dx(t)}{dt} = Ax(t - \tau), \qquad (13.8.27)$$

where A and τ are constants. If we assume a solution of the form

$$x(t) = e^{\lambda t}, \qquad (13.8.28)$$

then substitution in the differential-delay equation gives

$$\lambda = Ae^{-\lambda \tau} \qquad (13.8.29)$$

and

$$\lambda = \frac{W(\tau A)}{\tau}, \qquad (13.8.30)$$

and, finally,

$$x(t) = B \exp\left\{ \left[\frac{W(\tau A)}{\tau} \right] t \right\} \qquad (13.8.31)$$

where B is an arbitrary constant.

13.8.6   *Generalizations of Lambert-W Equations*

The standard Lambert-W function provides a solution to a vast number of equations arising in essentially all areas of the natural, engineering, and applied mathematical sciences (see, for example, the articles by Mezo and collaborators). Work on generalized W functions was begun in 2006 by Scott and his collaborators:

- T. C. Scott, R. Mann, and R. E. Martinez, "General relativity and quantum mechanics: Towards a generalization of the Lambert W function," *Applicable Algebra in Engineering Communications and Computing* **17**, 41 (2006).
- T. C. Scott, G. Fee, J. Grotendorst, and W. Z. Zhang, "Numerics of the generalized Lambert W function," *SIGSAM* **48**, 42 (2014).

The generalized W functions are defined as solutions to the equation

$$e^x \left[\frac{(x - t_1)(x - t_2) \cdots (x - t_n)}{(x - s_1)(x - s_2) \cdots (x - s_m)} \right] = a, \qquad (13.8.32)$$

where (n, m) are non-negative integers, and the (t_n, s_m) are constants. The corresponding solutions are denoted by the symbol

$$x = W \begin{pmatrix} t_1, t_2, \ldots, t_n \\ s_1, s_2, \ldots, s_m \end{pmatrix}; a \end{pmatrix} \qquad (13.8.33)$$

with this definition, it follows that

$$\begin{cases} W(\ ; a) = \mathrm{Ln}(a), \quad W(^0; a) = W(a), \\ W(_0; a) = -W\left(-\frac{1}{a}\right), \end{cases} \qquad (13.8.34)$$

and

$$\begin{cases} W(^t; a) = t + W(ae^{-t}), \\ W(_s; a) = s - W\left(-\frac{e^s}{a}\right). \end{cases} \qquad (13.8.35)$$

Th preprint article of Mezo and Keady gives a good listing of applications of the generalized Lambert functions to a broad range of issues in the physical sciences.

13.9   Analysis of Solutions to $y'' = y(y - zy')$

The investigation of the thermal explosion of a fluid from a long, thin heated tube is based on the following nonlinear diffusion partial differential equation

$$uu_t = u_{xx}; \quad u = u(x, t), \quad x > 0, \quad t > 0, \qquad (13.9.1)$$

$$u(x, 0) = 0, \quad x > 0, \qquad (13.9.2a)$$

$$u(\infty, t) = 0 \quad t > 0, \qquad (13.9.2b)$$

$$u_x(0, t) = -1, \quad t > 0. \qquad (13.9.2c)$$

See Dresner below in the reference listing.

The goal of this section is to use similarity techniques to derive on associated second-order, ordinary differential equation

$$y''(z) = y(z)[y(z) - zy'(z)], \qquad (13.9.3)$$

having the specified conditions

$$y'(0) = -b, \quad b > 0; \qquad (13.9.4a)$$

$$y(\infty) = 0. \qquad (13.9.4b)$$

All of the following results, after the first section, are based on unpublished work by Wilkins and Mickens.

The following references are relevant for the issues discussed in this section.

(1) R. P. Boas, "Signs of derivatives and analytic behavior," *American Mathematical Monthly* **78**, 1085 (1971).

(2) L. Dresner, *Similarity Solutions of Nonlinear Partial Differential Equations* (Pitman, Boston, 1983). See Section 4.7.

(3) J. D. Logan, *Nonlinear Partial Differential Equations* (Wiley-Interscience, New York, 1994). See Section 4.2 Similarity Methods, and Section 4.3 Nonlinear Diffusion Models, Example 5.

(4) J. Ernest Wilkins, Jr. and R. E. Mickens, "Properties of the solutions to $y'' = y(y - zy')$," (Clark Atlanta University; Atlanta, GA; January 21, 1997). Unpublished.

13.9.1 *Similarity Solution for* $uu_t = u_{xx}$

A good, concise introduction to similarity methods applied to nonlinear partial differential equation is given in the book by Logan. The basic idea is to transform a partial differential equation to an ordinary differential equation and, in general, the latter equation is much easier to analyze and/or solve.

The partial differential equation, Eq. (13.9.1), is invariant under the change of variables

$$\bar{x} = \epsilon^a x, \quad \bar{t} = \epsilon^b t, \quad \bar{u} = \epsilon^{b-2a} u, \qquad (13.9.5)$$

and this gives the result that $u(x,t)$ can be expressed in the form

$$u(x,t) = t^{1-\frac{2a}{b}} y(z), \quad z = \frac{x}{t^{a/b}}. \qquad (13.9.6)$$

In these expressions (a, b) are constants, to be determined from the initial and/or boundary conditions, and ϵ is a real parameter restricted to some open interval containing $\epsilon = 1$.

The (a, b) can be found in the following manner. First, calculate u_x and obtain

$$u_x(x, t) = t^{1-\frac{3a}{b}} y' \left(\frac{x}{t^{a/b}} \right).$$ (13.9.7)

The condition given by Eq. (13.9.2c) requires that

$$u_x(0, t) = t^{1-\frac{3a}{b}} y'(0) = 1.$$ (13.9.8)

Since the right-side does not depend on t, it follows that

$$\frac{a}{b} = \frac{1}{3}$$ (13.9.9)

and

$$u(x, t) = t^{1/3} y(z) = t^{1/3} y \left(\frac{x}{t^{1/3}} \right).$$ (13.9.10)

Following the notation of Dresner and Logan, and Wilkins and Mickens, we will modify the definition of z and use

$$z = \frac{z}{\sqrt{3}t^{1/3}} \Rightarrow u(x, t) = t^{1/3} f(z) = t^{1/3} y \left(\frac{x}{\sqrt{3}t^{1/3}} \right).$$ (13.9.11)

Note that a factor of $\sqrt{3}$ has been added to the denominator of the "old z." This means that with the "new z," the condition for $y'(0)$ is

$$y'(0) = -\sqrt{3}.$$ (13.9.12)

Continuing, we see that the condition in Eq. (13.9.2b) holds if

$$t^{1/3} f(\infty) = 0 \Rightarrow f(\infty) = 0.$$ (13.9.13)

Further,

$$\operatorname*{Lim}_{\substack{t \to 0 \\ x \text{ fixed}}} u(x, t) = \operatorname*{Lim}_{z \to \infty} \left(\frac{x}{\sqrt{3}z} \right) y(z)$$

$$= 0$$ (13.9.14)

since $y(\infty) = 0$. Consequently, the condition of Eq. (13.9.2a) holds.

If $u(x, t)$ from Eq. (13.9.11) is substituted into Eq. (13.9.1), then after simplification, we obtain the boundary-value problem

$$y'' = y(y - zy'), \quad z > 0,$$ (13.9.15a)

$$y' = -\sqrt{3}, \quad y(\infty) = 0.$$ (13.9.15b)

Our major task is to obtain a value for $y(0)$ and to do this in light of the fact that Eqs. (13.9.15) do not have an exact solution expressible in terms of a finite combination of the elementary functions.

The work to follow illustrates certain of the interesting relations involving solutions of Eqs. (13.9.15).

13.9.2 A Local Property and a Sum Rule

From the differential equation, it follows that at $z = 0$, we have

$$y''(0) = y(0)^2, \tag{13.9.16}$$

a *local relationship* which holds for all solutions.

We now derive a global relationship for the bounded solutions. To start, observe that

$$zyy' = \left(\frac{1}{2}\right)\frac{d}{dz}(zy^2) - \left(\frac{1}{2}\right)y^2. \tag{13.9.17}$$

Therefore,

$$
\begin{aligned}
y'' - y^2 + zyy' &= \left(\frac{d}{dz}\right)y' - y^2 - \left[\left(\frac{1}{2}\right)y^2 + \left(\frac{1}{2}\right)\left(\frac{d}{dz}\right)(zy^2)\right] \\
&= \frac{d}{dz}\left[y' + \left(\frac{1}{2}\right)\left(\frac{d}{dz}\right)(zy^2)\right] - \left(\frac{3}{2}\right)y^2 \\
&= 0.
\end{aligned}
\tag{13.9.18}
$$

Integrating each term, from $z = 0$ to $z = \infty$, gives

$$\int_0^\infty dy' - \left(\frac{3}{2}\right)\int_0^\infty y^2 dz + \left(\frac{1}{2}\right)\int_0^\infty d(zy^2) = 0 \tag{13.9.19}$$

or

$$[y'(\infty) - y'(0)] - \left(\frac{3}{2}\right)\int_0^\infty y^2 dz + \left(\frac{1}{2}\right)(zy^2)\Big|_0^\infty = 0. \tag{13.9.20}$$

The third term is zero at both boundary-values and $y'(\infty) = 0$. Therefore, the following sum-rule is obtained

$$-y'(0) = \left(\frac{3}{2}\right)\int_0^\infty y^2(z)dz, \tag{13.9.21}$$

and using $y'(0) = -\sqrt{3}$, gives

$$\frac{2}{\sqrt{3}} = \int_0^\infty y^2(z)dz. \tag{13.9.22}$$

The above relation is called *sum-rule* because it is a "sum" (integral) of $y^2(z)$ over the interval, $0 \leq z < \infty$, and thus is a global feature of the solution $y(z)$.

We will make use of this sum-rule, in Section 13.9.4, to determine estimates of $y(0)$.

13.9.3 *General Mathematical Properties of $y(z)$*

The results stated below have been proved (see Wilkins and Mickens) under the following assumptions:

1) $y(z)$ is a solution to the differential equation

$$y''(z) = y(z)[y(z) - zy'(z)], \quad 0 < z < \infty. \tag{13.9.23}$$

2) There exists a fixed positive number a, such that

$$y'(0) = -a. \tag{13.9.24}$$

3) $y(\infty)$ is defined at $z = \infty$ and has the value

$$y(\infty) = 0. \tag{13.9.25}$$

Result-1.

$$y'(z) < 0, \quad 0 \le z < +\infty, \tag{13.9.26}$$

$$y'(\infty) = 0. \tag{13.9.27}$$

Result-2.

$$y(z) > 0, \quad y''(z) > 0, \quad y'''(z) < 0, \quad y^{(iv)}(z) > 0. \tag{13.9.28}$$

Result-3.

$$y(z) < \frac{4}{z^2 + \frac{4}{y(0)}}. \tag{13.9.29}$$

Result-4.

$$\frac{2a}{3} = \int_0^\infty y^2(z)dz. \tag{13.9.30}$$

This is just the result given in Eq. (13.9.22) with $a = \sqrt{3}$.

Result-5.

$$y^3(0) < 2a^2, \tag{13.9.31a}$$

which is the same as

$$y^3(0) < 2[y'(0)]^2. \tag{13.9.31b}$$

Result-6.

$$a^{2/3} < y(0) < (1.084614\ldots)a^{2/3}. \tag{13.9.32}$$

Note that for $a = \sqrt{3}$, the following bounds are determined for $y(0)$,

$$1.442249\cdots < y(0) < 1.564285\ldots, \tag{13.9.33}$$

and the estimate

$$y(0) = 1.503267, \tag{13.9.34}$$

is in error by at most 4.2%.

In summary, we have obtained very tight bounds on the value of $y(0)$. Thus, if a numerical method is used to determine $y(0)$, such as a shooting method, then the range of initial guesses for $y(0)$ is quite limited in range. Logan used a shooting method and obtained the value $y(0) = 1.5111\ldots$.

13.9.4 *Analytical Approximation to* $y(0)$

The methodology used in the previous section was based on the use of rigorous mathematical techniques. We now consider a method which relies on assuming a particular explicit functional form for $y(z)$ and then using various exact mathematical constraints to calculate approximations to $y(0)$. The basic procedure is as follows:

(a) Select an explicit functional form for $y(z)$ containing a number of (for the moment) unknown parameters.

(b) Select several mathematical relations which are satisfied by the exact functional form for $f(z)$. The number of such relations should equal the number of unknown parameters.

(c) Substitute the assumed (approximate) solution into the constraint equations and solve for the unknown parameters.

We illustrate this procedure by means of the following elementary functional form for an approximate solution $\bar{y}(z)$ to $y(z)$:

$$\bar{y}(z) = Ae^{-bz}, \tag{13.9.35}$$

where A and b are the unknown parameters.

Take the derivative of $\bar{y}(z)$ to obtain

$$\bar{y}(z)' = -bAe^{-bz}, \tag{13.9.36}$$

and note that

$$\bar{y}(0) = A, \quad -\sqrt{3} = -bA, \tag{13.9.37}$$

and this means that

$$\bar{y}(z) = \bar{y}(0) \exp \left\{ - \left[\frac{\sqrt{3}}{\bar{y}(0)} \right] z \right\}. \tag{13.9.38}$$

Using the result in Eq. (13.9.16), we obtain

$$\frac{3}{\bar{y}(0)} = [\bar{y}(0)]^2, \tag{13.9.39}$$

and

$$[\bar{y}(0)]^3 = 3 \Rightarrow \bar{y}(0) = 31/3 = 1.4422, \tag{13.9.40}$$

a result which coincides with the lower bound in Eq. (13.9.33).

Let us now use a different constraint, namely, the one given by Eq. (13.9.22). Therefore, substitution of Eq. (13.9.35) into this relation gives

$$\frac{2}{\sqrt{3}} = \int_0^\infty A^2 e^{-2bz} dz = \frac{A^2}{2b} \tag{13.9.41}$$

with

$$\bar{y}'(0)' = -bA = -\sqrt{3}. \tag{13.9.42}$$

Therefore, using the fact that $A = \bar{y}(0)$, we obtain

$$b = \frac{\sqrt{3}}{\bar{y}(0)}, \quad \bar{y}(0)^3 = 4 \Rightarrow \bar{y}(0) = 4^{1/3} = 1.5874. \tag{13.9.43}$$

Note that this estimate for $y(0)$ is slightly more than the upper bound given in Eq. (13.9.33).

In summary, we have demonstrated that the use of even an elementary analytical functional form to approximate $y(z)$ provides a very reasonable result for $y(0)$. It should be expected that more complex functions for $\bar{y}(z)$ will produce more accurate estimates for both $y(z)$ and $y(0)$.

13.9.5   *Discussion*

The above work illustrates the power of the similarity methodology for deriving information on at least certain types of nonlinear partial differential equations. In particular, since a number of important physical and engineering systems can be modeled by one-space dimension evolution equations, these techniques need to be in the set of analytical tools of most scientists.

Two additional examples of equations on which extensive study have been done are the *Boltzmann problem*

$$\begin{cases} c_t = (cc_x)_x & c = c(t, x), \\ \int_{-\infty}^{\infty} c(t, x)dx = 1, \\ c(x, 0) = 0, & x \neq 0, \\ c(\infty, t) = 0, & t > 0; \end{cases} \qquad (13.9.44)$$

which models the problem of an instantaneous heat pulse in a plane at $x = 0$ at time $t = 0$ (see Dresner, pp. 25–27); and the problem of *transient heat transfer in superfluid helium* (see Dresner, pp. 41–45).

$$\begin{cases} T_t = \left[(T_x)^{1/3}\right]_x, & T = T(x, t), \\ T_x(0, t) = -1, & t > 0, \\ T(x, 0) = 0, & x > 0, \\ T(\infty, t) = 0, & t > 0. \end{cases} \qquad (13.9.45)$$

13.10 Method of Dominant Balance

Few equations can be exactly solved in terms of a finite sum of the elementary functions. A consequence of this fact is that, in general, only approximations to the actual solutions are available. While there are many techniques for the determination of approximate solutions, one procedure is often used for this purpose, namely, the method of dominant balance. Briefly, this procedure takes the following form (see Brenner, p. 18, below):

i) Assume that you have an equation consisting of several terms; for example,

$$A + B + C + D + E = 0. \qquad (13.10.1)$$

ii) Assume that two of the terms are "larger" (in some sense defined by the problem you are trying to solve/answer) than all the other terms. For the above, take these to be B and D.

iii) Now approximate the original equation by the reduced equation

$$B + D = 0, \qquad (13.10.2)$$

and solve it.

iv) Finally, verify that the reduced equation is *consistent* by substituting its solution into the original equation and checking that the neglected terms (A, C, E) are smaller than the kept two terms. If

(A, C, E) are indeed small compared with (B, D), then the obtained solution to Eq. (13.10.2) is consistent. However, for the case where the solution is *inconsistent*, then another method must be found or another set of terms must be tried.

Comments

There really are no uniformly agreed set of strict rules that characterizes the method of dominant balance. Consequently, for a particular equation, more than one type of dominant balancing procedure may be applicable. This means that the determination of a valid approximate solution is greatly influenced by "artistic" insight and mathematical knowledge.

Note that the method of dominant balance is always associated with obtaining solutions for some asymptotic limit. Thus this method is closely related to both perturbation procedures and asymptotic analysis. Within this framework, the previously stated steps for applying dominant balance can be expressed as follows:

(a) In the original equation, drop all terms thought to be small and replace the exact equation with an asymptotic relation.

(b) Next, replace this asymptotic relation with an equation by exchanging the asymptotic symbol $(\sim)$, with an equal, $(=)$, symbol.

(c) Solve this equation exactly.

(d) Finally, verify that the obtained solution is consistent with the approximations made in (a).

In the sections to follow, we will apply the method of dominant balance to four problems: an algebraic equation, a differential equation, an issue related to finite-time dynamics, and the determination of the front-behavior of the traveling wave solution of a nonlinear reaction-diffusion partial differential equation. The references listed below provide several views on the dominant balance methodology and also illustrate the broad applicability of the general technique:

(1) C. M. Bender and S. A. Orszag, *Advanced Mathematical Methods for Scientists and Engineers* (McGraw-Hill, New York, 1978). See, in particular, pp. 83–84.

(2) M. P. Brenner, *Physical Mathematics* (School of Engineering and Applied Science, Harvard University, Cambridge, MA, 2010). See Section 2.3 "Dominant Balance and Approximate Solutions."

(3) A. W. Bush, *Perturbations for Engineers and Scientists* (CRC Press, Boca Raton, FL, 1992). See Chapter 6, "The Dominant Balance and WKB Methods."

(4) R. Grimshaw, *Nonlinear Ordinary Differential Equations* (CRC Press, Boca Raton, FL, 1990).

(5) R. E. Mickens, *Oscillations in Planar Dynamic Systems* (World Scientific, Singapore, 1996).

(6) J. D. Murray, *Asymptotic Analysis* (Springer-Verlag, New York, 1984).

(7) A. H. Nayfeh, *Perturbation Methods* (Wiley, New York, 1973).

(8) W. Paulsen, *Asymptotic Analysis and Perturbation Theory* (CRC Press, Boca Raton, FL, 2014). See Section 1.5.

(9) J. G. Simmonds and J. E. Mann, Jr., *A First Look at Perturbation Theory*, 2nd Edition (Dover, Mineola, NY, 1998). See Chapter 2, "Roots of Polynomials."

13.10.1 *Algebraic Equations:* $x^2 - 2x + \epsilon = 0$ *and* $\epsilon x^2 - 2x + 1 = 0$

Consider the quadratic equation

$$x^2 - 2x + \epsilon = 0, \quad |\epsilon| \ll 1, \quad (13.10.3)$$

where ϵ is a small parameter, and let us pretend that we don't know how to obtain the two exact solutions. Note that the two solutions, $x^{(1)}(\epsilon)$ and $x^{(2)}(\epsilon)$, depend on ϵ. For $\epsilon = 0$, it follows that

$$x^2 - 2x = 0 \Rightarrow x^{(1)}(0) = 0, \quad x^{(2)}(0) = 2. \quad (13.10.4)$$

The simplest assumption to make is that for small ϵ, the two solutions have the representations

$$x^{(1)}(\epsilon) = 0 + a_1\epsilon + a_2\epsilon^2 + O(\epsilon^3), \quad (13.10.5)$$

$$x^{(2)}(\epsilon) = 2 + b_1\epsilon + b_2\epsilon^2 + O(\epsilon^3), \quad (13.10.6)$$

where only terms up to order ϵ^2 have been retained.

Now

$$(x^{(1)})^2 - 2x^{(1)} + \epsilon = (1 - 2a_1)\epsilon + (a_1^2 - 2a_2)\epsilon^2 + O(\epsilon^3)$$
$$= 0, \quad (13.10.7)$$

$$(x^{(2)})^2 - 2x^{(2)} + \epsilon = (2b_1 + 1)\epsilon + (b_1^2 + 2b_2)\epsilon^2 + O(\epsilon^3)$$
$$= 0, \quad (13.10.8)$$

which upon solving for (a_1, a_2) and (b_1, b_2) gives

$$a_1 = \frac{1}{2}, \quad a_2 = \frac{1}{8}; \quad b_1 = -\left(\frac{1}{2}\right), \quad b_2 = -\left(\frac{1}{8}\right), \tag{13.10.9}$$

and, consequently,

$$x^{(1)}(\epsilon) = \left(\frac{1}{2}\right)\epsilon + \left(\frac{1}{8}\right)\epsilon^2 + O(\epsilon^3), \tag{13.10.10}$$

$$x^{(2)}(\epsilon) = 2 - \left(\frac{1}{2}\right)\epsilon - \left(\frac{1}{8}\right)\epsilon^2 + O(\epsilon^3). \tag{13.10.11}$$

For purposes of comparison, the two exact solutions to Eq. (13.10.3) are

$$x^{(1)}(\epsilon) = 1 - \sqrt{1 - \epsilon}, \quad x^{(2)}(\epsilon) = 1 + \sqrt{1 - \epsilon}. \tag{13.10.12}$$

For $|\epsilon| \ll 1$, we have

$$\sqrt{1 - \epsilon} = 1 - \left(\frac{1}{2}\right)\epsilon - \left(\frac{1}{8}\right)\epsilon^2 + O(\epsilon^3). \tag{13.10.13}$$

(Note that the full series converges for $|\epsilon| < 1$.) Therefore, the substitution of Eq. (13.10.13) into the results presented in Eq. (13.10.12) give exactly the results in Eqs. (13.10.10) and (13.10.11). For this problem, the dominant balance solution is just the expansion of the solutions in a series of powers of ϵ and then determining the coefficients of the individual power of ϵ. It is clear that this procedure can be carried out to order ϵ^n, where n is any given positive integer.

Finally, it should be indicated that this problem is the one corresponding to a regular dominant balance procedure, i.e., the original equation, for $\epsilon \neq 0$, has two solutions and the result equation for $\epsilon = 0$, also has two solutions. The next equation does not have this property.

The quadratic equation

$$\epsilon x^2 - 2x + 1 = 0, \quad |\epsilon| \ll 1, \tag{13.10.14}$$

for $\epsilon \neq 0$, has two distinct solutions, $x^{(3)}(\epsilon)$ and $x^{(4)}(\epsilon)$, but at $\epsilon = 0$, it becomes

$$-2x + 1 = 0, \tag{13.10.15}$$

and one solution seemingly disappears. What's going on here?

Let's attempt to calculate a regular dominant balance series expansion in ϵ solution, i.e., assume $x^{(3)}(\epsilon)$ has the representation

$$x^{(3)}(\epsilon) = \frac{1}{2} + c_1\epsilon + c_2\epsilon^2 + O(\epsilon^3). \tag{13.10.16}$$

The substitution of this expression into Eq. (13.10.14) gives

$$\epsilon(x^{(3)})^2 - 2x^{(3)} + 1 = \left(\frac{1}{4} - 2c_1\right)\epsilon + (c_1 - 2c_2)\epsilon^2 + O(\epsilon^3)$$
$$= 0, \qquad (13.10.17)$$

which gives on solving for (c_1, c_2), the values

$$c_1 = \frac{1}{8}, \quad c_2 = \frac{1}{16}, \qquad (13.10.18)$$

and thus the following result for $x^{(3)}(\epsilon)$

$$x^{(3)}(\epsilon) = \frac{1}{2} + \left(\frac{1}{8}\right)\epsilon + \left(\frac{1}{16}\right)\epsilon^2 + O(\epsilon^3). \qquad (13.10.19)$$

So, what about $x^{(4)}(\epsilon)$? One way to proceed is to use the fact that if a quadratic equation has the roots, u and v, then it can be written as

$$(x - u)(x - v) = 0 \qquad (13.10.20)$$

or

$$x^2 - (u + v)x + uv = 0. \qquad (13.10.21)$$

Inspection of the latter expression shows that the "constant term" is always the product of the two roots, uv. Therfore, for $\epsilon \neq 0$, Eq. (13.10.14) becomes

$$x^2 - \left(\frac{2}{\epsilon}\right)x + \frac{1}{\epsilon} = 0, \qquad (13.10.22)$$

and we can conclude that

$$x^3(\epsilon)x^{(4)}(\epsilon) = \frac{1}{\epsilon}. \qquad (13.10.23)$$

Therefore,

$$x^{(4)}(\epsilon) = \frac{1}{\epsilon x^{(3)}(\epsilon)}$$
$$= \frac{1}{\epsilon\left[\frac{1}{2} + \left(\frac{1}{8}\right)\epsilon + \left(\frac{1}{16}\right)\epsilon^2 + O(\epsilon^3)\right]}$$
$$= \left(\frac{2}{\epsilon}\right)\frac{1}{1 + \left(\frac{1}{4}\right)\epsilon + \left(\frac{1}{8}\right)\epsilon^2 + O(\epsilon^3)}$$
$$= \left(\frac{2}{\epsilon}\right)\left[1 - \left(\frac{1}{4}\right)\epsilon + O(\epsilon^2)\right]$$
$$= \frac{2}{\epsilon} - \frac{1}{2} + O(\epsilon). \qquad (13.10.24)$$

Let us now compare these forms for $x^{(3)}(\epsilon)$ and $x^{(4)}(\epsilon)$ with the exact solution for $0 < |\epsilon| \ll 1$. We have, for Eq. (13.10.14), the results

$$x^{(3)}(\epsilon) = \left(\frac{1}{\epsilon}\right)\left[1 - \sqrt{1 - \epsilon}\right]$$

$$= \frac{1}{2} + \left(\frac{1}{8}\right)\epsilon + O(\epsilon^2), \tag{13.10.25}$$

$$x^{(4)}(\epsilon) = \left(\frac{1}{\epsilon}\right)\left[1 + \sqrt{1 - \epsilon}\right]$$

$$= \frac{2}{\epsilon} - \frac{1}{2} + O(\epsilon), \tag{13.10.26}$$

and these, expansions in ϵ representations, are in exact agreement with Eqs. (13.10.19) and (13.10.24).

These calculations indicate that when applying dominant balance to singular problems, one must be careful with how the procedure is actually carried out. In particular, additional knowledge of the equation (in this case, the fact that for a normalized quadratic equation, the constant term is equal to the product of the two roots) will be required to determine a valid singular solution.

13.10.2 *Differential Equation:* $y'' - (y/x) = 0$

We now demonstrate the construction of an approximation to the solution of the differential equation

$$\frac{d^2y}{dx^2} - \left(\frac{1}{x}\right)y = 0, \tag{13.10.27}$$

as x becomes large and positive, and we will abbreviate this condition as $x \to \infty$.

Assume $y(x)$ has the representation

$$y(x) = e^{\phi(x)}, \tag{13.10.28}$$

therefore,

$$y'(x) = \phi'(x)e^{\phi(x)}, \quad y''(x) = \phi''(x)e^{\phi(x)} + (\phi'(x))^2 e^{\phi(x)}. \tag{13.10.29}$$

Substituting these results into Eq. (13.10.27) gives

$$\phi'' + (\phi')^2 = \frac{1}{x}. \tag{13.10.30}$$

There are two cases for the dominant balancing. The first is

$$\phi'' \sim \frac{1}{x}, \tag{13.10.31}$$

and integrating once gives

$$\phi' \sim \text{Ln}(x). \qquad (13.10.32)$$

However, this cannot produce the needed balancing since consistency requires

$$(\phi')^2 = [\text{Ln}(x)]^2 \ll \phi'' = \frac{1}{x}, \qquad (13.10.33)$$

and this does not hold.

The second balancing relation is

$$(\phi')^2 \sim \frac{1}{x}, \qquad (13.10.34)$$

and solving for $\phi(x)$ gives

$$\phi' \sim (\pm)\frac{1}{\sqrt{x}}, \quad \phi \sim (\pm)2\sqrt{x}, \qquad (13.10.35)$$

with

$$\phi'' \sim -\left(\frac{1}{2}\right)\left(\frac{1}{x^{3/2}}\right). \qquad (13.10.36)$$

Note that for this case, for large x

$$|\phi''| = \left(\frac{1}{2}\right)\left(\frac{1}{x^{3/2}}\right) \ll (\phi')^2 = \frac{1}{x}. \qquad (13.10.37)$$

Therefore, we conclude

$$y(x) \sim e^{\pm 2\sqrt{x}+\eta(x)}, \qquad (13.10.38)$$

where

$$\eta(x) = o(\sqrt{x}). \qquad (13.10.39)$$

To determine $\eta(x)$, we start by calculating $y'(x)$ and $y''(x)$ from the expression given in Eq. (13.10.38); doing this gives

$$y'(x) \sim \left[(\pm)\left(\frac{1}{\sqrt{x}}\right) + \eta'\right]e^{(\cdots)}, \qquad (13.10.40a)$$

$$y''(x) \sim \left[(\pm)\left(\frac{1}{\sqrt{x}}\right) + \eta'\right]^2 e^{(\cdots)} + \left[(\mp)\left(\frac{1}{2}\right)\left(\frac{1}{x^{3/2}}\right) + \eta''\right]e^{(\cdots)}, \qquad (13.10.40b)$$

where $(\cdots) = \pm 2\sqrt{x} + \eta(x)$. Substituting these results into the differential equation and simplifying, produces the relation

$$\eta'' + (\eta')^2 \pm \left(\frac{2}{\sqrt{x}}\right)\eta' \sim (\pm)\left(\frac{1}{2x^{3/2}}\right). \qquad (13.10.41)$$

There are now three situations, i.e., dominant balancing cases, to consider.

Case 1

$$\eta'' = (\pm) \left(\frac{1}{2x^{3/2}} \right), \tag{13.10.42}$$

which gives upon integration

$$\begin{cases} \eta' = O\left(\frac{1}{\sqrt{x}}\right), & (\eta')^2 = O\left(\frac{1}{x}\right), \\ (\pm)\left(\frac{2}{\sqrt{x}}\right)\eta' = O\left(\frac{1}{x}\right). \end{cases} \tag{13.10.43}$$

If this is a valid dominant balancing procedure, then we should have

$$\left. \begin{array}{l} (\eta')^2 = O\left(\frac{1}{x}\right) \\ \left(\frac{2}{\sqrt{x}}\right)\eta' = O\left(\frac{1}{x}\right) \end{array} \right\} \ll |\eta''| = O\left(\frac{1}{x^{3/2}}\right); \tag{13.10.44}$$

however, inspection of the left-side items show them to be larger than the right-side, and we must conclude that the dominant balance in Eq. (13.10.42) is not correct.

Case 2

$$(\eta')^2 \sim (\pm)\left(\frac{1}{2x^{3/2}}\right) \Rightarrow \eta' \sim O\left(\frac{1}{x^{3/4}}\right) \tag{13.10.45}$$

and

$$\eta'' = O\left(\frac{1}{x^{7/4}}\right). \tag{13.10.46}$$

For this to be a valid dominant balancing relation, we must have

$$\left. \begin{array}{l} |\eta''| \\ \left(\frac{2}{\sqrt{x}}\right)|\eta'| \end{array} \right\} \ll (\eta')^2 = O\left(\frac{1}{x^{3/2}}\right). \tag{13.10.47}$$

However, for large x,

$$|\eta''| \ll (\eta')^2, \tag{13.10.48}$$

but

$$\left(\frac{2}{\sqrt{x}}\right)|\eta'| \gg (\eta')^2, \tag{13.10.49}$$

and this latter failure informs us that the dominant balancing of Eq. (13.10.45) is not correct.

Case 3

The last case is

$$\left(\frac{2}{\sqrt{x}}\right)\eta' \sim \frac{1}{2x^{3/2}}, \tag{13.10.50}$$

and for this, we have

$$\eta' = O\left(\frac{1}{x}\right), \quad \eta'' = O\left(\frac{1}{x^2}\right). \tag{13.10.51}$$

Therefore

$$\left.\begin{array}{l} \eta' = O\left(\frac{1}{x^2}\right) \\ (\eta')^2 = O\left(\frac{1}{x^2}\right) \end{array}\right\} \ll \left(\frac{2}{\sqrt{x}}\right)|\eta'| = O\left(\frac{1}{x^{3/2}}\right), \tag{13.10.52}$$

and we have a valid dominant balance representation. The function, $\eta(x)$, is given by the solution to the following first-order differential equation

$$\frac{d\eta}{dx} = \frac{1}{4x}. \tag{13.10.53}$$

Therefore,

$$\eta = \left(\frac{1}{4}\right)\text{Ln}(x) = \text{Ln}(x^{1/4}), \tag{13.10.54}$$

and $y(x)$ now takes the form

$$y(x) \sim \exp\left\{\pm 2\sqrt{x} + \text{Ln}(x^{1/4}) + \psi(x)\right\}, \tag{13.10.55}$$

where

$$\psi(x) = o[\text{Ln}(x)]. \tag{13.10.56}$$

Repeating the calculations, as done above for $\phi(x)$ and $\eta(x)$, we find that $\psi(x)$ satisfies the equation

$$\psi'' + \left(\frac{1}{2x} \pm \frac{2}{\sqrt{x}}\right)\psi' = \frac{3}{16x^2}, \tag{13.10.57}$$

which becomes for large x

$$\psi'' \pm \left(\frac{2}{\sqrt{x}}\right)\psi' \sim \frac{3}{16x^2}. \tag{13.10.58}$$

Therefore, two dominant balancing relations exist; they are

$$\begin{cases} \psi'' = \frac{3}{16x^2}, \\ \psi' = (\pm)\left(\frac{3}{32x^{3/2}}\right). \end{cases} \tag{13.10.59}$$

Carrying out the required calculations demonstrates that the second expression is the proper relation and this gives

$$\psi(x) = (\mp)\left(\frac{3}{16\sqrt{x}}\right). \tag{13.10.60}$$

Therefore, to this level of expansion $y(x)$ is

$$y(x) = \exp\left[\pm 2\sqrt{x} + \left(\frac{1}{4}\right)\mathrm{Ln}(x) \mp \left(\frac{3}{16\sqrt{x}}\right) + \epsilon(x)\right], \tag{13.10.61}$$

where

$$\epsilon(x) = o\left(\frac{1}{\sqrt{x}}\right). \tag{13.10.62}$$

Note, for large x, that

$$\exp\left[(\mp)\left(\frac{3}{16\sqrt{x}}\right)\right] = 1 + O\left(\frac{1}{\sqrt{x}}\right). \tag{13.10.63}$$

This means that $y(x)$ can be written in the form (for $x \to \infty$),

$$\begin{aligned}
y(x) = x^{1/4}\Big\{ Ae^{2\sqrt{x}}\left[1 + O\left(\frac{1}{\sqrt{x}}\right)\right] \\
+ Be^{-2\sqrt{x}}\left[1 + O\left(\frac{1}{\sqrt{x}}\right)\right]
\end{aligned} \tag{13.10.64}$$

where the arbitrary constants, A and B, are a consequence of the fact that the original differential equation is linear and of second-order. Therefore, two linearly independent solutions exist and they may be added together to form the general solution.

Finally, examining the explicit functional forms for (ϕ, η, ψ), i.e.,

$$\phi(x) = O(\sqrt{x}), \quad \eta(x) = O(\mathrm{Ln}(x)), \quad \psi(x) = O\left(\frac{1}{\sqrt{x}}\right), \tag{13.10.65}$$

it follows that

$$\eta(x) = o(\phi(x)), \quad \psi(x) = o(\eta(x)), \quad \epsilon(x) = o(\psi(x)). \tag{13.10.66}$$

If we change the notation to read: $\phi_0(x) = \phi(x)$, $\phi_1(x) = \eta(x)$, $\phi_2(x) = \psi(x)$, and $\phi_3(x) = \epsilon(x)$, then these functions are the first members in an asymptotic sequence

$$\begin{cases} \{\phi_0(x), \phi_1(x), \phi_2(x), \ldots, \phi_k(x), \ldots\} \\ \phi_{k+1}(x) = o[\phi_k(x)]. \end{cases} \tag{13.10.67}$$

13.10.3   *Traveling Wave Front Behavior*

In the discussions presented below, the following abbreviations are used:

- FE - Fisher equation
- FESRD - Fisher equation with square-root dynamics
- ODE - ordinary differential equation
- PDE - partial differential equation
- TW - traveling wave
- WF - wave front.

Relevant background materials on traveling waves and related issues, and on the major results presented in this section are given in the three references:

(1) J. E. Medina-Ramírez, "Computational treatment of traveling wave solutions of a population model with square-root dynamics," *Advanced Studies in Biology* **5**, 89 (2013).

(2) R. E. Mickens "Wave front behavior of traveling wave solutions for a PDE having square-root dynamics," *Mathematics and Computers in Simulation* **82**, 1271 (2012).

(3) J. D. Murray, *Mathematical Biology* (Springer-Verlag, Berlin, 1989). See Sections 11.1 to 11.4.

Consider a system having one-space dimension and modeled by the PDE

$$u_t + A(u)u_x = Du_{xx} + R(u), \quad u = u(x,t), \tag{13.10.68}$$

where D is a positive constant, the diffusion constant; $A(u)u_x$ corresponds to the effects of advection; Du_{xx} is a linear diffusion term; and $R(u)$ incorporates reactions. Assume $R(u)$ has two zeros which can be normalized such that they are $\bar{u}^{(1)} = 0$ and $\bar{u}^{(2)} = 1$. Further assume that

$$0 \le u(x,0) \le 1 \Rightarrow 0 \le u(x,t) \le 1, \tag{13.10.69}$$

and that the solutions

$$u^{(1)}(x,t) = 0, \quad u^{(2)}(x,t) = 1, \tag{13.10.70}$$

are, respectively, unstable and stable. This PDE is said to have a TW solution if

$$\begin{cases} u(x,t) = f(x - ct) = f(z) \quad z = x - ct, \\ \displaystyle\lim_{z \to -\infty} f(z) = 1, \\ \displaystyle\lim_{z \to +\infty} f(z) = 0, \\ f'(z) = \frac{df(z)}{dz} < 0, \qquad -\infty < z < \infty. \end{cases} \tag{13.10.71}$$

One of the best studied PDE's having a TW solution is the Fisher equation (FE)

$$u_t = u_{xx} + u - u^2. \qquad (13.10.72)$$

The book by Murray provides a full discussion of this equation. The major objective of this section is to investigate the wave front properties of a modified Fisher equation having square-root dynamics, i.e., Eq. (13.10.72) is changed to the form

$$u_t = u_{xx} + \sqrt{u} - u. \qquad (13.10.73)$$

Note that the TW solution, $f(z)$, to the FE is monotonic decreasing from $f(-\infty) = 1$ to $f(+\infty) = 0$, and has the property

$$0 < f(z) < 1, \quad -\infty < z < +\infty. \qquad (13.10.74)$$

In particular, for the FE

$$f(z) \sim e^{-az}, \quad z \text{ large}, \qquad (13.10.75)$$

where $a > 0$ is a positive constant. Thus, the WF behavior, given by the expression in Eq. (13.10.75), is exponentially small, but never zero.

However, before proceeding with our specific investigation, let us consider a general form of Eq. (13.10.73), with all its dimensional coefficients appearing, i.e.,

$$u_t = D u_{xx} + \lambda_1 \sqrt{u} - \lambda_2 u. \qquad (13.10.76)$$

This equation can be transformed into a dimensionless equation if we first rescale the independent and dependent variables as follows

$$t = T\bar{t}, \quad x = L\bar{x}, \quad u = W\bar{u}, \qquad (13.10.77)$$

where (T, L, W) are the time, length, and dependent variable scales, which are to be found. If these expressions are substituted in Eq. (13.10.76), we obtain

$$\frac{\partial \bar{u}}{\partial \bar{t}} = \left(\frac{TD}{L^2}\right) \frac{\partial^2 \bar{u}}{\partial \bar{x}^2} + \left(\frac{T\lambda_1}{\sqrt{W}}\right) \sqrt{\bar{u}} - (T\lambda_2)\bar{u}. \qquad (13.10.78)$$

Thus, the requirements

$$\frac{TD}{L^2} = 1, \quad \frac{T\lambda_1}{\sqrt{W}} = 1, \quad T\lambda_2 = 1, \qquad (13.10.79)$$

gives

$$T = \frac{1}{\lambda_2}, \quad L = \sqrt{\frac{D}{\lambda_2}}, \quad W = \left(\frac{\lambda_1}{\lambda_2}\right)^2, \qquad (13.10.80)$$

and, dropping the bars, gives the result expressed in Eq. (13.10.73).

A detailed examination of Eq. (13.10.72) leads to the following conclusions:

(i) Two fixed-points or constant solutions exist

$$u^{(1)}(x,t) = 1, \quad u^{(2)}(x,t) = 0. \tag{13.10.81}$$

(ii) The TW solution, $g(z)$, satisfies the following second-order ODE

$$g''(z) + cg'(z) + \sqrt{g(z)} - g(z) = 0. \tag{13.10.82}$$

(iii) The boundary conditions for this TW are

$$g(-\infty) = 1, \quad g(+\infty) = 1. \tag{13.10.83}$$

(iv) Also, we have

$$g'(z) < 0, \quad -\infty < z < +\infty. \tag{13.10.84}$$

(v) Finally, it should be noted that the speed of the TW solution cannot be determined by the analysis or techniques that we present.

The appearance of the $\sqrt{g}$ term, in Eq. (13.10.73), suggests that its TW solution may go to zero at a finite value of z, say z_0, and remain zero for $z > z_0$. (See Section 13.5.) Our task is therefore to determine the functional behavior in a neighborhood of this point, if it exists. This asymptotic behavior will be studied using a form of dominant balance.

We start our analysis by understanding that near the WF the function $f(z)$ is small, i.e.,

$$0 \leq g(z) \ll 1, \tag{13.10.85}$$

and this fact implies

$$\sqrt{g(z)} \gg g(z), \tag{13.10.86}$$

and we can drop the $g(z)$ term. Therefore, Eq. (13.10.82) becomes

$$g'' + cg' + \sqrt{g} \sim 0. \tag{13.10.87}$$

Now for purposes of applying the method of dominant balance, there are three cases to consider

$$\begin{cases} \text{Case I}: cg' + \sqrt{g} = 0, \\ \text{Case II}: g'' + \sqrt{g} = 0, \\ \text{Case III}: g'' + g' = 0. \end{cases} \tag{13.10.88}$$

The "third" case where

$$g'' + cg' = 0, \tag{13.10.89}$$

is not valid, in the sense that its solution

$$g^{(III)}(z) = A_1 + A_2 e^{-cz}, \tag{13.10.90}$$

is only "small" if $A_1 = 0$. However, inserting this $g(z) = A_2 e^{-cz}$ into $\sqrt{g}$ gives a function which is larger (as $z \to \infty$) than either $g'(z)$ or $g''(z)$, and, thus, provides a solution not consistent with dominant balance.

Similarly, the first-case has the solution

$$g^{(I)}(z) = \begin{cases} \left(\frac{1}{4}\right) \left[\frac{z-z_0}{c}\right]^2, & z \le z_0, \\ 0, & z > z_0. \end{cases} \tag{13.10.91}$$

However,

$$g^{(I)}(z)'' = O((z - z_0)^0) = O(1), \tag{13.10.92}$$

and this fact indicates an inconsistent dominant balance.

So, we are left with Case II. Assume a solution of the form

$$g^{(II)}(z) = A(z - z_0)^\alpha, \tag{13.10.93}$$

where A and α are to be determined. The substitution of this into $g'' + \sqrt{g} = 0$, gives

$$\alpha(\alpha - 1)A(z - z_0)^{\alpha-2} + A^{1/2}(z - z_0)^{\alpha/2} = 0, \tag{13.10.94}$$

and

$$\alpha = 4, \quad A = \frac{1}{144}. \tag{13.10.95}$$

The substitution of $g^{(II)}(z)$ into Eq. (13.10.73) shows that this is a consistent dominant balance.

To summarize, we have found the following results:

(a) Near its WF, the FESRD has the asymptotic mathematical structure

$$g(z) \sim \begin{cases} \left(\frac{1}{144}\right)(z - z_0)^4, & (z_0 - \epsilon) < z \le z_0, \\ 0, & z > z_0. \end{cases} \tag{13.10.96}$$

(b) At $z = z_0$, we have

$$g(z_0) = 0, \quad g'(z_0) = 0, \quad g''(z_0) = 0. \tag{13.10.97}$$

(c) An ad hoc, phenomenological derived functional form for an approximation to the traveling wave solution, over the full range of z values, i.e., $-\infty < z < +\infty$, is the following expression

$$u(x, t) = f(z)$$
$$= \left\{ 1 - \exp\left[-\left(\frac{z_0 - z}{2\sqrt{3}}\right)^4 \right] \right\} H(z - z_0), \tag{13.10.98}$$

where

$$H(w) = \begin{cases} +1, & w \le 0, \\ 0, & w > 0. \end{cases} \tag{13.10.99}$$

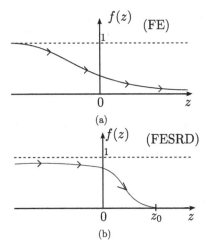

Fig. 13.10.1 (a) $f(z)$ for the TW solution to the Fisher equation. (b) $f(z)$ for the TW solution to the Fisher equation with square-root dynamics.

We conclude that the FESRD has a TW solution which starts at $z = -\infty$, with the value $f(-\infty) = 1$, and monotonic decreases to zero at $z = z_0$, i.e., $f(z_0) = 0$. For $z > z_0$, $f(z)$ is zero. See Fig. 13.10.1.

This problem illustrates the full power of the method of dominant balance.

Problems

Section 13.1

13.1.1 Derive the fact that $u(x, y)$ and $v(x, y)$ satisfy the Laplace equation from the Cauchy-Riemann relations.

13.1.2 Explain the existence of the functions $\phi_1(x)$ and $\phi_2(y)$ in Eqs. (13.1.5).

13.1.3 Prove Theorem-13.1.1.

13.1.4 Prove Theorem-13.1.2.

13.1.5 Use the following functions, each of which corresponds to the real part of a complex function, $f(z)$, to determine $f(z)$:

(i) $u(x, y) = \frac{x^3 - 3xy^2}{(x^2 + y^2)^3}$,

(ii) $u(x, y) = \cosh(x) \cos(y)$,

(iii) $u(x, y) = (x^2 + y^2)^{1/4} \cos\left[\left(\frac{1}{2}\right) \tan^{-1}\left(\frac{y}{x}\right)\right]$.

13.1.6 Derive the result given in Eq. (13.1.58).

13.1.7 Using the formulation of Section 13.1.7 to determine $f(z)$ given the following indicated functions:

  (i) $R(x, y) = (x^2 + y^2)^2$,

  (ii) $\theta(x, y) = y$,

  (iii) $\theta(x, y) = \tan^{-1}\left(\frac{y}{x}\right)$.

  A useful formula to know, in general, and one which be helpful for (iii) is

$$p = \tan(q) \Leftrightarrow q = \left(\frac{1}{2i}\right) \mathrm{Ln}\left(\frac{1 + ip}{1 - ip}\right).$$

  (iv) Prove the stated result in (iii).

Section 13.2

13.2.1 Explain the representation for $f(x)$ given by Eq. (13.2.20).

13.2.2 Using the Laplace transform of Eq. (13.2.26) and the condition given in Eq. (13.2.7) prove the following features:

  (i) $f(0) > 0$, and there exists a constant M such that $0 \le f(x) \le M$, for $0 \le x < \infty$.

  (ii) $(-1)^n f^{(n)} \ge 0$, $x > 0$; $n = 0, 1, 2, \ldots$.

  (iii) $\mathrm{Lim}_{x \to \infty} f(x) = 0$.

Section 13.3

13.3.1 Prove the results given in Eq. (13.3.2).

13.3.2 Plot $\mathrm{sign}(\sin \theta)$ and $\mathrm{sign}(\cos \theta)$ as a function of θ.

13.3.3 Prove Eqs. (13.3.14) and (13.3.16).

13.3.4 See if you can derive the Fourier series for $(\cos \theta)^{-1}$ using the standard definitions of the Fourier coefficients.

Section 13.5

13.5.1 Discuss the reasons for the condition given in Eq. (13.5.1).

13.5.2 Why should the three requirements of Eq. (13.5.9) be imposed on $f(\Delta, \lambda)^2$?

13.5.3 Prove that $(\Delta)^p$ is an odd function and $(\Delta)^{(1-p)}$ is an even function.

13.5.4 Solve and analyze in detail the solution of Eq. (13.5.24).

Section 13.6

13.6.1 In Eq. (13.6.1) what might be some other forces which could act on the object?

13.6.2 Let $v(t) = dx(t)/dt$ be given and assume that the initial condition for $x(t)$ is $x(0) = 0$. Demonstrate that $x(t)$ is given by the expression in Eq. (13.6.6).

13.6.3 Derive and explain the results for $v(t)$ and $x(t)$ given, respectively, in Eqs. (13.6.10) and (13.6.11).

13.6.4 Derive, explain, and discuss the solutions and plots of $v(t)$ and $x(t)$ for the α values: $0 < \alpha < 1$, $\alpha = 1$, $1 < \alpha < 2$, $\alpha = 2$, $\alpha > 2$.

Section 13.7

13.7.1 Solve Eq. (13.7.7) and show that the period is the expression given in Eq. (13.7.8).

13.7.2 Is the following mathematical expression dimensionlly correct

$$\frac{de^{ax^2}}{dx} = a^2 x e^{ax^2}?$$

If not, find the correct form.

13.7.3 Discuss why the arguments of functions such as $\cos(\cdots)$, $\exp(\cdots)$, $\text{Ln}(\cdots)$, must be dimensionless.

13.7.4 Consider the relation

$$F(x, a, b) = \int \frac{dx}{a + bx}.$$

By taking derivatives with respect to a and b, determine expressions for other integrals.

Hint: To start the process, first evaluate the integral.

13.7.5 Calculate the function $f(\theta)$ introduced in Section 13.7.5.

13.7.6 Discuss the validity of the formula $\bar{T}(R)$, given in Eq. (13.7.80). Are other selections possible? If so, construct an alternative to the one of Eq. (13.7.80).

13.7.7 For the July 16, 1945 Trinity nuclear bomb test, the following "data" exists for the radius of the fireball:

Time (ms)	Radius (m)
3.3	54
4.6	67
15	106
62	185

(a) Plot R vs t.

(b) Plot $\text{Ln}(R)$ vs $\text{Ln}(t)$.

(c) Which of the plots is more useful?

(d) Use the value of ρ, in MKS units, to estimate E_0.

(e) Calculate the speed of the wave front.

(f) Use dimensional analysis to determine the pressure as a function of the radius of the fireball.

Hint: Assume pressure $= P$ is a function of R, E_0, and ρ.

13.7.8 Use Eq. (13.7.97) to calculate the density of a BH having one solar mass. Compare this density with the average density of the sun.

13.7.9 Explain why d in Eq. (13.7.98) cannot be uniquely determined.

Section 13.8

13.8.1 Derive the following expression for the first derivative of the Lambert-W function

$$\frac{dW(x)}{dx} = \frac{1}{x + e^{W(x)}},$$

and explain why it does not exist at $x = -\left(\frac{1}{e}\right)$.

13.8.2 Show the following relation is true

$$\int_0^e W(x)dx = e - 1.$$

13.8.3 Show that

$$W(x \, \text{Ln} \, x) = \text{Ln}(x)$$
$$= W + \text{Ln}(W).$$

13.8.4 Prove that the radius of convergence for the series

$$W(x) = \sum_{k=1}^{\infty} \left[\frac{(-1)^{k-1}k^{k-2}}{(k-1)!} \right] x^k,$$

is $(1/e)$.

13.8.5 Show that

$$\int \frac{W(x)}{x} dx = \left(\frac{1}{2}\right)[1 + W(x)]^2 + c.$$

13.8.6 Calculate $W(1) = 0.567143290\ldots.$

Section 13.9

13.9.1 Give physical reasons why it should be expected that $y(z) \geq 0$ and $y'(z) \leq 0$ for $0 \leq z < \infty$.

13.9.2 Calculate the result given in Eq. (13.9.7).

13.9.3 Derive Eq. (13.9.15).

13.9.4 Given, for $0 \leq z < \infty$, $y(z) > 0$ and $y'(z) < 0$, use the differential equation for $y(z)$ to show that $y''(z) > 0$, $y'''(z) < 0$, and $y^{(iv)}(z) > 0$. What happens for $y^{(v)}(z)$?

13.9.5 Show that $y(z)$ has an exact solution of the form $y(z) = Az^\alpha$ and calculate A and α. Explain why this is not the solution we want.

13.9.6 Use the following approximating functions for $y(z)$ to estimate $y(0)$:

(a) $\bar{y}(z) = \frac{A}{1+Bz^2}$

(b) $\bar{y}(z) = \frac{A}{(1+Bz)^2}$

(c) $\bar{y}(z) = \frac{A}{1+Bz+Cz^2}$.

13.9.7 Use the results from Problem 13.9.5 to examine the possible solution behaviors of Eq. (13.9.15a). Are all solutions bounded? If not, explain why?

13.9.8 Construct a Taylor series for $y'' = y^2 - zyy'$. Show that with $y_0 = y(0)$ and $y'(0) = -\sqrt{3}$, the first four terms are

$$y(z) = y_0 - \sqrt{3}z + \left(\frac{y_0^2}{2}\right)z^2 - \left(\frac{\sqrt{3}y_0}{6}\right)z^3 + \cdots.$$

13.9.9 The Boltzmann problem can be solved exactly. Calculate this solution and discuss its properties.

Section 13.10

13.10.1 Is it possible for there to be a three or more terms dominant balance? If so, what does this mean?

13.10.2 Study the properties of the solutions for ϵ small for the algebraic equations

(i) $\epsilon x^3 + x - 1 = 0$

(ii) $\epsilon^2 x^2 + \epsilon x - 1 = 0$

13.10.3 Can dominant balance be applied to the equation

$$x = e^{-\epsilon x}$$

for ϵ small?

13.10.4 Derive the results given in Eq. (13.10.24).

13.10.5 Explain why $\eta(x)$ must satisfy the condition of Eq. (13.10.39).

13.10.6 Show that the result given in Eq. (13.10.63) is correct.

13.10.7 Determine the first three terms in the asymptotic expansion ($x \to \infty$) of the differential equation

$$y''(x) - \left(\frac{x^3}{1 + x^2} \right) y(x) = 0.$$

Hint: Use, for $x \to \infty$, the fact that

$$\frac{x^3}{1 + x^2} = \frac{x^3}{x^2 \left(1 + \frac{1}{x^2} \right)}$$

$$= \frac{x}{\left(1 + \frac{1}{x^2} \right)}$$

$$= x \left[1 - \frac{1}{x^2} + O \left(\frac{1}{x^4} \right) \right]$$

$$= x - \frac{1}{x} + O \left(\frac{1}{x^3} \right).$$

13.10.8 Derive the results given in Eq. (13.10.95) from Eq. (13.10.95).

13.10.9 Give arguments to suggest that $u(x,t) = f(z)$, as given in Eq. (13.10.99) is an approximation to the actual solution to the FESRD. Show that this $f(z)$ and its first derivative is continuous at $z = z_0$. Further show that $f'(z) \leq 0$ over the range, $-\infty < z < +\infty$.

Appendix A

Mathematical Relations

This appendix gives various mathematical relations that are used regularly in the calculations of the text. The references listed at the end of this appendix contain extensive tables of other useful mathematical relations and analytic expressions.

A.1 Trigonometric Relations

A.1.1 *Exponential Definitions of Trigonometric Functions*

$$\sin x = \frac{e^{ix} - e^{-ix}}{2i} \tag{A.1}$$

$$\cos x = \frac{e^{ix} + e^{-ix}}{2}. \tag{A.2}$$

A.1.2 *Functions of Sums of Angles*

$$\sin(x \pm y) = \sin x \cos y \pm \cos x \sin y \tag{A.3}$$

$$\cos(x \pm y) = \cos x \cos y \mp \sin x \sin y. \tag{A.4}$$

A.1.3 *Powers of Trigonometric Functions*

$$\sin^2 x = \left(\frac{1}{2}\right)(1 - \cos 2x) \tag{A.5}$$

$$\cos^2 x = \left(\frac{1}{2}\right)(1 + \cos 2x) \tag{A.6}$$

$$\sin^3 x = \left(\frac{1}{4}\right)(3\sin x - \sin 3x) \tag{A.7}$$

$$\cos^3 x = \left(\frac{1}{4}\right)(3\cos x + \cos 3x) \qquad (A.8)$$

$$\sin^4 x = \left(\frac{1}{8}\right)(3 - 4\cos 2x + \cos 4x) \qquad (A.9)$$

$$\cos^4 x = \left(\frac{1}{8}\right)(3 + 4\cos 2x + \cos 4x) \qquad (A.10)$$

$$\sin^5 x = \left(\frac{1}{16}\right)(10\sin x - 5\sin 3x + \sin 5x) \qquad (A.11)$$

$$\cos^5 x = \left(\frac{1}{16}\right)(10\cos x + 5\cos 3x + \cos 5x) \qquad (A.12)$$

$$\sin^6 x = \left(\frac{1}{32}\right)(10 - 15\cos 2x + 6\cos 4x - \cos 6x) \qquad (A.13)$$

$$\cos^6 x = \left(\frac{1}{32}\right)(10 + 15\cos 2x + 6\cos 4x + \cos 6x). \qquad (A.14)$$

A.1.4 *Other Trigonometric Relations*

$$\sin x \pm \sin y = 2\sin\left(\frac{x \pm y}{2}\right)\cos\left(\frac{x \mp y}{2}\right) \qquad (A.15)$$

$$\cos x + \cos y = 2\cos\left(\frac{x + y}{2}\right)\cos\left(\frac{x - y}{2}\right) \qquad (A.16)$$

$$\cos x - \cos y = -2\sin\left(\frac{x + y}{2}\right)\sin\left(\frac{x - y}{2}\right) \qquad (A.17)$$

$$\sin x \cos y = \left(\frac{1}{2}\right)[\sin(x + y) + \sin(x - y)] \qquad (A.18)$$

$$\cos x \sin y = \left(\frac{1}{2}\right)[\sin(x + y) - \sin(x - y)] \qquad (A.19)$$

$$\cos x \cos y = \left(\frac{1}{2}\right)[\cos(x + y) + \cos(x - y)] \qquad (A.20)$$

$$\sin x \sin y = \left(\frac{1}{2}\right)[\cos(x - y) - \cos(x + y)] \qquad (A.21)$$

$$\sin^2 x - \sin^2 y = \sin(x + y)\sin(x - y) \qquad (A.22)$$

$$\cos^2 x - \cos^2 y = -\sin(x + y)\sin(x - y) \qquad (A.23)$$

$$\cos^2 x - \sin^2 y = \cos(x + y)\cos(x - y) \qquad (A.24)$$

$$\sin^2 x \cos x = \left(\frac{1}{4}\right)(\cos x - \cos 3x) \qquad (A.25)$$

$$\sin x \cos^2 x = \left(\frac{1}{4}\right)(\sin x + \sin 3x) \tag{A.26}$$

$$\sin^3 x \cos x = \left(\frac{1}{8}\right)(2\sin 2x - \sin 4x) \tag{A.27}$$

$$\sin^2 x \cos^2 x = \left(\frac{1}{8}\right)(1 - \cos 4x) \tag{A.28}$$

$$\sin x \cos^3 x = \left(\frac{1}{8}\right)(2\sin 2x + \sin 4x) \tag{A.29}$$

$$\sin^4 x \cos x = \left(\frac{1}{16}\right)(2\cos x - 3\cos 3x + \cos 5x) \tag{A.30}$$

$$\sin^3 x \cos^2 x = \left(\frac{1}{16}\right)(2\sin x + \sin 3x - \sin 5x) \tag{A.31}$$

$$\sin^2 x \cos^3 x = -\left(\frac{1}{16}\right)(2\cos x + \cos 3x + \cos 5x) \tag{A.32}$$

$$\sin x \cos^4 x = \left(\frac{1}{16}\right)(2\sin x + 3\sin 3x + \sin 5x). \tag{A.33}$$

A.1.5 Derivatives and Integrals of Trigonometric Functions

$$\frac{d}{dx}\cos x = -\sin x \tag{A.34}$$

$$\frac{d}{dx}\sin x = \cos x \tag{A.35}$$

$$\int \cos x \, dx = \sin x \tag{A.36}$$

$$\int \sin x \, dx = -\cos x \tag{A.37}$$

$$\int \sin^2 x \, dx = \left(\frac{1}{2}\right)x - \left(\frac{1}{4}\right)\sin 2x \tag{A.38}$$

$$\int \cos^2 x \, dx = \left(\frac{1}{2}\right)x + \left(\frac{1}{4}\right)\sin 2x \tag{A.39}$$

$$\int \sin mx \sin kx \, dx = \frac{\sin(m-k)x}{2(m-k)} - \frac{\sin(m+k)x}{2(m+k)} \qquad m^2 \neq k^2 \tag{A.40}$$

$$\int \cos mx \cos kx \, dx = \frac{\sin(m-k)x}{2(m-k)} + \frac{\sin(m+k)x}{2(m+k)} \qquad m^2 \neq k^2 \qquad \text{(A.41)}$$

$$\int \sin mx \cos kx \, dx = -\frac{\cos(m-k)x}{2(m-k)} - \frac{\cos(m+k)x}{2(m+k)} \qquad m^2 \neq k^2 \qquad \text{(A.42)}$$

$$\int_{-\pi}^{\pi} \cos mx \cos kx \, dx = \pi \delta_{mk}; \qquad m, k \text{ integers} \qquad \text{(A.43)}$$

$$\int_{-\pi}^{\pi} \sin mx \cos kx \, dx = 0; \qquad m, k \text{ integers} \qquad \text{(A.44)}$$

$$\int_{-\pi}^{\pi} \sin mx \sin kx \, dx = \pi \delta_{mk}; \qquad m, k \text{ integers} \qquad \text{(A.45)}$$

$$\int x \sin x \, dx = \sin x - x \cos x \qquad \text{(A.46)}$$

$$\int x^2 \sin x \, dx = 2x \sin x - (x^2 - 2) \cos x \qquad \text{(A.47)}$$

$$\int x \cos x \, dx = \cos x + x \sin x \qquad \text{(A.48)}$$

$$\int x^2 \cos x \, dx = 2x \cos x + (x^2 - 2) \sin x. \qquad \text{(A.49)}$$

A.2 Factors and Expansions

$$(a \pm b)^2 = a^2 \pm 2ab + b^2 \qquad \text{(A.50)}$$

$$(a \pm b)^3 = a^3 \pm 3a^2 b + 3ab^2 \pm b^3 \qquad \text{(A.51)}$$

$$(a + b + c)^2 = a^2 + b^2 + c^2 + 2(ab + ac + bc) \qquad \text{(A.52)}$$

$$(a+b+c)^3 = a^3 + b^3 + c^3 + 3a^2(b+c) + 3b^2(a+c) + 3c^2(a+b) + 6abc \qquad \text{(A.53)}$$

$$a^2 - b^2 = (a - b)(a + b) \qquad \text{(A.54)}$$

$$a^2 + b^2 = (a + ib)(a - ib), \qquad i = \sqrt{-1} \qquad \text{(A.55)}$$

$$a^3 - b^3 = (a - b)(a^2 + ab + b^2) \qquad \text{(A.56)}$$

$$a^3 + b^3 = (a + b)(a^2 - ab + b^2). \qquad \text{(A.57)}$$

A.3 Quadratic Equations

The quadratic equation

$$ax^2 + bx + c = 0 \tag{A.58}$$

has the two solutions

$$x_1 = \frac{-b + \sqrt{b^2 - 4ac}}{2a}, \tag{A.59}$$

$$x_2 = \frac{-b - \sqrt{b^2 - 4ac}}{2a}. \tag{A.60}$$

A.4 Cubic Equations

The cube equation

$$z^3 + pz^2 + qz + r = 0 \tag{A.61}$$

can be reduced to the form

$$x^3 + ax + b = 0 \tag{A.62}$$

by substituting for z the value

$$z = x - \frac{p}{3}. \tag{A.63}$$

The constants a and b are given by the expressions

$$a = \frac{3q - p^2}{3}, \tag{A.64}$$

$$b = \frac{3p^3 - 9pq + 27r}{27}. \tag{A.65}$$

Let A and B be defined as

$$A = \left[-\left(\frac{b}{2}\right) + \left(\frac{b^2}{4} + \frac{a^3}{27}\right)^{1/2} \right]^{1/3}, \tag{A.66}$$

$$B = \left[-\left(\frac{b}{2}\right) - \left(\frac{b^2}{4} + \frac{a^3}{27}\right)^{1/2} \right]^{1/3}. \tag{A.67}$$

The three roots of (A.62) are given by the following expressions

$$x_1 = A + B, \tag{A.68}$$

$$x_2 = -\left(\frac{A + B}{2}\right) + \sqrt{-3}\left(\frac{A - B}{2}\right), \tag{A.69}$$

$$x_3 = -\left(\frac{A + B}{2}\right) - \sqrt{-3}\left(\frac{A - B}{2}\right). \tag{A.70}$$

Let

$$\Delta = \frac{b^2}{4} + \frac{a^3}{27}. \tag{A.71}$$

If $\Delta > 0$, then there will be one real root and two complex conjugate roots. If $\Delta = 0$, there will be three real roots, of which at least two are equal. If $\Delta < 0$, there will be three real and unequal roots.

A.5   Differentiation of a Definite Integral with Respect to a Parameter

Let $f(x,t)$ be continuous and have a continuous derivative $\partial f/\partial t$, in a domain in the x-t plane that includes the rectangle

$$\psi(t) \le x \le \phi(t), \qquad t_1 \le t \le t_2. \tag{A.72}$$

In addition, let $\psi(t)$ and $\phi(t)$ be defined and have continuous first derivatives for $t_1 \le t \le t_2$. Then, for $t_1 \le t \le t_2$, we have

$$\frac{d}{dt} \int_{\psi(t)}^{\phi(t)} f(x,t)dx = f[\phi(t),t]\frac{d\phi}{dt} - f[\psi(t),t]\frac{d\psi}{dt}$$

$$+ \int_{\psi(t)}^{\phi(t)} \frac{\partial}{\partial t} f(x,t)dx. \tag{A.73}$$

A.6   Eigenvalues of a 2 × 2 Matrix

The eigenvalues of a matrix A are given by the solutions to the characteristic equation

$$\det(A - \lambda I) = 0, \tag{A.74}$$

where I is the identity or unit matrix. If A is an $n \times n$ matrix, then there exists n eigenvalues λ_i, where $i = 1, 2, \ldots, n$.

Consider the 2 × 2 matrix

$$A = \begin{pmatrix} a & b \\ c & d \end{pmatrix}. \tag{A.75}$$

The characteristic equation is

$$\det \begin{pmatrix} a - \lambda & b \\ c & d - \lambda \end{pmatrix} = 0. \tag{A.76}$$

Evaluating the determinant gives

$$\lambda^2 - T\lambda + D = 0, \tag{A.77}$$

where

$$T \equiv \mathrm{trace}(A) = a + d,$$
$$D \equiv \det(A) = ad - bc. \tag{A.78}$$

The two eigenvalues are given by the expressions

$$\lambda_1 = \left(\frac{1}{2}\right)\left[T + \sqrt{T^2 - 4D}\right], \tag{A.79a}$$

$$\lambda_2 = \left(\frac{1}{2}\right)\left[T - \sqrt{T^2 - 4D}\right]. \tag{A.79b}$$

A.7 Routh-Hurwitz Theorem

Let $P_n(x)$ be an nth degree polynomial

$$P_n(x) \equiv x^n + a_1 x^{n-1} + a_2 x^{n-2} + \cdots + a_n, \qquad (A.80)$$

where the $(a_i : i = 1, 2, \ldots, n)$ are real numbers. Define the n numbers $(\Delta_i : i = 1, 2, \ldots, n)$ to be

$$\Delta_1 \equiv a_1, \quad \Delta_2 \equiv \begin{vmatrix} a_1 & 1 \\ a_3 & a_2 \end{vmatrix}, \quad \Delta_3 \equiv \begin{vmatrix} a_1 & 1 & 0 \\ a_3 & a_2 & a_1 \\ a_5 & a_4 & a_3 \end{vmatrix},$$

$$\cdots \Delta_n \equiv \begin{vmatrix} a_1 & 1 & 0 & 0 & \cdots & 0 \\ a_3 & a_2 & a_1 & 1 & \cdots & 0 \\ a_5 & a_4 & a_3 & a_2 & \cdots & 0 \\ \vdots & \vdots & \vdots & \vdots & & \\ a_{2n-1} & a_{2n-2} & a_{2n-3} & a_{2n-4} & \cdots & a_n \end{vmatrix}, \qquad (A.81)$$

where $a_r = 0$ for $r > n$.

The necessary and sufficient conditions for all the zeros of $P_n(x) = 0$ to have negative real parts are

$$\Delta_i > 0 \qquad (i = 1, 2, \ldots, n). \qquad (A.82)$$

A.8 Integration by Parts

$$\int u \, dv = uv - \int v \, du. \qquad (A.83)$$

A.9 Leibnitz's Relation

Let $u(x)$ and $v(x)$ have the nth derivatives, then

$$\frac{d^n(uv)}{dx^n} = \sum_{k=0}^{n} \binom{n}{k} u^{(n-k)} v^k. \qquad (A.84)$$

A.10 L'Hopital's Rule

Let $u(x)$ and $v(x)$ have a zero at $x = x_0$, i.e., $u(x_0) = 0$ and $v(x_0) = 0$. Assume that their first derivatives at $x = x_0$ are not both zero. Then

$$\lim_{x \to x_0} \left[\frac{u(x)}{v(x)} \right] = \frac{u'(x_0)}{v'(x_0)}. \qquad (A.85)$$

If $u(x_0) = 0$ and $v(x_0) = 0$, and their first n derivatives are zero at $x = x_0$ i.e.,

$$u^{(s)}(x_0) = 0, \qquad v^{(s)}(x_0) = 0, \qquad s = 1, 2, \ldots, n,$$

then provided that not both the $(n+1)$th derivatives are zero at $x = x_0$

$$\lim_{x \to x_0} \left[\frac{u(x)}{v(x)} \right] = \frac{u^{n+1}(x_0)}{v^{n+1}(x_0)}. \tag{A.86}$$

A.11   Special Determinants

A.11.1   Jacobian Determinants

Let $f_1(x), f_2(x), \ldots, f_n(x)$ be real-valued functions such that they are differentiable with respect to the n-variables $x = (x_1, x_2, \ldots, x_n)$. The Jacobian of these functions is defined to be the determinant

$$J_f(x) = \frac{\partial(f_1, f_2, \ldots, f_n)}{\partial(x_1, x_2, \ldots, x_n)}$$

$$\equiv \begin{vmatrix} \dfrac{\partial f_1}{\partial x_1} & \dfrac{\partial f_1}{\partial x_2} & \cdots & \dfrac{\partial f_1}{\partial x_n} \\[2mm] \dfrac{\partial f_2}{\partial x_1} & \dfrac{\partial f_2}{\partial x_2} & \cdots & \dfrac{\partial f_2}{\partial x_n} \\[2mm] \vdots & \vdots & & \vdots \\[2mm] \dfrac{\partial f_n}{\partial x_1} & \dfrac{\partial f_n}{\partial x_2} & \cdots & \dfrac{\partial f_n}{\partial x_n} \end{vmatrix}. \tag{A.87}$$

A.11.2   Hessian Determinants

The Hessian determinant involved the second derivatives of a scalar function of n variables, $\phi = \phi(x_1, x_2, \ldots, x_n)$. It is defined to be

$$H_\phi = \begin{vmatrix} \dfrac{\partial^2 \phi}{\partial x_1^2} & \dfrac{\partial^2 \phi}{\partial x_1 \partial x_2} & \dfrac{\partial^2 \phi}{\partial x_1 \partial x_3} & \cdots & \dfrac{\partial^2 \phi}{\partial x^1 \partial x_n} \\[2mm] \dfrac{\partial^2 \phi}{\partial x_2 \partial x_1} & \dfrac{\partial^2 \phi}{\partial x_2^2} & \dfrac{\partial^2 \phi}{\partial x_2 \partial x_3} & \cdots & \dfrac{\partial^2 \phi}{\partial x_2 \partial x_n} \\[2mm] \vdots & \vdots & \vdots & & \vdots \\[2mm] \dfrac{\partial^2 \phi}{\partial x_n \partial x_1} & \dfrac{\partial^2 \phi}{\partial x_n \partial x_2} & \dfrac{\partial^2 \phi}{\partial x_n \partial x_3} & \cdots & \dfrac{\partial^2 \phi}{\partial x_n^2} \end{vmatrix}. \tag{A.88}$$

Note that the Hessian of ϕ is the Jacobian of the first derivatives ($\partial \phi / \partial x_i$: $i = 1, 2, \ldots, n$).

A.11.3 *Wronskian Determinants*

Let $f_1(x), f_2(x), \ldots, f_n(x)$ be n functions of the single variable x and assume that each has an nth derivative in the open interval $a < x < b$. The Wronskian, $W_f(x)$, of these functions is defined as

$$W_f(x) \equiv \begin{vmatrix} f_1 & f_2 & \cdots & f_n \\ f_1^{(1)} & f_2^{(1)} & \cdots & f_n^{(1)} \\ f_1^{(2)} & f_2^{(2)} & \cdots & f_n^{(2)} \\ \vdots & \vdots & & \vdots \\ f_1^{(n-1)} & f_2^{(n-1)} & \cdots & f_n^{(n-1)} \end{vmatrix} \tag{A.89}$$

where $f_i^{(j)} \equiv d^i f_i / dx^j$.

A.12 Special Series

A.12.1 *Power Series*

$$(1+x)^q = 1 + qx + \left[\frac{q(q-1)}{2!}\right] x^2 + \left[\frac{q(q-1)(q-2)}{3!}\right] x^3$$
$$+ \cdots + \left[\frac{q(q-1)\cdots(q-k+1)}{k!}\right] x^k + \cdots . \tag{A.90}$$

This expansion converges for q real and $|x| < 1$. Particular cases:

$$(1+x)^{-1} = 1 - x + x^2 - x^3 + \cdots = \sum_{k=0}^{\infty} (-1)^k x^k \tag{A.91}$$

$$(1+x)^{-2} = 1 - 2x + 3x^2 - 4x^3 + \cdots = \sum_{k=0}^{\infty} (-1)^k (k+1) x^k \tag{A.92}$$

$$(1+x)^{1/2} = 1 + \left(\frac{1}{2}\right) x - \left(\frac{1 \cdot 1}{2 \cdot 4}\right) x^2 + \left(\frac{1 \cdot 1 \cdot 3}{2 \cdot 4 \cdot 6}\right) x^3$$
$$- \left(\frac{1 \cdot 1 \cdot 3 \cdot 5}{2 \cdot 4 \cdot 6 \cdot 8}\right) x^4 + \cdots \tag{A.93}$$

$$(1+x)^{-1/2} = 1 - \left(\frac{1}{2}\right) x + \left(\frac{1 \cdot 3}{2 \cdot 4}\right) x^2 - \left(\frac{1 \cdot 3 \cdot 5}{2 \cdot 4 \cdot 6}\right) x^3 + \cdots \tag{A.94}$$

A.13 Trigonometric and Hyperbolic Series

$$\sin(x) = x - \frac{x^3}{3!} + \frac{x^5}{5!} - \frac{x^7}{7!} + \cdots = \sum_{k=0}^{\infty} (-1)^k \frac{x^{2k+1}}{(2k+1)!} \tag{A.95}$$

$$\cos(x) = 1 - \frac{x^2}{2!} + \frac{x^4}{4!} - \frac{x^6}{6!} + \cdots = \sum_{k=0}^{\infty}(-1)^k\frac{x^{2k}}{(2k)!} \tag{A.96}$$

$$\sinh(x) = x + \frac{x^3}{3!} + \frac{x^5}{5!} + \frac{x^7}{7!} + \cdots = \sum_{k=0}^{\infty}\frac{x^{2k+1}}{(2k+1)!} \tag{A.97}$$

$$\cosh(x) = 1 + \frac{x^2}{2!} + \frac{x^4}{4!} + \frac{x^6}{6!} + \cdots = \sum_{k=0}^{\infty}\frac{x^{2k}}{(2k)!} \tag{A.98}$$

All of these series converge for $|x| < \infty$.

A.14   Exponential and Logarithmic Series

$$e^x = 1 + x + \frac{x^2}{2!} + \frac{x^3}{3!} + \frac{x^4}{4!} + \cdots = \sum_{k=0}^{\infty}\frac{x^k}{k!} \tag{A.99}$$

$$e^{-x} = 1 - x + \frac{x^2}{2!} - \frac{x^3}{3!} + \frac{x^4}{4!} - \cdots = \sum_{k=0}^{\infty}(-1)^k\frac{x^k}{k!} \tag{A.100}$$

Both of these series converge for $|x| < 1$.

$$\ln(1 + x) = x - \frac{x^2}{2} + \frac{x^3}{3} - \frac{x^4}{4} + \cdots = \sum_{k=1}^{\infty}(-1)^{k+1}\frac{x^k}{k} \tag{A.101}$$

This series converge for $-1 < x \le 1$.

A.15   Certain Standard Indefinite Integral Involving Exponential Functions

$$\int e^{ax}dx = \left(\frac{1}{a}\right)e^{ax} \tag{A.102}$$

$$\int a^x dx = \int e^{(\ln a)x}dx = \frac{a^x}{\ln a} \tag{A.103}$$

$$\int xe^{ax}dx = e^{ax}\left(\frac{x}{a} - \frac{1}{a^2}\right) \tag{A.104}$$

$$\int x^2 e^{ax}dx = e^{ax}\left(\frac{x^2}{a} - \frac{2x}{a^2} + \frac{2}{a^3}\right) \tag{A.105}$$

$$\int P_n(x)e^{ax}dx = \left(\frac{e^{ax}}{a}\right)\sum_{k=0}^{n}(-1)^k\frac{P_n^{(k)}(x)}{a^k}, \tag{A.106}$$

where $P_n(x)$ is an nth degree polynomial and $P_n^{(k)}(x)$ is the kth derivative of $P_n(x)$.

$$\int \frac{dx}{a + be^{cx}} = \left(\frac{1}{ac}\right) [cx - \ln(a + be^{cx}] \tag{A.107}$$

$$\int e^{ax} \sin(bx)dx = \left(\frac{e^{ax}}{a^2 + b^2}\right) [a\sin(bx) - b\cos(bx)] \tag{A.108}$$

$$\int e^{ax} \cos(bx)dx = \left(\frac{e^{ax}}{a^2 + b^2}\right) [a\cos(bx) + b\sin(bx)] \tag{A.109}$$

$$\int xe^{ax} \sin(bx)dx = \left(\frac{e^{ax}}{a^2 + b^2}\right) \left[\left(ax - \frac{a^2 - b^2}{a^2 + b^2}\right) \sin(bx)\right.$$
$$\left. - \left(bx - \frac{2ab}{a^2 + b^2}\right) \cos(bx)\right] \tag{A.110}$$

$$\int xe^{ax} \cos(bx)dx = \left(\frac{e^{ax}}{a^2 + b^2}\right) \left[\left(ax - \frac{a^2 - b^2}{a^2 + b^2}\right) \cos(bx)\right.$$
$$\left. + \left(bx - \frac{2ab}{a^2 + b^2}\right) \sin(bx)\right] \tag{A.111}$$

A.16 Partial Fractions

Let $N(x)$ and $D(x)$ be polynomials in x of degrees, respectively, m and n. Then define the rational function $R(x)$ given by

$$R(x) \equiv \frac{N(x)}{D(x)}. \tag{A.112}$$

If the degree of $N(x)$ is greater than or equal to the degree of $D(x)$, then division allows us to obtain the following form

$$R(x) \equiv \frac{N(x)}{D(x)} = P(x) + \frac{\bar{N}(x)}{D(x)}, \tag{A.113}$$

where $P(x)$ is a polynomial of degree $(m - n)$ and the degree of $\bar{N}(x)$ is less than the degree of $D(x)$. Consequently, for the partial fraction decomposition of $R(x)$, we only need consider the case where $m < n$. The general method to form the partial fraction decomposition of $R(x)$ is now given by the following steps.

(1) Factor $D(x)$ into its basic set of real factors, i.e., no factor should contain complex coefficients. The following possibilities may occur:

(i) linear factors with multiplicity one, $(ax + b)$;

(ii) linear factors with multiplicity ℓ, $(ax + b)^\ell$;

(iii) quadratic factors with multiplicity one, $(ax^2 + bx + c)$;

(iv) quadratic factors with multiplicity k, $(ax^2 + bx + c)^k$.

The coefficients (a, b, c) are assumed to be real numbers and it is further assumed that none of the quadratic factors can be written as a product of real linear factors.

(2) Each linear factor, $ax+b$, with multiplicity one, makes a contribution to the partial fraction decomposition having the form

$$\frac{A}{ax + b}, \tag{A.114}$$

where A is, for the moment, an undetermined constant.

(3) For each linear factor with multiplicity ℓ, include the following terms in the partial fraction decomposition

$$\frac{B_1}{ax + b} + \frac{B_2}{(ax + b)^2} + \cdots + \frac{B_\ell}{(ax + b)^\ell}, \tag{A.115}$$

where $(B_1, B_2, \ldots, B_\ell)$ are, for now, undetermined constant.

(4) For each quadratic factor with multiplicity one, include a term

$$\frac{C_1 x + D_1}{ax^2 + bx + c}, \tag{A.116}$$

where (C_1, D_1) are constants to be determined.

(5) For each quadratic factor with multiplicity k, include the terms

$$\frac{E_1 x + F_1}{ax^2 + bx + c} + \frac{E_2 x + F_2}{(ax^2 + bx + c)^2} + \cdots + \frac{E_k x + F_k}{(ax^2 + bx + c)^k}, \tag{A.117}$$

where $(E_i, F_i : i = 1, 2, \ldots, k)$ are constants to be determined.

(6) Set $R(x)$ equal to all the terms generated in (2), (3), (4), and (5).

(7) The various coefficients can be calculated by multiplying the expression for $R(x)$, obtained in (6), by the factored form of $D(x)$ and setting the coefficients of the corresponding powers of x equal to each other. There will be enough algebraic equations to obtain a unique solution for all coefficients.

A.17   Special Constants

$$\pi = 3.14\ 159\ 265\ 358\ 979$$

$$\frac{1}{\sqrt{2\pi}} = 0.39\ 894\ 228\ 040\ 143$$

$$e = 2.71\ 828\ 182\ 845\ 904$$

$$\log_{10}(e) = 0.43\ 429\ 448\ 190\ 325$$

$$\ln(10) = 2.30\ 258\ 509\ 299\ 404$$

$$\Gamma\left(\frac{1}{2}\right) = \sqrt{\pi} = 1.77\ 245\ 358\ 090\ 551$$

$$\Gamma\left(\frac{1}{4}\right) = 3.62\ 560\ 990$$

The Euler constant γ is defined as follows

$$\gamma \equiv \lim_{n\to\infty}\left[1 + \frac{1}{2} + \frac{1}{3} + \cdots + \frac{1}{n} - \ln(n)\right]$$

$$= 0.57\ 721\ 566.$$

Bibliography

[1] Chemical Rubber Company, *Standard Mathematical Tables* (Chemical Rubber Publishing Company, Cleveland, various editions).

[2] H. B. Dwight, *Tables of Integrals and Other Mathematical Data* (MacMillan, New York, 1961).

[3] A. Erdélyi, *Tables of Integral Transforms, Vol. I* (McGraw-Hill, New York, 1954).

[4] I. S. Gradshteyn and I. M. Ryzhik, *Table of Integrals, Series and Products* (Academic, New York, 1965).

[5] E. Jaknke and F. Emde, *Tables of Functions with Formulas and Curves* (Dover, New York, 1943).

[6] A. Jeffrey, *Handbook of Mathematical Formulas and Integrals* (Academic, New York, 1995).

[7] National Bureau of Standards, *Handbook of Mathematical Functions* (U.S. Government Printing Office; Washington, DC; 1964).

Appendix B

Asymptotics Expansions

B.1   Gauge Functions and Order Symbols

Let $f(\epsilon)$ be a function of the real parameter ϵ. If the limit of $f(\epsilon)$ exists as ϵ tends to zero, then there are three possibilities:

$$f(\epsilon) \to \begin{cases} 0, \\ A, \\ \infty, \end{cases} \qquad \text{(B.1)}$$

with $0 < |A| < \infty$. (The case where $f(\epsilon)$ has an essential singularity at $\epsilon = 0$, such as $\sin(\epsilon^{-1})$ or $\exp(\epsilon^{-1})$, is excluded.) In the first and last cases, the rate at which $f(\epsilon) \to 0$ and $f(\epsilon) \to \infty$ can be expressed by comparing $f(\epsilon)$ with certain known functions called *gauge functions*. The simplest and often most useful gauge functions are members of the set $\{\epsilon^n\}$ where n is an integer. Other gauge functions used are $\sin\epsilon$, $\sinh\epsilon$, $\log\epsilon$, etc.

The behavior of a function $f(\epsilon)$, as $\epsilon \to 0$, can be compared with a gauge function $g(\epsilon)$ by employing the symbols "O" and "o."

B.1.1   *The Symbol O*

The symbol O is defined as follows: Let $f(\epsilon)$ be a function of the parameter ϵ and let $g(\epsilon)$ be a gauge function. Let there exist a positive number A independent of ϵ and an $\epsilon_0 > 0$ such that

$$|f(\epsilon)| \leq A|g(\epsilon)|, \qquad \text{for all } |\epsilon| \leq \epsilon_0, \qquad \text{(B.2)}$$

then

$$f(\epsilon) = O[g(\epsilon)] \qquad \text{as } \epsilon \to 0. \qquad \text{(B.3)}$$

The condition given in (B.3) can be replaced by

$$\lim_{\epsilon \to 0} \left| \frac{f(\epsilon)}{g(\epsilon)} \right| < \infty. \tag{B.4}$$

Let $f(x, \epsilon)$ be a function of the variable x and the parameter ϵ, and let $g(x, \epsilon)$ be a gauge function. Then

$$f(x, \epsilon) = O[g(x, \epsilon)], \tag{B.5}$$

if there exists a positive number A independent of ϵ and an $\epsilon_0 > 0$, such that

$$|f(x, \epsilon)| \le A|g(x, \epsilon)|, \qquad \text{for all } |\epsilon| \le \epsilon_0. \tag{B.6}$$

If A and ϵ_0 are independent of x, the relationship is said to hold uniformly.

B.1.2   *The Symbols o*

The symbol o is defined as follows: Let $f(\epsilon)$ be a function of ϵ and let $g(\epsilon)$ be a gauge function. Let there exist an $\epsilon_0 > 0$, and let, for every positive number δ, independent of ϵ, the following condition hold

$$|f(\epsilon)| \le \delta|g(\epsilon)|, \qquad \text{for } |\epsilon| \le \epsilon_0, \tag{B.7}$$

then

$$f(\epsilon) = o[g(\epsilon)] \qquad \text{as } \epsilon \to 0. \tag{B.8}$$

The condition given by (B.8) can be replaced by

$$\lim_{\epsilon \to 0} \left| \frac{f(\epsilon)}{g(\epsilon)} \right| = 0. \tag{B.9}$$

Let $f(x, \epsilon)$ be a function of x and ϵ, and let $g(x, \epsilon)$ be a gauge function. Then

$$f(x, \epsilon) = o[g(x, \epsilon)] \qquad \text{as } \epsilon \to 0, \tag{B.10}$$

if for every positive number δ, independent of ϵ, there exists an ϵ_0 such that

$$|f(x, \epsilon)| \le \delta|g(x, \epsilon)|, \qquad \text{for } |\epsilon| \le \epsilon_0. \tag{B.11}$$

If δ and ϵ_0 are independent of x, then (B.11) is said to hold uniformly.

B.1.3 Combination of Order Relations

The following relations involve various combinations of one or more of the two order symbols. Based on the definitions of the order relations, the proofs of these formulas are easy to construct.

$$O(O(f)) = O(f) \tag{B.12}$$

$$O(o(f)) = o(O(f)) = o(o(f)) = o(f) \tag{B.13}$$

$$O(f)O(g) = O(fg) \tag{B.14}$$

$$O(f)o(g) = o(f)o(g) = o(fg) \tag{B.15}$$

$$O(f) + O(f) = O(f) + o(f) = O(f) \tag{B.16}$$

$$o(f) + o(f) = o(f). \tag{B.17}$$

If $f = O(g)$ and $a > 0$, then

$$|f|^a = O(|g|^a). \tag{B.18}$$

If $f_i = O(g_i)$, $(i = 1, 2, \ldots, k)$ and $(a_1, a_2, \ldots, a_k)$ are constants, then

$$\sum_{i=1}^{k} a_i f_i = O\left(\sum_{i=1}^{k} |a_i||g_i|\right). \tag{B.19}$$

If $f_i = O(g_i)$, $(i = 1, 2, \ldots, k)$, then

$$\prod_{i=1}^{k} f_i = O\left(\prod_{i=1}^{k} g_i\right). \tag{B.20}$$

B.2 Asymptotic Expansion

Let $\{\delta_n(\epsilon)\}$ be a sequence of functions such that

$$\delta_n(\epsilon) = o[\delta_{n-1}(\epsilon)] \qquad \text{as } \epsilon \to 0. \tag{B.21}$$

Such a sequence is called an *asymptotic sequence*.

Consider the series

$$\sum_{m=0}^{\infty} a_m \delta_m(\epsilon), \tag{B.22}$$

where the a_m are independent of ϵ and $\{\delta_m(\epsilon)\}$ is an asymptotic sequence. This is an asymptotic sequence, denoted by

$$y \sim \sum_{m=0}^{\infty} a_m \delta_m(\epsilon) \qquad \text{as } \epsilon \to 0, \tag{B.23}$$

if and only if

$$y = \sum_{m=0}^{n} a_m \delta_m(\epsilon) + O[\delta_{n+1}(\epsilon)] \qquad \text{as } \epsilon \to 0. \tag{B.24}$$

The expansion given by (B.23) may diverge. However, if the series is an asymptotic expansion, then although (B.23) may diverge, for fixed n, the first n terms in the expansion can represent y with an error that can be made arbitrarily small by taking ϵ sufficiently small. Thus, the error made in truncating the series after n terms is numerically less than the first neglected term, namely, the $(n+1)$th term.

Note, however, that given a function $y(\epsilon)$, the asymptotic expansion is not unique. In fact, $y(\epsilon)$ can be represented by an unlimited number of asymptotic expansions, since there exists an unlimited number of possible asymptotic sequences that can be used. However, once a particular asymptotic sequence is selected, the representation of $y(\epsilon)$ in terms of this sequence is unique. If $y(\epsilon)$ has the asymptotic expansion

$$y(\epsilon) \sim \sum_{m=0}^{\infty} a_m \delta_m(\epsilon) \qquad \text{as } \epsilon \to 0, \tag{B.25}$$

for a particular sequence $\{\delta_m(\epsilon)\}$, then the coefficients a_n are given uniquely by

$$a_n = \lim_{\epsilon \to 0} \frac{y(\epsilon) - \sum_{m=0}^{n-1} a_m \delta_m(\epsilon)}{\delta_n(\epsilon)}. \tag{B.26}$$

B.3   Uniform Expansions

Let y be a function of x and ϵ. The asymptotic expansion of y in terms of the asymptotic sequence $\{\delta_m(\epsilon)\}$ is

$$y(x, \epsilon) \sim \sum_{m=0}^{\infty} a_m(x) \delta_m(\epsilon) \qquad \text{as } \epsilon \to 0, \tag{B.27}$$

where the coefficients a_m are functions of x only. This expansion is said to be uniformly valid if

$$y(x, \epsilon) = \sum_{m=0}^{n} a_m(x) \delta_m(\epsilon) + R_{n+1}(x, \epsilon), \tag{B.28}$$

where

$$R_n(x, \epsilon) = O[\delta_n(\epsilon)], \tag{B.29}$$

uniformly for all x of interest. If these conditions do not hold, then the expansion is said to be nonuniformly valid.

For the expansion to be uniformly valid, the term $a_m(x)\delta_m(\epsilon)$ must be small compared with the preceding term $a_{m-1}(x)\delta_{m-1}(\epsilon)$ for each m. Since

$$\delta_m(\epsilon) = o[\delta_{m-1}(\epsilon)] \qquad \text{as } \epsilon \to 0, \tag{B.30}$$

then $a_m(x)$ can be no more singular than $a_{m-1}(x)$ for all values of x of interest, if the expansion is to be uniform. This means that $a_m(x)/a_{m-1}$ is bounded. Consequently, each term in the expansion must be a small correction to the preceding term irrespective of the value of x.

B.4 Elementary Operations on Asymptotic Expansions

B.4.1 *Addition and Subtraction*

In general, asymptotic expansions can be added and subtracted. For example, let $\{\delta_m(\epsilon)\}$ be an asymptotic sequence and consider the asymptotic expansions of the two functions $f(x, \epsilon)$ and $g(x, \epsilon)$, where the expansions are defined for the same intervals of x and ϵ; that is

$$f(x, \epsilon) \sim \sum_{m=0}^{\infty} a_m(x)\delta_m(x), \tag{B.31}$$

$$g(x, \epsilon) \sim \sum_{m=0}^{\infty} b_m(x)\delta_m(x). \tag{B.32}$$

Then, for constants A and B

$$Af(x, \epsilon) + Bg(x, \epsilon) \sim \sum_{m=0}^{\infty} [Aa_m(x) + Bb_m(x)]\delta_m(\epsilon). \tag{B.33}$$

B.4.2 *Integration*

If $f(x, \epsilon)$ and $a_m(x)$ are integrable functions of x, then

$$\int_{x_1}^{x} f(\bar{x}, \epsilon)d\bar{x} \sim \sum_{m=0}^{\infty} \delta_m(\epsilon) \int_{x_1}^{x} a_m(\bar{x})d\bar{x}. \tag{B.34}$$

If $f(x, \epsilon)$ and $\delta_m(\epsilon)$ are integrable functions of ϵ, then

$$\int_{0}^{\epsilon} f(x, \bar{\epsilon})d\bar{\epsilon} \sim \sum_{m=0}^{\infty} a_m(x) \int_{0}^{\epsilon} \delta_m(\bar{\epsilon})d\bar{\epsilon}. \tag{B.35}$$

B.4.3   *Multiplication*

In general, two asymptotic expansions cannot be multiplied to form another asymptotic expansion. This is because in the formal product of $f(x, \epsilon)$ and $g(x, \epsilon)$, all products of the form $\delta_n(\epsilon)\delta_m(x)$ occur and it may not be possible to arrange them so as to obtain an asymptotic sequence. Multiplication of asymptotic expansions is only justified if the asymptotic sequence $\{\delta_m(\epsilon)\}$ is such that $\delta_n(\epsilon)\delta_m(x)$ either form an asymptotic sequence or possess asymptotic expansions. An important asymptotic sequence that does have this property is $\{\epsilon^n\}$, the collection of (non-negative) powers of ϵ. If

$$f(x, \epsilon) \sim \sum_{m=0}^{\infty} a_m(x)\epsilon^m, \tag{B.36}$$

$$g(x, \epsilon) \sim \sum_{m=0}^{\infty} b_m(x)\epsilon^m, \tag{B.37}$$

then

$$f(x, \epsilon)g(x, \epsilon) \sim \sum_{m=0}^{\infty} c_m(x)\epsilon^m, \tag{B.38}$$

where

$$c_m(x) = \sum_{n=0}^{m} a_n(x)b_{m-n}(x). \tag{B.39}$$

B.4.4   *Differentiation*

In general, it is not justified to differentiate asymptotic expansions with respect to either x or ϵ. When differentiation is not justified, nonuniformities occur. The following theorem is useful:

Theorem   *Let $f(x, \epsilon)$ be an analytic function of x and ϵ, where S and T are, respectively, the domains of analyticity of x and ϵ. Let $f(x, \epsilon)$ have a uniformly valid asymptotic expansion*

$$f(x, \epsilon) \sim \sum_{m=0}^{\infty} a_m(x)\epsilon^m, \tag{B.40}$$

for all x in S. Under these conditions, the $a_m(x)$ are analytic for x in S and

$$\frac{\partial f(x, \epsilon)}{\partial x} \sim \sum_{m=0}^{\infty} \frac{da_m(x)}{dx}\epsilon^m, \tag{B.41}$$

uniformly in every compact proper subset T_1 of T.

B.5 Examples

This section provides illustrations of some of the concepts introduced in earlier sections of this appendix. For a fuller discussion and additional examples, the references listed at the end of this appendix can be consulted.

B.5.1 *The Symbol O*

As $\epsilon \to 0$,

$$\sin \epsilon = O(\epsilon) \qquad \cos \epsilon = O(1)$$
$$\sinh \epsilon = O(\epsilon) \qquad \cosh \epsilon = O(1)$$
$$\tanh \epsilon = O(\epsilon) \qquad \cot \epsilon = O(\epsilon^{-1})$$
$$\sin \epsilon^2 = O(\epsilon^2) \quad 1 - \cos^2 \epsilon = O(\epsilon^2)$$

and

$$\sin(x + \epsilon) = O(1) \qquad \text{uniformly as } \epsilon \to 0$$
$$\sin(\epsilon x) = O(\epsilon) \quad \text{nonuniformly as } \epsilon \to 0.$$

B.5.2 *The Symbol o*

As $\epsilon \to 0$,

$$\sin \epsilon = o(1) \qquad \cos \epsilon = o(\epsilon^{-1/2})$$
$$\sinh \epsilon = o(1) \qquad 1 - \cos \epsilon = o(\epsilon)$$
$$\coth \epsilon = o(\epsilon^{-3/2}) \qquad \sin^2 \epsilon = o(\epsilon)$$

and

$$\sin(x + \epsilon) = o(\epsilon^{-1/3}) \qquad \text{uniformly as } \epsilon \to 0$$
$$e^{-\epsilon x} - 1 = o(\epsilon^{1/2}) \quad \text{nonuniformly as } \epsilon \to 0.$$

B.5.3 *The Function* $\sin(x + \epsilon)$

Consider the function $\sin(x + \epsilon)$ in more detail. For $\epsilon \to 0$

$$\sin(x + \epsilon) = \sin x \cos \epsilon + \cos x \sin \epsilon$$
$$= \left(1 - \frac{\epsilon^2}{2!} + \frac{\epsilon^4}{4!} + \cdots \right) \sin x + \left(\epsilon - \frac{\epsilon^3}{3!} + \frac{\epsilon^5}{5!} + \cdots \right) \cos x$$
$$= \sin x + \epsilon \cos x - \left(\frac{\epsilon^2}{2!} \right) \sin x - \left(\frac{\epsilon^3}{3!} \right) \cos x + \cdots . \qquad \text{(B.42)}$$

For all values of x the coefficients of all powers of ϵ are bounded. Consequently, the expansion is uniformly valid.

B.5.4   *The Function* exp(−ϵx)

For a nonuniformly valid expansion, consider the expansion of $\exp(-\epsilon x)$ for small ϵ:

$$\exp(-\epsilon x) = \sum_{m=0}^{\infty} (-1)^m \frac{(\epsilon x)^m}{m!}. \tag{B.43}$$

This function can be accurately represented by a finite number of terms only if ϵx is small. Since ϵ is assumed small, this means that $x = O(1)$. Note that if x is as large as $O(\epsilon^{-1})$, then ϵx is not small, and a finite number of terms cannot give an accurate representation of $\exp(-\epsilon x)$. To obtain a satisfactory expansion for all x, all terms in (B.43) must be retained.

B.6   Generalized Asymptotic Power Series [3]

Many calculational procedures are based on expansions that take the following form

$$f(x, \epsilon) = f_0[\nu(\epsilon)x] + f_1[\nu(\epsilon)x]\epsilon + \cdots e + f_n[\nu(\epsilon)x]\epsilon^n + O(\epsilon^{n+1}). \tag{B.44}$$

Observe that ϵ occurs not only in the powers of ϵ^m, but also appears in the coefficients of these powers. A requirement is that this expansion be uniformly valid for all x. This will be the case if all the coefficients remain bounded as $\epsilon \to 0$, for fixed x. Such an expansion is called a generalized asymptotic power series. This is a special case of the generalized asymptotic expansion

$$
\begin{aligned}
f(x, \epsilon) = {}& \bar{f}_0(x, \epsilon)\delta_0(\epsilon) + \bar{f}_1(x, \epsilon)\delta_1(\epsilon) + \cdots \\
& + \bar{f}_n(x, \epsilon)\delta_n(\epsilon) + O[\delta_{n+1}(\epsilon)],
\end{aligned} \tag{B.45}
$$

where $\{\delta_n(\epsilon)\}$ is an asymptotic sequence.

Bibliography

[1] N. G. de Bruijn, *Asymptotic Methods in Analysis* (Interscience, New York, 1958).

[2] A. Erdély, *Asymptotic Expansions* (Dover, New York, 1956).

[3] J. A. Murdock, *Perturbations: Theory and Methods* (Wiley-Interscience, New York, 1991). See Sections 1.8 and 4.2.

[4] A. H. Nayfeh, *Perturbation Methods* (Wiley, New York, 1973). See Chapter 1.

[5] J. G. Van der Corput, *Asymptotic Expansions* (Lecture Notes, Stanford University, 1962).

[6] W. Wason, *Asymptotic Expansion for Ordinary Differential Equations* (Interscience, New York, 1965).

Index

Series on Advances in Mathematics for Applied Sciences

Editorial Board

Series on Advances in Mathematics for Applied Sciences

Aims and Scope

This Series reports on new developments in mathematical research relating to methods, qualitative and numerical analysis, mathematical modeling in the applied and the technological sciences. Contributions rlated to constitutive theories, fluid dynamics, kinetic and transport theories, solid mechanics, system theory and mathematical methods for the applications are welcomed.

This Series includes books, lecture notes, proceedings, collections of research papers. Monograph collections on specialized topics of current interest are particularly encouraged. Both the proceedings and monograph collections will generally be edited by a Guest editor.

High quality, novelty of the content and potential for the applications to modern problems in applied science will be the guidelines for the selection of the content of this series.

Instructions for Authors

Submission of proposals should be addressed to the editors-in-charge or to any member of the editorial board. In the latter, the authors should also notify the proposal to one of the editors-in-charge. Acceptance of books and lecture notes will generally be based on the description of the general content and scope of the book or lecture notes as well as on sample of the parts judged to be more significantly by the authors.

Acceptance of proceedings will be based on relevance of the topics and of the lecturers contributing to the volume.

Acceptance of monograph collections will be based on relevance of the subject and of the authors contributing to the volume.

Authors are urged, in order to avoid re-typing, not to begin the final preparation of the text until they received the publisher's guidelines. They will receive from World Scientific the instructions for preparing camera-ready manuscript.

Series on Advances in Mathematics for Applied Sciences

*To view the complete list of the published volumes in the series, please visit:
http://www.worldscibooks.com/series/samas_series.shtml

Printed in the United States
By Bookmasters